Harry C. Trexler Library
Muhlenberg College

Pharmaceutical Chemicals in Perspective

Pharmaceutical Chemicals in Perspective

Bryan G. Reuben
South Bank Polytechnic
London, England

Harold A. Wittcoff
The Chem Systems Group
Tarrytown, New York

A WILEY-INTERSCIENCE PUBLICATION
John Wiley & Sons
NEW YORK / CHICHESTER / BRISBANE / TORONTO / SINGAPORE

Library of Congress Cataloging in Publication Data:

Reuben, B. G.

Pharmaceutical chemicals in perspective / Bryan G. Reuben, Harold A. Wittcoff.

p. cm.

"A Wiley-Interscience publication."

Includes bibliographies and index.

1. Pharmaceutical industry—United States. 2. Drugs—United States. I. Wittcoff, Harold. II. Title.

[DNLM: 1. Drug Industry. 2. Drugs. QV 55 R442p]

HD9666.5.R48 1989

338.4′76151′0973—dc 19

DNLM/DLC

for Library of Congress 88-36857

CIP

ISBN 0-471-84363-6

Printed in the United States of America

10 9 8 7 6 5 4 3 2 1

To Catherine and Dorothy

Preface

The pharmaceutical industry makes a profit by providing that most desirable of all benefits—good health. It is an important and prosperous sector of the chemical and allied products industries and employs a range of scientists from synthetic organic chemists who synthesize new drugs, and molecular biologists who elucidate how they work, to pharmacologists and life scientists who test them and evaluate their safety. It is a highly innovative industry and operates at the frontiers of chemical and medical technology.

This book aims to provide an overview of the pharmaceutical industry and its products. The first part provides a background to the industry. A brief history is followed by a chapter dealing with economic and technical aspects. Chapter 3 discusses the people who consume pharmaceuticals—the patients—and those who prescribe them—the doctors. Chapter 4 provides a summary of the principles of pharmacology.

The second part describes the most important pharmaceuticals, their mode of action, and synthetic routes by which they may be obtained. There are about 1100 chemical entities on the market and selection presents a problem. We have focused on the 100 most frequently prescribed pharmaceuticals in the United States and give the syntheses of virtually all of them. We have expanded our discussion to include some drugs of historical significance, some which are dispensed in hospitals but not widely prescribed by general practitioners, and some which are new and have not yet achieved a place in the Top-100 but may do so in the future.

The third part deals with drugs that are not in the Top-100, perhaps because the diseases at which they are targeted are not prevalent in the United States, as with drugs against tropical diseases, and perhaps because they offer particular hope for the future, as with the prostaglandins and drugs against cancer and virus diseases.

In general, we have divided drugs according to the diseases they are intended to cure; there are chapters on antibacterials, drugs affecting the central nervous system, cardiovascular drugs, and so on. But we have also described some drugs in significant chemical groups, such as steroids and prostaglandins, where a single group of chemical compounds appears to have a range of medical applications.

The industry rests on a solid chemical basis and chemical aspects are stressed in this book. The book is suitable for anyone who wishes to gain a perspective on the pharmaceutical industry and is suitable as a textbook for students who have completed a course in organic chemistry.

We have attempted to make our text up to date but the rate of innovation in the pharmaceutical industry makes this an impossible target. We have added a final chapter to cover some of the drugs introduced after the completion of the main manuscript. Even so, since over 20 new chemical entities are introduced each year in the United States and over 40 worldwide, there are problems of selection and emphasis. We stress, therefore, that our objective is rather to provide a perspective on the industry and to give a feel for the way it has evolved and where it is likely to go in the future.

BRYAN G. REUBEN
HAROLD A. WITTCOFF

London, England
August 1988

Just after completing the first draft of this book, I fell disastrously while skiing in the French Alps. In addition to breaking many bones, I tore the aortic valve of my heart (see Figure 7.1) and required open-heart surgery. As an additional dedication, I wish to thank the rescue teams at La Plagne, and the medical staff at Bourg-St-Maurice and Grenoble hospitals, and especially Paul Pilichowski, M.D., surgeon; and Nicole Crepel, M.D., anesthetist. Without their skills and alertness, this book would be a posthumous publication.

B.G.R.

Acknowledgments

We are grateful for the invaluable help of Ms Maryann M. Grande, Ms Elizabeth Coles and Ms Karen Rose, whose library skills provided us with many data. We also thank Ms Lydia Aretakis, Ms Deborah Speyer and Ms Lucy Rusi for lending their computer skills to the formulation of the index, and Mr Anthony Jacob Reuben for help with proofreading. A special thanks to The Chem Systems Group for its invaluable encouragement and support.

The following material is reprinted or adapted with permission from Wittcoff, H. A. and Reuben, B. G., *The Pharmaceutical Industry: Chemistry and Concepts*, ACS Audio Course C-78; American Chemical Society: Washington, DC, 1987. Copyright 1987 American Chemical Society.

Figures: 1.1, 3.8, 4.4, 4.5, 4.7, 6.2, 6.3, 6.4, 6.8, 6.9, 6.10, 6.12, 6.13, 6.16, 6.17, 6.19, 6.20, 6.21, 6.22, 6.23, 6.24, 7.1, 7.3, 7.5, 7.6, 7.8, 7.12, 7.13, 7.14, 7.15, 7.16, 7.17, 7.18, 7.19, 7.22, 7.23, 8.1, 8.2, 8.4, 8.5, 8,7, 8.8, 8.9, 8.12, 8.13, 8.14, 8.15, 8.18, 8.19, 8.23, 8.24, 8.25, 8.26, 8.29, 8.30, 9.4, 9.5, 9.8, 10.1, 10.2, 10.3, 10.5, 10.6, 10.8, 10.10, 10.11, 10.12, 11.1, 11.3, 11.5, 11.6, 12.1, 12.9, 12.10, 13.1, 13.2, 14.1, 14.2, 15.1, 15.4, 15.6, 16.1, 17.3, 17.6, 17.7, 18.1, 18.2, 18.3, 18.4, 18.5, 19.2, 20.1, 20.3, 21.1, 21.2, 21.4, 21.5.

Tables: 1.1, 1.2, 1.3, 1.4, 4.1, 5.4, 17.1, 19.1.

Structures on pages: 37, 57, 58, 60, 61, 65, 67, 116, 138, 151, 154, 155, 156, 157, 162, 163, 164, 165, 169, 170, 179, 181, 183, 184, 187, 188, 199, 210, 216, 227, 229, 230, 233, 248, 252, 253, 254, 265, 266, 278, 285, 287, 294, 297, 300, 301, 305, 309, 310, 315, 319, 336, 349, 358, 363, 364, 366, 379, 386, 404, 405, 406, 410, 411, 415, 416, 417, 419, 422, 423, 427, 453, 456, 458.

Contents

PART I THE BACKGROUND TO PHARMACEUTICALS

Part I

The Background to Pharmaceuticals

1

History

The growth of the modern pharmaceutical industry has parallelled that of the organic chemical industry and can be said to have started in 1935 with the introduction of the sulfonamide antibacterials. Before 1935 it was unusual for a physician to be able to prescribe a drug to cure a specific disease. In general he could only recommend that the patient stay in bed, take plenty of hot drinks and aspirin to alleviate the symptoms, and allow nature to take its course.

This is not to say that the industry does not have a long and honorable prehistory. Records of medication go as far back as an Egyptian papyrus of 1150 B.C. and the literature is full of herbal and folk remedies that were occasionally effective but more often were not. Table 1.1 shows some of the important dates in the development of the pharmaceutical industry, subdivided into four categories: alkaloids, anesthetics, inoculation, and the role of bacteria in disease.

Early medicines were mainly alkaloids and other plant extracts—materials such as digitalis, morphine, strychnine, quinine, atropine, papaverine, and cocaine—and the isolation of these provides most of the significant dates in the first half of the nineteenth century.

Other significant dates in that period derive from the use of simple although dangerous general anesthetics in surgery: at the end of the century, local anesthetics such as benzocaine and procaine were discovered.

Inoculation against smallpox goes back to Lady Mary Wortley Montagu in 1718 and Jenner introduced vaccination in 1796, but most of the landmarks in inoculaton were in the early years of this century.

TABLE 1.1. Important Dates in the Development of Drugs

1618	First London Pharmacopeia published
	Alkaloids and other plant extracts
1785	Digitalis used against heart disease by Withering
1803	Morphine extracted from *Papaver somniferum*
1818	Strychnine isolated
1820	Quinine isolated from cinchona bark
1831	Atropine isolated from deadly nightshade
1848	Papaverine isolated from opium
1860	Cocaine isolated from *Erythoxylon coca*
1884	Cocaine introduced to medical practice
	Anesthetics
1842	Ether introduced to surgery by Long and Morton (1846)
1844	Nitrous oxide
1847	Chloroform
1909	Procaine prepared by Einhorn and used as a local anesthetic
1910	Benzocaine, identified 1890, used as a local anesthetic
	Inoculation
1718	Smallpox inoculation introduced to Europe from Turkey by Lady Mary Wortley Montagu
1796	Smallpox vaccination used by Jenner by infection with cowpox
1890	Antirabies inoculation demonstrated by Pasteur
1896	Antityphoid vaccine demonstrated
1905	Anticholera vaccine first used successfully in India
1928	Yellow fever vaccine
1942	Mass diphtheria immunization introduced to Britain
	Role of Bacteria in Disease
1847	Semmelweis demonstrates contagious nature of childbed fever
1864	Pasteur demonstrates bacterial cause of infectious diseases
1865	Lister introduces carbolic acid (phenol) as the first antiseptic
1877	Pasteur and Joubert demonstrate microbial antagonism
1882	Koch isolates tubercle and later cholera bacilli
1884	Laffler isolates diptheria bacillus
1889	Kitasato isolates bacilli causing tetanus and anthrax

In the 19th century, a scientific approach to disease began to bear fruit. The dates in the table cluster in the second half of the century and are milestones in the recognition of the microbial basis of infectious disease. They heralded the chemotherapeutic revolution.

Important dates in that revolution are indicated in Table 1.2. Note the early use of amyl nitrite for heart disease, salicylic acid for rheumatism, and antipyrine for fevers. Then, at the turn of the century, three important drugs were synthesized notably barbital, aspirin, and the general anesthetic procaine.

TABLE 1.2. The Chemotherapeutic Revolution

1867	Brunton introduces amyl nitrite for the treatment of angina.
1876	Salicylic acid introduced for rheumatism.
1887	Knorr synthesizes antipyrine.
1898	Barbital first used as a sedative.
1899	Dreser synthesizes aspirin and Bayer markets it.
1904	Stolz and Dakin synthesize epinephrine (adrenalin).
1910	Ehrlich discovers salvarsan for use against syphilis.
1916	Ehrlich discovers the use of suramin against trypanosomiasis (African sleeping sickness).
1916	Heparin found to be a natural anticoagulant.
1921	Banting and Best isolate insulin for use against diabetes.
1926	Bayer introduces pamaquine against malaria.
1928	Fleming discovers penicillin but is unable to isolate it.
1933	Haworth and Reichstein synthesize vitamin C.
1935	Prontosil and then sulfanilamide introduced against the hemolytic streptococcus.
1935	Reichstein isolates cortisone.
1937	Bovet and Staub discover first antihistamines.
1939	Waksman and Dubos discover peptide antibiotics.
1940	Florey and Chain demonstrate use of penicillin against bacterial infections.
1945	Streptomycin discovered for tuberculosis, followed by *p*-aminosalicylic acid in 1946 and isoniazid in 1952.
1948	Chlortetracycline discovered, followed by oxytetracycline in 1950 and tetracycline in 1952.
1948	Vitamin B_{12} isolated.
1948	Partial synthesis of cortisone.
1948	Chloramphenicol discovered.
1950	Chlorpromazine, the first major tranquilizer, introduced.
1952	Reserpine, the first antihypertensive agent, isolated from *Rauwolfia serpentina.*
1954	Meprobamate, the first minor tranquilizer, introduced.
1957	Isolation of the penicillin nucleus, 6-aminopenicillanic acid.
1957	The tricyclic imipramine recognized as an antidepressant.
1957	Iproniazid, the first antidepressant monoamine oxidase inhibitor, introduced.
1959	First semisynthetic penicillin introduced.
1960	Chlordiazepoxide, the first benzodiazepine minor tranquilizer, introduced.
1960	Methyldopa introduced as an antihypertensive.

The seminal work of Ehrlich in the early years of the twentieth century provided in 1910 the first true chemotherapeutic agent—the arsenobenzyl compound salvarsan active against syphilis. Heparin, extracted from animals, was used as an anticoagulant from 1916. Insulin, also from animals, saved the lives of diabetics from 1921. Suramin was introduced against trypanosomiasis in 1916 and pamaquine against malaria in 1926. Various vitamins, whose uses were understood, became available, and Table 1.2 lists vitamin C in 1933 and vitamin B_{12} in 1948. An intravenous saline drip turned out to be a fairly effective answer to cholera. All the same, in the 1920s, six drugs, singly or in combination, made up 60% of all British prescriptions. The physician's armory was tiny compared with what it is today.

TABLE 1.3. Some Drugs Available in 1932[a]

Vaccines	*Stimulants*
Paratyphoid vaccine	Strychnine
Smallpox vaccine	Caffeine (Section 8.5.2)
Tuberculin vaccine	
Antidysentery serum	*Antisyphilitics*
	Arsphenamine* (Salvarsan) (Figure 1.1)
Hypnotics	Neoarsphenamine* (Figure 1.1)
Chloral hydrate* (Section 8.1)	Sulfarsphenamine* (Figure 1.1)
Barbital* (Figure 8.1)	
Phenobarbital* (Figure 8.1)	*Purgatives*
Paraldehyde* (Figure 1.1)	
Potassium bromide	Castor oil
Sulfonal* (Figure 1.1)	Calomel
	Cascara sagrada
Analgesics and Antipyretics	Syrup of Senna
	Syrup of figs
Aspirin* (Section 10.2)	Phenolphthalein* (Figure 1.1)
Phenacetin* (Section 10.2.2)	
Antipyrine* (Figure 10.10)	*Local Anesthetics*
Codeine* (Figure 10.1)	
Quinine (Section 18.2.1)	Cocaine* (Section 8.5.2)
Amidopyrine* (Pyramidon)	Procaine*
Morphine (Figure 10.1)	Amylocaine*
	Benzocaine*
Cardiovascular Drugs	
	Other Important Alkaloids
Digitalis (Section 7.2)	
Quinidine (Section 7.3)	Diamorphine
Epinephrine (Figure 7.14)	Papaverine
Nitroglycerol* (Section 7.5.1)	Pilocarpine (Section 4.3.2)
Amyl nitrite* (Section 7.5.1)	Atropine (Section 14.1)
Erythrityl tetranitrate*	Hyoscyamine (Section 14.1)
	Hyoscine (Section 14.1)

[a]Figure references indicate where structures may be found.

*Made by chemical synthesis.

Table 1.3 gives some idea of the more effective drugs that were available in 1932. There is a selection of vaccines, hypnotics, analgesics and antipyretics, cardiovascular drugs, stimulants, antisyphilitics, purgatives, local anesthetics, and assorted alkaloids. The structures of most of these drugs are given either in Figure 1.1 or later in the book.

This pharmaceutical armory, such as it was, owed little to the synthetic chemist. The drugs in Table 1.3 that were made synthetically are starred. The British Pharmacopeia of 1932 contained just 32 synthetic drugs including aspirin, phenacetin, and barbital, all three of which had been discovered in Germany before 1900.

In 1935 the first of the sulfonamides was introduced. They will be described in more detail later. Suffice it now to say that they were the first effective

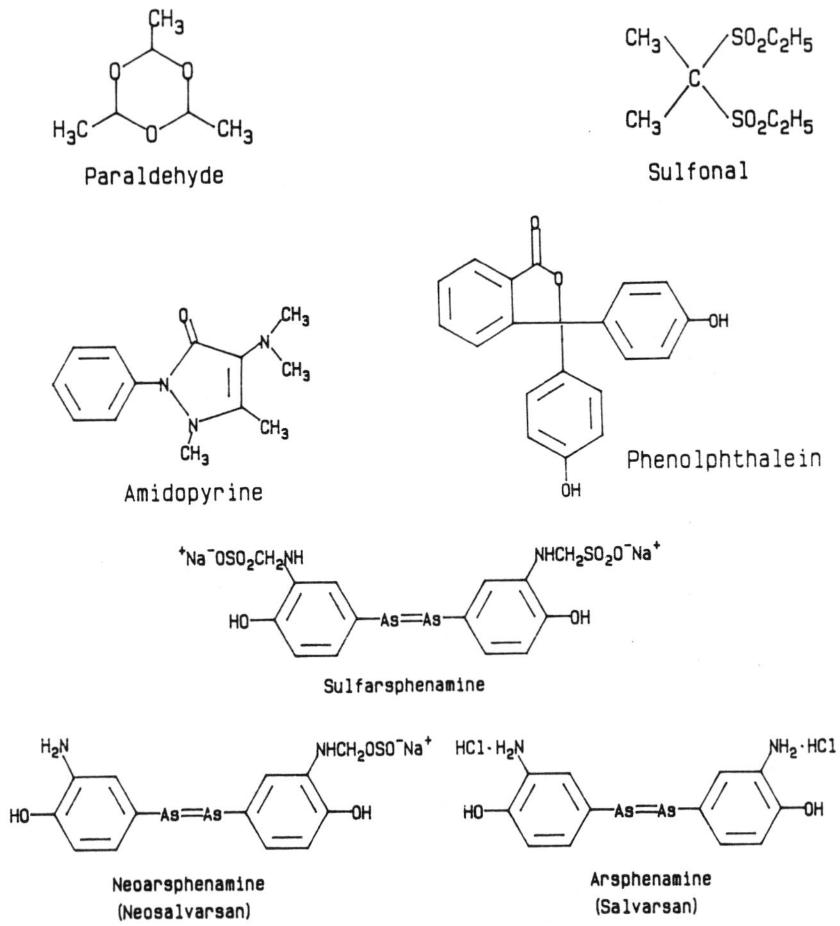

Paraldehyde

Sulfonal

Amidopyrine

Phenolphthalein

Sulfarsphenamine

Neoarsphenamine (Neosalvarsan)

Arsphenamine (Salvarsan)

Figure 1.1. *Some synthetic drugs available in the 1932 British Pharmacopoeia*

antibacterials and they made possible for the first time the cure of infectious diseases other than syphilis. The effect of these drugs was not only medical. They caused the various strands of pharmaceutical activity to draw together to found the modern pharmaceutical industry.

The sulfa drugs were especially active against hemolytic streptococcus. The worst manifestation of this kind of bacterial infection was puerperal fever, an infection of the genital tract after childbirth. Figure 1.2 shows the decrease in puerperal fever and maternal mortality in the United Kingdom from 1860 onward. Note the break in the curve after the introduction of sulfanilamide in 1935.

Several years before the discovery of the sulfonamides, Sir Alexander Fleming had observed the antibacterial properties of a penicillin mold but it was not until the pressures of World War II and the more modern isolation

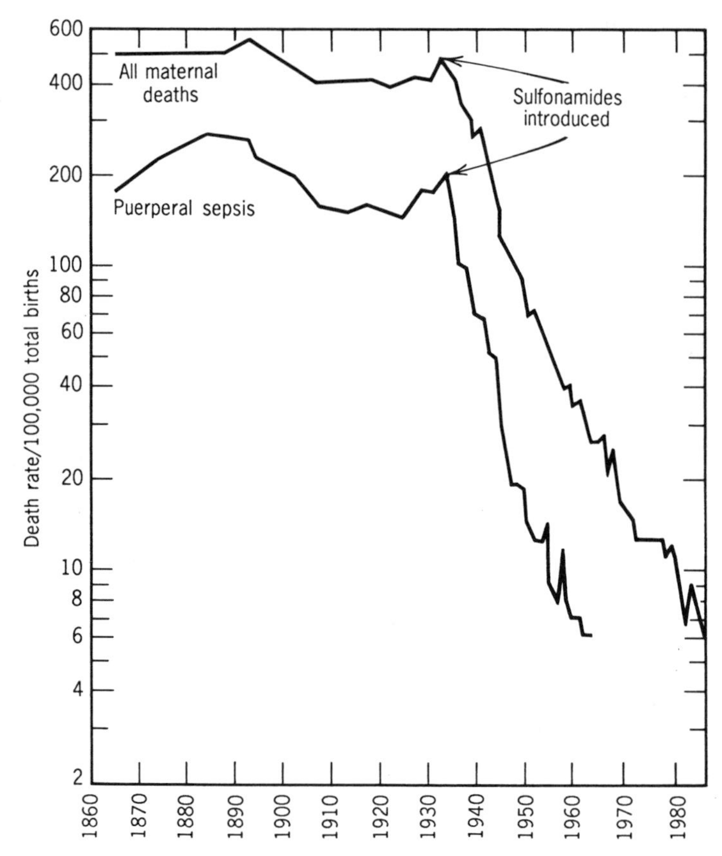

Figure 1.2. *Maternal mortality and puerperal fever (England and Wales: 1860– 1985). Ten-year averages, 1861– 90; five-year averages, 1891– 1930; annual rates 1931– 80. Source: Registrar General's Statistical Reviews for England and Wales. Graph updated from B. G. Reuben and M. L. Burstall, The Chemical Economy, Longman, London, 1973, and Office of Health Economics publications.*

techniques then available that penicillin became a useful drug (Section 6.3). Penicillin replaced salvarsan as the drug of choice in syphilis and sulfonamides as the drug of choice in lobar pneumonia, a form of pneumonia that attacked all age groups and led to a mortality of between 5 and 20%. Figure 1.3 shows mortality from all pneumonia and influenza. The death rate declined slowly from 1900, more rapidly from the 1930s with the introduction of sulfa drugs, and very rapidly after the development of penicillin.

The death rate from tuberculosis of the lung is shown in Figure 1.4. Tuberculosis is a sensitive measure of public health. The death rate declined from 1900 onward and highlights the important point that much of the increased expectation of life in this century has come not from better drugs but from public health measures, that is, better hygiene and housing condi-

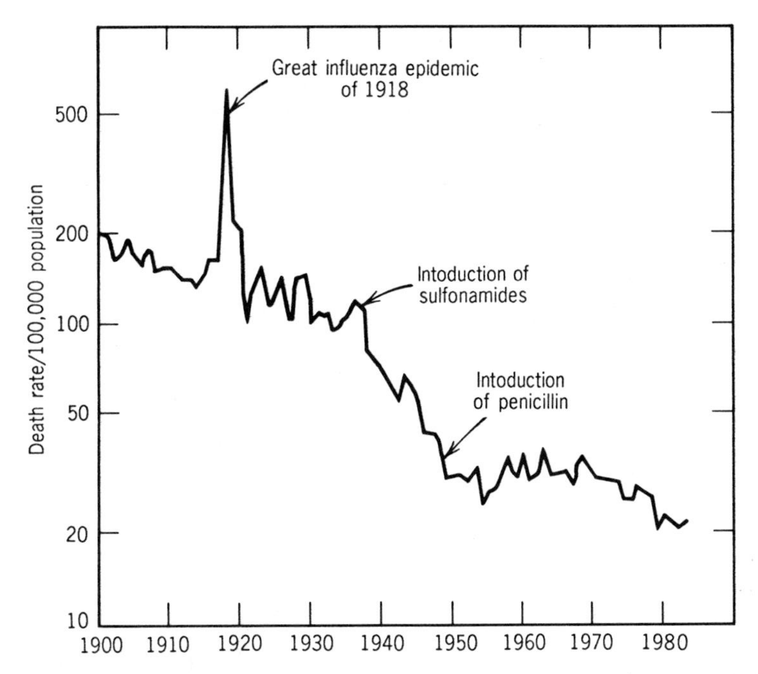

Figure 1.3. *Death rate from pneumonia (excludes pneumonia of newborn) and influenza (United States: 1900– 1983). Source: Historical Statistics of the United States, U.S. Department of Commerce, Washington, DC, 1975.*

tions, improved nutrition, and provision of clean water and adequate sewage systems. It has been said with some truth that the toilet is one of the greatest inventions of mankind.

Thus, most of the increased life expectancy until 1935 was due to public health measures. No antituberculosis drugs were available until 1945, yet tuberculosis declined by about 3% per year in Britain between 1880 and 1945. In 1944 the antituberculosis antibiotic streptomycin was discovered and only a year later was made available. Unfortunately, patients taking it made an apparent recovery but frequently relapsed because of the appearance of streptomycin-resistant strains of bacteria. In 1946, the simple molecule *para*-aminosalicylic acid was shown also to be active against the tubercle bacillus and in 1952 isoniazid, another small molecule, turned out to be a third antituberculosis drug. Given together, these three drugs could prevent or at least delay the appearance of resistant strains. In addition, a program of immunization against the disease was started in Britain. The mortality rate from tuberculosis began to drop by 15% per year and by 1960 was a tenth of what it had been in 1945. Tuberculosis, described by the poet John Milton in the seventeenth century as "the Captain of the Men of Death," had almost ceased to be a threat.

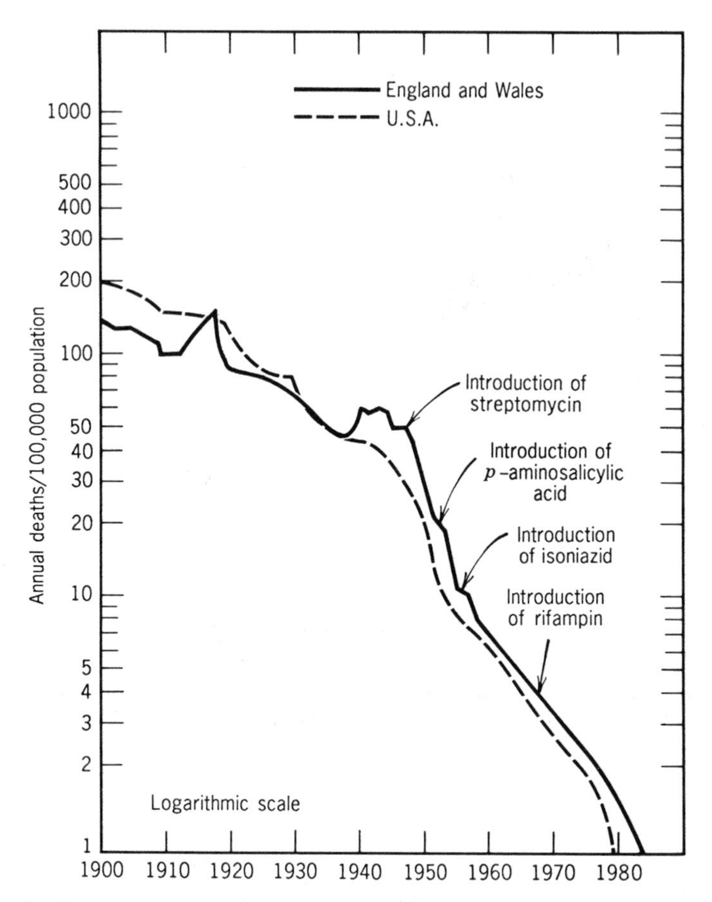

***Figure 1.4**. Mortality rate from tuberculosis (United States, England, and Wales: 1900–1984). 1900–1945: Mortality declined at 3% per year mainly because of improved living conditions. 1945–1960: Mortality declined at 15% per year mainly because of antibiotics. Source: U.S. Department of Health and Social Services, quoted by N. Wells, Medicines, 50 Years of Progress, Office of Health Economics, London, 1980.*

After streptomycin, exhaustive screening programs were undertaken to find other antibiotics. In 1948 chloramphenicol was discovered and so was chlortetracycline, the first tetracycline. By 1950 most of the major groups of antibiotics had been identified. The next breakthrough was the isolation of the penicillin nucleus, 6-aminopenicillanic acid, in 1957. Adding side chains to this nucleus led to the first semisynthetic penicillin in 1959. Antibiotics had largely vanquished infectious disease as a cause of death in developed countries and had also ameliorated many uncomfortable illnesses such as tonsillitis.

The most spectacular effect of the decline in mortality from infectious diseases was a drop in the child death rate. Figure 1.5 shows the child death rate in Britain from 1931/35 to 1977 from major infectious diseases. Data are

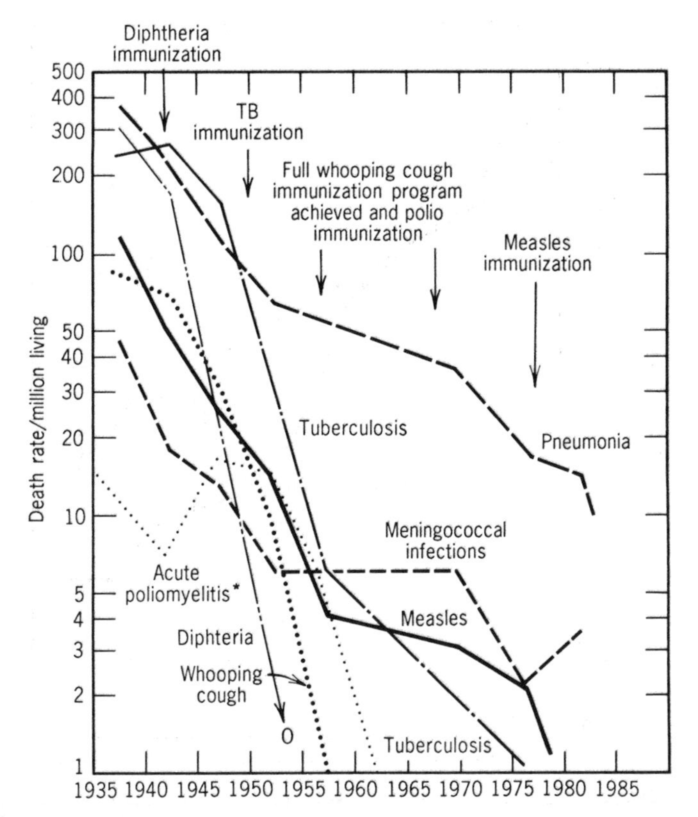

Figure 1.5. *Childhood (ages 1 – 14 years) deaths by selected causes (England and Wales: 1935 – 1984). (*Death rate for acute poliomyelitis embraces all children in England and Wales under 15 years.) Source: N. Wells, Medicines, 50 Years of Progress, Office of Health Economics, London, 1980.*

given for diphtheria, tuberculosis, whooping cough, polio, measles, pneumonia, and meningococcal infections. For all these diseases the childhood death rate has dropped by at least an order of magnitude and frequently more than two.

How much of the declining child mortality was due to better public health and how much to the new chemotherapy? Figure 1.6 attempts to provide an answer. The dotted line shows the child death rate from 1901 to 1935 extrapolated to 1985 while the full line shows the actual death rate. If the death rate had continued to decline on the 1901–1935 trend because of improved living conditions, the shaded area between the lines would represent the decrease in mortality that can be ascribed to pharmaceuticals. On the extrapolated death rate the 1985 figure would have been expected to be about 800 per 100,000 living; in fact it is about a third of that. Stated another way, there are about 300,000 people alive in England and Wales today who would have died in childhood but for the new drugs.

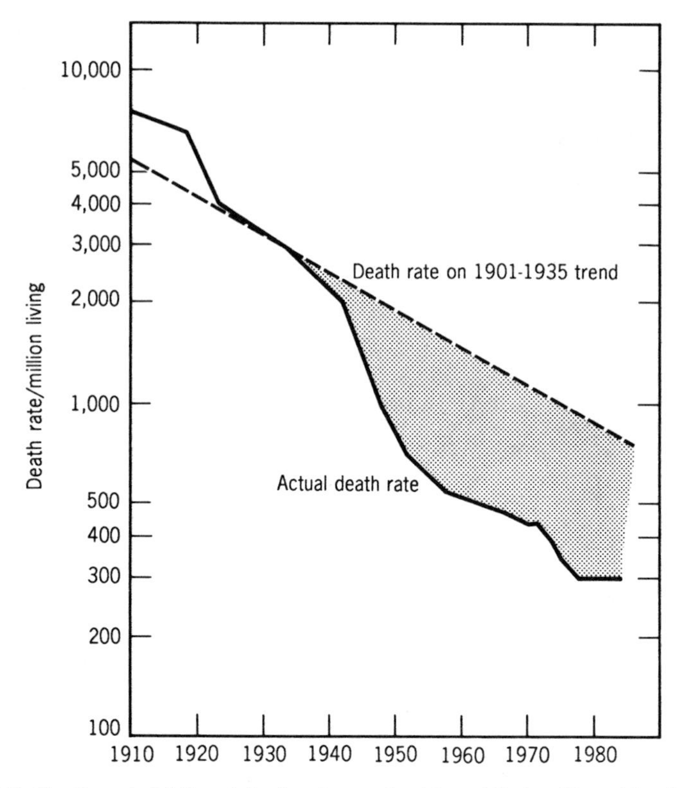

Figure 1.6. *Decline of childhood death rate ascribed to public health and to pharmaceuticals (1–14 years). Note: Five-year averages 1911/1915–1961/65; annual rates 1966–1985. Source: N. Wells, Medicines, 50 Years of Progress, Office of Health Economics, London, 1980. UK Annual Abstract of Statistics.*

As Table 1.3 indicates, if the 1940s were the decade of antibiotics, the 1950s were the decade of psychotropic medicines. The first major tranquilizer, chlorpromazine, was discovered in 1950, followed in 1954 by the minor tranquilizer, meprobamate, and in 1960 by another minor tranquilizer, chlordiazepoxide, better known by its trade name Librium. By 1960 there were also available two groups of antidepressants, the monoamine oxidase inhibitors and the tricyclics. For the first time there were medicines for the treatment of schizophrenia, anxiety, and acute depression. These will be discussed in Chapter 8.

Another area where the advances have been slower but also impressive is heart disease. Reserpine was discovered in 1952 and methyldopa in 1960. The golden age for heart drugs, however, was the late 1960s and early 1970s. An impressive range of materials is now available.

Mortality from heart disease dropped in the United States and United Kingdom from 1900 to 1922 when it reached a minimum and it then started to

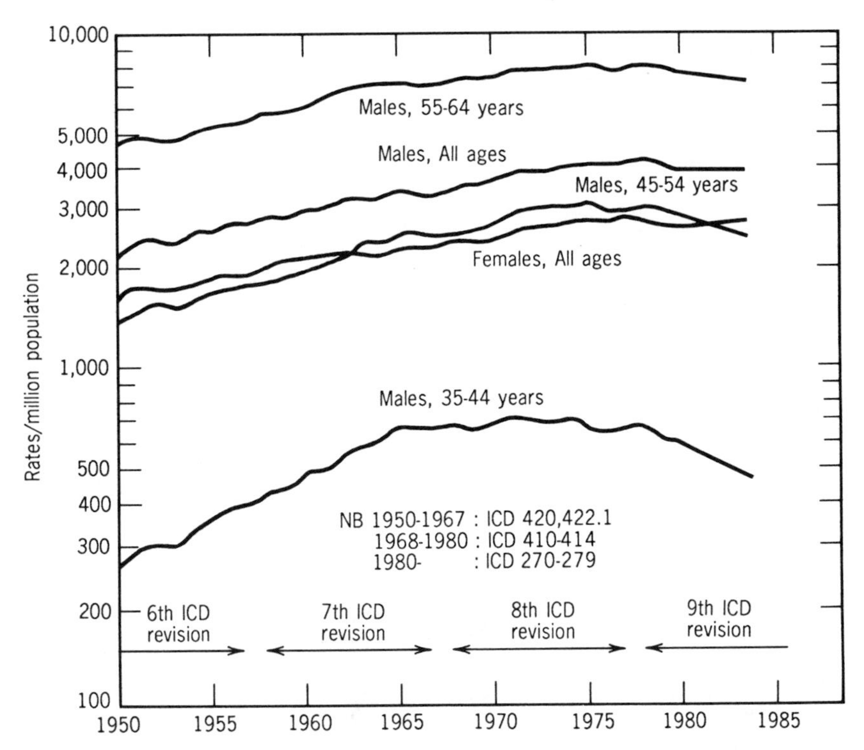

Figure 1.7. *Mortality from heart disease. Deaths from arteriosclerotic/ischemic heart disease in England and Wales, 1950–1984. After 1900, mortality rates dropped, reaching a minimum in both the United States and United Kingdom in 1922, and then rose. Source: United Kingdom Office of Population Census and Surveys.*

rise. British data since 1950 are shown in Figure 1.7. The death rate had leveled by 1985 and preliminary 1988 data suggest it has now turned decisively down. Women are only one-third as prone to heart disease as men. As Figure 1.8 indicates, the pattern of mortality varies from country to country, with the Japanese escaping lightly and the British and Americans being particularly at risk. The most striking occurrence, however, has been the drop in U.S. mortality since 1968 so that it is now below the British and Swedish figures. Although many factors are involved, changes in life-style and diet seem to have played a major part in the improvement, and presumably other countries will be able to build on the U.S. experience. Indeed, Finland has achieved a similar drop. Almost everywhere, the rise in mortality up to the 1970s has now been checked and there are grounds for optimism about the future.

Cancer is a major killer disease and chemotherapeutic treatment, despite decades of effort, is still only a limited success. Cancer mortality and cancer drugs will be discussed in Chapter 21.

The various advances in public health and chemotherapy have been reflected by an increase in life expectancy. Figure 1.9 shows the increase in life

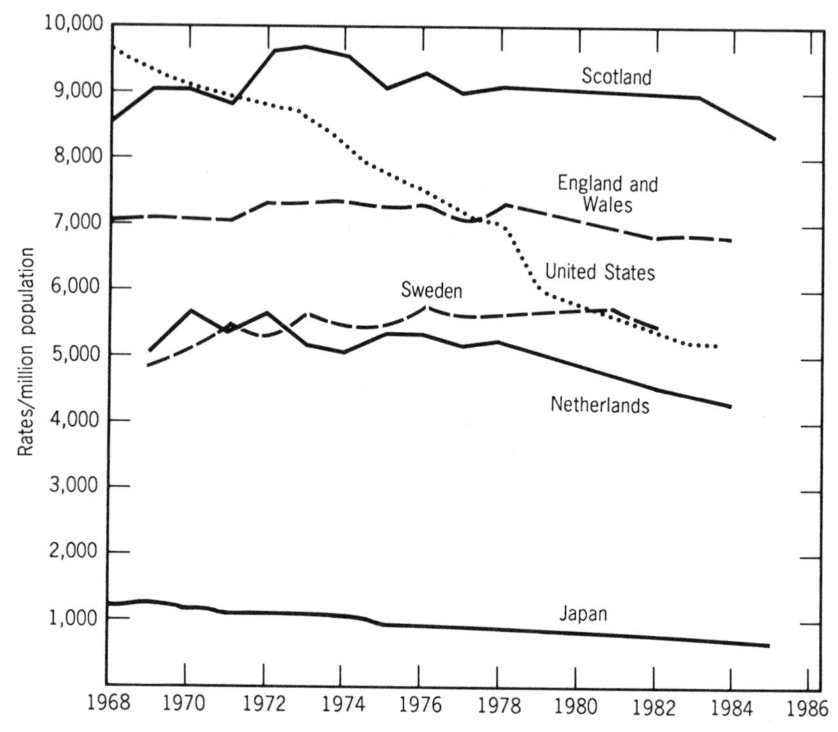

Figure 1.8. *Deaths from ischemic heart disease among males aged 55–64 years in selected countries, 1968–1985. Source: World Health Organization and Office of Health Economics.*

expectancy at birth for men and women in the United States between 1920 and 1982. Figure 1.10 shows the expectancies for men at different ages in England and Wales. The most impressive increase in life expectancy has been at birth where there has been an increase of about 30 years in the past century. Old people's life expectancy has risen less spectacularly but there has been an increase of about four years for 65-year-olds.

Table 1.4 shows the diseases that have largely been conquered in the past 50 years, those that have been alleviated, and those that are largely untouched. These changes have taken place within the memory and experience of people now in their sixties and represent the most remarkable period in medical history. The benefits brought by the changes are clear. Nonetheless, these advances have also brought problems in their wake. A full discussion of these is outside the scope of this book, but, along with the successes of the pharmaceutical industry, we shall mention drugs with unfortunate side effects.

The period from 1935 to the 1960s has been termed the first chemotherapeutic revolution. Many of the drugs discovered were life-saving. The regulatory system was none too strict. Drugs were discovered by a sort of molecular roulette involving the screening of millions of more or less arbitrarily chosen

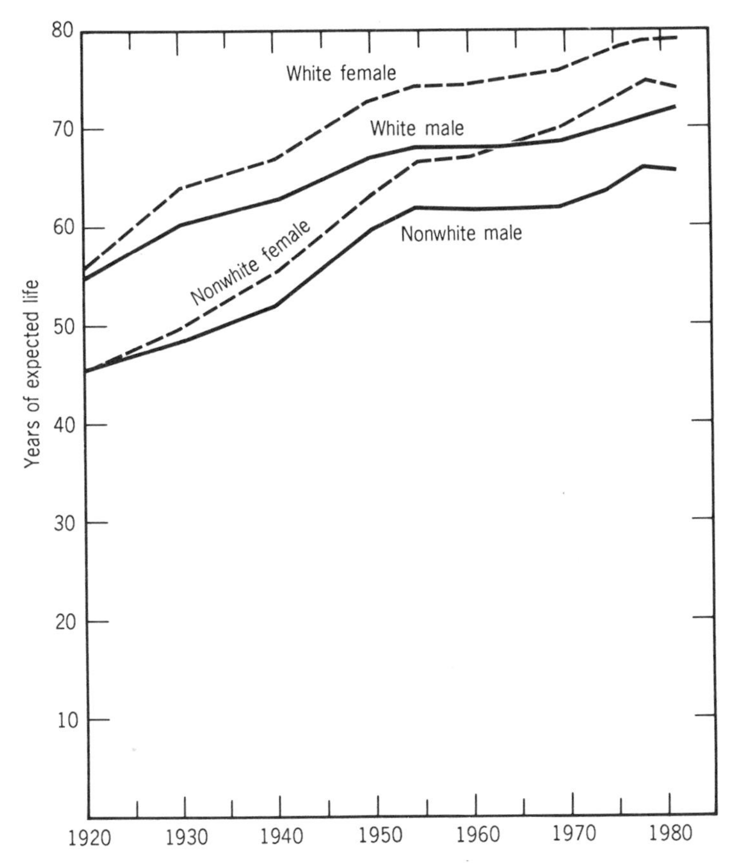

Figure 1.9. *Life expectancy at birth (United States: 1920– 1982). Source: Statistical Abstract of the United States, U.S. Department of Commerce, Bureau of the Census, Washington, DC, 1982.*

compounds. Paul Ehrlich, the father of chemotherapy, synthesized 606 compounds before he found one, salvarsan, that was reasonably effective against syphilis. His search was labeled the quest for the magic bullet and he visualized many magic bullets, each effective against a different microorganism. Similarly, May and Baker synthesized 693 compounds before coming up with sulfapyridine. This "batting average" is still good when compared with modern researchers who, until recently, synthesized of the order of 8000 compounds for every useful one produced.

The empirical approach was remarkably effective and, of course, the compounds were not chosen entirely at random. Chemists had some idea of the sort of structures liable to be biologically active and, once having found one that exhibited a degree of activity, they had well-known lines of development for increasing activity. Nonetheless they often had little idea of how drugs worked.

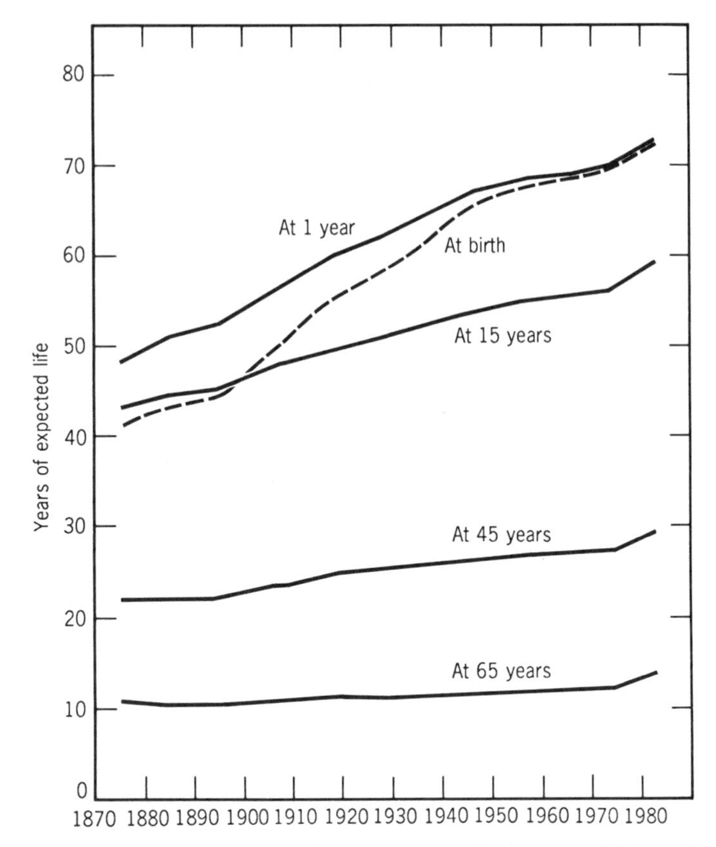

Figure 1.10. *Life expectancy at selected ages for males (England and Wales, 1871 – 1982). Source: N. Wells, Medicines, 50 Years of Progress, Office of Health Economics, London, 1980. Office of Population Censuses and Surveys.*

These pioneering days have now passed. New drugs are being developed within a much stricter regulatory framework. Their function frequently is more to improve the quality of life than to save life. Most significant of all, new drugs are being designed rather than discovered; they are being sought on the basis of an understanding of the biochemistry of the systems they are intended to influence. Although drug design still relies on the intuitive feel of the scientist, developments of receptor theory (the modern equivalent of Ehrlich's "magic bullet" hypothesis) give a way of escape from molecular roulette.

This new approach gathered momentum in the 1970s and the results are seen in some of the pharmaceutical innovations of the 1970s and in most of the innovations of the 1980s. Much of this book is concerned with this seminal period. It truly represents a second chemotherapeutic revolution.

TABLE 1.4. Status of Some Diseases

1. Diseases Largely Cured or Avoided
Cholera
Diphtheria
Erysipelas
Lobar pneumonia
Measles
Meningococcal meningitis
Pertussis (whooping cough)
Plague
Poliomyelitis
Rheumatic fever
Scarlet fever
Smallpox
Staphylococcal septicemia
Subacute bacterial endocarditis
Tuberculosis
Typhoid fever
Vitamin deficiency
2. Diseases Alleviated
Asthma
Diabetes
Heart disease
Schizophrenia
Syphilis and other veneral diseases
3. Diseases Which Still Present Challenges
AIDS
Alzheimer's disease
Arthritis
Cancer
Cirrhosis
Common cold
Genetically transmitted diseases
Genital herpes
Huntingdon's chorea
Influenza
Multiple sclerosis
Parkinson's disease
Pulmonary fibrosis
Senility, geriatric problems

2

The Characteristics of the Pharmaceutical Industry

The pharmaceutical industry is unique in both its technology and its economics. This chapter deals with the way the industry works and the characteristics that distinguish it from industry in general and the bulk organic chemicals industry in particular.

2.1 THE WORLD PHARMACEUTICAL INDUSTRY

The pharmaceutical industry is international in scope and is dominated by large multinational companies. Such are the costs of new drug development that no other pattern would be economic. Worldwide in 1986 $100 billion worth of pharmaceuticals were manufactured, of which the United States was responsible for about one-quarter, and about one-sixth of this was shipped overseas, making pharmaceuticals one of America's most important exports. Apart from the United States, three other countries—the United Kingdom, West Germany, and Switzerland—are leaders in pharmaceutical manufacture and innovation. These countries have a large positive balance of trade in pharmaceuticals and high sales as shown in Table 2.1. France still is in the big league but seems to be slipping while Japan and Sweden are trying very hard to join. Indeed Japan introduced more new pharmaceutical entities to the market than any other country in most years in the 1980s and is expected to make further headway in the next decade.

Most pharmaceutical companies outside the big four are subsidiaries of multinationals based in these countries and few important drugs originate elsewhere. Not many countries and companies have the financial, technical,

TABLE 2.1. World Pharmaceutical Production, Consumption, and Balance of Trade (1986)

Country or Area	Consumption (millions of $US)[a]	Production (millions of $US)[a]	Balance of trade
France	6,150	7,086	936
Germany	7,830	9,254	1,424
Italy	5,120	4,820	−300
Spain	1,810	1,751	−59
United Kingdom	3,280	4,530	1,250
Other EEC[b]	2,840	3,118	278
Total EEC	27,030	30,559	3,529
Sweden	690	829	139
Switzerland	790	2,450	1,660
Other non-EEC Europe	1,120	700	−420
Total non-EEC Europe	2,600	3,979	1,379
United States	23,850	25,029	1,179
Canada	1,780	1,425	−355
Australia	590	383	−207
Japan	17,750	16,539	−1,211
Other developed countries outside Europe	900	n.a.	n.a.
Total developed countries outside Europe	44,870	44,276	−594
Developing America	4,750	4,594	−156
Developing Africa	1,750	small	−1,750
Developing Asia	5,000	2,798	−2,202
Middle East / OPEC	1,200	small	−1,200
Total developing countries	12,700	7,063	−5,637
Total market economies	87,200		
Soviet Union	6,000	n.a.	n.a.
China	3,500		
Other command economies	3,000		
Total command economies	12,500		
World total	99,700	99,700	0

[a]At manufacturers' prices and 1986 average exchange rates.
[b]EEC, European Economic Community.

Source: Authors' estimates based on M. L. Burstall (Bibliography 2 / 12) and UN trade statistics. UN data are not always internally consistent and figures do not necessarily add up.

and medical resources to develop and test new drugs, and even among developed countries there are few that succeed. It is surely of significance that since 1945, to the best of our knowledge, no nationalized drug company nor any drug company in a centrally planned economy has discovered a major new pharmaceutical.

So much for the countries that produce pharmaceuticals. Where are they consumed? The second column of Table 2.1 provides the answer. The United States consumed about a quarter of the total. Japan, West Germany, France,

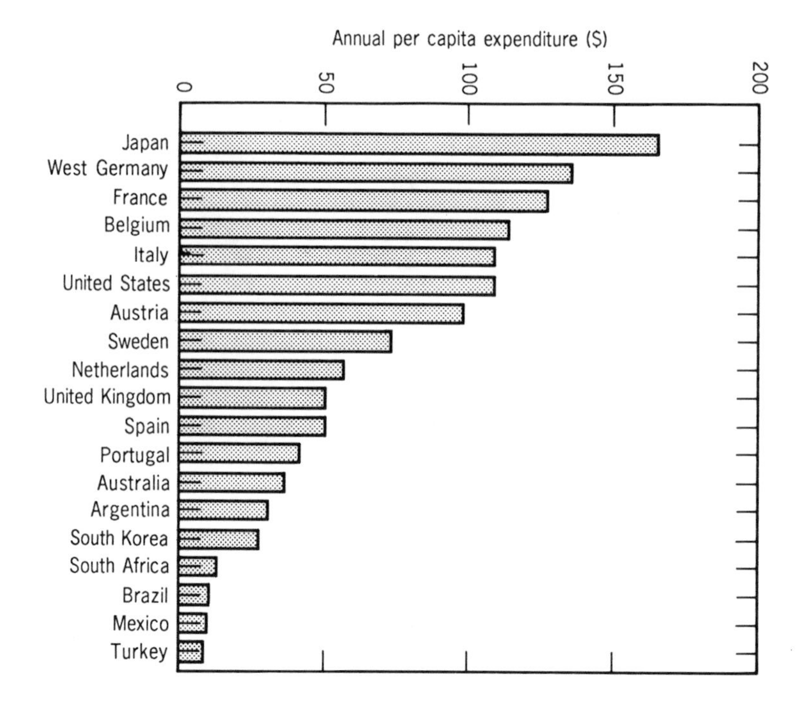

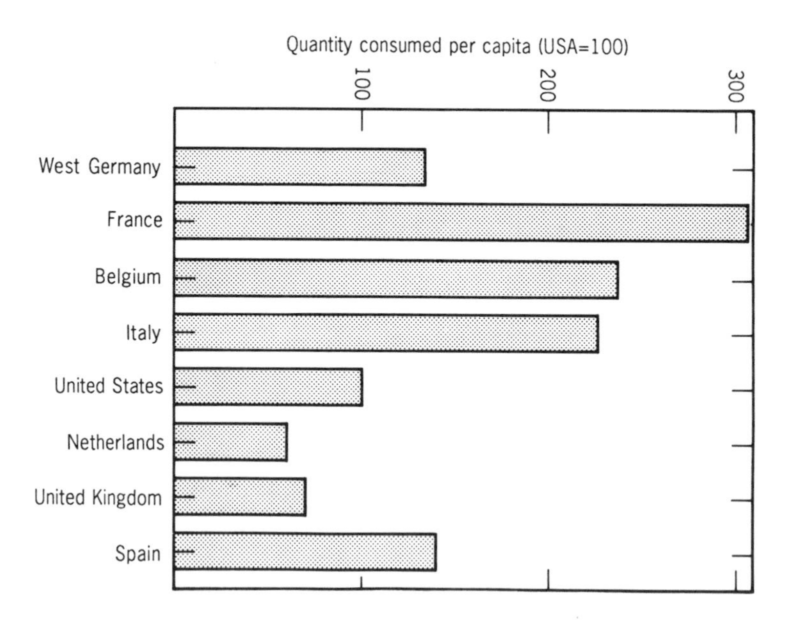

Figure 2.1. *Per capita expenditure on medicines, 1986.*

Italy, and the United Kingdom were also major consumers. Western Europe, North America, Japan, and the other developed countries outside the "command economies" together accounted for about three-quarters of all drug consumption.

Figure 2.1 shows the pharmaceutical expenditure per head of the population for a range of countries. The Japanese, West Germans, French, and Belgians all spend more per head on drugs than the North Americans, and the Italians spend as much. This does not necessarily mean that the inhabitants of these countries consume a greater volume of pharmaceuticals because there is also the question of price. In Japan and West Germany, as in the United States, drug prices are fixed by market forces and are relatively high. In most other countries, government pressure is applied to the drug companies and prices are lower. Division of amount spent by relative price indices leads to the second part of Figure 2.1 and the conclusion that the French are the world leaders in the volume of pharmaceuticals consumed. The Swiss and Japanese also swallow large numbers of tablets, but pharmaceutical price indices are not available.

In developing countries, by contrast, expenditure on pharmaceuticals is small, mainly because of poverty but partly because the inhabitants do not have access to medical treatment or prescription. The per capita consumption of drugs in Africa is only $3, but even this expenditure saves many lives because it is used for powerful and relatively cheap drugs such as antibiotics to cure diseases that in the past were fatal. Furthermore, as indicated in Table 2.2, it amounts to about the same percentage of GDP as is spent by the developed countries.

Table 2.3 shows the 25 largest pharmaceutical companies in the world together with their sales and country of origin. Many of them, including Hoechst in second place and Bayer in sixth, are not specifically pharmaceutical firms but the pharmaceutical subsidiaries of chemical companies.

Of the 25 companies listed, 12 are American, 4 are British, 3 are West German, 3 are Swiss, 1 is French, and 2 are Japanese, illustrating the point made above about the "big four."

TABLE 2.2. Consumption of Pharmaceuticals as % GDP

Geographical Area	% GDP
"Developed" countries, average	0.74
American industrially less developed countries	0.83
Asian industrially less developed countries	0.78
Middle Eastern industrially less developed countries	0.42
African industrially less developed countries	0.79
Average industrially less developed countries	0.70
World average	0.73

Source: D. Taylor, *Medicines, Health and the Third World*, Office of Health Economics, London 1982.

TABLE 2.3. The 25 Largest Pharmaceutical Companies in the World in 1987

Rank	Company	Country of Origin	Pharmaceutical Sales ($m)	Profit margin (%)
1	Merck & Co.	USA	4,630	28.4
2	Hoechst	W. Germany	4,278	n.a.
3	American Home Products	USA	3,788	27.2
4	Ciba-Geigy	Switzerland	3,726	n.a.
5	Glaxo	UK	3,525	40.4
6	Bayer	W. Germany	3,391	n.a.
7	Sandoz	Switzerland	3,202	n.a.
8	Takeda	Japan	3,189	13.0
9	Pfizer	USA	3,118	26.7
10	Lilly, Eli	USA	2,895	23.2
11	Rhône-Poulenc	France	2,679	9.3
12	SmithKline Beckman	USA	2,640	26.1
13	Roche-Sapac	Switzerland	2,436	n.a.
14	Sankyo	Japan	2,346	8.7
15	Abbott	USA	2,333	30.0
16	Warner-Lambert	USA	2,288	26.5
17	Bristol-Myers	USA	2,217	21.8
18	Boehringer-Ingelheim	W. Germany	2,121	6.6
19	ICI	UK	2,085	27.7
20	Upjohn	USA	2,068	20.7
21	Beecham	UK	2,044	31.4
22	Johnson & Johnson	USA	1,993	32.8
23	Schering-Plough	USA	1,865	22.3
24	Wellcome	UK	1,846	15.6
25	Squibb	USA	1,843	24.3

The fact that innovation has been so successful in the past 30 years has also meant that the pharmaceutical industry has grown rapidly, typically by 9–15% per year.

2.2 THE PHARMACEUTICAL INDUSTRY IN THE U.S. ECONOMY

The pharmaceutical industry is responsible for about one-sixth of the sales of the Chemical and Allied Products Industries in the United States. In 1988 sales of pharmaceutical products at the manufacturer's level were about $41 billion out of total Chemical and Allied Products Industry's shipments of $238 billion, that is 17.4%. The Chemical and Allied Products Industry is what the U.S. Census Bureau calls the group of industries that manufacture and formulate chemicals (Table 2.4). It includes, for example, the companies that make monomers for polymers, the companies that convert the monomers into

TABLE 2.4. The Chemical and Allied Products Industry (USA 1985 and 1988)

Industry	Sales 1985[a] ($ billion)	%	Value Added 1985[a] ($ billion)	%	Sales 1988[b] ($ billion)	%
Chemical and Allied Products (total)	197.3	*100*	95.3	*100*	237.7	*100*
Pharmaceuticals	31.3	*15.9*	27.7	*29.1*	41.3	*17.4*
Industrial organic chemicals	41.8	*21.2*	14.9	*15.6*	45.4	*19.1*
Soaps, detergents and toilet preparations	29.6	*15.0*	17.3	*18.2*	35.4	*14.9*
Industrial inorganic chemicals[c]	20.3	*10.3*	11.1	*11.6*	19.7	*8.3*
Plastics and Resins	33.5	*17.0*	12.2	*12.8*	40.4	*17.0*
Coatings	11.6	*5.9*	5.2	*5.5*	11.9	*5.0*
Agricultural chemicals and fertilizers	14.8	*7.5*	5.1	*5.2*	15.3	*6.4*

[a]Bureau of the Census, Annual Survey of Manufacturers
[b]Estimates based on US Industrial Outlook 1989
[c]Excluding pigments

polymers (which may of course be the same companies), and the companies that in turn convert those polymers to plastics, elastomers, and fibers.

The pharmaceutical industry is one of the more important industries in this group and only one sector—industrial organic chemicals—had larger sales in 1988.

If the industries are listed by value added, however, the pharmaceutical industry is much more impressive. Value added is the sales of an industry less the cost of raw materials, utilities, and bought-in services and is a better measure of industry size than overall sales. The pharmaceutical industry share of the value added by the Chemical and Allied Products Industries was about 29% in 1985 and no other sector was larger.

2.3 PROFITABILITY

The pharmaceutical industry is undoubtedly the most profitable sector of the Chemical and Allied Products Industries. Indeed, in 1985, it had the highest return on sales of any industry. Table 2.3 gives the operating, pre-tax profits of the largest U.S. companies in 1987. Operating profit varies from a low of 6.6% to a high of 40.4%. The chemical industry would find it hard to match any but the poorest of these returns.

The industry's profitability, however, is in danger of declining because of the increasingly stringent government regulation, which increases the cost of introduction of new drugs, and legislation about generic prescribing, which decreases profits. Generics will be discussed in Section 2.11. There is also government pressure, particularly in Europe, to reduce drug prices. Certainly, the prices of drugs have increased less than those of other commodities. The consumer price index for all manufactured goods in 1984 was 311 on the basis of 100 for 1967, that is, prices of goods in general had more than tripled. For drugs, however, the 1984 index was only 234. The index for medical care overall was 380. Thus, drugs dropped from 13.3% of total health expenditure in 1950 to 6.7% in 1984 in spite of the greater range and effectiveness of drugs available.

It is clear, however, that profitability, although variable, remains high by the standards of other industries. Declining profits have been obviated by the industry's creativity and this should continue for at least a decade. The complaints of the industry appear to be a symptom more of the high risks associated with it. It takes perhaps ten years to discover a drug and bring it to market. Right up to the launch, and indeed after it, side effects might appear, which can wreck the whole enterprise. The industry feels it gets a bad press, that it is blamed for anything that goes wrong, while getting little credit for the lives it saves.

2.4 CAPITAL INVESTMENT

The pharmaceutical industry makes small tonnages of a large number of complicated organic chemicals in batch equipment. Consequently it is not particularly capital intensive by chemical industry standards. Table 2.5 shows that the pharmaceutical industry took only about 12.7% of the capital investment in the Chemical and Allied Products Industries, but, as shown earlier, it was responsible in 1985 for 15.9% sales and 29.1% value added.

TABLE 2.5. Investment in the Chemical and Allied Products Industry (1988)

Industry	Investment by Industry ($ billion)	% of Investment by Manufacturing Industry	% of Investment by Chemical and Allied Products Industry
All manufacturing	168.0	100	—
Chemical Process Industries[a]	51.9	30.9	—
Chemical and Allied Products	20.4	12.1	100
Drugs	2.6	1.5	12.7
Plastics	3.7	2.2	18.1
Industrial organic chemicals	6.9	4.1	33.8
Industrial inorganic chemicals	2.3	1.4	11.2
Agricultural chemicals	1.5	0.9	7.4
Soap, cleaners and toilet preparations	1.5	0.9	7.4

[a]Includes Chemical and Allied Products; Pulp and Paper; Rubber and Plastics; Stone, Clay, and Glass; Petroleum Refining; Non-Ferrous Metals; Coal Products; Food and Beverages; Textiles.

TABLE 2.6. Research Expenditures in Industry (USA 1987)

Industry	$ Million
All industry	96,050
Chemical and Allied Products	9,950
Pharmaceuticals	5,550
Industrial and other chemicals	4,400

Source: Chemical & Engineering News, 27 July 1987.

2.5 RESEARCH EXPENDITURE

The pharmaceutical industry may not be capital intensive but it is certainly research intensive. Table 2.6 shows the research expenditure within the Chemical and Allied Products Industries. The drug industry accounts for 56% of all funds spent on research by these industries.

Table 2.7 shows the expenditure of four chemical and four pharmaceutical companies in 1986. The average large chemical company in 1986 spent 3.1% of its sales on research and development, whereas a typical pharmaceutical company spends 15%, up from 11.3% ten years earlier.

The reason is not hard to find. Bringing a drug to market is a lengthy and expensive process. In 1970, the last date for which such figures are available, member companies of the Pharmaceutical Manufacturers' Association prepared or isolated 126,000 substances and pharmacologically tested 703,900, many obtained from outside sources. A total of 1013 compounds reached the stage of clinical trials, while only 16 new compounds actually appeared on the market and not all of these were a success.

There was broad agreement between the Pharmaceutical Manufacturers' Association and the FDA (the Food and Drug Administration) in 1987 that

TABLE 2.7. Research Expenditures by Various Companies (1986)

Company	$ Million	% Sales
	Chemical Companies	
Dow Chemical	605	5.4
Union Carbide	148	2.4
Monsanto	596	8.7
Hercules	71	2.7
	Pharmaceutical Companies	
Ciba-Geigy	341	15.0
SmithKline Beckman	251	10.9
Warner-Lambert	162	8.7
Bristol-Myers	242	13.8

TABLE 2.8. Cost of Developing a New Chemical Entity (United States, millions of 1987 dollars)

Testing Phase	Average Cost for a Single Successful Drug ($ million)	Average Cost Including Failures ($ million)	Expenditures Capitalized to Point of Marketing Approval at Alternative Interest Rates	
			5%	10%
Discovery	4.213	33.704	50.17	75.07
Preclinical animal toxicity	0.371	2.972	4.23	5.97
Phase I clinical	0.636	5.086	6.92	9.47
Phase II clinical	3.374	13.482	17.19	21.91
Phase III clinical	5.921	9.008	10.11	11.34
Long-term animal testing	1.609	3.738	4.53	5.45
Totals	16.124	67.990	93.15	129.21

the cost of developing a new pharmaceutical had risen to over $100 million. This figure is so large that it is worth some explanation. The relevant data appear in Table 2.8. The first column shows the stages in the development of a new drug and will be explained further in Section 2.10. A new chemical entity (NCE) is a new drug, a novel therapeutically active substance, not a new dosage form of an established drug. The second column of the table gives the average costs for a single NCE. The discovery and the phase II and III clinical tests are the most expensive parts of the process and the total is a modest $16 million. Not all drugs that are developed, however, complete the testing process successfully; indeed only one in eight of the drugs that start clinical trials receives marketing approval. Column three includes the cost of the failed NCEs. Note that the ratio of the third column to the second decreases as one descends the table, reflecting the fact that the proportion of failures diminishes with an increasing number of tests. The cost of failures approximately quadruples the cost per NCE.

Failures, however, are not the only problem. New drugs take about ten years to develop and test before they bring any return. The money for discovery and testing must either be borrowed, in which case interest must be paid, or it must come from a company's cash flow in which case it could otherwise have been invested and would have earned interest. The cost of developing an NCE should therefore be capitalized up to the point of marketing approval. This is done by a discounted cash flow calculation and gives the "opportunity cost," that is, the interest that is foregone because the money is spent on discovery and testing. The values obtained depend on the interest rates assumed. The value that should be taken is the real interest rate, that is, the current interest rate (the interest receivable on a safe investment) minus the inflation rate. The calculation is shown in column 4 for 5% and

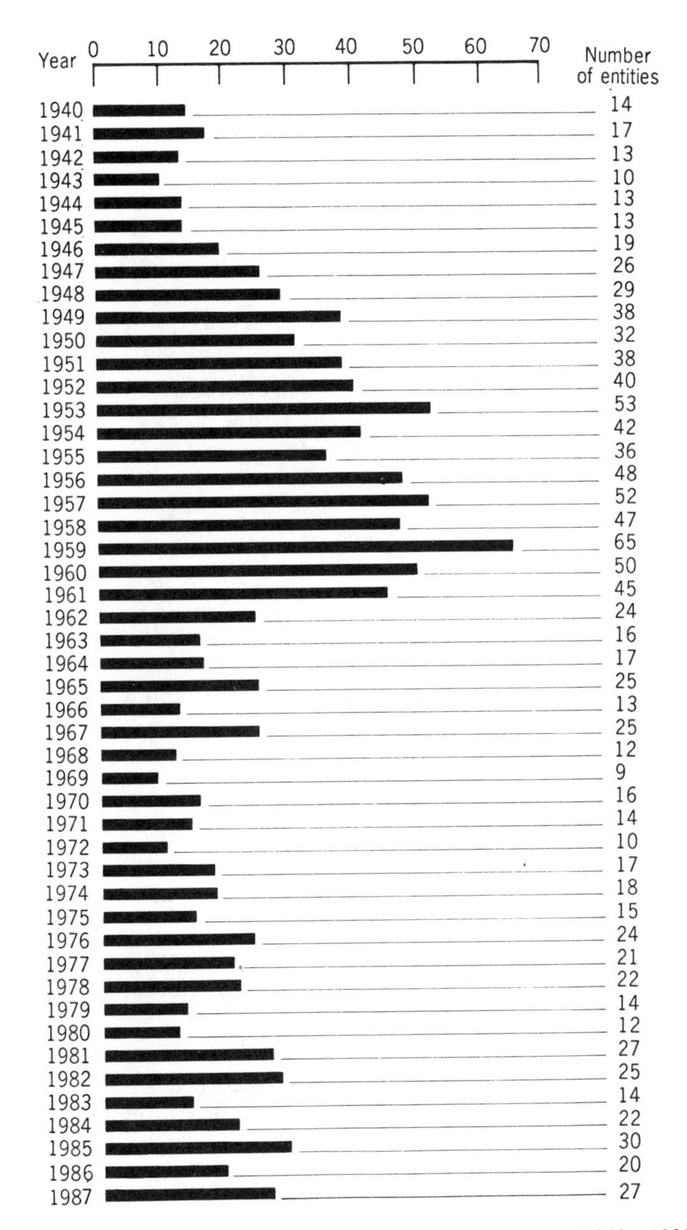

Figure 2.2. *New single entity drug introductions to U.S. Market (1940– 1986). Source: 1940– 1975 New Chemical Entity Data Base; Pharmaceutical Manufacturer's Association; 1975– 1986 U.S. Food and Drug Administration.*

column 5 for 10% and it is these totals that represent the true NCE development cost.

Figure 2.2 shows the introductions of NCEs from 1940 to 1986. Numbers rose steadily until the late 1950s, the height of the first chemotherapeutic revolution. The stringent new regulations in the wake of the thalidomide disaster (1960) (Section 8.1.2) reduced the rate of innovation and new drug introductions in the United States fell 80% between 1959 and 1980. A streamlining of procedures and greater cooperation between the FDA and the drug companies has increased the number of approvals, as has the advent of the second chemotherapeutic revolution. The current rate of approvals—about 25 per year—is determined by the speed at which the FDA can process applications and is only secondarily related to the innovation rate.

The effectiveness of the new drugs that are developed is reflected by the fact that, of the drugs available in the United States in 1969, 92% originated after 1945, 45% originated after 1960, and 20% originated after 1965. Of the drugs available in 1977, 95% originated after 1950. Of the 50 most widely used drugs in 1987, only 18 had appeared on the list eleven years earlier.

2.6 EMPLOYMENT

Because research spending is high in the pharmaceutical industry, many scientists are employed. Data are given in Table 2.9. About 610,000 scientists and engineers were employed in research and development in the United States in 1987. Of these, 66,000, or one in nine, were employed in the Chemical and Allied Products Industries. Of these 66,000, 29,000, or 44% were employed in the pharmaceuticals sector despite the fact that pharmaceuticals account for only about 17% of the sales of the industries. The pharmaceutical industry employs 18% of all chemists.

TABLE 2.9. R & D Scientists and Engineers in Industry (USA, 1987)

Industry	Thousands	%
Chemical and Allied Products	66.2	10.9
Industrial chemicals	27.1	4.4
Pharmaceuticals	29.0	4.8
Other chemicals	10.1	1.7
Motor vehicles[a]	34.8	6
Electrical equipment[a]	116.1	20
Aircraft and missiles[a]	104.5	18
Machinery[a]	81.2	14
Other industry	174.0	30
Total	610	

Source: Chemical & Engineering News, 20 June 1988.

[a]1986.

Not only are pharmaceutical industry employees frequently scientists but management too is often technically oriented. In the chemical industry, top managers generally have business, accountancy, or legal training. In the pharmaceutical industry, a far higher proportion of top management started their careers as scientists. One reflection of this is the pharmaceutical industry's continuing commitment to research.

2.7 PATENT PROTECTION

Because innovation is so important in the pharmaceutical industry, patent protection is always sought for new drugs. This grants an inventor the sole right to exploit an invention for a period of 17 years in the United States, 20 years in Britain, and similar periods elsewhere. Although patent protection is weak in some countries such as Spain or absent (as in Italy before 1978), effective patent protection is frequently possible in the pharmaceutical industry. The inventor of a successful drug can still expect rich returns providing not too much of the patent time is taken up with development work. In 1986, two drugs—the anti-stomach ulcer compounds cimetidine and ranitidine—both brought in over one billion dollars in revenue worldwide.

The length of time taken to develop a new drug has shortened—frequently more than halved—the time available for a company to recoup its development costs. In 1987 the processes of invention and testing took, on average, 10.5 to 12 years. In 1962 the tests were less elaborate and took only two years. This has been a constant complaint of the pharmaceutical industry. In particular, a delay by the FDA in assessing evidence or processing documents might cost a company millions of dollars in lost revenue. The legislators had intended inventors to have a 17-year monopoly and not, as now, only eight or nine years.

The 1984 amendment to the Federal Food, Drug and Cosmetic Act went some way toward meeting the industry's complaints. It extended the patent term of pharmaceuticals and biologicals by *half* the time from which the first investigational new drug application (IND) became effective until the date of filing of the new drug application (NDA) plus *all* the time the FDA takes to review the new drug application up to a maximum of 5 years.

A limitation is that a company cannot have more than a 14-year patent life by this method and time can also be subtracted if the FDA feels the company is not pursuing the project with sufficient diligence.

Furthermore, marketing exclusivity for 5 years is granted to a company that markets a chemical entity for the first time even if the product is not patentable. Three years' exclusivity may be granted to a new mixture, but there is always the possibility of a full 17 years if the idea is sufficiently novel.

The Japanese government has also legislated for patent term restoration and the European Community has moved a long way down the path of a unified 20-year patent law. Sweden and Switzerland who are outside the Community have also acceded to the European Patent Convention.

2.8 COMPETITION AND PROMOTION

The pharmaceutical industry is highly competitive and spends a great deal of money on promotion. Competition is of three kinds. First, there is competition between different companies to make the same drug, for example, a tetracycline no longer covered by patents. Price and the reputation of the company may be selling points; the situation is little different from consumer goods generally. Second, there is the competition between "me-too" drugs, that is, patent-protected drugs that differ from one another chemically but do more or less the same job medically. Many antiarthritic drugs fall into this category. Finally, there is competitive innovation where a company tries to find a new and better drug that will displace another patent-protected drug from the market or treat a hitherto intractable disease. This last is likely to bring the largest profit but also to involve the highest risk. The efforts of research chemists to make obsolete an established and successful drug is one of the main themes of pharmaceutical industry research.

In all these situations, the pharmaceutical companies see intensive advertising to the medical profession as indispensable. General practitioners receive five or ten advertising leaflets each day and are visited regularly by sales representatives, the so-called detail people, who in the United States often have medical training. The job of the sales representatives is to keep physicians up-to-date with what their company is doing and to persuade them to prescribe their products. The inducements offered to physicians to prescribe particular drugs are often substantial. It is the job of the industry associations to monitor the various schemes and to take action when the bounds of ethical behavior have been overstepped.

2.9 MANUFACTURE OF PRODUCTS

Three hundred and fifty billion pounds of organic chemicals and polymers are manufactured annually in the United States, compared with a total pharmaceutical production of about 300 million pounds. Ethylene is the largest volume organic chemical with 37 billion pounds manufactured in 1986. The largest tonnage chemical exclusively for pharmaceutical use is aspirin and United States production is about 32 million pounds. Thus, the pharmaceutical industry produces by volume only one-thousandth of the total organic chemicals and polymers. Compare this with the proportion of its sales and employment of scientists and engineers.

TABLE 2.10. The 18 Largest Volume Pharmaceuticals Consumed (Data for United States, Western Europe, and Japan, 1985)

Compound	Activity	Value ($ million)	Consumption (millions of pounds)	
			Medicinal	Total
Ascorbic acid	Vitamin	325	36.4	66.1
Penicillin G	Antibiotic	240	4.2	18.3
Vitamin E	Vitamin	230	7.2	14.3
Ampicillin	Antibiotic	150	3.7	3.7
Aspirin	Analgesic	115	55.5	55.6
Acetaminophen	Analgesic	100	31.3	31.3
Penicillin V	Antibiotic	100	1.5	6.8
Riboflavin	Vitamin	90	1.8	4.7
Sulfonamides	Antibiotics	90	8.0	11.0
Thiamine	Vitamin	85	3.6	5.5
Niacin and derivatives	Vitamin	65	4.8	19.4
Calcium pantothenate	Vitamin	60	1.8	8.8
Caffeine[a]	Stimulant	45	4.4	9.9
Chlortetra-cycline	Antibiotic	40	0.3	2.5
Pyridoxine	Vitamin	40	2.1	2.6
Tetracycline	Antibiotic	35	0.8	2.9
Oxytetracycline	Antibiotic	30	1.6	2.6
Theophylline	Bronchodilator	15	3.0	3.0
Total		1855	172.1	269.0

[a]Synthetic caffeine only.
Source: S.C. Stinson, *Chem. and Eng. News,* 16 September 1985

Numbers of prescriptions are taken throughout this book as a measure of a drug's importance but it is worthwhile at this stage to record, instead, the drugs that are made on the largest scale. Table 2.10 shows the 18 largest tonnage pharmaceutical chemicals. They account for 82% of the total tonnage of pharmaceuticals but only 40% of value, the smaller tonnage materials naturally being more expensive. If only the medical uses of the chemicals are included (i.e., excluding uses as food additives, feedstocks for other chemicals, etc.) they account for only 53% of pharmaceutical tonnage.

At the other extreme from the compounds in Table 2.10 are the drugs whose worldwide consumption is tiny. Three hundred pounds of the anti-cancer drug methotrexate are consumed annually; 300 pounds of the antihypertensive drug clonidine are sold. The steroids represent a large number of drugs taken widely both for medicinal purposes and in oral contraceptives. Yet they are so potent that world consumption is only 300,000 pounds for all the steroids together.

The majority of the large volume drugs are antibiotics, analgesics, and vitamins. Except for the antibiotics, they are available without prescription and are consumed in large quantities for this reason.

2.10 DRUG TESTING

How does the pharmaceutical industry know if a drug is effective and safe? The answer is through elaborate and expensive preclinical and clinical trials that must satisfy the U.S. Food and Drug Administration (FDA), the Council for the Safety of Medicines (CSM) in the United Kingdom, and similar regulatory bodies in other countries. In the nineteenth century there was virtually no control. The Food and Drugs Act of 1906 required manufacturers to meet certain standards, primarily that the material be properly labeled but not necessarily that it be safe.

In 1937 S. E. Massengill, a small company in Tennessee, prepared an elixir of sulfanilamide in diethylene glycol, a toxic solvent. Ninety-three people died. The chief chemist certainly felt responsible, for he committed suicide. The law was so weak, however, that it was possible to convict Massengill only of "adulteration and misbranding" for calling his medicine an elixir which, strictly speaking, should have been alcohol-based. The company was fined a total of $16,800 but the episode precipitated amendments to the Food and Drug Act requiring that manufacturers test candidate drugs for safety in animals prior to human trials.

The view gained ground in the 1950s that it was insufficient for a drug to be proved safe and that evidence of efficacy should also be required. In the aftermath of the thalidomide disaster (Section 8.1.2), amendments to the Food, Drug and Cosmetic Act in 1962 formalized the practice of proving efficacy and introduced the investigational new drug procedure (see below). The pattern established by the 1962 amendments has remained broadly unchanged; the 1984 amendments were discussed in Section 2.7.

Figure 2.3 shows the pattern of evaluation. A program of chemical synthesis in the laboratory, inspired perhaps by understanding of the biochemistry of the systems involved, produces a drug with a degree of the desired physiological activity in animals. The activity is refined by structural modifications until a chemical structure believed to be close to the optimum is achieved. Meanwhile, the biochemical metabolism of the various chemical structures is studied, together with the other activities on the left of the figure. A single candidate drug emerges from this process that has the appropriate physiological effect on animals.

Massive doses are then given to animals—typically mice, rats, rabbits, and guinea pigs—for three or four days and their organs are examined. These are the acute toxicity tests and if they show severe toxicity the candidate drug is rejected. If the drug passes the acute tests, it is subject to subacute tests in which animals are given smaller doses for 30–90 days. Acceptable results encourage the company to apply to the FDA for investigational new drug status (IND) to permit the drug to be tried by humans.

Clinical trials on humans have four phases. In phase I, tests are performed on normal people to prove that the drug does not harm them and to determine how they metabolize it. The phase II trials begin with a limited number of sick

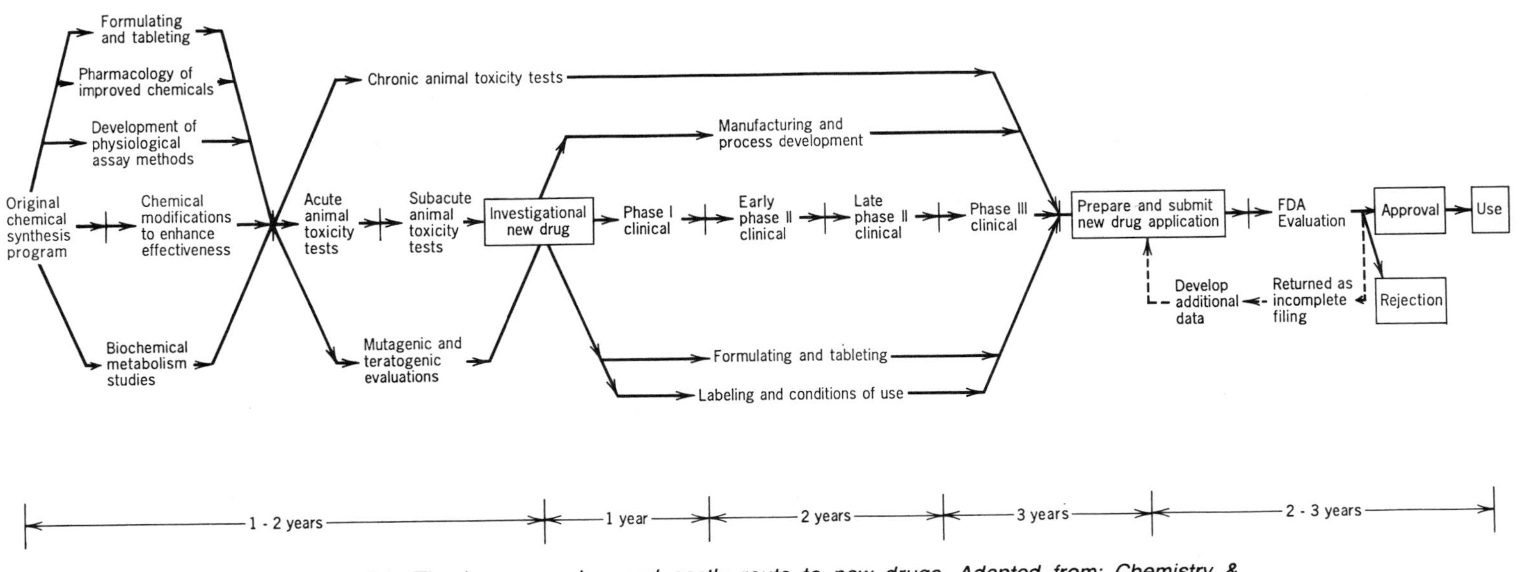

Figure 2.3. *The long, complex, and costly route to new drugs. Adapted from: Chemistry & Medicine— The Legacy and the Responsibility, Commitee on Chemistry and Public Affairs, Washington DC, Copyright 1977 American Chemical Society. Figures also from Pharmaceutical Manufacturers' Association, Facts at a Glance, 1987.*

people to see if the drug genuinely cures disease and has few and acceptable side effects. These are open trials and the physicians and sometimes even the patients know whether they are receiving the new drug, an older approved drug, or a placebo. The later phase II trials involve much larger numbers of people and double-blind testing, in which neither physicians nor patients know what is being administered. It is the phase II trials that establish whether a new invention has taken place and the drug will genuinely cure disease. Double-blind tests with placebos and batteries of statistical techniques are employed to ensure that the highly expensive phase III tests are not started needlessly. Parallel with the clinical testing, long-term chronic animal toxicity tests are undertaken. The FDA encourages an end-of-phase II conference with them.

Successful phase II testing leads to phase III tests, which are designed to confirm statistically the results of phase II, discover side effects undetected in the smaller samples, show interactions with other drugs, and so on. The tests are expensive and involve on the average about 2000 patients in several countries.

Companies then prepare a new drug application (NDA) with:

1. A full report of investigations to show the drug is safe and effective.
2. A full list of components.
3. A full statement of the composition of the drug.
4. A full description of the methods, facilities, and controls used for manufacturing, processing, and packaging the drug.
5. Samples of the drug, components, and standards that may be required.
6. Specimens of the labeling.

An NDA runs to thousands of pages and the FDA is required to complete its initial evaluation in 6 months. It can accept it, reject it, or declare an "incomplete filing" and ask further questions or request further data. It will normally classify drugs as offering an important therapeutic gain or little or no therapeutic gain and give them corresponding priority. There is also a distinction between new molecular entities (NME), which seem to be identical with the new chemical entities (NCE) described in other literature, and new combinations, formulations, and dosage forms.

Once the NME has been approved, marketing can go ahead, but the cautious company will monitor performance and watch for side effects that could not be detected even during phase III. There is, thus, an informal phase IV involving postapproval testing. Companies pursue aggressive marketing policies and the buildup of sales to a plateau level will usually take only 1–2 years.

The approval of an NME is a lengthy procedure. There is currently pressure for phase III to be shortened and phase IV to be made formal to permit drugs to be brought to market more quickly. In 1981 the FDA instituted a stream-

TABLE 2.11. Failures in Present Methods of Toxicity Testing

1. Effects observed in Humans but Not in One or More of the Species Commonly Used in Toxicity Testing

Chemical[a]	Effect	Species in Which Effect Not Observed
Thalidomide	Fetal abnormalities	Most rat strains[b]
Practolol	Skin, eye, and intestinal lesions	All species examined
Halothane	Liver failure	All species examined
Oral contraceptive pill	Thrombosis	All species examined
Dinitrophenol	Cataracts	All mammals examined
Azauracil	CNS effects	All mammals examined
Clioquinol	Optic and peripheral neuritis	All mammals examined
TCDD	Chloroacne	All except hairless mouse
Benzene	Leukemia[c]	All mammals examined
β-Naphthylamine	Bladder cancer[c]	Rats and rabbits

2. Effects Observed in One or More Common Laboratory Animals but Not in Humans

Chemical	Effect	Species in Which Effect Observed
Oral contraceptive pill	Bleaching of retina and cancer	Monkey and mice
DDT, dieldrin, and phenobarbital	Liver tumors	Mice
Penicillin	Rapid death after single dose	Guinea pig
Aspirin	Fetal abnormalities	Rats

Reproduced by permission from J. V. Bridges and S. A. Hubbard, Principles, Practice, Problems and Priorities in Toxicology, In *Current Approaches in Occupational Health*, Vol. 2, W. Gardener (Ed.), Wright, Bristol, 1982.

[a]Some of these drugs were not subjected to the rigorous animal testing now required.
[b]Thalidomide does cause fetal abnormalities in rabbits.
[c]Although these specific cancers are not observed, both materials are carcinogenic in animals.

lined program of review and approval. Nonetheless, the mean approval time for the 141 NMEs approved between 1981 and 1986 was 32.2 months—almost three years. This figure is coupled with a standard deviation of 21.3 months indicating a wide scatter in approval times.

Preclinical and clinical trials are expensive and drawn out. Worse, the results of animal tests are difficult to evaluate. The short-term effects of large doses of a chemical on mice may be quite different from the long-term effects of small doses on humans. Table 2.11 gives examples of drugs for which the method breaks down. The first section of the figure lists some drugs that have deleterious effects on humans but not on some test animals, while the second half lists some that are toxic to some test animals but not to humans.

It is ironic that aspirin causes fetal abnormalities in rats while thalidomide does not. A Pfizer statistician in the mid 1980s found that of 19 known human carcinogens, only 7 had been shown to cause cancer in rodents, and he concluded that flipping a coin was as satisfactory a method of determining carcinogenicity in humans. This is obviously an exaggeration, and we doubt that anybody would buy drugs that had not been tested first on animals, but improved methods of testing are desperately needed.

Such a dilemma is a challenge to the scientific community and there are now indications that a breakthrough has been achieved. The new technique is known as receptor technology. Receptors are the sites in the body to which drugs must bind to exert their effects. This will be discussed in more detail in Section 4.2. It is now possible to isolate cell membrane receptors from organs such as the brain and peripheral nerves, the heart, blood vessels, lungs, and intestines. It thus becomes possible to screen compounds in test tubes instead of in animals. Scientists can determine quickly whether a compound is active in the body and, if so, where it acts. Receptor technology is much faster and cheaper than animal testing. In animal screening, several days may be required to test one compound. Receptor technology can even be automated so that hundreds of compounds can be assessed daily for their effects on scores of receptors. As an added bonus, the number of animal deaths can be reduced, which is a cause for satisfaction not only to the animal lobby but throughout the pharmaceutical industry. The number of animal experiments per year was estimated in 1985 as 100 million worldwide.

The FDA has reacted sympathetically to the new technique and has given speedier-than-normal approval to at least one drug because receptor technology reinforced other evidence for the drug's safety and efficacy.

2.11 DRUG NOMENCLATURE AND GENERIC DRUGS

The preceding sections dealt especially with exciting new drugs, but there are many old drugs whose patents have expired but which are still of value. They are newsworthy at present because governments and consumer organizations are trying to reduce drug costs, whereas drug companies are correspondingly reluctant to sacrifice profits. Out-of-patent drugs are frequently available under a bewildering variety of names and we begin this section by describing drug nomenclature.

Every drug has at least three names: a chemical name, a generic name, and a trade or brand name. The systematic chemical names of the complex chemicals that are used by the drug industry are long and unwieldy. Accordingly, the industry decided to institutionalize the use of trivial or so-called generic names such as amphetamine or tetracycline. Any individual manufacturer may submit a generic name to the U.S. Council on adopted names sponsored by the American Medical Association, the U.S. Pharmacopeial Convention, and the Food and Drug Administration. If the Council agrees to the generic name, it is published in the *Trademark Bulletin* of the Pharmaceutical Manufacturers' Association. Anyone who believes that the generic name is too close to that of another drug or who has any other objections to the new name may voice them at this point. If no objections are raised, the name is submitted to cooperative organizations worldwide. These include the British Pharmacopeia Council, the French Codex Commission, and the Nordic Pharmacopeia Council. When everyone has agreed, the final step is the publication of the new name in the "New Names" section of the *Journal of the American Medical Association*. Thereafter, any company manufacturing this drug uses the generic name. To indicate that the drug is manufactured by a particular company, however, the manufacturer will attach to it a trade or brand name, the third name by which the drug is known.

As a simple example, the compound 1-phenyl-2-aminopropane is a central nervous system stimulant. It has the generic name amphetamine and one of its many trade names is Benzedrine. Note that the name is capitalized but the generic name is not.

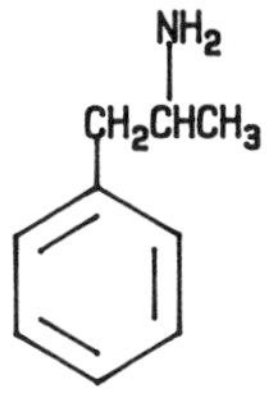

Generic name: amphetamine

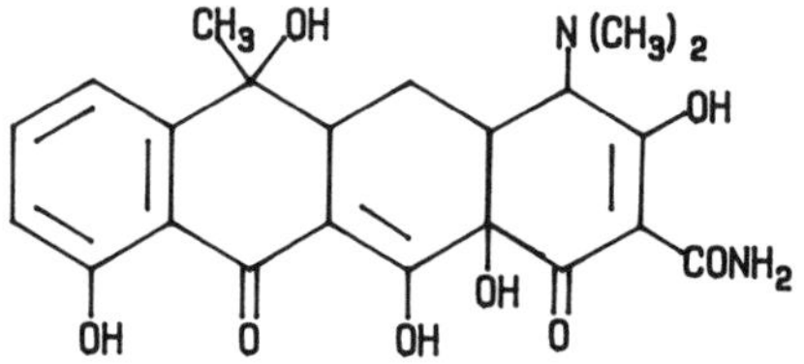

Generic name: tetracycline

A more complex example: The generic name of one of the most important antibiotics is tetracycline (above). Its chemical name is 4-(dimethylamino)-1,4,4a,5,5a,6,11,12a-octahydro-3,6,10,12a-pentahydroxy-6-methyl-1,11-dioxo-2-naphthacenecarboxamide. This is certainly more of a mouthful than the generic name and is the justification for it. Tetracycline has numerous brand names including Achromycin, Sumycin, Tetracyn, Hostacyclin, Steclin Panmycin, Tetrabon, and Ambramicina. The *Merck Index* lists the generic names of drugs and adds numerous brand names.

In spite of the elaborate arrangements for generic names to be internationally recognized, confusion still occurs. For example, the hormone epinephrine is known in Europe as adrenaline; the analgesic acetaminophen is known as paracetamol, and the antiasthmatic albuterol is known as salbutamol. Because salbutamol was discovered in Europe and licensed for use in the United States almost a decade later than in Europe, the United States generic name scarcely occurs in the literature.

A standard complaint against the drug industry is that even physicians, let alone the general public, become confused when a single chemical entity is sold under a host of brand names. The antidepressant, imipramine, appears round the world under about 24 different names. Aspirin has over 200. The average pharmaceutical is said to have 8 brand names. The multiplicity of names had tragic consequences in the thalidomide disaster because when the birth-deforming effects of Contergan, the German brand name for thalidomide, were reported in Sweden, the Swedes did not immediately recognize it as Neurosedyn, the Swedish brand name.

The multiplicity of names, however, serves a commercial purpose. Chemically identical drugs, with perhaps minor differences of formulation and packaging, are sold in different countries under different names at different prices (Section 2.1). In general, drugs are more expensive in rich countries such as Germany and cheaper in poor ones such as Spain. Thus, to some extent, the richer countries subsidize the poorer ones.

If drugs had the same brand name in all countries, it would pay pharmaceutical wholesalers to set up systems of parallel imports to the detriment of the pharmaceutical manufacturers. They would buy drugs cheaply in Spain, ship them to Germany and resell them at a profit. The product differentiation brought about by the different brand names makes this impossible.

This differentiation is threatened within the European Community by the removal of non-tariff trade barriers in 1992. The legislation is intended to remove price distortions so that uniform prices prevail throughout Western Europe as they do throughout the United States. If this degree of integration is imposed, then the number of brand names will decrease. Probably the Germans will pay less for their drugs while the Spaniards pay more. The Spaniards may benefit in other ways from European integration but the issue is essentially a political one.

2.11.1 Generics versus Branded Pharmaceuticals

While a drug is patent protected, there will be only one brand name in a given country. Whether the physician prescribes by generic or brand name, the consumer will get the patent holder's product. When the drug moves out of patent, however, a number of manufacturers may start to produce it under different brand names. Until recently, if the physician prescribed by brand name, the consumer would get a particular company's product; if the physician prescribed by generic name, the consumer might decide which brand to buy since there was often a disparity between the prices of different brands. For example, a major national pharmacy chain in 1987 charged \$31.94 for 100 tablets of Diabinese, the brand name for a drug for diabetics. The same chain charged \$7.59 for the same quantity of chlorpropamide, the generic version of Diabinese.

Brand and generic names have invaded the American consumer's life in the past few years because legislation has been passed in all fifty states that allows pharmacists to substitute a less expensive generic version even when the physician prescribes by brand. Indeed, the physician must make a positive effort to insist on a brand name drug. In some states "prescribe as written" must appear on the prescription form; in others the physician must sign on one line of the form rather than on another.

Generic prescribing was also encouraged by the Waxman–Hatch Act of 1984. Prior to that, generic copies of products that first appeared after 1962 were treated as new drugs by the FDA and full NDAs had to be submitted at prohibitive cost. Hence, generics were confined to pre-1962 drugs, mainly antibiotics. The Waxman–Hatch Act of 1984 instituted the abbreviated new drug application in which applicants need only prove to the FDA that they have formulated the product correctly, using the same amount of the identical active ingredient as the brand name version. The evidence required is a statement of chemical composition and bioavailability (Section 4.9).

Legislation encouraging the use of generic names encourages small manufacturers with low overheads and virtually no research costs to enter the drug business, manufacturing products on which the patents have expired. They then are able to compete on price. The large, research-based drug companies are, of course, bitterly opposed to generic prescribing and, according to *Consumer Reports*, many of them have used less-than-ethical methods to counteract it (see bibliography). The core of their arguments is that generic drugs are not identical, that the quality control of generic firms is inadequate, and that the bioavailability of the generic drugs is open to question.

Their case is bolstered by two episodes in Britain. In the early 1960s, the British National Health Service, always hard-pressed for money, started to buy patent-protected tetracyclines cheaply from unlicensed manufacturers in Italy and Poland, countries that did not subscribe to patent conventions. This was possible because certain clauses in the Patents Act did not apply to government bodies. In 1965 it was shown that many of the cheap tetracyclines had been poorly manufactured and had deteriorated to the point where only a fraction of their activity remained. Possibly some of the material had been defective in the first place. At any rate, the National Health Service returned to buying from the American and British patent holders and their licensees.

The second episode occurred in 1972 and became known as the digoxin affair. It is described in Section 4.9. It was shown that chemically identical drugs were absorbed at dramatically different rates by the body because of apparently trivial differences in preparation and formulation. Quality compliance with official standards did not guarantee identical pharmaceutical effects.

The force of these arguments is diminished by the FDA's insistence on bioavailability testing. Drugs such as digoxin where bioavailability problems may arise are clearly identified.

There is apparently no reason to fear recurrence of these episodes and the attacks on ill-equipped generic manufacturers are vitiated by the fact that some 80% of generics are made by the research-based pharmaceutical companies themselves.

Penetration of generics in the United States market is the highest in any country. About a quarter of all prescriptions were written for generic products in 1987 13.6% of first prescriptions. The discrepancy is due to the fact that patients who require frequent repeat prescriptions are more likely to make efforts to reduce their costs. Because of generic substitution, however, about 43% of first prescriptions were dispensed generically.

In Table 5.1, we list the top 100 drugs in the United States by numbers of prescriptions. Only one of the first 16 compounds on this list and about 10 in the top fifty are still under patent.

In no European country is there legislation permitting generic substitution, although in the United Kingdom there is a "blacklist" of brand name drugs where only the generic form may be prescribed on the National Health Service. Partly as a result, the percentage of the sales value of the top 100 drugs attributable to patented products has dropped from 73% in 1973 to 49% in 1985. In 1984, generics claimed 16.9% of the U.K. market by number of prescriptions and 4.6% by value. Under 3% by value of the European market is attributable to generics but both this and the American figure seem likely to increase. West Germany also has a "blacklist". France, Italy, and Spain have positive lists in which only named drugs qualify for reimbursement.

The legislation on generic names is motivated by the American belief in free competition. The patent holder had a right to a monopoly until the patent expired. The Waxman-Hatch Act improved the companies' patent position. Now, the customer has the right to make a choice, assuming, of course, that government standards assure equivalent products. Free competition will keep prices low. On the other hand, competition will reduce the profitability of the research-oriented drug companies. One can scarcely be surprised if they complain.

3

Patterns of Illness and Health Care

Medical care is expensive. Total health care expenditure in the United States comes second only to expenditure on food, drink, and tobacco and amounted in 1985 to $425 billion. It has risen from 4.5% of the GNP in 1950 to 10.7% in 1985. The chemotherapeutic revolution has also cost money. Expenditure on pharmaceuticals in the United States has increased from $1.6 billion in 1950 to $28.5 billion in 1985. At first sight this is a huge increase, but measured in constant 1985 dollars it is an increase only from 7.15 to 28.5 billion. As a percentage of total United States expenditure on health care it has dropped from 13.3 to 6.7%. As a percentage of gross national product, it has increased from 0.60 to 0.71%, a small price to pay for the enhanced effectiveness of modern drugs.

3.1 ETHICAL AND OVER-THE-COUNTER DRUGS

Pharmaceuticals are manufactured by the pharmaceutical industry but they are consumed by people. Sometimes the patients themselves are permitted to buy drugs and sometimes drugs must be prescribed for them by a physician.

Pharmaceuticals regarded as harmless may be bought directly by the patient and are described as proprietary or over-the-counter drugs; pharmaceuticals that are potentially dangerous are described as ethical pharmaceuticals and must be prescribed by a physician.

Ethical pharmaceuticals are frequently more active pharmacologically than proprietary pharmaceuticals. Proprietary pharmaceuticals are often based on traditional recipes or on chemicals such as aspirin, generally regarded as safe.

They are backed by advertising to the general public, although the claims for them may be limited by law. Conversely, ethical pharmaceuticals are often patent-protected and are advertised only to the medical profession.

In 1986, sales of ethical drugs in the United States were $15.2 billion while sales of over-the-counter drugs were $4.8 billion, a 76–24 split.

Because patients are the final consumers, it is their illnesses that determine the therapeutic group of the pharmaceuticals that are purchased. Within the group, however, choice of the appropriate compound rests for 70% of the time with the physician. Thus, the pattern of pharmaceutical industry sales is dictated by patients and their illnesses, and physicians and their preferences.

3.2 PHYSICIANS

In the developed world, it is taken for granted that physicians will be adequately trained and that most people will have access not only to physicians but to general medical care. This is not the case worldwide. In Table 3.1, the first two columns of figures show the number of people per physician and per nurse in 50 countries. Developed countries in general have fewer people per physician than developing ones. The pattern is broadly what one would expect but not entirely so and cultural factors clearly influence the amount spent on health care and its efficacy. The Soviet Union and East European countries are well supplied with physicians, partly because of general Communist ideology and partly because the study of medicine provides an outlet for enterprising people who, in the West, might have channeled their energies into capitalist endeavors. Israel, which economically is only on the fringe of the developed countries, has the fifth fewest people per physician. This is perhaps of cultural significance since the Jews have held the practice of medicine in esteem since medieval times, and even earlier. The appearance of Spain in fourth place—up from twelfth place in 1977—is less easily explicable.

At the end of Table 3.1 come the sad countries of the third world with no medical services outside the large towns and scarcely any within them. The range is spectacular, Ethiopia having 246 times as many people per physician as the Soviet Union.

The spread in the number of people per nurse is less spectacular and Bangladesh, Afghanistan, and Nepal are exceptional. This presumably reflects the relative ease of providing a trained or partly trained nurse compared with training a physician.

Certain countries, for example, Spain and Italy, have many physicians but few nurses and this may reflect the ease of qualifying as a physician in those countries as well as the possibility that physicians have to do many jobs done elsewhere by medical auxiliaries.

How far does a population benefit from higher numbers of physicians and nurses? Comparisons of levels of health care in different countries are difficult, but a sensitive indicator is infant mortality. Column 3 in Table 3.1 shows the

TABLE 3.1. People per Physician (~ 1980) and per Nursing Person, Infant Mortality (1980 – 1985), and National Income per Head (1986) in 50 Selected Countries

Country	People per Physician	People per Nurse[a]	Infant Mortality[b]	National Income per Head ($) (1986)
USSR	290	n.a.	25	4,110[c]
Czechoslovakia	354	139	16	1,290[c]
Hungary	390	179	21	1,940
Spain	390	295	12	4,360
Israel	403	127	15	4,920
Switzerland	409	152	8	16,380
West Germany	442	181	13	10,940
Sweden	454	107	7	11,890
Italy	490	330[c]	14	6,520
East Germany	494	n.a.	12	6,430[c]
Argentina	530	n.a.	36	2,130
United States	549	144	12	16,400
Australia	556	116	11	10,840
Poland	610	214	21	2,120
France	610	151	10	9,550
Kuwait	686	193	30	14,270
United Kingdom	750	267	12	8,390
Japan	761	209	8	11,330
Egypt	850	771	113	680
Venezuela	930	345	39	3,110
Cuba	1110	369	20	1,410[c]
Hong Kong	1163	379	12	6,220
Peru	1550	918	99	960
Chile	1620	419	40	1,440
Ecuador	1620	n.a.	77	1,160
Turkey	1631	1250	110	1,130
Brazil	1700	2440	71	1,640
Mexico	1820	1396	53	2,080
China	1908	1175	38	310
Paraguay	2150	870	45	940
Guatemala	2490	n.a.	68	1,240
India	2546	1857	118	250
Saudi Arabia	2604	1390	103	8,860
Pakistan	3175	4492	120	380
Honduras	3290	400	82	730
Philippines	6711	2591	50	600
Thailand	6849	1104	51	830
Sudan	8696	1343	118	330
Bangladesh	8929	14922	133	150
Nigeria	9615	1179	114	760
Kenya	11630	3417	94	290
Afghanistan	13514	9111	205	170[c]
Ivory Coast	15220	1475	122	1,040[c]
Zaire	15530	1574	107	170
Tanzania	17550	n.a.	n.a.	270
Uganda	22222	2009	98	290[c]
Nepal	28571	7448	144	160
Chad	41940	4434	143	110[c]
Burkina Faso	55555	3567	149	140[c]
Ethiopia	71430	4116	143	110

[a]Nurses include midwives and nursing auxiliaries.
[b]Per 1,000 live births.
[c]1979.

rates of infant mortality between 1980 and 1985 in the various countries. Some countries such as the United States, Sweden, Australia, the United Kingdom, Japan, and Hong Kong seem to get full value from their medical personnel. The Soviet Union, Hungary, Kuwait, and Argentina are doing notably worse.

The fourth column of Table 3.1 lists national income per head. In a simple world, countries would presumably be in the same rank order for national income as they are for people per physician. This again is clearly not so and the cultural factors mentioned earlier come into play and determine the proportion of national income devoted to health care.

What happens to sick people who either cannot afford to visit a physician or do not have access to one? Some manage to consult medical auxiliaries, pharmacists, or witch doctors. Some die when suitable treatment would have saved them. Some would have died anyway. In many countries where physicians are scarce, the laws governing sale of ethical pharmaceuticals are lax and sick people are able to buy powerful drugs on the recommendation of friends, sometimes in collaboration with a pharmacist. Even Switzerland, where doctors abound, is relaxed about what citizens may buy and swallow.

The conclusion is that access to physicians, medical care, and drugs varies widely among countries and that these patterns are of importance to an international drug company.

3.3 PATIENTS

One might think of patients as being the consumers of pharmaceuticals but this is not strictly true. The products of the pharmaceutical industry are consumed not only by people who are ill but also by people in robust health. Vaccination and immunization against diphtheria, polio, tetanus, whooping cough, and measles are carried out on healthy subjects. Oral contraceptives are taken by healthy women as a preventive measure, although it is pregnancy not illness that is feared. Other pharmaceuticals taken by healthy people include laxatives, multivitamin pills, and a host of proprietary tonics. These are of questionable value although, for example, the consumption of massive doses of vitamin C as a cold preventive has influential supporters in the scientific community.

The majority of pharmaceuticals are nonetheless consumed by sick people. What are the patterns of sickness within a community and how do these influence the demand for pharmaceuticals? Measures of community sickness include mortality, disability, visits to the physician, and numbers of prescriptions. These will be discussed in turn.

3.3.1 Mortality

Figure 3.1 shows the causes of death in the United States, the United Kingdom, and Egypt in the 1980s. They are subdivided according to the

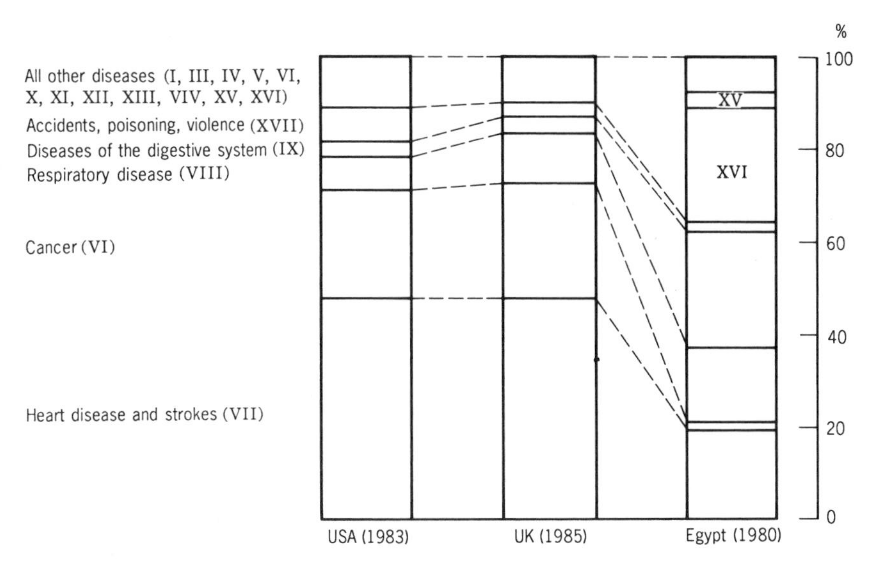

Figure 3.1. *Causes of death in the United States, the United Kingdom, and Egypt.*

International Classification of Diseases which is what the Roman numerals refer to. This classification is given in Appendix 2 together with a reference to a book containing a more detailed list.

We have chosen Egypt as our developing country for purposes of comparison; the comparison is not meant as criticism. Egypt is trying harder than most developing countries to meet the problems of a high birth rate, and, as a spin-off from this, publishes detailed mortality statistics. How unusual this is may be judged from the fact that only one other country in Africa—the tiny Mauritius—also does so. In Asia, only Japan, the Philippines, Thailand, Israel, and Kuwait do so.

Of all of the causes of death recognized by the international classification, only the major ones have been displayed in Figure 3.1. The five categories individually listed make up about 90% of deaths in the United States and the United Kingdom, but only 64% in Egypt. The difference is due to the omission of ICD groups XV and XVI. Category XV is "certain causes of perinatal mortality" which is responsible for 374 deaths per 100,000 in the under 1 age group in Egypt. Perinatal mortality consists of stillbirths and deaths within four weeks of birth. High perinatal and infant mortality is a characteristic of developing countries. Nine percent of Egyptian children die before their first birthday and another 2% before their fourth birthday. Terrible as this figure is, it is half to two-thirds the figure in third world countries that have not made Egypt's efforts in public health. It is a mere fifth of the figure in working-class industrial slums in England in 1840 when one in two children died in his or her first year.

Category XVI represents senility and ill-defined causes and contains a quarter of Egyptian deaths. In the West, a physician would be reluctant not to define a cause of death, and old age is scarcely recognized as a reason for demise. Hence, the numbers in category XVI are negligible in the United States and the United Kingdom.

About half the deaths in the United States and the United Kingdom are from cardiovascular illness—heart disease and strokes. Between a quarter and a fifth are due to cancer. That leaves about 30% of deaths to be attributed to other causes.

Over twice as high a proportion of Americans die in accidents, poisoning, and violence as do British citizens and, in spite of the vastly higher murder rate in the United States, about half these deaths are road accidents. The British are more prone to car accidents than the Americans, but there are more cars per person in the United States and they are driven more miles, so the death rate per 100,000 persons is higher. A death rate per 100,000 cars would be lower for the United States.

Another transatlantic difference is that the U.K. death rate from respiratory disease is 50% higher than that in the United States. The main component of this is bronchitis deaths where the UK rate, 45.3 per 100,000, is about four and a half times the U.S. rate. Bronchitis is a common British disease and is due to a combination of smoking, air pollution, and climatic and hereditary factors.

The addition of diseases of the digestive system, especially cirrhosis of the liver, which oddly enough is included in this category, gives a total of about 90% of U.S. and U.K. deaths.

The main causes of death in Egypt are quite different. Heart disease accounts for 17% of deaths and cancer for a trivial 1.6%. The reason is partly that Egyptians frequently die of something else first. The death rate from heart disease in any given age group in Egypt is higher than in the West. The overall rate is lower because the age balance of the population is different in Egypt than in the United States or the United Kingdom.

This is illustrated in Figure 3.2. The United Kingdom has an aging society, with 15% of citizens aged 65 or over, which is almost as many as the 19.2% who are 14 and under. The United States is slightly younger, with 11.6% 65 and over, and 22.2% under 14. In Egypt, however, almost 40% of the population are under 14 and a mere 3.6% are 65 and over. These figures may be compared with the crude birth and death rates in Table 3.2.

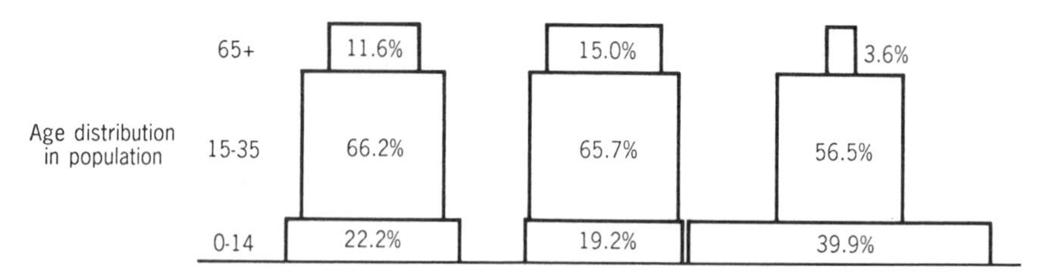

Figure 3.2. *Age distribution in populations of the United States (1983), the United Kingdom (1985), and Egypt (1980).*

TABLE 3.2. Crude Birth and Death Rates in the United States, the United Kingdom, and Egypt

	USA (1982)	UK (1985)	Egypt (1980)
Birth rate (per 1000)	15.7	12.8	36.9
Death rate (per 1000)	8.7	11.4	10.3

The U.S. birth rate is higher than that in the United Kingdom, hence more young people. The Egyptian birth rate is very high indeed. The U.S. death rate is lower than the U.K. death rate because there are fewer old people. The Egyptian death rate is lower than the U.K. death rate because, in spite of the much shorter expectation of life in Egypt, the death of even middle-aged people reduces the population at risk by an even greater amount. Since about 17% of Egyptian deaths are attributed to senility, it would be interesting to know how that is defined and how much "old age" is part of it.

If heart disease is no more likely at any given age in the West than in Egypt, the same is not true of cancer. Cancer deaths in Egypt are lower even when analyzed by age group. This may be related to dietary and environmental factors.

Accidents, poisoning, and violence in Egypt are surprisingly low, and the death rate from traffic accidents reflects the scarcity of automobiles rather than the skill with which they are driven. The death rate per car on the road is high.

The "killer" diseases in Egypt, in contrast to the West, are diseases of the respiratory system and the digestive system. To anyone who knows Egypt, however, the pattern shown by these statistics is misleading. Visiting physicians in Egypt have estimated that about 50% of all Egyptian deaths have bilharzia, otherwise known as schistosomiasis, as a prime or contributory cause (see Section 18.3.2). It involves infestation of the veins with worms and slow destruction of the liver and kidneys. There is an ICD category involving bilharzia and hardly any deaths are classified into it, presumably because a disease that is so widespread receives little comment. Many bilharzia deaths are probably classified as deaths due to diseases of the digestive system or from ill-defined causes. This view is supported by the high published Egyptian death rate from cirrhosis of the liver, a disease that is usually alcohol-induced and is unlikely to be significant in a Moslem country, however broad-minded.

These points underline the great care that must be exercised in interpreting official statistics. Not only may statistics from developing countries be unreliable, but there are also problems of diagnosis in the developed world. For example, there is a growing feeling that Alzheimer's disease—a form of senile dementia—is involved in far more deaths in the United States and the United Kingdom than has been imagined, and statistics showing it as the cause of death may increase sharply as the condition becomes more widely recognized.

The data nonetheless show that the major causes of death in Egypt and in other developing countries are quite different from those in the United States and even the causes of death in the United Kingdom differ significantly from the American pattern. Tropical diseases will be discussed in Chapter 18. Meanwhile, this section has indicated which illnesses are most important in developed and developing countries in terms of the deaths resulting from them.

3.3.2 Disability

Potentially fatal diseases, however, are not the only ones that cause distress. There are also the diseases that are disabling rather than fatal. The numbers of disabled people are difficult to estimate because degree of disability is a continuum and a threshold of disability is hard to define. A survey of Great Britain in 1985 produced a figure of 13.5% of all adults. This is in reasonable agreement with a Social Security Administration study that suggested that one American in seven of working age can be considered disabled, with members of racial minorities and persons of low educational status being especially at risk.

The British study defined ten categories of disability based on scores for limitation of locomotion, reaching/stretching, dexterity, personal care, continence, seeing, hearing, communication, behavior, and intellectual functioning. The numbers in each severity category are shown in Figure 3.3, together with "pen pictures" of typical cases in each category. About half the people in category 10 and over a fifth in category 9 are cared for in communal establishments. The establishments cope with about 1% of the population and 7% of the disabled, but the vast majority of disabled are cared for at home. Indeed, 14% of households have a member suffering from at least one disability.

The incidence of disability rises inexorably with age, slowly at first then accelerating after 50 and rising very steeply after about 70. Almost 70% of disabled adults are 60 + and nearly a half are 70 + .

The disabled represent a huge store of human misery, which can sometimes be alleviated by the use of pharmaceuticals. They also represent a tempting market for the pharmaceutical industry in that many of them are likely to consume drugs consistently over a period of years.

The main causes of disability in adults as estimated in the 1985 study are shown in Figure 3.4. The total number of sufferers in each ICD class add up to more than the total disabled (6.2 million) because many of the disabled suffer from multiple disabilities. Arthritis is overwhelmingly the main disabling illness followed by deafness, blindness, heart disease, and stroke. Bronchitis is peculiarly British. Mental disorders are less prevalent than might have been expected. Nonetheless, among patients in establishments, 56% are handicapped mentally with 26% suffering from senile dementia, this being the largest single problem.

Severity Category	Typical Cases	Cumulative Rate per 1000 Population
1	Man aged 59, deaf in one ear. Difficulty hearing talking in quiet room.[a]	142
2	Woman aged 75, angina, eye problem. Cannot walk 200 yards without stopping or severe discomfort.[b] Difficulty reading newspaper print or recognizing friend across road.[c]	114
3	Man aged 47, spinal arthritis. Has difficulty putting either hand behind back, (b). Can only walk up and down 12 steps if he holds on.[d] Difficulty getting in and out of bed.[e] Difficulty following conversation against background noise.[f]	95
4	Man aged 25, deaf in both ears. Cannot hear alarm clock etc., use telephone,[g] follow TV program,[h] (a). Very difficult to understand strangers.[i]	78
5	Woman aged 75, phlebitis. Loses control of bladder at least once per day[j] (b) and (d).	62
6	Woman aged 40, epileptic. Has seizures between 1 and 4 times a year. Cannot count well enough to handle money, write a short letter without help, read a short article in a newspaper, watch and recollect a half-hour TV program.[k] Thoughts muddled and slow,[l] gets confused about time of day. Feels need to have someone present all the time,[m] sits for hours doing nothing,[n] often feels aggressive or hostile to other people. Difficult for strangers to understand.[o]	46
7	Man aged 31, addicted to tablets. Gets so upset that he hits other people[p] and breaks or rips things up, (m) and (n). Finds relationships outside family difficult, impossible for strangers to understand and quite difficult for those who know him.	33
8	Woman aged 77, "old age," (p), (k), and (l). Finds it difficult to stir herself to do things. Often forgets what she is supposed to be doing, loses track in middle of conversation, forgets names of family and friends seen regularly. Cannot pass on a message correctly. Cannot squeeze out a sponge with either hand and can only turn a tap with one hand. Difficulty wringing out light washing or using scissors.	22
9	Man aged 79, arthritis of spine and deafness. Finds it impossible to understand people who know him well, (e), (h), (g), (j), (c). Has difficulty hearing loud voice in quiet room. Has difficulty serving from pan with spoon or ladle or unscrewing top of coffee jar. Cannot touch knees and straighten up again and can only manage 12 steps if he stops and takes a rest.	13
10	Man aged 55, stroke. Cannot walk at all. Cannot feed themselves, get in and out of chair or bed, wash hands and face, dress and undress, or go to toilet without help. Cannot do anything involving gripping, holding or turning, cannot put hat or jacket on or tuck shirt in. Difficulty shaking hands with anyone, (o) and (c). Loses control of bladder at least once a month.	5

Figure 3.3. *Numbers of disabled by severity category. Disabilities are given a superscript letter and only the letter is quoted when the disability recurs. Cumulative rates are in reverse order, that is, 142 per 1000 is category 1 or worse, 114 per 1000 category 2 or worse, and so on.*

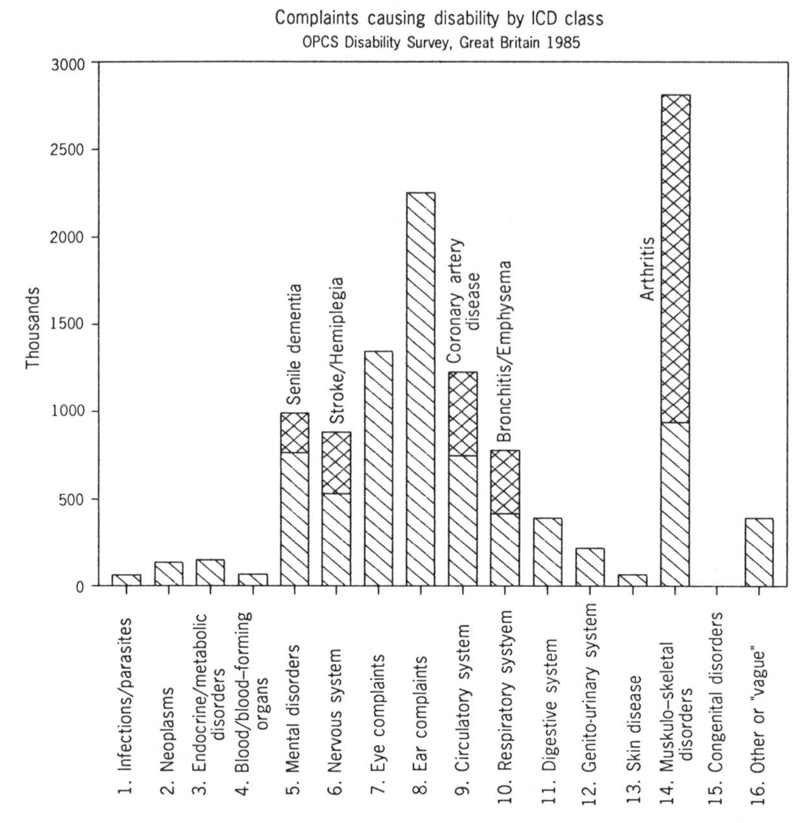

***Figure 3.4**. Complaints causing disability by ICD class (OPCS Disability Survey, Great Britain 1985). Cross-hatching indicates sufferers from a particular disease within an ICD category.*

3.3.3 Visits to the Physician

In addition to disabling or potentially fatal illnesses, there are illnesses that are merely inconvenient, such as colds or headaches, or ones that might have been fatal before antibiotics, such as chest infections. These illnesses may warrant a visit to the physician's office and, indeed, general practitioners may well spend most of their time dealing with minor illnesses. On the one hand, they may well be able to alleviate the inconvenience of the illness and on the other they may need to reassure the patient that the illness is indeed minor and will either cure itself in a few days or prove susceptible to treatment.

The average United States male visited his physician 4.0 times in 1980, while the average female visited hers 5.4 times. The corresponding U.K. figures for 1981–1982 are said to be 2.7 and 4.0 (See bibliography) but other data suggest figures 60% higher than these. It is interesting to note, however, that in spite of free visits to general practitioners in the United Kingdom, the consultation rate is little if any higher than in the United States.

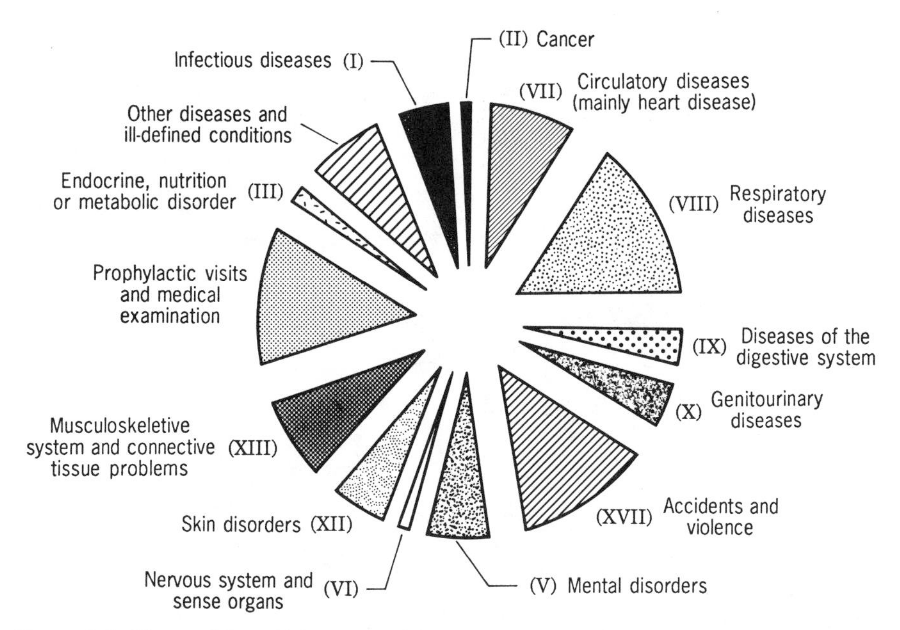

Figure 3.5. *Illnesses for which people visit their physician (United Kingdom 1981 – 1982).*

The excess of female over male visits is a feature of medical consultations worldwide and is only partly due to "female" problems. A series of reasons has been proposed, including greater willingness by women to seek medical advice and greater possibilities within their life-styles of attending the doctor's office. Combined with the lower male life expectancy, this has given rise to the adage that "women get sick but men die."

The illnesses for which people in Britain visit their general practitioners are shown in Figure 3.5. One visit in six is for respiratory diseases such as coughs, colds, and bronchitis. This figure may be lower in the United States but is still probably the most common cause for visiting the physician's office. One visit in seven relates to prophylaxis—vaccination, provision of oral contraceptives, and general medical checkups. After that come groups of illnesses rating between 6 and 10% of the visits, including arthritic diseases in ICD group XIII, which account for 8.5%, and heart disease in group VII, which accounts for 8.9%. Finally, there are the diseases of which the general practitioner sees relatively little, although in several cases they are major causes of mortality. Cancer rates only 1.3% of visits and a typical general practitioner is likely to see only about 10 new cancer cases per year.

This pattern is quite different from the patterns of death and disability and reflects the point that one of the major roles of the general practitioner is to sort out those patients who are seriously ill from those who will get better anyway.

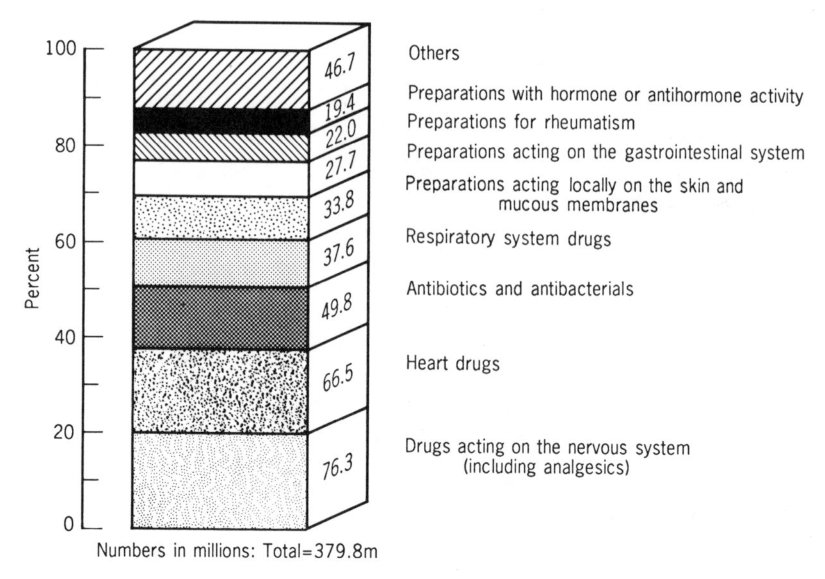

Figure 3.6. *Distribution of prescriptions (millions) by therapeutic class (Great Britain, 1985).*

3.3.4 Prescribing Drugs

Having examined a patient and attempted a diagnosis, the physician comes to the point at which it must be decided which drug, if any, is going to be prescribed for the patient. Prescribing patterns vary widely from physician to physician. Some prescribe virtually every time a patient comes to see them. Others do so on as little as one consultation in four. In the United Kingdom, the average is three prescriptions per four consultations, and it is probably slightly higher in the United States. It is highest of all in Italy where 19 consultations out of 20 end in prescription.

The figure is seen by many authorities as much too high and it is claimed that physicians succumb too readily to drug companies' advertising and to patients' demands. This is a subject beyond the scope of this book but there are a few references in our bibliography. It is significant, however, that 5% of British prescriptions are never presented, apparently because the patients are aware that their illness will get better of their own accord or because they have no faith in the efficacy of the drug. Among depressed patients this proportion can rise as high as 20%.

What diseases are the subject for prescription? Figures 3.6 and 3.7 show the distribution of prescriptions by therapeutic class in Britain in 1985. A fifth of all prescriptions are for central nervous system drugs—tranquilizers, sleeping pills, and antidepressants. In these statistics, analgesics are also classified under central nervous system drugs. Even so, the figures reflect a high level of real or imagined mental illness and do not entirely correlate with the relatively small proportion of patients visiting their physician with mental disorders.

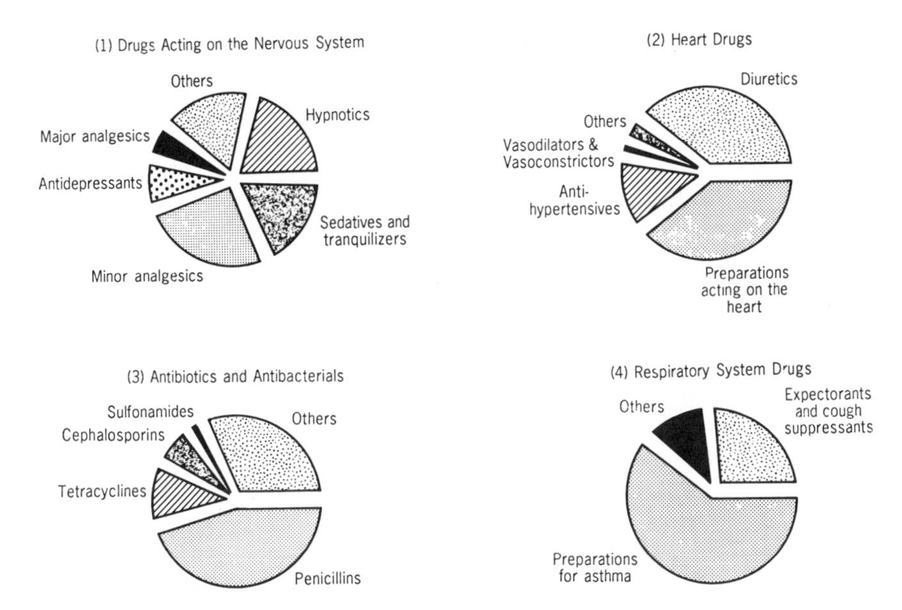

Figure 3.7. *Distribution of prescriptions by therapeutic class (Great Britain, 1985).*

The next largest group of prescriptions is for heart drugs, which accounts for about 16.7% of prescriptions. This category reflects long-term consumption of drugs by people who have had an episode of heart disease or who show symptoms of being especially at risk.

Third is antibiotics and antibacterials, the drugs that counter infectious diseases. If one bears in mind that a course of antibiotics lasts for less than a week and compares this figure with the data in Figure 3.5 for visits to the physician, it is clear that antibiotics are prescribed frequently for respiratory infections. Since something like 60% of such infections are viral rather than bacterial, there is a strong presumption that antibiotics are being prescribed as prophylactics rather than to counter identified infections. Except in certain relatively rare conditions, they are unsuited to this role.

Fourth is preparations acting on the respiratory system and these are largely made up of antiasthma drugs and cough mixtures. British cough mixture consumption, as might be guessed from our previous data, is high but is still far lower than that in Spain and Italy where one might have supposed that the equable climate would have diminished the need for it. Fifth is the topical preparations applied to the skin or the mucous membranes. This category may be expected to diminish because hydrocortisone ointments are now available over the counter in the United Kingdom as they have been for some time in the United States. Sixth is the antacids, antispasmodics, and antiulcer drugs used for gastrointestinal disturbances. Seventh, and surprisingly low on the list, is the analgesic preparations used to soothe rheumatic and arthritic pain. The availability of aspirin over the counter means that this class of drugs appears less important than it really is. Recent legislation

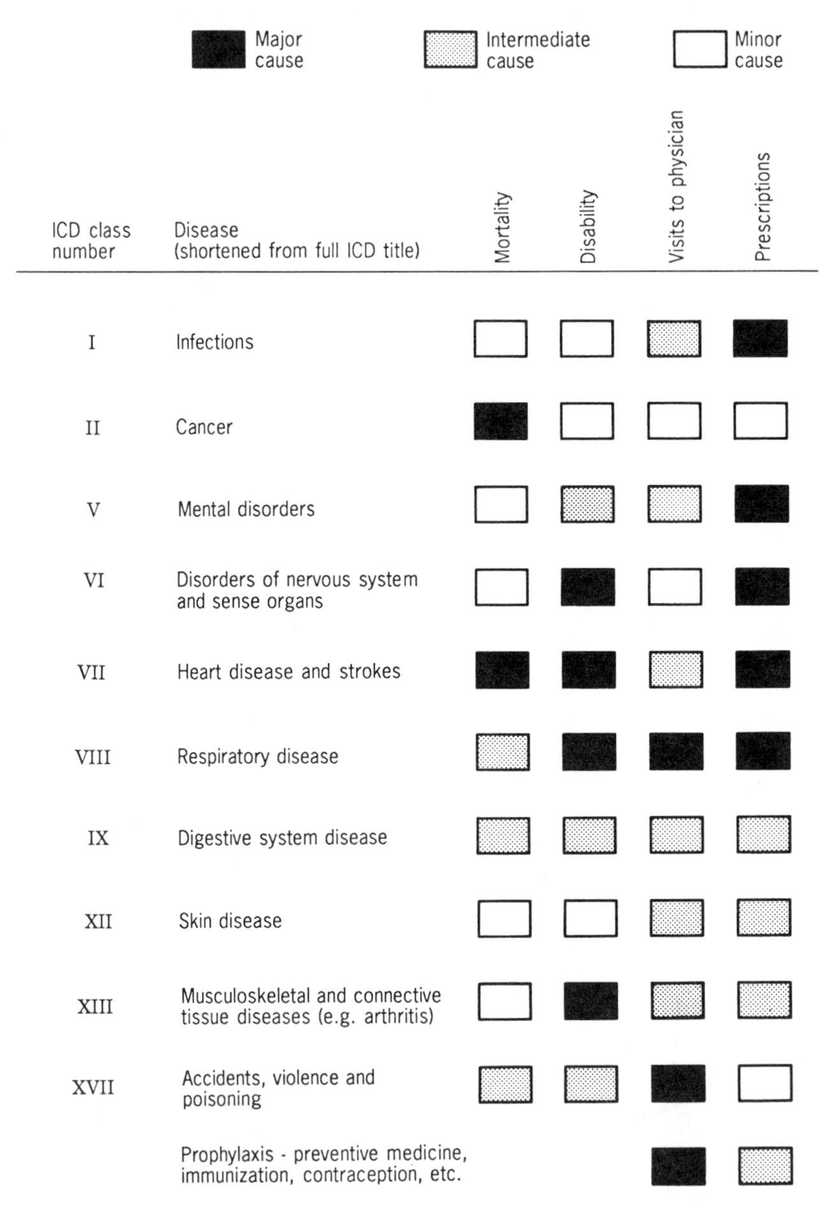

Figure 3.8. *Correlation between mortality, disability, visits to the physician, and prescriptions for various diseases (United Kingdom).*

permitting ibuprofen over the counter will lead to still fewer prescriptions in the future. Eighth is the hormone preparations, which are mainly oral contraceptives.

3.3.5 Correlation

Figure 3.8 is an attempt to correlate the above data. The relation between death, disability, visits to the physician, and drugs prescribed is a complex one but it is possible to pick out various diseases that show markedly different patterns.

Infections, if respiratory tract infections are included, lead to few deaths or disability, but they are the cause of many visits to the physician and many prescriptions are issued to counter them. Drugs to counter infectious diseases are among the most effective drugs in the physician's armory.

Cancer, on the other hand, is responsible for many fatalities but is not the cause of many visits to the physician, nor of a great deal of disability. Drugs to counter it are mainly used in hospitals so they do not appear high on the list of prescriptions.

Large numbers of drugs are prescribed for mental disorders and disorders of the nervous system even though they account for only a moderate number of visits to the physician, and even fewer deaths. Heart disease is responsible for many fatalities and much disability. It occupies a moderate amount of a physician's time and many drugs are consumed because effective ones are available and they have to be taken over a long period.

Respiratory disease scores high on everything except mortality where it is an intermediate cause. Arthritis and rheumatoid arthritis, which make up most of ICD category XIII, do not cause many fatalities. They lead to about the same number of visits to the physician as heart disease. They cause a huge amount of disability, and many drugs are taken against them. Many of the analgesics for these diseases are available over the counter.

These patterns are of great importance to the pharmaceutical industry. Drug companies, after all, are in business to make money and they can do so only if they direct their research effort toward drugs that will be prescribed. Drugs that keep people alive who would otherwise have died, that enable the disabled to be more independent, and that alleviate symptoms of inconvenient though minor illnesses are all important and there will be many examples of all of them in the rest of the book.

4

Pharmacological Concepts

Pharmacology is the science of drugs; it embraces their physical and chemical properties, their compounding, biochemical, and physiological effects, and their mechanism of action, absorption, distribution, biotransformation, and excretion. This chapter can do no more than scratch the surface of such a vast subject. It is intended at least to explain the vocabulary that is used and give some examples.

The first part of the chapter deals with the molecular basis for drug action and explains why drugs behave as they do; the middle section deals with what happens when a drug passes through the body; and the final section summarizes the properties sought in an ideal drug.

4.1 RECEPTOR THEORY

Some drugs are structurally nonspecific and act throughout the body. Examples include the gaseous anesthetics and alcohol. Nonspecific drugs are few, however, and they will not be discussed here. Most drugs are specific and the action of specific drugs is usually explained in terms of receptor theory. The idea of receptor sites goes back to Ehrlich's idea of a magic bullet that would cure disease by attacking bacteria but not human tissues, just as certain dyes had an affinity for specific cells but left others unstained. His ideas led him in 1909 to the discovery of salvarsan, shown in Figure 1.1, an arsenobenzyl compound effective against syphilis.

The receptor theory states that there are chemically receptive sites in the body that combine with complementary functional groups in a chemical either produced by the body naturally or introduced into it as a drug. These complementary functional groups may be basic nitrogen atoms, aromatic rings, hydroxyls, and indeed almost any of the functional groups known to the organic chemist. Drug–receptor interactions can involve electrostatic effects, such as charge–charge and charge–dipole interactions; hydrophobic interactions, such as the formation of hydrophobic or hydrophilic areas; secondary force interactions, such as dipole-induced dipole interactions and hydrogen bonding; and charge transfer processes, such as pi–pi interactions. The presence of the appropriate chemical groups, however, is not enough. The groups must be in a particular configuration to be effective. This is demonstrated by the classic example of diethylstilbestrol:

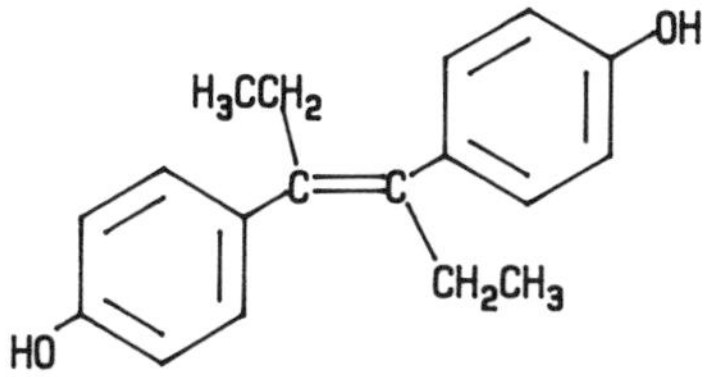

trans-Diethylstilbestrol

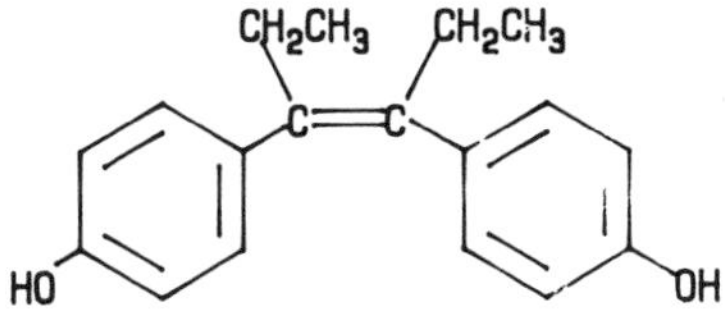

cis-Diethylstilbestrol

The trans isomer is estrogenic; that is, it behaves like a female sex hormone. The cis isomer, on the other hand, has only 7% of the activity of the trans isomer. In subsequent chapters, many examples will be given where small differences in structure cause large changes in physiological action.

Several bonding sites have been recognized within the body. The most important of them are the active sites on enzymes, nucleic acids, and permeases, all described below. In addition, receptors may be small molecules.

4.1.1 Enzymes

Enzymes are proteins that are highly specific biological catalysts. Essentially all biochemical reactions are enzyme-catalyzed. The reaction system on which an enzyme works is called the enzyme substrate. The enzyme attaches itself to the substrate by means of chemical bonds located in an active site. Some enzymes work only in the presence of a specific nonprotein organic molecule called a coenzyme.

It was early recognized that drugs may combine with enzymes just as enzyme substrates do. The active site or receptor did not bind the drug very tightly as a rule and never changed it chemically. Thus, the drugs were not comparable to enzyme substrates but to the less tightly bound coenzymes. Many drugs interfere with enzyme action and are called antimetabolites. They can either inhibit the reaction normally brought about by the enzyme or they can cause the enzyme to produce substances different from those it usually produces, and these new substances may have unique biochemical properties. An example of an antimetabolite is the antibacterial compound sulfanilamide:

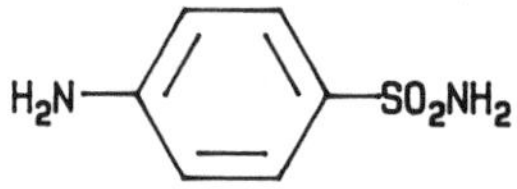

H_2N — SO_2NH_2

Sulfanilamide

Sulfanilamide occupies the active site on a bacterial enzyme, dihydrofolate synthetase, that would otherwise take up *p*-aminobenzoic acid for biosynthetic reactions in the bacterium. Without these biosynthetic reactions the bacterium cannot survive.

4.1.2 Nucleic Acids

A receptor may also be the active site on a nucleic acid. A nucleic acid consists of long chains of nucleotides combined through phosphate diester linkages. A nucleotide is composed of a purine or a pyrimidine base attached to a sugar (D-ribose or 2-deoxy-D-ribose) by a glycosidic linkage; the sugar is then combined with phosphoric acid. Nucleotides containing D-ribose are referred to as ribonucleic acid (RNA) and those containing 2-deoxy-D-ribose as deoxyribonucleic acid (DNA). The structure in Figure 4.1 shows the tetranucleotide portion of one strand of DNA composed of adenine, thymine, cytosine, and guanine deoxynucleotides. The drug enters the helical portion of the nucleic acid or adsorbs onto the nonhelical portion. Examples (see p. 60) are the antimalarial drug chloroquine, which affects the DNA of the malaria parasite, and the anticancer drug adriamycin, which affects the DNA of cancer cells.

The antibacterial agent, proflavine, is planar and is believed to act by insertion of the flat aromatic rings of the drug between the bases of the nucleic acids, as shown in Figure 4.1. This process is called intercalation.

(a)

Deoxyribonucleic acid, if represented two-dimensionally as in (a), has a "ladder" structure. The sides are chains of deoxyribose joined by phosphate groups. The rungs are pairs of bases, thymine (T) and adenine (A), or guanine (G) and cytosine (C). The bases are bonded covalently to the sides of the ladder but the rungs are joined by multiple hydrogen bonds.

(b) Represents the three-dimensional structure. The ladder is twisted into a spiral — the double helix. The base pairs fill the inside of the spiral and interact via their aromatic pi-bonds. The spacing between the pairs is 0.34 nm, similar to the spacing between the layers in graphite. The spiral repeats every 3.4nm, that is, every ten base pairs. The spiral has major and minor grooves, which can act as receptor sites (Figure 4.3(b)).

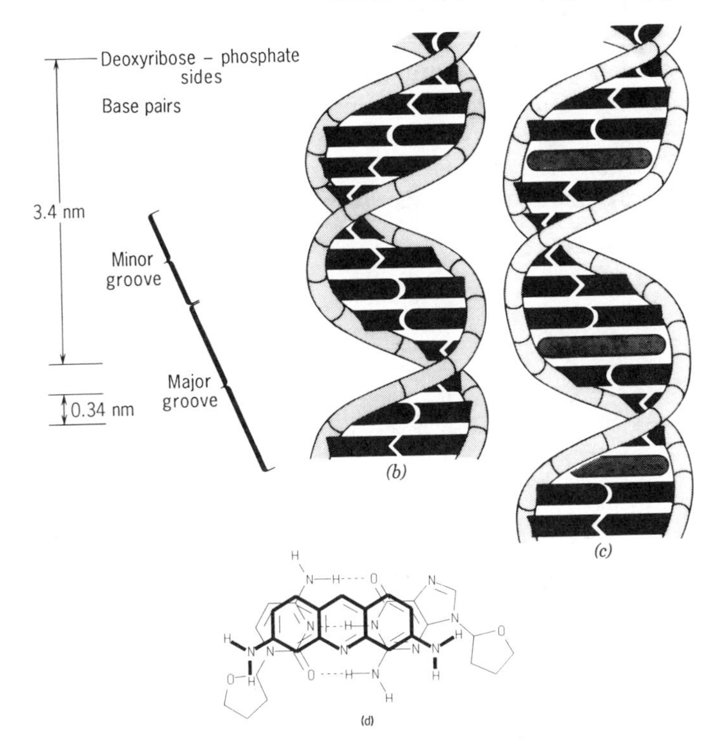

Planar molecules such as proflavine (3,6-diaminoacridine) interact with DNA by intercalation, that is, by insertion between the base pairs. The spiral has to distort and unwind to accommodate them as shown in (c).

(d) is a view along the axis of the spiral and shows the superposition of the aromatic rings of proflavavine over the pyrimidine ring of cytosine and the purine ring of guanidine.

Figure 4.1. *Intercalation of proflavine. Adapted from Taylor and Kennewell, see bibliography.*

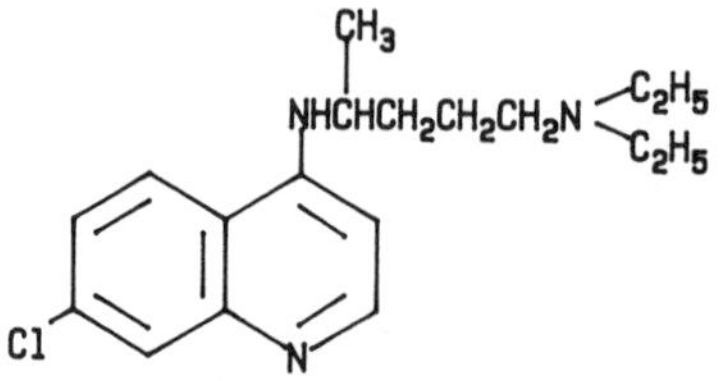

Chloroquine

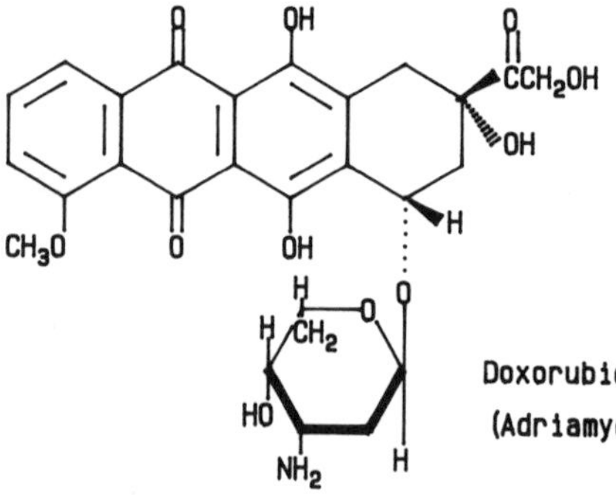

Doxorubicin (Adriamycin)

4.1.3 Permeases

A third type of bonding site is the active site on a permease. Permeases are proteins and are similar to enzymes, but, instead of catalyzing chemical reactions, they regulate the transport of molecules—usually inorganic ions—through cell membranes. They have also been shown to carry receptors and they combine with neurotransmitters such as acetylcholine (Figure 7.12). Verapamil (Section 7.5.2) is a drug that blocks the transfer of calcium ions through cell walls and reduces intracellular free calcium ion levels:

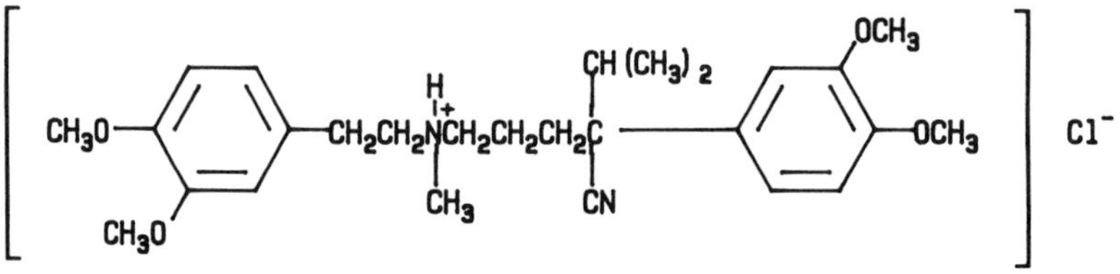

Verapamil

This reduction is desirable in certain types of heart disease.

Small molecules can also be receptors. The 8-hydroxyquinoline antibacterials act by chelating a variable valence metal such as iron or copper and causing it to become more toxic to bacteria. The receptor is the small molecule thioctic acid $HSCH_2CH_2CH(SH)(CH_2)_4COOH$, which is the essential coenzyme for the oxidative decarboxylation of pyruvic acid. It is oxidative destroyed by the chelate. An example of such an antibacterial, although it is no longer used in the United States because of doubts about its side effects, is iodochlorohydroxyquin, a drug against amoebic dysentery:

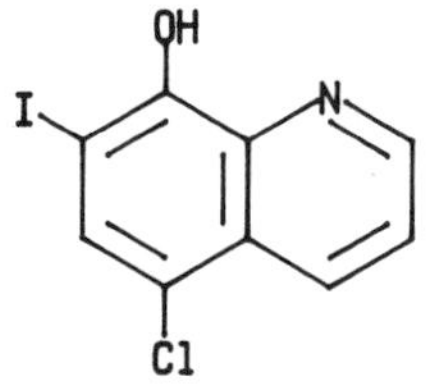

OH
I
N
Cl

Iodochlorohydroxyquin

8-Hydroxyquinoline, a related compound, is used analytically to chelate metals.

4.2 RADIOLIGANDS

Receptor molecules are present in the body in very small amounts. For example, the receptor proteins in the brain account for less than one-millionth of total brain protein. Receptor molecules cannot be isolated as yet, but in the last 15 years or so it has become possible to show binding to receptors by the use of radioactive ligands with very high specific activities of at least 20 Curies/nanomol. These are tritium- (^{3}H)- labeled analogues of potential drugs that have recently become commercially available.

A small amount of tissue—about one milligram—is first incubated for 15–20 minutes with a measured concentration of radioactive ligand in a buffered medium. The tissue is separated and washed and the amount of ligand that has been bound is measured by liquid scintillation counting of the whole sample.

Under these conditions, the ligand is bound in two different ways: first, specific bonding to the receptor under investigation, and second, nonspecific bonding to other biopolymers in the tissue. The number of specific sites is limited, whereas the number of nonspecific sites is effectively infinite in terms of the amount of ligand added. The experiment is therefore repeated with the radioactive ligand in the presence of a large amount of nonradioactive ligand. Statistically this is so large that effectively no radioactive ligand will bond to receptor sites, while the nonspecific bonding, because there are unlimited sites, will remain unchanged.

The difference in binding values for these two experiments therefore gives the specific binding. If the experiment is carried out at increasing concentrations of ligand, a graph like that shown in Figure 4.2 is obtained. The amount of specific binding increases with concentration until it reaches a plateau at which point all the receptor sites are occupied. Analysis of the curve gives the number of receptor sites and the dissociation constant. Receptor theory explains that there are chemicals that block receptor sites, and if they are added in an experiment such as this, the amount of blocker that occupies 50% of the sites can be determined. Such values are called IC_{50} or

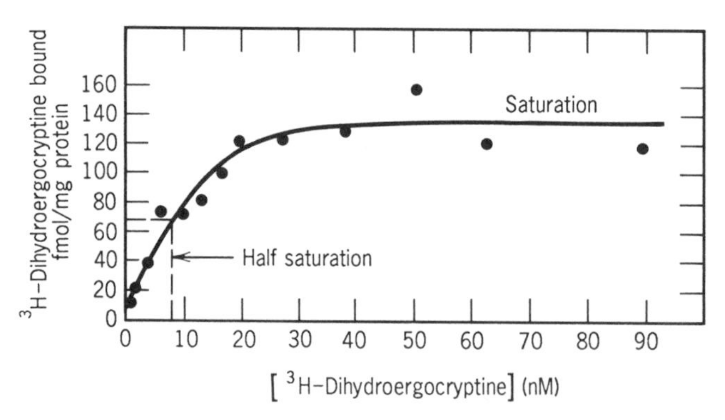

Figure 4.2. *Specific binding of ^{3}H-dihydroergocryptine to rabbit uterus membranes as a function of ^{3}H-dihydroergocryptine concentration. Source: L. T. Williams and R. J. Lefkowitz, Receptor Binding Sites in Adrenergic Pharmacology, Raven Press, New York, 1978, p. 59.*

inhibitory concentration levels. In Figure 4.2, saturation occurs at 140 femtomoles (140×10^{-15} mol) ^{3}H-dihydroergocryptine/mg protein and half-saturation occurs at a ^{3}H-dihydroergocryptine concentration of 8 nanomoles (8×10^{-9} mol).

The technique is experimentally difficult and was originally used to demonstrate the existence of receptors and their malfunction in disease states. Recently, it has become the basis of a technique known as receptor technology which can be used to test a wide range of potential drugs and receptors to see which of them interact and to what extent. It has proved possible to automate the process so that evaluation is rapid (Section 2.10) and, hence, many more compounds can be screened and a drug can be tailored much more closely to fit its receptor.

Receptor technology not only offers speed and a reduction in animal experiments, but also it has other advantages. Because it estimates a drug's activity at its binding site, it prevents the loss of promising materials that, when administered to an animal, are metabolized before reaching their receptors. By experiments with different tissues, it shows whether the drug interacts selectively with one receptor or with more than one and hence gives some idea of what side effects are likely to arise. The drawback is that the technique shows only whether or not a drug binds to a receptor; it does not show what effect it has on it. Thus, animal trials have not been superseded.

Receptor technology tools are steadily being improved. A recent advance is the development of positron emission tomography—a medical imaging technique that studies brain metabolism—to map receptor sites in the brain. One such study indicated that neuroleptics such as the phenothiazines (Section 8.2) rapidly bind to brain dopamine receptors even though the effect of the drugs is delayed by several weeks. That suggests that the benefit of the drugs may be related to their ability to regulate the number of receptors.

(a) Inhibitor (dotted surface) bound in the active site cleft of the enzyme thermolysin.

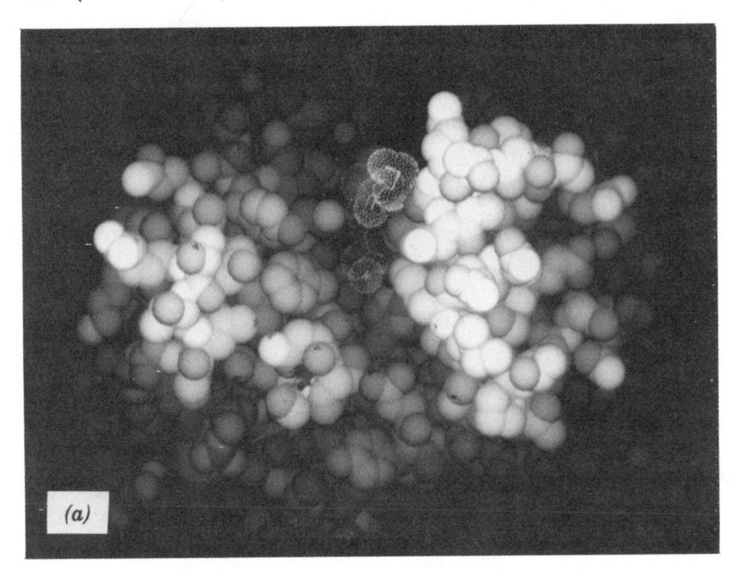

(b) The anticancer / antimicrobial drug netropsin (dotted surface):

CH_3

$H_2NCNHCH_2CONH$ — N — $CONHCH_2CH_2CNH_2$

NH — CONH — NH

N

CH_3

bound in the minor groove of DNA.

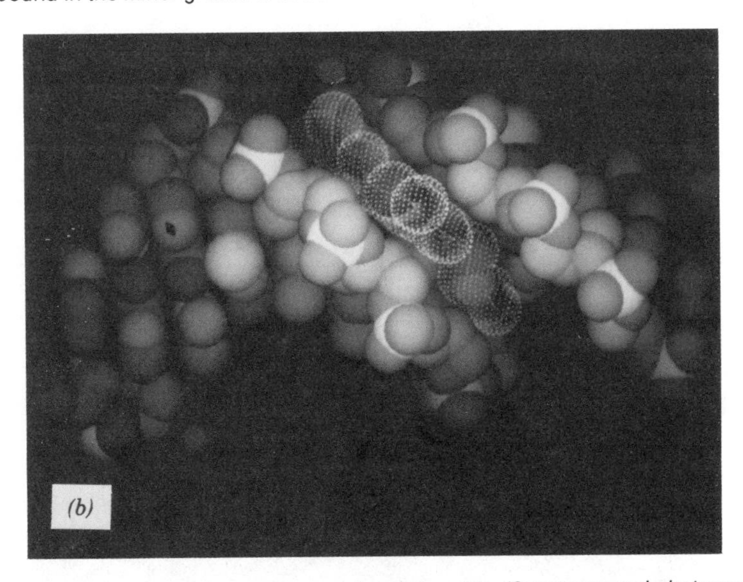

Figure 4.3. *Receptor sites on enzymes and nucleic acids. (Structures and photographs by Mike Hann, Glaxo Pharmaceuticals, reproduced by permission.)*

4.2.1 Receptor Geometry

Much work has been expended on the isolation and purification of the macromolecules that serve as receptors. Many successes have been reported, including the identification of receptors for acetylcholine, opiates, insulin, cholera toxin, cardiac glucagon, and the beta-adrenergic receptor of the heart. No atomic resolution structures of receptors are yet known but many enzyme structures are known and are often used as models of the way in which receptors work. A typical example is the complex between dihydrofolate reductase and the anticancer drug methotrexate.

Enzymes are proteins, which are linear sequences of aminoacids; these chains are folded into elaborate three dimensional structures stabilized by hydrogen bonding and other interchain interactions. Receptor sites are invariably crevices or grooves in the protein surfaces. Enzyme binding sites are usually lined with a combination of polar and less polar amino acids. Many contain metal ions. Nucleic acids such as DNA (which also have receptor

(c) Ethidium bromide:

NH_2

H_2N N^+ CH_2CH_3

binding as an intercalator to DNA (cf. proflavine, Fig 4.1)

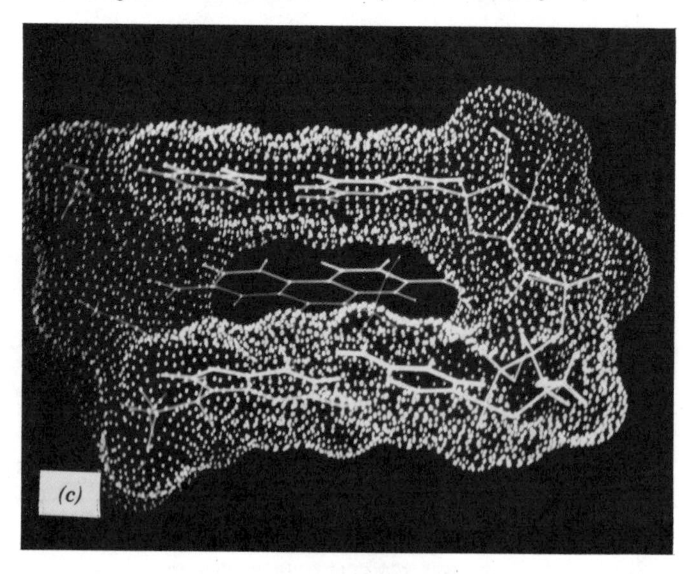

Figure 4.3. *Continued*

sites) bind other molecules by intercalation or via helical "grooves" in their structure. The latter usually involves strong electrostatic interactions with negatively charged phosphate groups on the nucleic acid backbone as well as various types of hydrogen bonding. This is illustrated in Figure 4.3.

4.3 REPLACEMENTS, REDESIGNED AGONISTS, AND ANTAGONISTS

Specific drugs may be classified according to their action on receptors. They fall into three classes: replacement drugs or agonists, redesigned or modified agonists, and antagonists. There is also a subgroup called pseudoagonists which are a sort of antagonist.

4.3.1 Agonists

Replacement drugs or agonists are chemicals that are identical with substances the body needs but cannot itself synthesize. The reason may be an illness such as diabetes where the body cannot produce sufficient insulin or it may be inherent such as the inability of the body to synthesize vitamins and certain essential amino acids and unsaturated fatty acids. In the case of diabetes, the body has lost the ability to synthesize the chemical, insulin. In the case of scurvy, the body was never able to synthesize vitamin C and the disease occurs because the chemical is not available from outside sources. Insulin (Section 13) is a replacement drug for diabetes, while vitamin C (Section 17.7.1) is a replacement drug for scurvy. Replacement drugs for chemicals the body cannot itself synthesize are also known as diet supplements.

Replacement drugs or agonists interact with receptors in the same way as the body's natural materials. It is sometimes convenient, however, to modify a natural agonist to prolong its action, slow down its elimination, or cause it to act at a specific site. Such drugs are called redesigned or modified agonists. They share the affinity for the receptor of the natural material and have a certain amount of its efficacy in persuading the receptor to act.

Norethindrone is a modified agonist mimicking the effect of the hormone progesterone:

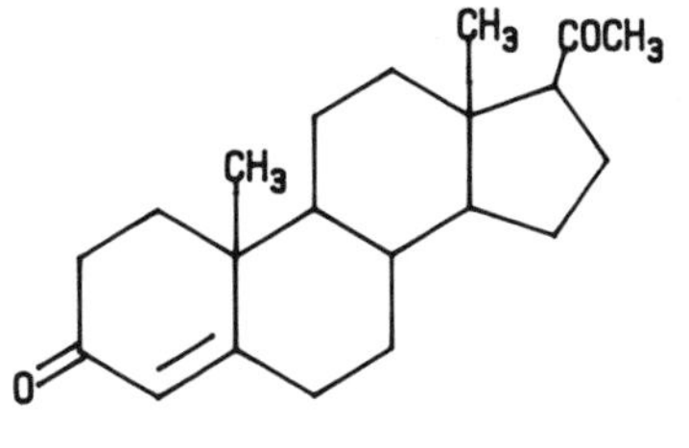

CH_3 $COCH_3$ CH_3 O

Progesterone

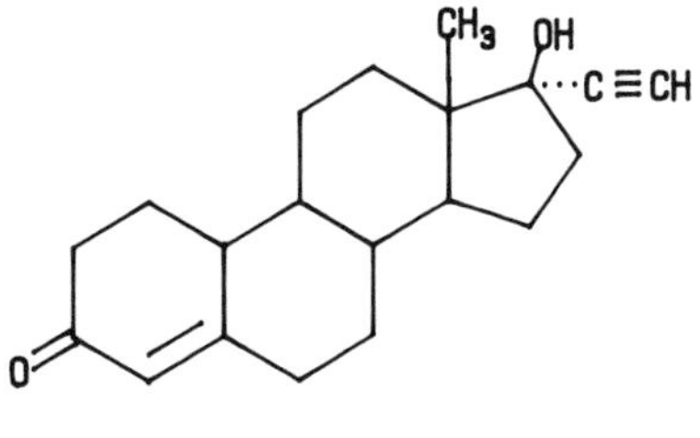

CH_3 OH $C\equiv CH$ O

Norethindrone

Under certain circumstances, progesterone will block ovulation; hence, it should be useful in oral contraceptives. The difficulty is that it is broken down in the liver and is therefore inactive when taken by mouth. Norethindrone, on the other hand, has a similar effect on ovulation and may be taken orally.

4.3.2 Antagonists

Although replacement drugs and redesigned agonists account for a number of useful drugs, by far the majority of modern pharmaceuticals are receptor antagonists. Agonists have an affinity for receptors and are also efficacious; that is, they mimic the effect of the natural substance. Antagonists share the affinity for the receptors but lack the efficacy. That is to say, they adsorb on receptors and block their action by preventing natural materials from gaining access to them.

Receptor antagonists are by their nature toxic but the useful ones are selectively toxic. The idea of medicines as poisons that heal is an old one. Moses Maimonides, the twelfth century Jewish philosopher and physician, wrote that "Medicine is the science of weighing up errors and the art of choosing between risks and ills." Six centuries later, in 1775, William Withering, a physician in Birmingham, England, conducted the first documented British clinical trial in Britain when he evaluated the use of an ancient herbal remedy concocted from the foxglove as a remedy for the dropsy, an old name for a type of heart disease. Digoxin, extracted from the foxglove, is still used for this purpose, as we shall see. Thus, Withering's work provided the oldest drug still widely used today. Withering noted that excessive doses of the herbal remedy could prove fatal and in his account he commented, "Poisons in small doses are the best medicines, and useful medicines in too large doses are poisons."

The aim of the pharmacologist is to find a material that will damage one kind of living matter without harming any other kind, even though the two are in intimate contact. A material may be sought that will permanently damage certain cells, for example, bacteria or cancer cells. This would be called a chemotherapeutic agent. Alternatively, chemicals may be sought that will exert a gently graded and finally reversible effect, such as anesthetics that block the mechanism by which pain is felt. These are called pharmacodynamic agents. Their effect should be completely reversible with time, and toxicity to other tissues should be negligible.

Allopurinol, a drug used to relieve gout, is an example of a receptor anatagonist. Gout is a painful disease caused by the accumulation of uric acid in joints in the body. Uric acid is produced by two enzymatic conversions, both brought about by xanthine oxidase, in which hypoxanthine is oxidized in two positions in the body (Figure 4.4). The antigout drug allopurinol has a structure, also shown in the figure, that is close to that of hypoxanthine. The enzyme receptor mistakenly "recognizes" allopurinol as hypoxanthine but is unable to convert it to uric acid. Consequently, the receptor is blocked, and far

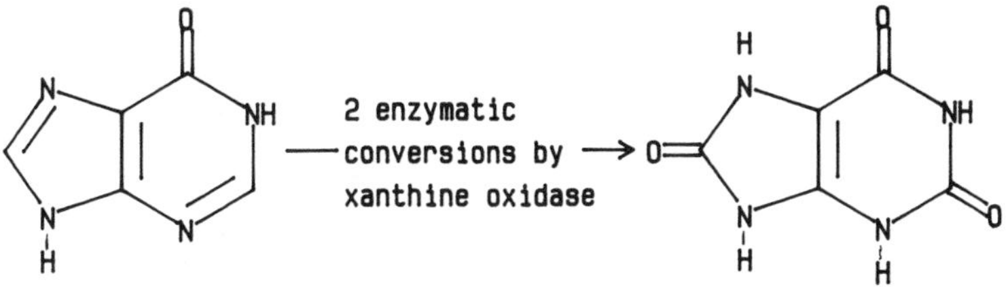

Hypoxanthine —2 enzymatic conversions by xanthine oxidase→ Uric acid

Allopurinol (keto form) blocks xanthine oxidase — incapable of enzymatic conversion

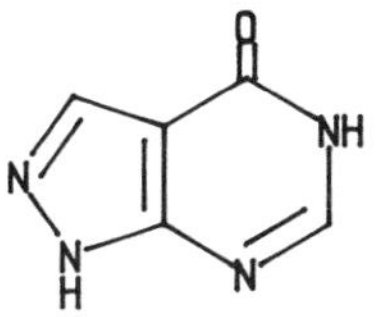

Figure 4.4. Allopurinol as a receptor antagonist.

less uric acid is produced by the body. Three particularly important groups of blocking drugs are the β blockers and calcium antagonists, which will be described in Section 7.5 on heart drugs, and the H_2 blockers cimetidine and ranitidine, which will be described in Section 9.6 on antihistamines.

Pseudoagonists are not truly a fourth group of drug but are nonetheless worthy of mention. They are drugs that inhibit enzymes that would otherwise destroy agonists, so, although they are enzyme antagonists, their effect is that of agonists and they are called pseudoagonists. An example is pilocarpine, an alkaloid used for glaucoma:

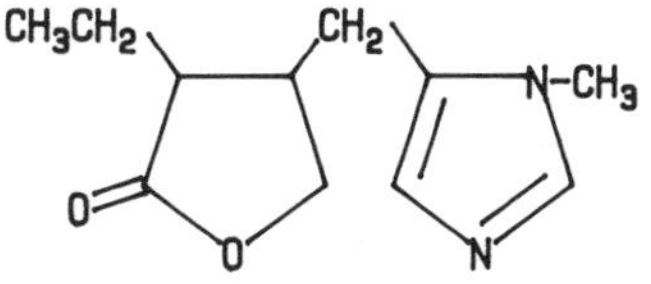

Pilocarpine

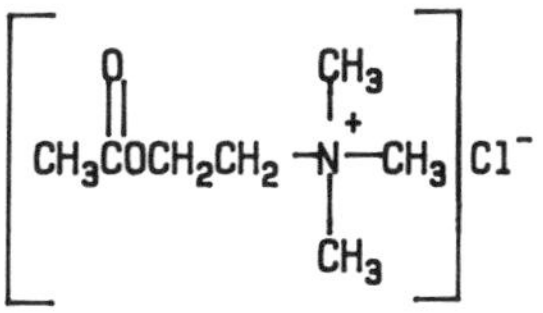

$[CH_3COCH_2CH_2N^+(CH_3)_3]Cl^-$

Acetylcholine Chloride

1. Suppose there is a receptor R controlling a muscle M. The receptor is like a lock. In the body there is a chemical A the key that fits the lock.

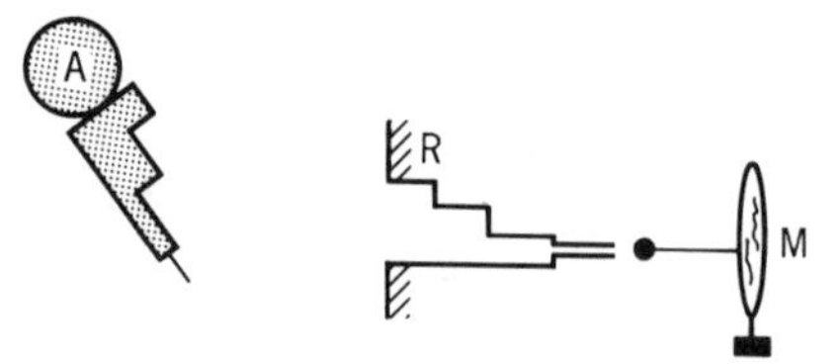

2. When the chemical interacts with the receptor, the receptor sends an electrical message to the muscle M, which contracts. The chemical A is said to be a natural agonist.

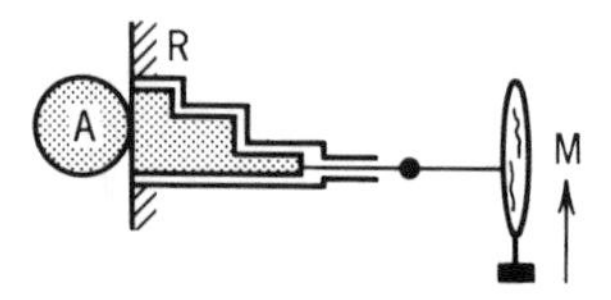

3. If the body runs short of A for some reason, then A can be given as a drug. It is called a replacement drug and has exactly the same effect in the body as would natural supplies of A. It is an agonist.
4. It is sometimes convenient to modify an agonist in order to prolong its action, slow down its elimination, or cause it to act specifically at a particular site. It fits the lock but perhaps more or less tightly. It is called a redesigned agonist.

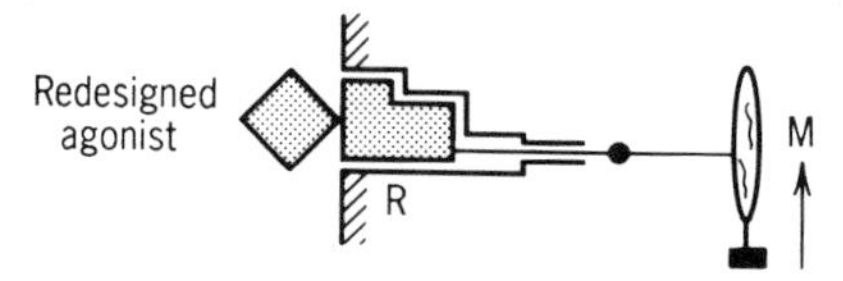

5. If there is too much of the chemical A in the body, it is sometimes possible to prevent the overstimulation of the muscle by use of a drug which combines with the receptor more strongly than the natural agonist but does not cause the muscle to respond. Such drugs are called receptor antagonists or blockers and they block the lock so that none of the keys can operate it.

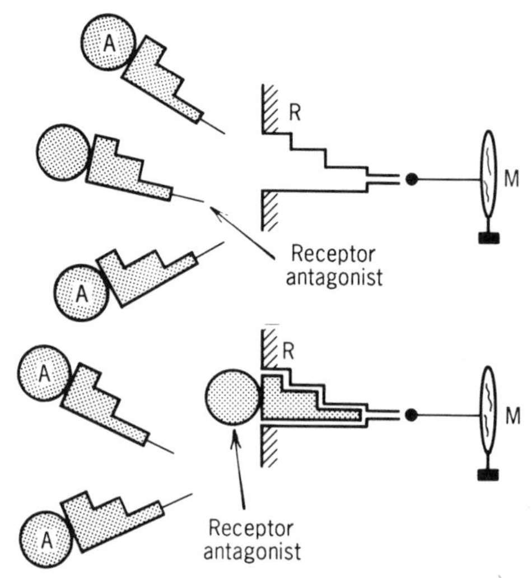

Figure 4.5. *Replacements, redesigned agonists, and antagonists.*

Glaucoma is relieved by acetylcholine, shown above as the hydrochloride, and acetylcholine is destroyed within the body by the enzyme acetylcholinesterase. Pilocarpine blocks this enzyme and thus the patient's acetylcholine level rises.

Figure 4.5 illustrates the differences between replacements, modified agonists, and antagonists by use of the lock-and-key analogy. This was introduced in the nineteenth century by Emil Fischer and gives a vivid, simple picture. As a rigorous model of what does on, however, it has been replaced by Koshland's induced fit hypothesis in which the substrate is believed to *induce* the required orientation of groups in the active site required for binding and catalysis. This is more difficult to show pictorially. The *reversibility* of the binding process is another aspect of receptor behavior not shown by the lock-and-key analogy.

4.3.3 Drug Design

The examples given of drugs that are agonists and antagonists give rise to a serious question. Both modified agonists and antagonists have a close resemblance to natural agonists that adsorb on receptor sites. Why does one group stimulate receptors and the other group block them? Insofar as there is a clear answer, it is that drugs behave in one way if they resemble the natural agonist sufficiently well to take its place, as does norethindrone, but have the opposite effect if they differ from the agonist enough to interfere with its stimulation of the receptor as do the β blockers and H_2 blockers mentioned above.

Both kinds of drugs require a certain common structure in order to bond to the receptor, but the design of antagonists is not as difficult as that of agonists because they have only to block the receptor whereas agonists have to make it respond. Indeed there are some antagonists, as will be seen in Section 9.6 on antihistamines, that have structures widely different from natural agonists. Meanwhile, one method often used to create an antagonist is to modify an agonist by increasing its molecular weight, perhaps by lengthening a hydrocarbon chain. This usually results in stereochemical hindrance at the receptor site so that the receptor no longer responds. It also usually lengthens the time that the molecule adheres to the receptor site.

It was at one time widely believed that drug receptor interactions could be described simply in terms of the relative strengths of bonding of the drug and the natural agonist. This is now the subject of debate and a summary of current theories is given by Taylor and Kennewell and Roberts and Price, cited in the bibliography.

Compared with a decade ago, tremendous progress has been made toward a rational basis for drug design, even though the industry's approach is still empirical in many respects. It is known that the fundamental requirements for a particular kind of drug action are dictated by the structure and location of the receptor and much has been learned about the topography of receptors. New techniques are available to match potential drug structures to them.

An effective drug has to satisfy three requirements. First, it must have a size, shape, and electron distribution complementary to that of the receptor. Second, it must have a lipophilic–hydrophilic balance that will ensure that it reaches the tissues where the receptor is located. Third, it must be selectively toxic and not a destructive poison. That means it must accumulate selectively in the cells one hopes to influence or destroy, or it must influence a biochemical system important for the cells, or it must interact with a receptor found only in those cells.

4.3.4 Structure – Activity Relations

The above considerations lead to the establishment of relations between the biological activity of a drug and its physical and chemical properties. These are called structure–activity relations (SAR) and, if they can be expressed numerically, they become quantitative structure–activity relations (QSAR).

Structurally similar compounds often produce similar effects. For example, most penicillins have antibiotic action. Even when compounds in a particular class change from agonists to antagonists, if, for example, the length of a side chain is altered, the change is a gradual one and the compounds are clearly acting at the same receptors.

Structurally diverse compounds may also be connected by SARs. They may produce a common biological effect despite the fact that their structures differ widely. This is because they may possess a common physical property such as partition coefficient, degree of ionization, vapor pressure, or size and shape. Thus, it is believed that a variety of apparently unrelated anesthetics produce the same effect because their partition coefficients are similar. Compounds with different structures may motivate the same biological effect even though they act at different sites, because the end biological effect is the same. This is the case with blood-pressure-lowering drugs (Section 7.6). Another possibility is that structurally diverse chemicals are metabolized to similar intermediates. Perhaps the most interesting of all is the topographical effect. Morphine and the enkephalins are believed to have the same analgesic activity because they have the same topography or geometric characteristics, although certainly different mechanisms are involved.

The biological activity of a compound is usually expressed in terms of the biological response and the dose required to elicit that response. Two types of activity are recognized, called dichotomous and quantitative. Dichotomous data answer the question of response versus no response, activity versus no activity, or toxicity versus no toxicity. Quantitative data require a correlation between some physicochemical variable and a biological variable such as the time or the dose to achieve a response. ED_{50} is the dose required for 50% response and LD_{50} is the dose giving 50% death. An alternative approach is to measure response achieved for a given dose of material, but the "dose for a fixed response" method is usually preferred. A graph of biological response versus dose will normally give an S-shaped curve. If the response can be expressed as a percentage P of a maximum response, then the S-shaped curve can be transformed to a straight line by the relation

$$\mathrm{BA} = (\mathrm{MW}/d)\log[P/(1-P)]$$

where d is the dose, MW the molecular weight of the compound, and BA the biological activity.

How QSARs are derived and used is beyond the scope of this book, but is well described in Burger (see bibliography). Several examples are given including the use of computer-oriented methods for obtaining three-dimensional conformations. This was used to determine the pharmacore (the pattern described by a group of atoms and their interatomic distances) of the spasmophoric group in acetylcholine and anticholinergic agents (Chapter 14).

To illustrate the type of relations obtained, we take the series of β-adrenergic blocking drugs (Section 7.5.6) of general formula

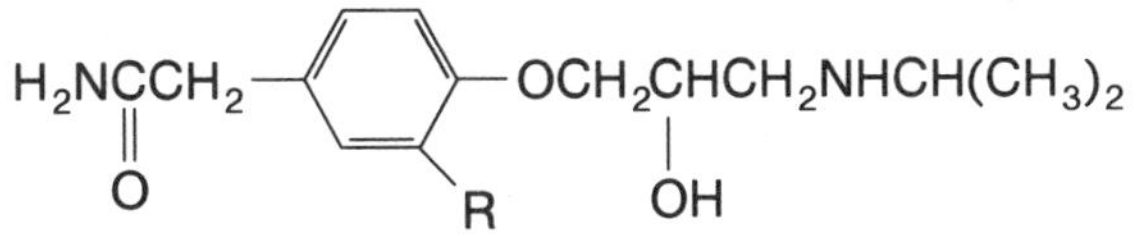

$H_2NC(=O)CH_2$–$C_6H_3(R)$–$OCH_2CH(OH)CH_2NHCH(CH_3)_2$

These compounds inhibit overrapid heartbeat (tachycardia) and their activity is measured as follows. Isoprenaline is injected into anesthetized rats where it induces tachycardia, the extent of which is measured. The rats are then given the test compound and half an hour later receive the same dose of isoprenaline as previously. The extent to which tachycardia is induced may be measured and it is possible, after sufficient experiments, to calculate the dose of test compound which would reduce by 50% the expected tachycardia. This is the ED_{50}.

The physicochemical property is π, the substituent constant for hydrophobic effects, given by

$$\pi = \log P_X - \log P_H$$

where P_X is the partition coefficient of the substituted compound and P_H the partition coefficient of the unsubstituted compound (R = H above). The model water-immiscible solvent for partition coefficient measurements is usually *n*-octanol (see Section 4.6.2).

The results are shown in Figure 4.6. Two orders of potency are observed, one containing the halogen substituents and the other, of shallower slope, containing the alkyl substituents. Clearly another factor is relevant besides partition coefficients. Meanwhile, the halogenated compounds affected receptors outside the heart, so the compound that was in fact chosen was the most active on the alkyl line, notably atenolol (R = H).

Although many other factors are important, there are numerous examples of relations between the biological activity of a compound and its partition

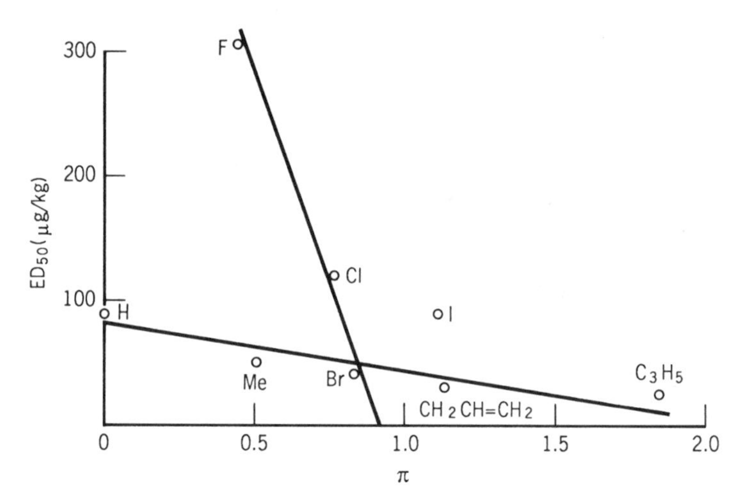

Figure 4.6. *QSAR for atenolol analogues. Source: D. J. le Count in Chronicles of Drug Discovery, Vol. 1, J. S. Blindra and D. Lednicer (Eds.), Wiley, New York, 1982.*

coefficient. Hansch (see bibliography) has demonstrated many of these and shown the effect of different substituent groups on the partition coefficients of various drugs.

4.4 WHAT HAPPENS TO DRUGS IN THE BODY

The foregoing discussion has covered the molecular basis for drug action. This section deals with the problems of absorption, distribution, biotransformation, and excretion of drugs. The four processes are illustrated in Figure 4.7 for aspirin taken by mouth. The aspirin (acetylsalicylic acid) is absorbed rapidly from the stomach and intestines so that the concentration in the stomach half an hour after a dose is taken is only about 5% of its initial value. Once absorbed in the body (mainly in the bloodstream) it is converted to salicylic acid (Figure 4.11). Thus, its concentration in the bloodstream reaches a maximum in less than half an hour and then declines. As described in Section 19.1.1, it also blocks the formation of inflammatory prostaglandins and hence reduces inflammation, fever, and pain.

Meanwhile, the concentration of the salicylic acid metabolite reaches a peak after about an hour. It is eliminated quite slowly via an amide formed with glycine in the liver (Figure 4.11). This combination of one molecule with another large molecule is known to pharmacologists as *conjugation* but it has nothing to do with the alternating single and double bonds of organic molecules. The salicyclic acid–glycine conjugate is excreted in the urine over the next two days.

Aspirin is a simple and well-known case. The individual stages will now be described in more detail.

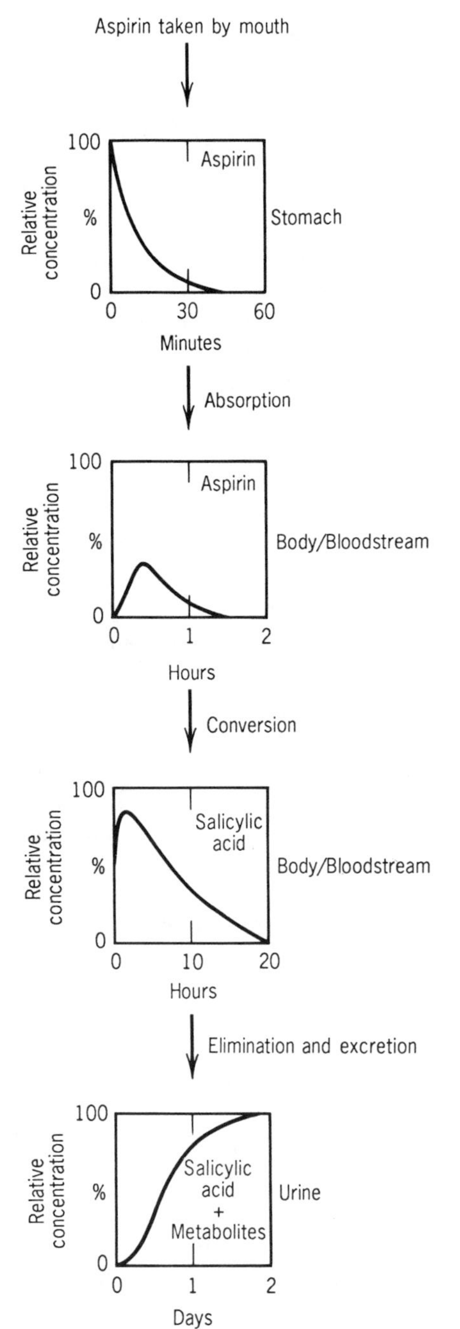

Figure 4.7. *The passage of aspirin through the body.*

4.5 DRUG ABSORPTION AND DOSAGE FORMS

A few drugs are not absorbed from the gut or urine and act locally in the gut or bladder. Some drugs applied topically also act only locally. Most drugs, however, must enter the bloodstream to reach receptors. They reach the bloodstream by one of four routes: enteral, percutaneous, inhalation, and parenteral.

When a drug is taken by mouth or injection (enterally or parenterally), its level in the blood is in the therapeutic range for perhaps 40–60% of the time. Sometimes, as discussed in Section 4.9, it may reach toxic levels. There is therefore great interest in controlled-release dosage forms, also termed sustained, prolonged, timed, programmed, or extended release. With such products, drug levels can be brought into the therapeutic range for 80–90% of the time and toxic levels avoided. An additional advantage of controlled release is possible long-term action. Some implanted drugs last for a month or even a year. Furthermore, if a controlled release drug is implanted near its intended site of action, it can be more effective and produce fewer undesired side effects.

There are several techniques for controlled release. Most are based on the fact that the skin and mucous membranes are semipermeable so that a drug may migrate through them to the dermis, which contains a network of blood capillaries where the drug may be absorbed.

Controlled-release drugs are currently available for the relief of cold symptoms, glaucoma, angina, motion sickness, and for birth control. Research is targeted on drugs to treat narcotics addiction, diabetes, malaria, pollen allergies, tooth decay, and high blood pressure. Anticancer drugs would be particularly valuable in a controlled-release form because they are highly toxic at little above therapeutic levels.

Various controlled-release techniques will be described under the four different methods of administration.

4.5.1 Enteral Dosage

The enteral route, whereby the patient swallows a tablet and the drug is absorbed through the gastrointestinal tract, is the simplest and most widely used method, but is also the slowest. The more soluble the drug, the faster it will be absorbed whatever the method of dosage. Rapid absorption from the gut is also aided by high lipid solubility. For example, the antibiotic streptomycin, is absorbed much more quickly in an oily vehicle than from an aqueous solution. Drug absorption from the gut is affected by a host of factors such as whether the stomach is full and how the tablet is formulated. There are marked variations from person to person and time to time. This is discussed further in Section 4.9.

Drugs may be given enterally in the form of tablets, soft gelatin capsules, hard capsules, liquids, powders, and solutions. The rectum is also part of the gastrointestinal tract, so suppositories inserted into it also come under the enteral classification. Suppositories are a useful dosage form for young children or for patients who are vomiting continually.

More will be said about metabolic transformation in Section 4.7, but it is relevant to note here that if a drug is readily metabolized, much of it may be lost on the first pass through the liver (the first-pass effect) before it has a chance to reach its point of action.

There are other problems with enteral drug administration. Some drugs are destroyed by the low pH of the stomach or by digestive enzymes, and others irritate the stomach and cause vomiting. Some tablets are made with outer layers that resist enzymes in the mouth but break up in the stomach. Others have enteric coatings that protect the active ingredient until it reaches the intestine.

The first controlled release system was an over-the-counter cold remedy marketed by Smith, Kline and French under the name "Spansules." A Spansule comprises a hard gelatin capsule containing hundreds of beads, each of which has a core containing the drug. Each core is surrounded by a layer of wax and these layers are of varying thicknesses. Thus some beads release their contents rapidly, others more slowly. By this means, drug delivery can be sustained over a 12-hour period.

Ion exchange resins may be used to control drug absorption. A basic drug such as nicotine will react with a sulfonic acid cationic ion exchange resin and, if the latter is incorporated into chewing gum, the nicotine is slowly released as the gum is chewed. This is the basis for the drug nicotine polacrilex (Section 17.4), which helps smokers break the habit.

Figure 4.8 shows the OROS tablet, which is typical of recent developments. An osmotic core (made, for example, of an ethylene-vinyl acetate copolymer) containing the active drug is surrounded by a semipermeable membrane which has a laser-produced orifice in it. Once the tablet has been swallowed, water penetrates the membrane by osmosis. It is more attracted to the core than is the contained drug, which is consequently forced out through the orifice.

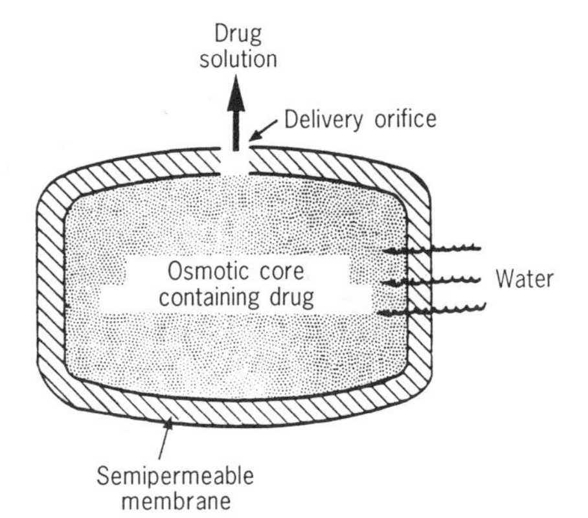

Figure 4.8. *The OROS tablet.*

4.5.2 Percutaneous Dosage

The second form of drug administration is called percutaneous; that is, absorption takes place through the skin, as in the case of ointments, or through the mucous membranes, as in the case of vaginal or sublingual tablets. Absorption through the skin may be fast or slow and the effect may be topical and local rather than systemic. It is usually fast through the mucous membrane. Thus, sublingual tablets are used for the administration of ergotamine tartrate for migraine and nitroglycerol for angina, where rapid relief is essential.

A recent contribution to percutaneous dosage forms is films in which drugs are dispersed and from which they are released slowly. The film is placed next to a semipermeable membrane such as the eye, the uterus, or even the skin, and the membrane absorbs the drug as the film releases it. This may continue over a long period, making repeated doses unnecessary. Figure 4.9 shows the Ocusert system. The rate-controlling membrane is an ethylene–vinyl acetate copolymer and it releases the drug from a so-called reservoir.

The first drug to be administered in this way was pilocarpine for the treatment of glaucoma. If pilocarpine is dropped into the eye, the tissue is flooded, and an excess of the drug is present. The new system avoids flooding and supplies the drug over a sustained period. A product with a similar membrane and a progesterone reservoir, when placed in the uterus, can provide contraception for one year.

Diffusion of drugs through the skin rather than membranes is usually slow. Nonetheless, a new and widely prescribed mode of dispensing nitroglycerol to angina sufferers is through a patch stuck to the chest. The construction of

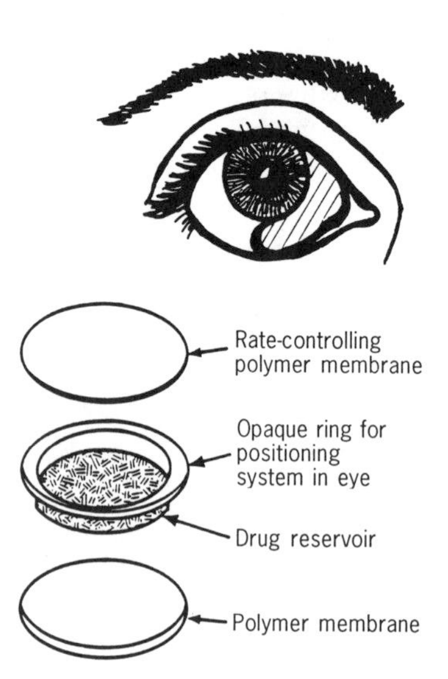

Figure 4.9. *The Ocusert system.*

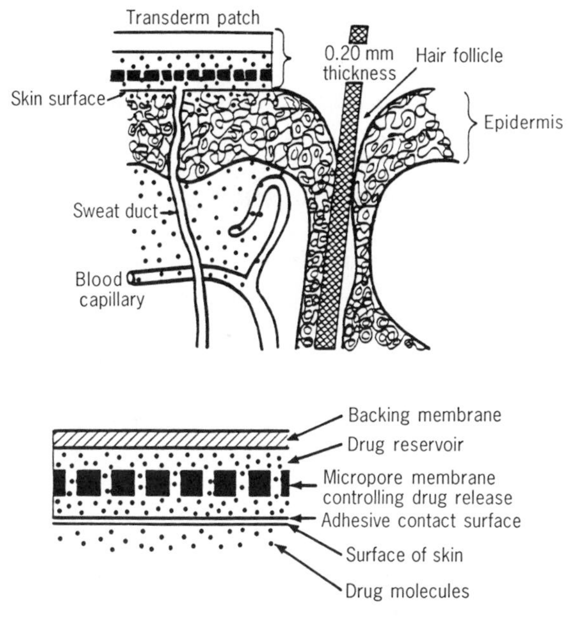

Figure 4.10. *Transderm patch.*

the patch and the way the material diffuses through the epidermis are shown in Figure 4.10. The drug is contained in an assembly comprising an ethylene–vinyl acetate copolymer semipermeable membrane, a backing membrane of an aluminized plastic, the drug reservoir, and a membrane that controls the drug release. A porous silicone adhesive allows this patch to be attached to the skin, after which the drug migrates through the membrane and then through the skin into the blood capillaries. A variation involves dispersal of the drug in a gel containing glycerol, water, lactose, poly(vinyl alcohol), poly(vinyl-2-pyrrolidone), and sodium citrate.

4.5.3 Inhalation

The third method of dosage is by inhalation. The drug is breathed into the lungs, where it is absorbed from the so-called pulmonary epithelium. This has a very high surface area because it is involved in gas exchange in the lungs and it is also well-supplied with blood vessels so absorption is very fast. Inhalation is used especially for antiasthma drugs and anesthetics. Antiasthma drugs may be administered as liquids from atomizers or aerosols, or as solids from insufflators, which disperse the solid and enable it to be inhaled. Probably some of these drugs act locally and not systemically.

4.5.4 Parenteral Dosage

The fourth method of dosage is by injection and is called parenteral administration. An injection may be made directly into a nerve for local effect, as when a dentist deadens the pain from the nerve of a tooth, or into the spinal fluid, as in spinal anesthesia. But the great majority of injections are intravenous (into a vein) or intramuscular (into a muscle) or subcutaneous (under the skin).

Intravenous injections have an almost instantaneous effect, as mainlining drug addicts know. The drug reaches a high concentration in the blood very quickly. The snag is that the dosage has to be measured exactly or toxic levels might be reached. The drug has to be soluble and the injection has to take place under sterile conditions. Thus patients can dose themselves only after careful training. Many drug addicts contract hepatitis or even AIDS as a result of using infected needles. Injections are usually reserved for situations where a rapid effect is vital or there is no other way of ensuring adequate absorption.

Local conditions may affect the rate of absorption of an intramuscular or subcutaneous injection. For example, a patient in a state of shock may fail to respond to an intramuscular injection because peripheral circulation is too poor to carry the drug to a useful site.

Some intramuscular injections are now given in prolonged release forms. These are called "depot" injections and the drug is slowly absorbed from the injection site over hours, days, and even weeks. The contraceptive medroxyprogesterone acetate (Section 11.3) is the most widely used drug in this dosage form.

A variation on depot injections is the use of implantation. Progesterone, the female sex hormone, is destroyed when taken enterally. It is insoluble, so it cannot be injected intravenously. It must either be injected into a muscle, which requires frequent injections, or be implanted in a short tube in the uterus, from which it is released slowly.

Most drugs cannot be implanted as easily as this and there are many controlled-release systems. Some are diffusion controlled with reservoirs and membranes like the Ocusert system and transdermal patch described above. The drug, usually in the form of a powder, is surrounded by a nonbiodegradable polymer through which it diffuses. The diffusion rate depends on the drug and polymer properties and must be carefully controlled. Among the membranes in use are the ethylene–vinyl acetate copolymer already mentioned, hydrogels, and polydimethylsiloxane (Silastic).

A typical hydrogel is formed from poly(hydroxymethyl methacrylate), lightly cross-linked. The water-containing hydrogel does not irritate surrounding tissue; the drug it encapsulates, even if it has a high molecular weight, escapes the entangled polymer structure and enters the bloodstream. Such a system, attached to a tooth, can dispense fluoride continuously for 6 months, although this concept has not yet been commercialized.

Polydimethylsiloxane is used in Finland and Sweden to encapsulate the contraceptive levonorgestrel. The system is implanted under the skin of a woman's arm whence it is released over a five-year period.

A drawback of reservoir systems is that the membrane must be removed once the drug is consumed. Another problem is that, if the membrane ruptures, a large and possibly toxic dose of drug may be released into the body.

An alternative to the reservoir system is the matrix system in which the drug is absorbed inside a porous biocompatible polymer matrix. Body water dissolves the drug but the solution must then thread its way through tortuous channels to reach the polymer surface, so that a time-release system can be designed. Since the drug is not localized, problems of membrane rupture are obviated. Silicone rubber, ethylene–vinyl acetate copolymers, and ethycellulose are used as matrices. Even high molecular weight drugs such as insulin and interferon may be released by this method. It is believed but has not yet been proved that enough insulin to sustain a diabetic for a year could be implanted by this method.

Matrices may also be used with narcotic antagonists (Section 10.1.3). Drug addicts can be given an implant which lasts for thirty days. During that time they are relieved of the agonizing daily decision of whether or not to take the medicine. The development is still under clinical investigation.

An application of this technique to farming involves a sex hormone dispersed in a matrix consisting of a lightly cross-linked copolymer of hydroxyethyl methacrylate and ethylene glycol dimethacrylate. This is implanted in the ears of cattle and assures oestrus synchronization so that the cattle may all be impregnated at one time, either naturally or artificially. The implant must be removed after nine days.

4.5.5 Other Dosage Forms

Matrix systems obviate the problems of membrane rupture but the problems of matrix removal remain. One solution is the use of polymers that gradually decompose, that is, bioerodible polymers. The drug is dispersed through the polymer and is released as the latter disintegrates. Such systems have the added advantage over other controlled-release systems that the drug does not need to be water soluble.

The most frequently used bioerodible polymers are poly(lactic acid), poly(glycolic acid), and lactic and glycolic acid copolymers. Poly(lactic acid) was already well known as the material which replaced catgut in surgery to provide biodegradable sutures.

Bioerodible polymers have been used with narcotic antagonists (see above) and have also been tested for birth control drugs and for luteinizing-hormone-releasing hormone (Section 11.1), which acts on the brain's hypothalamus and might serve as a contraceptive for both men and women.

Another bioerodible polymer under study is poly(ϵ-caprolactone), which offers high permeability to steroids. It is 10,000 times more permeable to progesterone than is poly(lactic acid), a related polymer, and can thus deliver steroid hormones more efficiently.

There are also problems with taking matrix drugs by mouth because the capsule may pass beyond the digestive system while the drug is being released and discharge it in the wrong place (see Section 12.2). One possible solution is the bioadhesive matrix. The drug, for example, the diuretic chlorothiazide, is encapsulated in a water-dispersible polymer such as carboxymethylcellulose or an acrylic acid–divinyl glycol copolymer. Because the matrix is hydrophilic, it attaches to a mucosal lining, such as that of the stomach, and the drug is slowly released.

Further novel delivery systems include drugs which themselves are in a polymeric form or attached to polymers. The body is generally passive to polymeric materials and it might be possible to obtain great specificity at the desired site. A polymeric antibiotic might perhaps not be absorbed from the gut and thus be more effective against urinary tract infections. There is much research in progress on organometallic polymers, which could perhaps be implanted and have prolonged action against cancerous tumors.

Red blood cells could be useful drug carriers and, if properly modified, concentrate in the liver and the spleen. Potentially they provide a method for delivering drugs to these organs. Liposomes are multilayered vesicles that form when phospholipids are dispersed in water. These, too, provide a potential means for drug delivery.

The attachment of drugs to antibodies is another possibility and is exemplified by the Antibody Directed Enzyme Pro-drug Therapy (ADEPT). Antibodies can be made which will bind to antigens on cancer cells. However, when a drug is attached to an antibody, it does not reach all cancer cells because some of them do not have antigens and because antibodies are large molecules which cannot penetrate closely packed tumors.

The first step in the ADEPT treatment is to produce a pro-drug consisting of an anticancer drug rendered inactive by a small chemical "tail" which can be removed by an enzyme not found in the body. This enzyme (carboxypeptidase G2) is coupled with a cancer-seeking antibody and injected into the patient. A day later, when the antibody has bonded to the cancerous cells, the pro-drug is injected. When it reaches the enzyme, it reacts to give high concentrations of anticancer drug (a benzoic acid mustard) near the cancerous cell. The drug itself can diffuse through the spaces between the cancer cells much more readily than the antibody can. Hence the technique, which is still in the pioneering stages, should be particularly suitable for large tumors.

Implantable pumps, which deliver drugs at a predetermined rate, are currently available on an experimental basis for various drugs including morphine, insulin, heparin, and anticancer agents.

On the horizon are even more sophisticated systems. Research is underway on the use of magnetic beads, which can be set in motion by a magnetic field and thereby "squeeze" more of a drug such as insulin through a semipermeable membrane when the body needs it. This could make possible the administration of variable doses. Other work has involved the compounding of anticancer drugs with iron oxide particles. The mixture is encapsulated in protein microspheres and injected into the bloodstream. A magnet is held over the tumor to be treated; the magnetic iron oxide is attracted to the magnet and the drug carried to the place where it will be the most useful.

It should not be assumed that drugs will be administered in all these ways in the foreseeable future. The examples illustrate, however, that much research is in progress on the way drugs can be delivered to sites where they will be most effective, and that improvement of delivery methods is an important part of pharmaceutical research.

4.6 DRUG DISTRIBUTION

The above discussion indicates that different methods of drug dosage will give different rates of absorption into the bloodstream. A drug's effectiveness, however, depends not on its concentration in the bloodstream but on its concentration near the receptors with which it interacts. It therefore has to have the ability to travel in the bloodstream and to gain access to the appropriate receptors. There are wide differences in behavior among different drugs, and the factors affecting drug distribution are discussed in the following sections.

4.6.1 Binding to Plasma Protein and Blood Cells

As a drug enters the circulatory system, it may bind to plasma protein or to red blood cells. This is a reversible equilibrium:

$$\text{drug(unbound)} + \text{protein(bound)} \rightleftharpoons \text{drug–plasma protein(complex)}$$
$$\text{drug(unbound)} + \text{red blood cells(unbound)} \rightleftharpoons \text{drug–red blood cells(complex)}$$

It is set up rapidly in the case of plasma protein, slowly in the case of red blood cells.

Plasma protein and red blood cells are macromolecules and are unable to diffuse through tissues. Consequently, since only the free drug can diffuse freely between blood and tissues, the availability of the drug in the tissues depends on the position of equilibrium in the drug–protein and drug–red blood cell interactions. The complex also acts as a depot for the drug and slows its elimination. Thus, in general, strong binding to plasma and blood cells means that a drug will reach lower concentrations in tissues but the levels will be maintained longer. Suramin, the antitrypanosomiasis drug (Section 18.2.2), binds so strongly to blood protein that a single dose is effective for 3 months.

4.6.2 Physical Properties of the Drug

The second factor governing drug distribution is the physical properties of the drug. All drugs must have or acquire some minimum solubility in the polar extracellular fluids, otherwise they cannot be transported to cell surfaces. Some drugs are intrinsically soluble; others are solubilized by enzyme or other chemical action.

Two main properties, the pK_a values and the appropriate partition coefficients, govern the distribution of the drug between extracellular body water and tissues. The pH of the system is also important.

Highly hydrophilic drugs cannot penetrate cell walls and remain confined to body water outside cells. Drugs with a balanced lipophilicity can penetrate cell walls and enter the water inside cells. Highly lipophilic drugs become widely distributed in tissues and may become concentrated in body fat. Barbiturates and chlorinated compounds generally, including DDT, are examples.

In Section 4.3.4 the importance of partition coefficients in QSARs was noted. In choosing compounds for synthesis, the pharmaceutical chemist will often concentrate on those likely to have partition coefficients in a defined range.

4.6.3 Other Factors

A third factor governing drug distribution is binding to tissue. Sites in the tissues away from the circulatory system may bind drugs strongly. Chlorpromazine (Section 8.2) binds to melanin granules in eyes and skin which can lead to side effects in long-term users. Lead accumulates in the bones. Tetracycline antibiotics (Section 6.5) are stored in bones and teeth. Barbiturates (Section 8.1.1) are stored in fatty tissue. Quinacrine, an obsolete antimalarial, is stored in the liver. Drugs with flat molecules, such as the antibacterial proflavine, can intercalate between the base pairs of DNA. This is illustrated in Figure 4.1.

The fourth factor governing distribution is perfusion, defined as the blood flow per gram of tissue to a particular site. Tissues such as the liver, kidneys, lungs, brain, and heart have a high perfusion rate. Others, such as muscles, skin, and body fat are poorly perfused. Lipophilic drugs equilibrate with tissues at a rate governed by the perfusion rate. Consequently, high concentrations are reached quickly in highly perfused tissues but may be achieved slowly or not at all in poorly perfused tissues.

A fifth factor is the ability of a drug to cross special membranes in the body. The general problem of crossing cell walls is covered by the factors mentioned in Section 4.6.2, but the body has special barriers to protect sensitive organs, such as the brain, and these are more difficult for chemicals to penetrate. Such membranes with special features may affect the distribution of drugs in the body. With a few exceptions, only un-ionized and highly lipophilic drugs can cross the blood–brain barrier. Levodopa, used in

Parkinson's disease (Chapter 16), is able to do so even though it is ionic. Sulfadiazine is able to cross the cerebrospinal fluid barrier and thus is of use in meningococcal meningitis. Lipophilic substances, such as steroids, narcotics, anesthetics, some antibiotics, and established teratogens, can cross the placenta and this may cause problems in pregnancy.

There are many factors omitted from this discussion. For example, drug distribution varies in individual cases because of personal factors including age, pregnancy, disease states, and interactions with other drugs the patient may be taking. Peripheral circulation may be much slower in an old person so that perfusion rates are altered. Obese people can store much larger quantities of drugs in their body fat, and so on.

4.7 BIOTRANSFORMATION OF DRUGS

In terms of the story here, the journey of a drug to its receptors is complete and it will exert its therapeutic effect. Meanwhile, the first instinct of the body is to try to eliminate the intruding molecules by excreting them. Thus, the drug concentration will reach a peak some time after administration but will afterward drop as the drug is metabolized by the body. This process is called biotransformation. Biotransformation is brought about by enzyme systems in all the body cells and the blood but occurs especially in the liver. All drugs given orally pass from the gastrointestinal tract to the bloodstream via the hepatoportal vein which collects blood from the intestines and delivers it to the liver. After passing through the liver, material enters the systemic circulation and is delivered to the rest of the body. It is during this and subsequent passages through the liver that the bulk of metabolic transformations take place.

There are two classes of reactions that transform drugs, known as synthetic and nonsynthetic biotransformation. In the synthetic case, the drug or its metabolite is coupled or *conjugated* with another substance to give a soluble inactive product that is readily excreted. In the nonsynthetic case, the enzymes cause a chemical reaction, such as hydrolysis, oxidation, or reduction, leading to inactivation or reduced activity of the drug.

In both cases, the metabolic systems act to reduce the lipophilicity of the drug, usually by increasing its polarity. An interesting consequence of this is that when drugs are subject to high-pressure liquid chromatographic analysis, the metabolites are eluted *before* the parent molecule in a reverse phase system and *after* it on a normal phase column. After the drug has been metabolized, it can be eliminated dissolved in the urine. Unless the metabolic system is able to do this, the lipophile will remain in the body for a long time concentrated in the fatty tissues. This accounts for the gross difference in toxicity between benzene and toluene. Benzene has no functional group that can be attacked by an enzyme. Toluene, on the other hand, undergoes oxidation to benzoic acid which subsequently couples either with glucuronic acid or glycine, both of which are readily excreted.

Toluene → Benzoic acid → (H_2NCH_2COOH glycine) Hippuric acid; (D-glucuronic acid) Benzoyl glucuronic acid

To carry out the above functions, the liver is equipped with a series of enzymes known as mixed-function oxidases. They bring about a range of elegant chemical reactions. Figure 4.11 shows a selection of metabolic transformations occurring throughout the body, all involving drugs that will be discussed in later chapters. Hydroxylation is the commonest metabolic transformation, and dealkylation is thought to take place via hydroxylation of the alkyl group followed by cleavage of the carbon–oxygen bond. Hydrolyses are brought about by esterases and amidases, and many occur outside the liver. Reductive reactions also frequently take place outside the liver and are often brought about by the bacteria in the gut.

In infants these enzyme systems are immature and in old people they may have degenerated. Such patients are particularly at risk and may require much lower doses of drugs than healthy adults to produce the same effect. One of the problems with the antiarthritis drug benoxaprofen (Section 12.1), which had to be withdrawn in 1982, was that it was eliminated by old people at about one-fifth of the normal rate. In younger people it was removed by conjugation with glucuronic acid (Figure 4.11), but older patients were not as well equipped to bring about this reaction.

Drugs may interact in the course of their distribution. Consider, for example, the anticoagulant drug warfarin (Section 7.7) and the antirheumatic drug phenylbutazone (Section 12.3). Both drugs bind to plasma proteins. If both are given at once, there is competition for plasma proteins and the phenylbutazone wins. The level of free warfarin in the bloodstream is higher than it should be and the patient may suffer from severe bleeding rather than benefiting from mild anticoagulant action.

Most psychotropic drugs interact with ethyl alcohol and the patient who drinks while taking tranquilizers may become very drunk indeed. Barbiturates are particularly dangerous. Both they and ethyl alcohol are eliminated by the

Hydroxylation

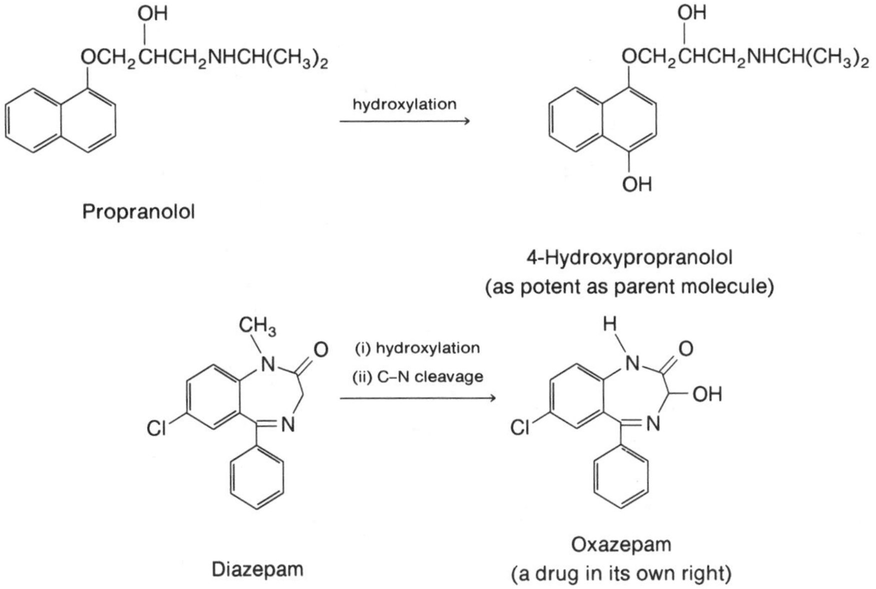

Propranolol —hydroxylation→ 4-Hydroxypropranolol (as potent as parent molecule)

Diazepam —(i) hydroxylation, (ii) C–N cleavage→ Oxazepam (a drug in its own right)

Dealkylation

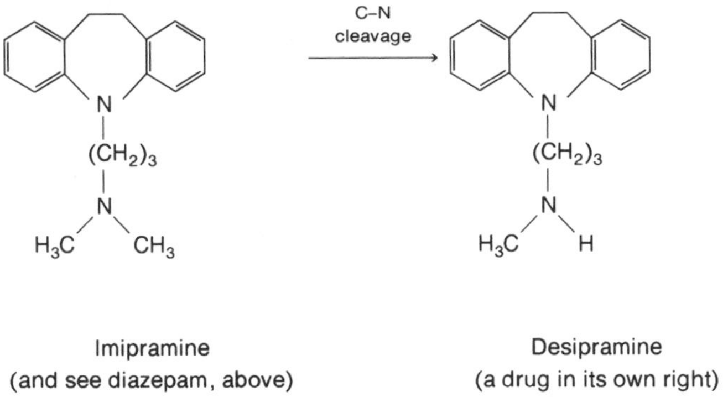

Imipramine (and see diazepam, above) —C–N cleavage→ Desipramine (a drug in its own right)

Figure 4.11. *Commonly observed metabolic pathways.*

same system in the liver; consumption of the two at the same time can lead to liver failure and death.

Not all drug interactions are unfavorable, however. With some drugs there is a synergistic effect and the one potentiates the other. This is exemplified by the antibacterial mixture cotrimoxazole, which will be discussed in Section 6.2. Finally, some drugs are simply rendered ineffective. For example, tetracycline antibiotics (Section 6.5) form chelate compounds with calcium and magne-

Oxidation

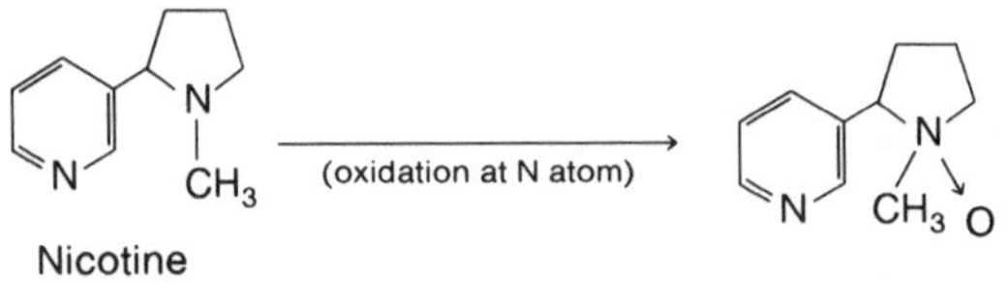

N CH$_3$ (oxidation at N atom) N CH$_3$ O

Nicotine

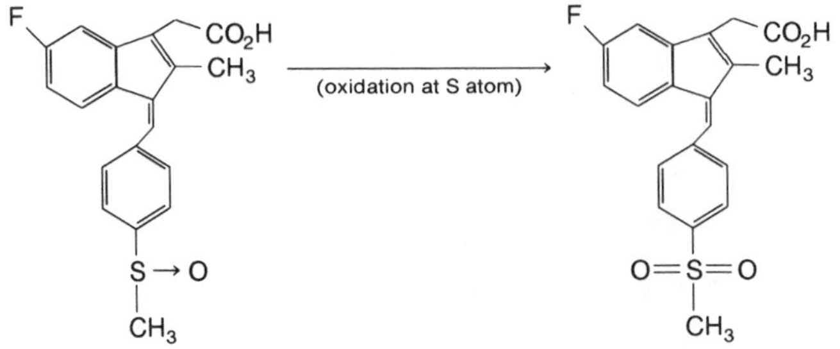

F CO$_2$H CH$_3$ (oxidation at S atom) F CO$_2$H CH$_3$

S → O CH$_3$ O=S=O CH$_3$

Sulindac

Hydrolysis

$C_6H_5CHCOHN$ NH_2 S N O $CO_2CH_2O_2CC(CH_3)_3$

Hydrolysis

$C_6H_5CHCOHN$ NH_2 S N O CO_2H

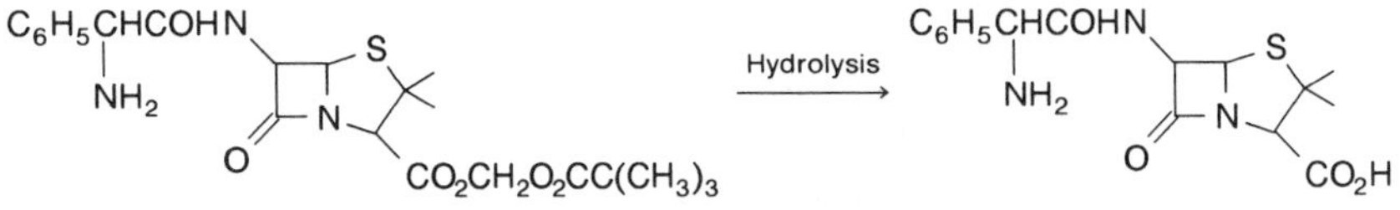

Pivampicillin

Ampicillin

Reduction

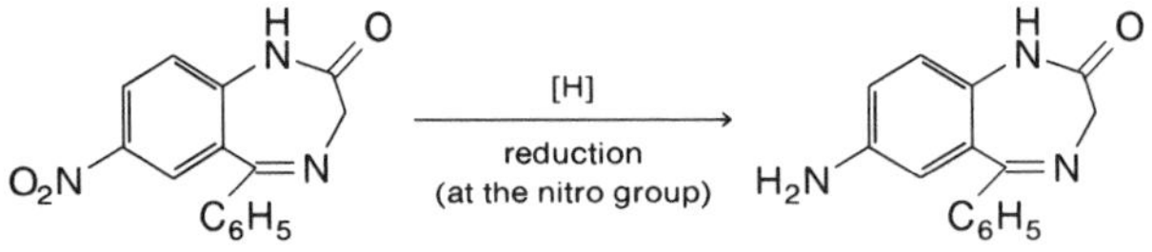

H N O O_2N N C_6H_5

[H] reduction (at the nitro group)

H N O H_2N N C_6H_5

Nitrazepam

HO C_6H_5 O $CHCH_2CCH_3$ O O

[H] reduction (at the carbonyl group)

HO C_6H_5 OH $CHCH_2CHCH_3$ O O

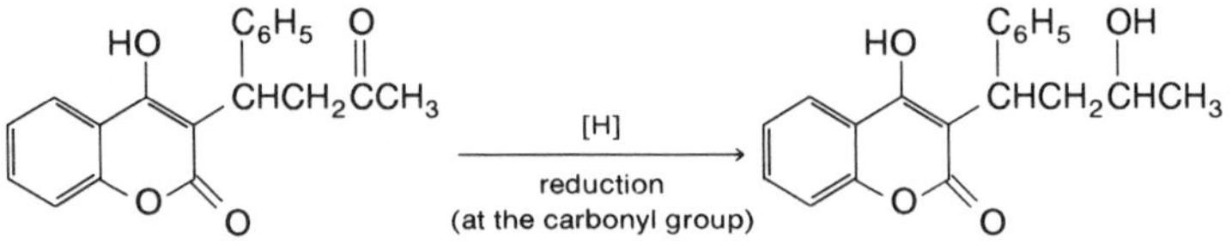

Warfarin

Figure 4.11. *Continued.*

sium. If they are taken with milk, which contains calcium, or indigestion remedies, many of which contain magnesium, they are sequestered and may never reach an effective level in the bloodstream.

There is an implication above that biotransformation invariably renders a drug less effective. This is not always true. Sometimes a metabolite is a drug in its own right and the discovery of the metabolic pathway may lead to the

Esterification ("conjugation")

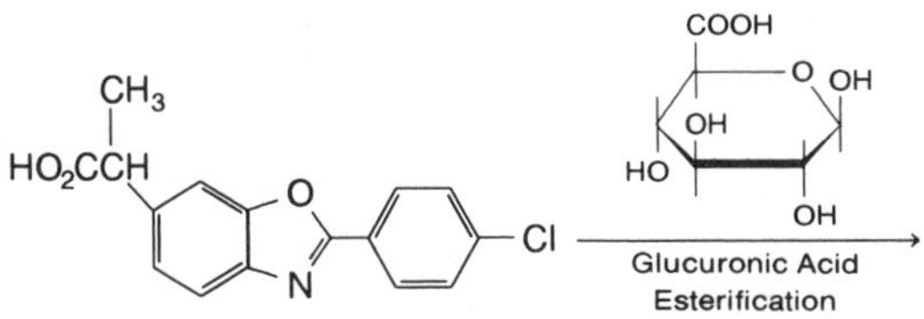

Benoxaprofen

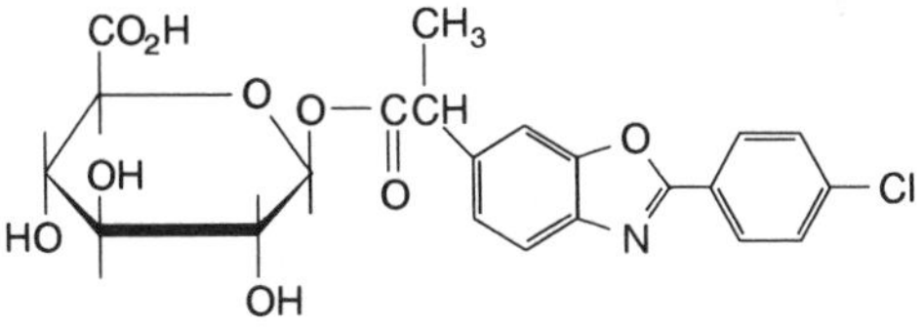

Glucuronic Acid Esterification

Glucuronide ester

Amidification ("conjugation")

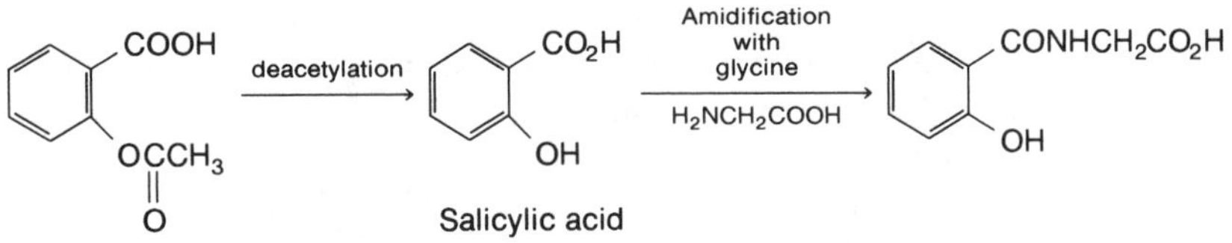

deacetylation

Amidification with glycine H_2NCH_2COOH

Aspirin

Salicylic acid

Acetylation

CH_3COOH

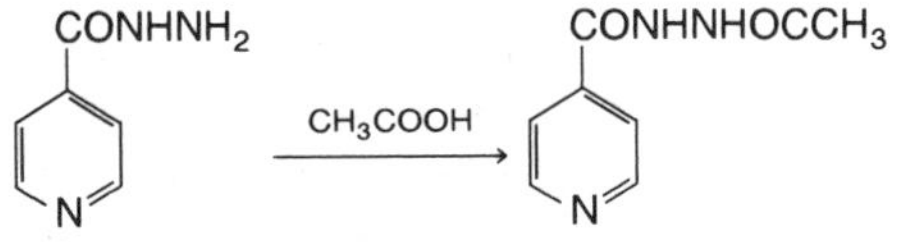

Isoniazid

Figure 4.11. *Continued.*

discovery of another drug. For example, oxazepam (Figure 4.11) is a metabolite of diazepam, has a similar but shorter lasting activity, and is marketed in its own right. Similarly, the antidepressant imipramine (Figure 4.11) is metabolized to desipramine, which has a more specific antidepressant activity and is also separately marketed.

Sometimes a drug must be converted to another chemical in the body before it is effective. Such drugs are called pro-drugs. One advantage of a pro-drug is that it may be more lipophilic than the drug and hence be better absorbed by the tissues. Also, the pro-drug may be less irritating to the gastrointestinal tract. The penicillin esters (e.g., pivampicillin, Section 6.3.6) illustrate both these advantages. They hydrolyze in the body to give the active drug ampicillin. Pro-drugs may sometimes be given in a different dosage form. For example, tetracycline is sparingly soluble and cannot be injected. The pro-drug rolitetracycline (Section 6.5.2) is adequately soluble and hydrolyzes to tetracycline in the bloodstream. Pro-drugs may be designed to decompose slowly so as to maintain a steady level in the body and to reduce the number of doses required. An example is insulin protamine zinc compound which decomposes to release insulin slowly into the bloodstream. Finally, the pro-drug may be more stable than its active metabolite and therefore be more suitable as a dosage form.

4.8 EXCRETION

Having been metabolized in the body, a drug is eventually excreted as its metabolites. The main organ involved in this process is the kidneys, which are adept at handling water-soluble materials. The metabolites then appear in the urine. Other organs that may be concerned include the bile system, the lungs, and the sweat, salivary, and other glands. Some drugs are excreted in breast milk and may affect the baby. Ethyl alcohol is eliminated through several channels and may be detected in the urine and in the breath where part of it has been metabolized to acetaldehyde (ethanal). This is, or course, the basis of the Breathalyzer.

The elimination of a drug administered in therapeutic doses usually follows first-order kinetics; that is, the rate of elimination is proportional to the concentration of the drug. Such drugs are said to be eliminated linearly and the half-life in the body is constant and independent of dose. A few drugs, for example, aspirin (Section 10.2) and phenytoin (Section 8.1.3), exhibit dose-dependent elimination, the half-life increasing with dose. The kinetics are of an order less than one and approaching zero, and elimination is said to be nonlinear. Nonlinear elimination is usually due to saturation of the elimination process. Other reasons include the inhibition of elimination by a metabolite of the drug.

4.9 BIOAVAILABILITY AND THE DIGOXIN AFFAIR

The past few sections have indicated some of the factors that govern the rates of absorption and distribution of drugs. These are the processes that carry the drugs to their target receptors and while this is going on there are simultaneous processes of biotransformation and excretion. The concentration of drug that reaches the receptor and the length of time for which it is maintained is the biological availability or bioavailability of the drug.

The bioavailability of a drug is governed by some factors related to the patient and others related to the drug. The patient-related factors include age, sex, body weight, the presence of food in the stomach, the pH of the stomach,

TABLE 4.1. Some Drug-Related Factors Affecting the Bioavailability of the Drug

Particle size of the drug; its crystal form
Chemical nature — Is it in an un-ionized form, a salt, or a hydrate?
Excipients — Diluents, fillers, lubricants, disintegrants, surface-active agents, and wetting agents are or may be added
Manufacturing process — Compressive force used in making a tablet
Dosage form — Gelatin capsules will have a different bioavailability from tablets, enteric coated tablets, slow release formulations, etc.
Surface-to-volume ratio of the tablet
Additives — May alter solubility by formation of eutectics, ion pairs, etc.
pH — Buffer type and amount
Flavoring and coloring agents, antioxidants, and preservatives
Impurities, contaminants, or allergenic substances in product
Storage factors — Time, heat, light, and vibration
Type of container and stopper — Glass? Impervious?
Package — Dehydration of cotton in package, amount of cotton, etc.

and the circadian rhythm. Some factors related to the drug are shown in Table 4.1. They mainly affect its rate of absorption. All the factors listed plus other minor but possibly significant ones must be studied by a pharmaceutical company before a drug can be marketed. Because of the variety of factors, responses to drugs are difficult to predict.

Until the 1960s, it was assumed nonetheless that two medicines containing identical amounts of active ingredients would produce identical responses in a given patient if taken under similar circumstances. At that time a routine chemical analysis was considered a satisfactory level of control for active ingredients. Only for biological preparations, such as vaccines, was measurement of therapeutic activity considered necessary. During the 1960s, various physicians reported disturbing discrepancies in patients' reactions to apparently identical drugs and the terms "bioavailability" and "bioequivalence" were coined. Bioavailability is the ease with which a drug is available from the dosage form in given circumstances; bioequivalence is the equivalent physiological availability. Thus, one must determine how much of a drug, when injected, will have the same effect as the same drug taken orally or in other dosage forms. Also one needs to know what doses of similar drugs have the same physiological effect.

Figure 4.12 shows the possible consequences to a patient of differing bioavailability. The plasma concentration of a drug is plotted against time after the dose. In each of the cases, A, B, and C, the same total quantity of active drug is eventually absorbed but in case A there will be a toxic response and in case C the therapeutically effective plasma concentration is never reached. Only in case B does the drug have the desired effect.

The importance of bioavailability was brought forcibly to the attention of prescribers by the so-called digoxin affair. Digoxin has been used for the treatment of congestive heart failure for 200 years and is still frequently prescribed (Section 7.2). Of all the widely used drugs it is the one that offers

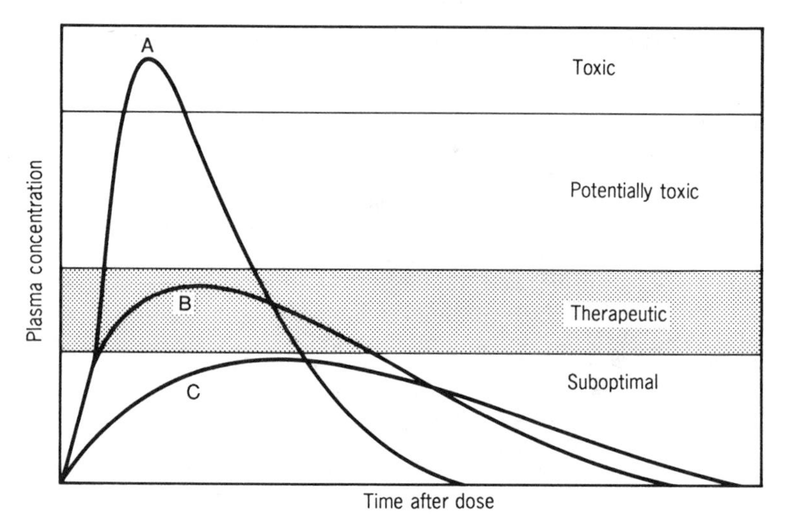

Figure 4.12. *Influence of drug delivery rate on therapeutic activity. Source: Brand Names in Prescribing, Office of Health Economics, 1976. (See also A. H. Beckett, Postgraduate Medical Journal 50, 125 (1974)).*

the largest accumulation of experience. Its discoverer, William Withering, warned in 1785 of the dangers of overdose and the margin between the effective and toxic doses was known to be small.

Yet in 1972 digoxin was found to produce widely different results in patients. Hospital teams in the United States and the United Kingdom confirmed these fluctuations. The leading brand of digoxin was made by

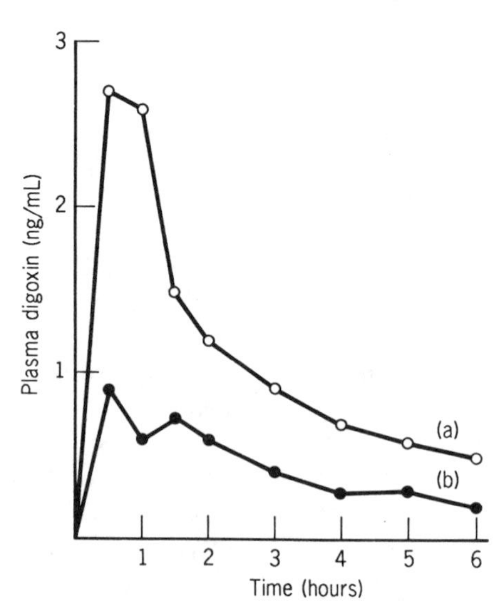

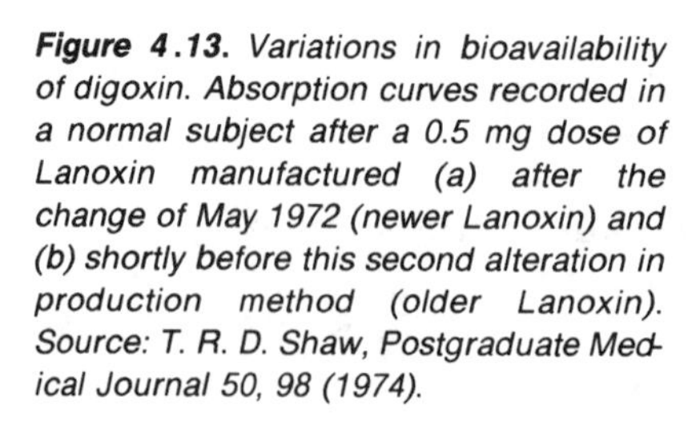
Figure 4.13. *Variations in bioavailability of digoxin. Absorption curves recorded in a normal subject after a 0.5 mg dose of Lanoxin manufactured (a) after the change of May 1972 (newer Lanoxin) and (b) shortly before this second alteration in production method (older Lanoxin). Source: T. R. D. Shaw, Postgraduate Medical Journal 50, 98 (1974).*

Burroughs Wellcome, a company of high moral reputation whose profits go mainly to the Wellcome Foundation, which is a charitable institution. It transpired that they had modified their method of processing digoxin in May 1972 and the new material had a different particle size from the old. Consequently, its rate of dissolution was quite different. Studies of other brands also showed wide differences in dissolution rates.

Figure 4.13 shows the plasma levels in a normal person after administration of digoxin made by the new method after May 1972 and after the return to the older production method. Amid embarrassment and consternation, Burroughs Wellcome made the facts public, warned the medical profession, and recalled ten million tablets.

Although the digoxin affair was the most publicized case of bioavailability problems, it was by no means the only one. For example, some formulations of chlorothiazide–potassium chloride tablets were found to cause intestinal perforations, a few of which were fatal. One brand of enteric coated aspirin had no affect on arthritis pain while another, containing apparently the same dose, led to toxic symptoms. And while diphenylhydantoin capsules diluted with calcium sulfate were satisfactory, other tablets diluted with lactose led to toxic symptoms.

Unfortunate though it was, the digoxin affair gave a boost to the large pharmaceutical companies who had long insisted that the only way a physician could be sure of consistent properties from a drug formulation was to use consistently a branded pharmaceutical from a reputable manufacturer, in whose laboratories bioavailability *in vivo* as well as product purity would be carefully monitored.

4.10 PROPERTIES OF AN IDEAL DRUG

The pharmacological concepts described above illustrate the many factors that can limit a drug's usefulness. An ideal drug would be without any such drawbacks and a convenient way to summarize this chapter is to list the properties of an ideal drug.

1. A drug should be of low toxicity and without side effects. Many drugs are selective poisons and carry some toxic hazard. The fatal dose, if the drug is indeed poisonous, should be many times the therapeutic dose, that is the therapeutic index should be high. In addition, individual patients may have acute allergic reactions to drugs or become sensitized to them. These risks have to be weighed against the potential benefit to the patient. The antibiotic, lincomycin, for example, is prescribed for certain infections by penicillin-resistant bacteria. Usually it also causes diarrhea, which is not considered an excessive price to pay if the drug is effective. On the other hand, chloramphenicol is a cheap, excellent, broad spectrum antibiotic that causes fatal aplastic anemia in about one case in 50,000. Consequently, its use is reserved

in the United States and the United Kingdom for diseases such as typhoid fever where it is the only effective treatment.

2. The physical quantity of the formulation to be taken should be of an acceptable size and the drug should not require too frequent administration over too long a period. That is the reason for the attractiveness of ivermectin (Section 18.3.1) and is one advantage that the antiulcer drug famotidine has over its competitor cimetidine (Section 9.6).

Patients who forget to take drugs are a major medical problem. Physicians find that most people can remember to take one tablet a day. Fewer can remember four tablets at three-hour intervals. Furthermore, many patients stop taking a drug once they feel better. This encourages resurgence of illness and with infections it encourages the development of antibiotic-resistant strains of bacteria. Nonetheless, at present, many heart disease and arthritis patients are forced to take large numbers of tablets several times a day for the rest of their lives.

3. The efficacy of a drug should not be seriously reduced by body fluids, exudates, plasma proteins, and tissue enzymes.

4. A drug should be capable of being stored for long periods in extreme climates and it should be possible to dispense it in a variety of dosage forms. It is useless to give patients who are not in a hospital a drug that has to be injected several times a day, unless they are trained to do it as are diabetics requiring insulin. Equally, patients who are vomiting cannot use a drug that has to be given by mouth.

5. A drug should be selective. An antihistamine should counter an allergy without putting the patient to sleep. An anesthetic should put the patient to sleep without depressing breathing. An antibiotic should eliminate the bacteria responsible for the disease leaving other bacteria in the body undisturbed. Eliminating all the bacteria leaves the patient prone to superinfection where a few antibiotic-resistant bacteria find conditions ideal for multiplication. A second infection therefore takes over from the first. It should nevertheless be noted that a broad-spectrum antibiotic is convenient and cost effective. It can be prescribed in a single visit to the physician and obviates the need for extensive microbiological investigation. A general practitioner's choice of antibiotic is most often an empirical one and a broad-spectrum antibiotic reduces the margin of error.

6. Preferably, a drug should cure disease rather than deal only with the symptoms. Regrettably, this is rarely possible. Among the vast range of pharmaceuticals, only the anti-infectives truly cure disease. Even with these there is a distinction between bactericides, which kill bacteria, and bacteriostats, which prevent their multiplication. An ideal antibacterial is bactericidal but most real ones are only bacteriostatic.

7. An ideal drug should be one to which patients will develop neither tolerance nor addiction. Regular users of barbiturates, for example, will need

larger and larger doses to induce sleep and this is an undesirable feature of the drug. Most sleeping tablets lead to some form of physical dependence over a sustained period and, of course, the classical drugs of abuse, such as heroin and morphine, are rapidly addictive.

In the case of antibiotics, it is not the patients who develop tolerance but the bacteria that develop resistant strains. Often these produce enzymes, such as β-lactamases, that destroy the drug. Resistant strains can develop in two ways. In any large population of a billion or more drug-sensitive bacteria, there are probably a few spontaneous mutants resistant, for example, to streptomycin. If the population is exposed to streptomycin, the mutants continue to multiply while the rest of the population dies. Antibiotic resistance in bacteria, however, seems usually to have resulted from a second process involving transfer of ready-made resistance genes from one bacterium to another. These genes are DNA molecules known as episomes or plasmids. The problem of resistant bacteria is discussed further in Section 6.12.

8. A drug should not contain impurities. One of the problems with the early penicillins was the removal of foreign proteins that caused toxic reactions. Similarly, levodopa, a drug used in treating Parkinson's disease, must not contain its optical isomer D-dopa because the latter causes a blood disorder in 25% of patients.

9. Finally, the absorption, distribution, biotransformation, and excretion of the drug should be such that therapeutically active levels in the appropriate parts of the body are rapidly reached and maintained for long periods. There should be no problems of bioavailability. Patients should react consistently to the drug and individual idiosyncratic reactions should not occur.

4.11 CONCLUSION

No drug ever possesses all these virtues, but the points mentioned in this chapter provide useful targets for pharmaceutical research. They should be borne in mind when considering the advantages and drawbacks of a given pharmaceutical.

The passage of a drug through the body has been described at some length to stress that, not only must a drug be designed such that it will interact with the necessary receptors and be selectively toxic, but it must also be formulated in such a way that it is transported to the site where it is to act in correct concentrations and for the appropriate time. If these conditions are fulfilled, it is the result of a careful balancing of rates of absorption, distribution, biotransformation, and excretion.

The design of a formulation may present just as many problems as the design of a drug. The development of a dosage form that need only be given once a day by mouth rather than having to be injected or taken several times a day is a legitimate aim of pharmaceutical research. Indeed even the packaging of a drug may require research. For example, it is easy enough to devise a child-proof bottle to contain acetaminophen to be taken by an adult for relief of symptoms of a cold, but how does one produce a child-proof package for arthritics whose manual dexterity has been impaired by their illness? The pharmaceutical industry has to pay attention even to problems apparently as trivial and nonmedical as this one.

Part II

The Top-100 Drugs

5

The Production of Pharmaceuticals

In the previous chapters, the background to the pharmaceutical industry was discussed. The remaining sections of this book describe where drugs come from and how they are made. There are immediate problems of selection and classification.

5.1 THE TOP-100 DRUGS

The problem of selection comes first. The *Physicians's Desk Reference* (see bibliography) describes over 1000 commonly prescribed chemical entities, over 2000 branded products, and about 4000 formulations. Which are the important ones? How is "important" defined? Is it the drugs that save lives in a few cases or those that ease discomfort in many? Is it the drugs at the forefront of medical knowledge and research or the bread-and-butter compounds that make a living for their manufacturers year in and year out? Is it the drugs that are manufactured in the largest quantities or that produce the largest revenue? There is no valid answer to these questions and we have taken a pragmatic decision to concentrate our attention on the drugs that are prescribed most frequently. One hundred chemical entities account for about half of all prescriptions filled in U.S. pharmacies and these will be the basis of the discussion. The list for 1986 is shown in Table 5.1A.

Many important drugs are omitted, for example, anesthetics and injectable antibiotics, which are administered mainly in hospitals, and drugs for tropical diseases not prevalent in the United States. We shall diverge from our selection to mention some of these. Hospital drugs make up about 15% by

TABLE 5.1A. The 100 Most Widely Prescribed Ethical Drugs in the United States (1986) (Plus the Next 35)

Rank	Generic Name	Typical Trade Name[a]	Therapeutic Class
1	Hydrochlorothiazide	Hydrochlorothiazide	Diuretic
2	Amoxycillin	Amoxil	Antibiotic
3	Codeine	Codeine	Analgesic
4	Triamterene	Dyazide	Diuretic
5	Erythromycin	E.E.S.	Antibiotic
6	Penicillin V	V-cillin	Antibiotic
7	Digoxin	Lanoxin	Cardiac glycoside
8	Potassium chloride	Slow-K	Potassium supplement
9	Furosemide	Lasix Oral	Diuretic
10	Propranolol	Inderal	β blocker
11	Ibuprofen	Motrin	Nonsteroid anti-inflammatory
12	Norethindrone	Ortho-Novum 10/11-28	Birth control
13	Ampicillin	Penbritin	Antibiotic
14	Cimetidine	Tagamet	H_2 blocker
15	Diazepam	Valium	Anxiolytic
16	Ethynylestradiol	Ortho-Novum 7/7/7-28	Birth control
17	Atenolol	Tenormin	β blocker
18	Naproxen	Naprosyn	Nonsteroid anti-inflammatory
19	Alprazolam	Xanax	Anxiolytic
20	Tetracycline	Achromycin-V	Antibiotic
21	Nitroglycerol	Nitrostat	Coronary vasodilator
22	Cephalexin	Keflex	Antibiotic
23	Propoxyphene	Darvocet	Analgesic
24	Mestranol	Ortho-Novum 1/35-28	Birth control
25	Cotrimoxazole	Bactrim DS	Antibiotic
26	Theophylline	Theo-Dur	Antiasthma
27	Conjugated estrogenic hormone	Premarin Oral	Steroid
28	Norgestrel	Ovral-21	Birth control
29	Methyldopa	Aldomet	Antihypertensive
30	Ranitidine	Zantac	H_2 blocker
31	L-Thyroxine, Na salt	Synthroid	Thyroid
32	Metoprolol	Lopressor	β blocker
33	Salbutamol	Proventil	Antiasthma
34	Amitriptyline	Triavil	Antidepressant
35	Prednisone	Decortisyl	Antiflammatory
36	Dipyridamole	Persantine	Coronary vasodilator
37	Nifedipine	Procardia	Calcium blocker
38	Phenylpropanolamine	Naldecon	Nasal decongestant
39	Insulin	Iletin I NPH	Hypoglycemic
40	Phenytoin	Dilantin sodium	Anticonvulsant
41	Triazolam	Halcion	Hypnotic

TABLE 5.1A. Continued.

Rank	Generic Name	Typical Trade Name[a]	Therapeutic Class
42	Cefaclor	Ceclor	Antibiotic
43	Caffeine	Caffeine	Stimulant
44	Phenobarbital	Phenobarbital	Anticonvulsant
45	Diltiazem	Cardizem	Calcium blocker
46	Oxycodone	Percocet-5	Analgesic
47	Captopril	Capoten	ACE inhibitor
48	Piroxicam	Feldene	Nonsteroid anti-inflammatory
49	Lorazepam	Ativan	Anxiolytic
50	Prazosin	Minipress	Antihypertensive
51	Terfenadine	Seldane	H_1 blocker
52	Clorazepate	Tranxene	Anxiolytic
53	Butalbital	Fiorinal	Hypnotic
54	Isosorbide dinitrate	Isordil	Coronary vasodilator
55	Timolol	Timoptic	β blocker
56	Clonidine	Catapres	Antihypertensive
57	Glyburide	Micronase	Hypoglycemic
58	Flurazepam	Dalmane	Hypnotic
59	Metaproterenol	Alupent aerosol	Antiasthma
60	Sulindac	Clinoril	Nonsteroid anti-inflammatory
61	Miconazole	Monistat	Fungicide
62	Atropine	Donnatal	Anticholinergic
63	Doxepin	Sinequan	Antidepressant
64	Nadolol	Corgard	β blocker
65	Thyroid	Thyroid	Thyroid deficiency
66	Temazepam	Restoril	Anxiolytic
67	Chlorpheniramine	Ornade	H_1 blocker
68	Warfarin	Coumadin Oral	Anticoagulant
69	Indomethacin	Indocin	Nonsteroid anti-inflammatory
70	Amiloride	Moduretic	Diuretic
71	Chlorpropamide	Diabinese	Hypoglycemic
72	Clotrimazole	Lotrimin	Antifungal
73	Cyclobenzaprine	Flexeril	Antimuscle spasm
74	Chlorthalidone	Hygroton	Diuretic
75	Haloperidol	Haldol	Neuroleptic
76	Potassium clavulanate	Augmentin	Antibiotic
77	Betamethasone	Valisone	Anti-inflammatory
78	Iodinated glycerol	Tussi-Organidin	Cough mixture
79	Chlordiazepoxide	Librium	Anxiolytic
80	Medroxyprogesterone	Provera	Anti-inflammatory
81	Hydrocodone	Vicodin	Analgesic
82	Multivitamin / fluoride	Poly-Vi-Flor	Vitamin
83	Trazodone	Desyrel	Antidepressant
84	Guaifenesin	Entex LA	Expectorant
85	Carbamazepine	Tegretol	Anticonvulsant
86	Nitrofurantoin	Macrodantin	Antibacterial
87	Thioridazine	Mellaril	Neuroleptic
88	Meclizine	Antivert	Antinausea
89	Allopurinol	Zyloprim	Antiurate
90	Metoclopramide	Reglan	Antiemetic

TABLE 5.1A. Continued.

Rank	Generic Name	Typical Trade Name[a]	Therapeutic Class
91	Cefadroxil	Duricef	Antibiotic
92	Nicotine polacrilex	Nicorette	Antismoking
93	Verapamil	Calan	Calcium blocker
94	Carbidopa / levodopa	Sinemet	Parkinson's disease
95	Diflunisal	Dolobid	Nonsteroid anti-inflammatory
96	Polymyxin	Cortisporin Otic	Antibiotic
97	Fenoprofen	Nalfon	Nonsteroid anti-inflammatory
98	Glipizide	Glucotrol	Hypoglycemic
99	Sucralfate	Carafate	Antiulcer
100	Pentoxifylline	Trental	Hemorheologic
101	Enalapril	Vasotec	Antihypertensive
102	Clemastine fumarate	Tavist-D	H_1 blocker
103	Doxycycline	Vibramycin	Antibiotic
104	Azatadine	Trinalin	H_1 blocker
105	Pseudoephedrine HCl	Trinalin	Decongestant
106	Prochlorperazine	Compazine	Neuroleptic
107	Terbutaline	Brethine	Antiasthma
108	Hydroxyzine	Atarax	Anxiolytic
109	Promethazine	Phenergan	H_1 blocker
110	Diphenoxylate	Lomotil	Antidiarrhea
111	Fluocinonide	Lidex	Anti-inflammatory
112	Methylprednisolone	Medrol Oral	Anti-inflammatory
113	Sulfisoxazole	Pediazole	Antibiotic
114	Desipramine HCl	Norpramin	Antidepressant
115	Minocycline	Minocin	Antibiotic
116	Tolazamide	Tolinase	Hypoglycemic
117	Oxazepam	Serax	Anxiolytic
118	Tolmetin	Tolectin DS	Nonsteroid anti-inflammatory
119	Dicyclomine	Bentyl	Muscle spasm
120	Meclofenamic acid	Meclomen	Nonsteroid anti-inflammatory
121	ZnO, BiO, etc.	Anusol HC	Antihemorrhoids
122	Procainamide	Procan SR	Antiarrhythmic
123	Bumetanide	Bumex	Diuretic
124	Amantadine HCl	Symmetrel	Parkinson's disease and antiviral
125	Benztropine mesylate	Cogentin	Parkinson's disease
126	Clidinium bromide	Librax	Anticholinergic
127	Dextromethorphan	Dextromethorphan	Cough mixture
128	Tretinoin	Retin-A	Antiacne
129	Chlorzoxazone	Parafon forte	Muscle spasm
130	Loperamide	Imodium	Antidiarrheal
131	Dihydrocodeine	Synalgos DC	Analgesic
132	Labetalol HCl	Normodyne	Antihypertensive
133	Flunisolide	Nasalide	Anti-inflammatory
134	Prazepam	Centrax	Anxiolytic
135	Hydrocortisone cream	Hydrocortisone cream	Anti-inflammatory

[a] The trade name given may be a mixture of which the specified drug is only one constituent.

TABLE 5.1B. Top-20 Hospital Drugs in the United States (1987)[a] by Value

Rank		Theraputic Class
1	Cefoxitin	Cephalosporin antibiotic
2	Cefazolin	Cephalosporin antibiotic
3	Vancomycin	Glycopeptide antibiotic
4	Clindamycin	Antibiotic
5	Albumin	Blood product
6	Ceftazidime	Cephalosporin antibiotic
7	Total parenteral nutrition	Intravenous nutrition
8	Rantitidine	Antiulcer, H_2 antagonist
9	Dextrose	Intravenous nutrition
10	Heparin	Anticoagulant
11	Cisplatin	Anticancer agent
12	Cefoperazone	Cephalosporin antibiotic
13	Cefotaxime	Cephalosporin antibiotic
14	Ticarcillin	Penicillin antibiotic
15	Tobramycin	Aminoglycoside antibiotic
16	Dobutamine	Cardiac stimulant
17	Ceftriaxone	Cephalosporin antibiotic
18	Imipenem-cilastin	Carbapenem antibiotic
19	Amino acids	Intravenous nutrition
20	Cimetidine	Antiulcer, H_2 antagonist

[a]Based on Scrip 1310 / 18.

value of the U.S. drug market and 28% of the non-Communist world market. The top twenty hospital drugs by sales value are shown in Table 5.1B. Nonetheless, from the point of view of the pharmaceutical industry, the top 100 prescription drugs are the most important in the sense that they provide the bulk of its cash flow, keep it in business, and finance its research programs.

Appendix 1 describes how Table 5.1A has been derived. The method produced a list of 135 chemical entities. The Top-100 are in reasonably reliable rank order and are the more important. The 35 entities left over are also recorded in Table 5.1A. They, together with other interesting drugs outside the Top-100, will also be mentioned. To aid identification of the important drugs, their position in the ranking will be indicated by the symbol #. Thus, hydrochlorothiazide is #1 and amoxycillin #2.

5.2 CLASSIFICATION OF DRUGS

Having selected the drugs for discussion, we are faced with the second problem, that of classification. Should drugs be classified by their chemical structure, by the diseases they cure, or by the body system on which they act? Classification by structure is important to those doing pharmaceutical research. An expert in steroid or prostaglandin chemistry will be primarily concerned with the various chemical manipulations that can be performed on these molecules and will rely on someone else to screen them for effect against the various illnesses susceptible to steroid or prostaglandin therapy. Classification by chemical class is difficult, however, and such classes are difficult to define. An impressive attempt has been made in the book by Wilson et al. (see Bibliography).

Classification of drugs by activity is important to a physician. Penicillin and streptomycin have no chemical relation to one another but both are antibiotics. Similarly, cortisone and indomethacin are not related chemically but both are anti-inflammatories. To the physician, their chemical structures are irrelevant. This is the approach adopted by Goodman and Gilman (see Bibliography).

Finally, classification of drugs by the body system on which they act (drugs affecting the central nervous system or the musculoskeletal system, for example) forms a useful bridge with the International Classification of Diseases (Appendix 2).

TABLE 5.2. Drug Classification by General Anatomical Group[a]

Code	Heading	Examples
A	Alimentary tract and metabolism	Anti-peptic ulcerants (9.6, 17.6, 19.1.1), anticholinergics (14), antidiarrheals (10.1.1), antiemetics (17.6), vitamins (17.7), anorectics (8.5), hypoglycemics (13)
B	Blood and blood forming organs	Anticoagulants (7.7), thrombolytics (7.9), hypolipemics (7.8)
C	Cardiovascular system	Cardiovascular drugs (7)
D	Dermatologicals	Antifungals (6.11), antibiotics (6), corticosteroids (11.4), antiacne (17.5)
G	Genitourinary system and sex hormones	Antibacterials (6), corticosteroids (11.4), sex hormones (11, 21.3)
H	Other systemic hormonal preparations	Glucocorticoids (11.4), thyroid therapy (17.1)
J	General systemic anti-infectives	Antibacterials and antibiotics (6), antivirals (20)
L	Antineoplastic and immunosuppressive drugs	Antineoplastics (21)
M	Musculoskeletal system	Corticosteroids (11.4), antigout agent (17.3), nonsteroid anti-inflammatory agents (12), muscle relaxants (17.2)
N	Central nervous system	Central nervous system (8), analgesics (10), anti-Parkinson drugs (16)
P	Antiparasitic products	Drugs for tropical diseases (18)
R	Respiratory system	Antihistamines (9.1 – 9.4), antiasthmatics (15), cough and cold preparations (9.5)
S	Sensory organs	
V	Various	

[a]Numbers in parentheses indicate sections in this book.

In principle there should be relationships between structure and activity and this remains a goal of pharmaceutical research. Certain structure–activity relations have been found and these are reviewed in Burger, Volume 1, Chapter 10 (see Bibliography).

The most powerful and useful system developed so far is a compromise between the methods, known as the anatomical–therapeutic–chemical system (ATC). The system divides products into 13 general groups (Table 5.2) according to the body system on which they act: A, alimentary system; B, blood and blood-forming organs, and so on. This is followed by a number that describes the main group of drugs, a letter that describes the subgroup, and a number describing therapeutic class. The system is explained in detail by the Nordic Council for Medicines (see Bibliography).

In this text, we have classified drugs primarily by the anatomical system on which they act (e.g., Chapter 8 discusses drugs affecting the central nervous system) and secondarily by the diseases they cure (e.g., Section 8.3 discusses anxiolytics and Section 8.4 discusses antidepressants). We have, however, deviated from this principle with especially significant chemical groups, such as the steroids and the prostaglandins.

Table 5.3 shows the categories of drugs in the Top-100 according to this system. They fall into 10 broad pharmacological groups plus a miscellaneous

TABLE 5.3. Categories of Drugs in the Top-100

ATC Category	Type	Number of Drugs
C	Cardiovascular drugs	24
N[a]	Drugs affecting the central nervous system	18
J	Antibiotics and antibacterials	12
	Antihistamines[b]	5
R6	H_1 blockers	3
A2	H_2 blockers	2
R5	Cough and cold preparations	3
N2	Analgesics	4
G3[c]	Steroid drugs	8
M1A	Nonsteroid anti-inflammatory drugs	7
A10	Hypoglycemics	4
A3	Anticholinergic drugs	1
D1	Antifungals	2
R3A	Antiasthma drugs	3
	Miscellaneous drugs	9

[a] Includes anticonvulsants; excludes drugs for Parkinson's disease.

[b] The three H_1 blockers are used as antiallergics or in cough and cold preparations. The H_2 blockers are antiulcer drugs, which, together with one miscellaneous drug, make up three antiulcer drugs.

[c] Includes two anti-inflammatory steroids that the ATC system classifies under a variety of headings, depending on application.

category and the steroids, which we discuss as a chemical class rather than by their activity.

The largest group by number of chemical entities is cardiovascular drugs, that is, drugs for heart disease. Since heart disease is the number one killer in the United States, this is entirely appropriate. Almost as well represented are the psychotropic drugs, that is, drugs affecting the central nervous system. This group includes tranquilizers, sleeping pills, appetite suppressants, and antidepressants. The mental disorders for which they are prescribed are rarely fatal and the level of prescribing of such mind-affecting drugs has given rise to misgivings which we discuss later.

The third well-represented category is antibiotics and they represent human triumph over bacterial infections, a group of illnesses that caused many fatalities before the 1930s but that are rarely fatal today. The 12 antibiotics and antibacterials are particularly important because 3 of them occur in the top 6 drugs and 7 in the top 25. On the basis of number of prescriptions in the United States, they are second only to heart drugs.

On a worldwide basis, until the mid 1980s, more antibiotics were sold than any other category. Bacterial infections were widespread and dangerous. Antibiotics administered by mouth or by injection in field clinics played a major role in keeping people alive. Outside the developed world, heart disease and mental illness appeared less prevalent, perhaps because they were diagnosed and treated less often and perhaps because they were, in fact, less common. Infections are still widespread and dangerous, but the low price of antibiotics and antibacterials and the increasing availability of sophisticated and expensive heart drugs has meant that, by value of sales, cardiovascular drugs are now the most important, accounting for some 18% of the world pharmaceutical market. Antibiotics and antibacterials come second with 16%. Drugs for the alimentary tract, including the expensive H_2 blockers (Section 9.6), come third with 15.6%, and central nervous system drugs come fourth with 11%.

Among hospital drugs (Table 5.1B) the pattern is quite different. Only the two antiulcer drugs, cimetidine and ranitidine, are common to both lists. The hospital list also contains eleven antibiotics (including six injectable cephalosporins), two cardiovascular drugs, an anticancer agent, a blood product, and three intravenous nutrition products. The list reflects the presence in hospitals of cancer sufferers, patients with intractable infections, and patients who might have had operations and cannot eat normally.

5.3 SOURCES OF DRUGS

Since this chapter deals with drug production, the question of raw materials should next be considered. If its subject were bulk organic chemicals, there would be no problem. Bulk organics are derived from oil or natural gas via fewer than ten major building blocks (see Bibliography). Pharmaceuticals are

TABLE 5.4. Sources of Drugs

Source	Examples
Fermentation	All antibiotics except chloramphenicol, vitamin B_{12}, one step in cortisone from diosgenin, and one step of vitamin C; biotechnology.
Chemical synthesis	Heart drugs, psychotropic drugs, antihistamines, analgesics. Also chemical modification of materials from other sources.
Animal extracts	Heparin, thyroid, insulin, bile acids.
Biological sources	Vaccines and serums (e.g., against diphtheria, polio, whooping cough).
Vegetable extracts	Alkaloids (e.g., papaverine, atropine, codeine, quinidine) steroids, glycosides (e.g., digoxin).

much more complicated molecules and the picture is far from simple. There are five different sources of pharmaceuticals, listed in Table 5.4. They are fermentation, chemical synthesis, animal extracts, biological sources, and plant extracts. By value of products, fermentation is probably the most important since almost all antibiotics are produced by this method. Even the chemically modified antibiotics require a fermentation stage to provide their precursors. In addition, the new biotechnology processes may also be classified under fermentation and a material identical with human insulin is currently being made via recombinant DNA.

By tonnage, chemical synthesis is undoubtedly the most important source of pharmaceuticals. It provides the great majority of the heart drugs, the drugs affecting the central nervous system, the antihistamines, and the analgesics. The methods are sophisticated and are closer to conventional laboratory organic synthesis than most industrial chemical processes.

Purity is of greater importance than high yield. The purity of most drugs is greater than 99.9% and the content of the remaining 0.1% is usually specified in the Pharmacopoeia. Penicillin unexpectedly turns up as a contaminant in many preparations because the *Penicillium* microorganism is ubiquitous. Its concentration must be kept below 1 ppm to avoid allergic reactions in penicillin-sensitive patients. Absorbent cotton fibers may present a problem, and ointments should contain less than 1 fiber cm^{-3}, the fiber being less than 1 mm long. Bacteria insensitive to an antibiotic may grow in a medium containing it, and most countries demand the absence of harmful varieties (e.g., salmonella) and specify a limit for the total bacterial count. Thus the U.S. Pharmacopoeia quotes 5000 cm^{-3} for a gelatin base.

The third source of drugs is animal extracts, primarily packing house wastes. Bile acids are a minor source of steroids. Insulin is obtained from the pancreas of pigs and cows, and L-thyroxine is obtained from the thyroid glands of domestic animals. Conjugated estrogenic hormone comes from the urine of pregnant mares, and is exceptional in that the animal does not have to be slaughtered.

Vaccines and serums are obtained from biological sources and are extremely important for control of diphtheria, polio, measles, whooping cough, and a range of other ailments. Influenza vaccine is becoming more effective and widely used. These materials are typically given as a single dose, which is effective for at least a year and at best for life.

Plant extracts are the final source in Table 5.4. Stigmasterol and sitosterol from soybeans and diosgenin from the Mexican yam, genus *Dioscorea*, are the major steroid precursors. Alkaloids also come from plants and our Top-100 contains codeine and atropine. Alkaloids are much less important in medicine than they were. An encyclopedia published early in the 1930s gives plants as the only significant source of drugs and most of these were alkaloids. Such compounds as morphine and quinine are less important now than they were then and do not find a place in the Top-100 although they are still used.

5.4 MANUFACTURING PROCESSES

The pharmaceutical industry manufactures a large number of specific chemical entities to rigorous specifications. Each entity will probably be made by only a few manufacturers or perhaps only one. The product, especially if it is under patent, can be sold, like other specialty chemicals,at a price that reflects its value-in-use rather than its manufacturing cost.

The heavy chemicals industry, in contrast, manufactures commodity chemicals which are made by a large number of producers and have to be sold at the ruling market price. The individual company, having little control over prices, is therefore preoccupied with reducing costs. A major source of cost reduction is economies of scale and these in turn are related to the use of continuous rather than batch processes. Indeed the former are well-nigh universal in the heavy chemicals sector.

5.4.1 Quality Control

Economies of scale are relatively inaccessible to the pharmaceutical industry. Instead, it is preoccupied with quality control, and this is more easily achieved in batch equipment. Batches can be given numbers and, if problems arise, a single batch can be easily identified and recalled. The process must always be operated in such a way that the quality of the final product is maintained, and this leads to certain design features not normally found in chemical manufacturing. Where the purified product is being handled, walls and floors are rounded off to allow efficient cleaning, air may be filtered and certain areas may be maintained under aseptic conditions, where only suitably clothed personnel are allowed access. Packaging is carried out in individual cubicles to

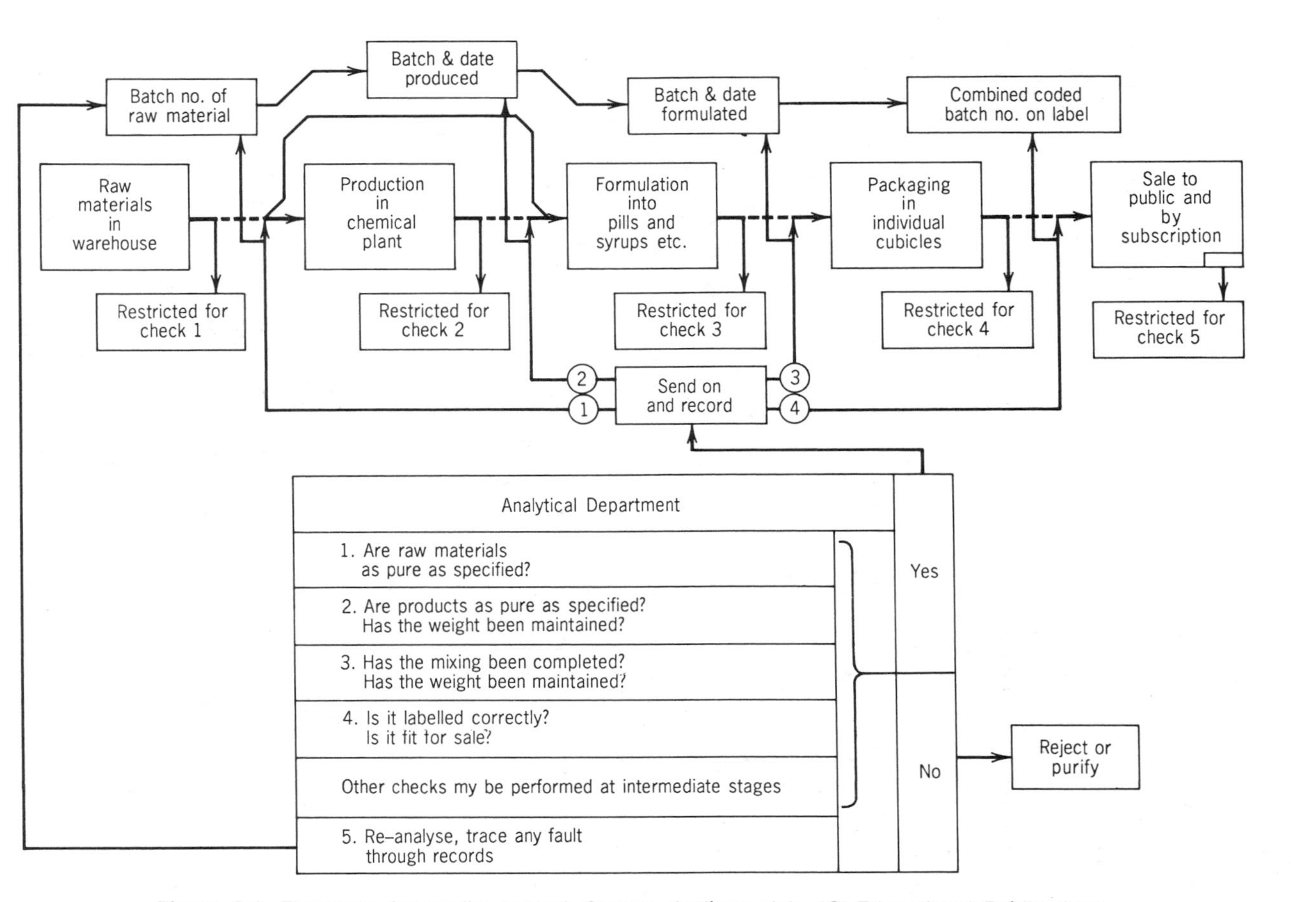

Figure 5.1. *Flowchart for quality control. Source: Antibacterials, ICI Educational Publications, 1973. Reprinted with permission.*

avoid cross-contamination if there is spillage. An analytical department intervenes at every stage to monitor the progress of a drug through the system. A flowchart for quality control is shown in Figure 5.1.

5.4.2 Choice of Reagents

The special characteristics of the pharmaceutical industry also affect choice of reagents and plant design. In general, processing costs are higher than in the chemical industry as a whole, and capital costs are lower. Because of the high unit cost of pharmaceuticals, expensive chemicals are practicable. For example, lithium aluminum hydride might be used as a reducing agent while, in the petrochemical industry, hydrogen is the only economic reagent. The use of some exotic chemicals may lead to hazards. Nitrations would probably be carried out in a special nitrating plant with equipment designed to prevent runaway reactions; phosgene would be handled only under carefully controlled conditions because of its toxicity. Experience is similarly needed with organofluorine compounds, reactions at cryogenic temperatures, and fermentation routines for steroids and antibiotics. A firm that has been engaged in such operations for many years accumulates know-how and this gives it a great advantage over new entrants to the industry. The Bhopal disaster, due surprisingly to methyl isocyanate and not to the phosgene also on site, illustrates the problems that can arise with toxic specialty chemicals.

A modest way in which the industry tries to keep costs within bounds is by the use of standard intermediates that can be raw materials for a number of products. For example, all the different chemically modified penicillins are based on 6-aminopenicillanic acid, which can consequently be manufactured on a much larger scale than any of the individual penicillins. Such intermediates are known as synthons.

5.4.3 Plant Design

As far as equipment is concerned, the pharmaceutical industry rarely follows the chemical industry pattern of purpose built plant, carefully optimized and with the larger sections fabricated on site. In general, equipment is bought "off the shelf" and assembled like an erector set. Optimization is sacrificed to availability of plant items. A secondary benefit to emerge from this is that lead times (i.e., the time taken to build a plant) are lower than in the chemical industry.

The equipment supplier will therefore aim to offer versatile, multipurpose equipment that can be used in a range of processes. Most of it will be stainless steel or glass-lined to avoid problems with contamination or corrosion. In spite of its versatility, however, the plant, once built, will rarely be changed. In the past, legal accredition of a pharmaceutical depended on the plant remaining unchanged and, if the manufacturer wanted to add a recycle stream, legal clearance had first to be obtained.

Use of a plant for different products is inhibited by worries about contamination, and fermentation equipment is always dedicated. Nonetheless, the situation has eased somewhat in recent years, and more multiuse equipment is being installed. There is still not the flexibility in equipment use that one might expect in comparison, say, with the dyestuffs industry, but there is a trend in that direction.

Furthermore, the availability of on-line microcomputers and microprocessor-controlled equipment is extending many of the benefits of continuous processing to the batch processor. The pharmaceutical industry has been a leader in the employment of these new techniques.

The processes themselves resemble laboratory preparations carried out on a large scale. The petrochemical industry's concern with, for example, the saving of small packages of heat by the use of heat exchange networks, is rarely appropriate in equipment of one-thousandth the capacity. Problems of heat and mass transfer are certainly more important than in the laboratory, but can usually be overcome by scaled-up laboratory techniques.

In downstream processing, too, pharmaceutical industry practise differs from that in the petrochemical industry. The small, simple molecules of the petrochemical industry can usually be separated from impurities by distillation. This may be true for pharmaceutical intermediates and solvents but is rarely true for end products. In fermentation, it is never true, and the extraction of penicillin described in Section 6.3.1 is an example of the way unstable molecules may be purified.

Even drugs produced by chemical synthesis, however, may comprise complex molecules that degrade on distillation; hence, favored methods of separation are the liquid–liquid and liquid–solid separation processes such as solvent extraction, leaching, ion exchange, freeze-drying, and crystallization.

5.4.4 Cost Structure

Equipment and buildings for making high-technology ethical pharmaceuticals costs tens of millions of dollars. Glaxo's proposed new ranitidine unit (Section 9.6) will cost $65 million, which represents the top end of the market. Twenty-five million dollars, on the other hand, would provide a multiproduct facility to make between three and eight chemical entities of lesser complexity.

Formulation is carried out separately. A small tableting machine makes 500 tablets/minute and costs about $16,000. A large one produces 10,000/minute and costs about $110,000. These costs are trivial, however, in comparison with the costs of housing the machines. A large, high-technology factory with 10–20 large machines making 40–50 products might cost $32 million but a small factory making simple over-the-counter pharmaceuticals on 3–6 machines could be built for only about $8 million.

Dosage forms other than tablets are more expensive. Hard gelatin capsules, for example, present problems. Gelatin is a natural material and variable in its properties. The machines to fill capsules are bigger and more difficult to run

TABLE 5.5. Typical Cost Structure for Pharmaceutical Manufacture

Expenditure	Percentage
Manufacturing, including raw materials	42.1
Scientific information	12.5
Research and development	11.4[a]
Marketing	9.0
Advertising	4.3
Administration	6.4
Licensing fees	1.7
Interest payments	1.8
Taxes	1.5[b]
Miscellaneous	9.3

[a]R & D money is laid out many years before there is any return. If the expenditure is capitalized to the point of launch (Section 2.5), this figure is more or less doubled.
[b]This does not include company taxation.

than tableting machines. Only two companies (Eli Lilly and Parke-Davies) make hard gelatin capsules and they cost between one and two cents each. For an expensive product where a rapid launch is essential, it pays to use them, but in the long run tablets are much cheaper. Tableting costs usually work out at about 5–10% of total manufacturing costs.

The high unit value of drugs means that pressure to reduce costs is not as great as in the heavy chemicals industry. That is not to say that the cost of manufacture of drugs is unimportant, but that it is less important. An estimate for large West German drug firms in the early 1980s suggested that manufacturing costs were about 35% of total sales (40% of total costs) compared with almost 90% for heavy organics. Instead, money is spent on research and development, quality control, and testing and promotion of products. A typical cost analysis is shown in Table 5.5. The markup on average is about 25%; that is, costs are about 80% of sales.

The above figures are, of course, guesstimates and there will be wide fluctuations from product to product, company to company, and year to year. Nonetheless, they should provide a feel for the orders of magnitude of expenditure demanded by the pharmaceutical industry.

6

Antibacterials and Antibiotics

Antibacterials are compounds that attack microorganisms. The useful ones are those that attack illness-causing bacteria without harming the host. Some antibacterials actually kill bacteria but most merely inhibit their reproduction, that is, they are bacteriostatic rather than bactericidal. Frequently they are antimetabolites or enzyme anatagonists (Section 4.1.1) and interfere in some way with the biosynthetic processes in the bacteria.

Antibiotics are antibacterials that come from microorganisms, or compounds that have a molecular structure similar to materials produced by microorganisms. The term antibiotic is derived from the Greek *anti* (against) and *bios* (life). Antibiotics are based on the idea of microbial antagonism which was first observed in 1877 by Pasteur and Joubert. They found that if "common bacteria" were introduced into a pure culture of anthrax bacilli, the anthrax bacilli died. Similarly, an otherwise deadly injection of anthrax bacilli into an animal was harmless if common bacteria were injected simultaneously.

Antibiotics are produced by the enzymes contained in bacteria, fungi, and molds. Penicillin G, for example, is produced by a mold, *Penicillium chrysogenum*. A potent source of antibiotics is the microbes found in the soil, especially those of the order *Actinomycetes* and among these especially the genus called *Streptomyces*. Over 14,000 antibiotics have been identified. About 50 new ones are announced each year but, unlike penicillin, most of them are useless because they are toxic to humans.

Sulfa drugs and the quinolones are examples of antibacterials that are not antibiotics. They are not proliferated by microorganisms but are made by chemical synthesis. Chloromycetin is also made by chemical synthesis but is an antibiotic because it is a natural product and was originally found in the soil.

Aztreonam is made by chemical synthesis but is still an antibiotic because it contains the β-lactam ring that characterizes the naturally occurring penicillins and cephalosporins.

Some writers do not adhere to these definitions and antibiotics are sometimes treated as a separate class from antibacterials rather than as a subset of them.

TABLE 6.1. Antibiotics, Antibacterials, and Antifungals

Antibiotics and Antibacterials
Sulfonamides
Sulfamethoxazole (#25)
Acetylsulfisoxazole (#113)
Pyrimidines
Trimethoprim (#25)
Penicillins
Amoxycillin (#2)
Phenoxymethylpenicillin (Penicillin V) (#6)
Ampicillin (#13)
Clavulanic acid (#76)
Cephalosporins
Cephalexin (#22)
Cefaclor (#42)
Cefadroxil (#91)
Tetracyclines
Tetracycline (#20)
Doxycycline (#103)
Minocycline (#115)
Aminoglycosides
Streptomycin
Tobramycin
Macrolides
Erythromycin (#5)
Peptides
Polymyxin (#96)
Quinolones
Nalidixic acid
Miscellaneous
Chloramphenicol
Nitrofurantoin (#86)
Clindamycin
Vancomycin
Antifungals
Miconazole (#61)
Clotrimazole (#72)

Note: Cotrimoxazole (#25) is a mixture of sulfamethoxazole and trimethoprim. Pediazole (#104) is a mixture of erythromycin and sulfisoxazole. Clavulanic acid is a β-lactamase inhibitor used with amoxycillin or ticarcillin (q.v.).

TABLE 6.2. World Antibiotics Market (1986)

Antibiotic	$ Million	Predicted Growth %
Cephalosporins	4,510	3.0
Penicillins	2,420	4.1
Macrolides (erythromycin)	893	−2.0
Aminoglycosides	591	−6.3
Tetracyclines	555	2.9
Sulfonamides	493	−3.6
Quinolones	466	23.3
Antituberculosis	167	0.0
Others	919	5.6
Total	11,014	3.7

Source: Based on Robert Fleming Securities, *Chem. Ind.*, 16 Nov. 1987.

The Top-100 (Figure 5.1) contains 14 antibacterials and these are listed in Table 6.1 together with a few less widely used but still important compounds and two antifungals which counter infection by fungi rather than bacteria. Other less widely prescribed antibacterials, for example, the antituberculosis drugs, will be introduced under the individual headings.

Table 6.1 also indicates the chemical classes to which the main antibiotics and antibacterials belong. The penicillins, cephalosporins, and erythromycin are evidently the most important. The world market by value of sales is shown in Table 6.2.

6.1 CLASSIFICATION OF BACTERIA

Before antibacterial compounds are described, bacteria should be mentioned because it is they at which the drugs are aimed. Three characteristics are shown in Figure 6.1.

Bacteria were arbitrarily divided into two classes by the Danish physician Hans Gram in the 1880s. He decided to call positive those that were stained blue by a mixture of crystal violet and iodine. To this day such bacteria are called gram-positive. Bacteria that are not stained at all or turn red or pink are termed gram-negative.

Bacteria are unicellular organisms having a sturdy peptide/carbohydrate cell wall surrounding a fragile plasma membrane. Gram-negative bacteria possess a second outer cell wall made up of lipid and polysaccharide. It is more difficult to find dyes or drugs that will interact with this second, high lipid, cell wall; hence, there is generally a relation between the Gram reaction of a bacterium and ease of treatment of infections caused by it.

Bacteria are also classified by shape. They occur as rods known as bacilli, spheres known as cocci, spirals known as spirilla, and incomplete spirals,

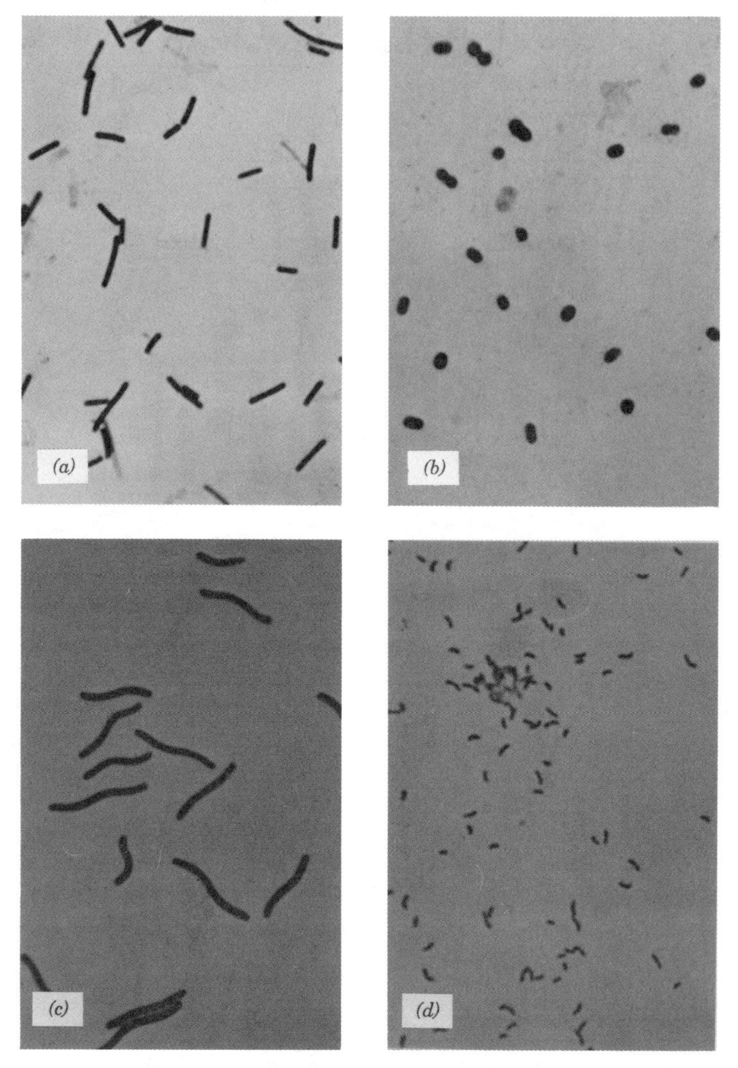

Figure 6.1. Classification of bacteria. Photographs supplied by the National Collections of Industrial and Marine Bacteria, Aberdeen, Scotland. Magnification ×2500.
(a) Bacillus (rods)
(b) Acinetobacter (very short rods– coccobacilli)
(c) Aquaspirillus (spirals)
(d) Desulfovibrio (incomplete spirals)

which are called vibrios. Cocci are about one micrometer in diameter. Bacilli have the same width as cocci but are 2 to 4 times as long, whereas spirilla appear as fine threads. Depending on the shape of the organisms, the cells may be arranged singly, in pairs designated diplo, in packets of four designated tetrads, in packets of eight designated sarcinae, in chains designated strepto,

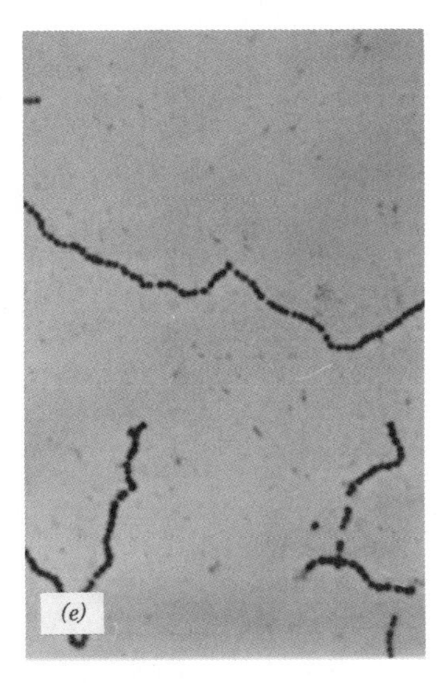

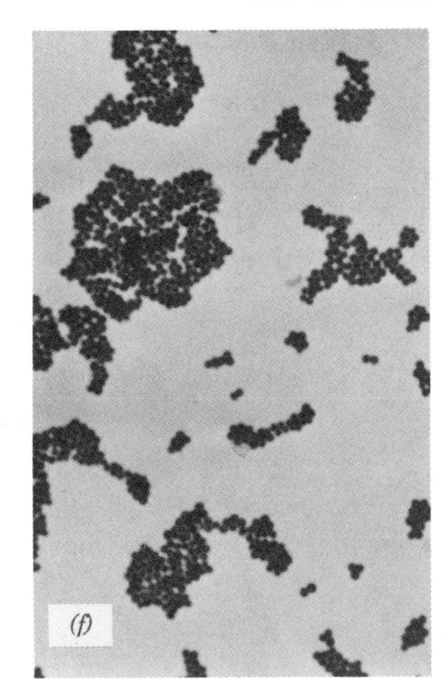

Figure 6.1. *Continued.*
(e) Streptococcus (chains of spheres)
(f) Staphylococcus (grapelike clusters of spheres)

and in grapelike clusters designated staphylo. As an example, bacteria that appear in chains of spherical cells would be called streptococci.

6.2 SULFONAMIDES

For historical reasons, the sulfonamides are a suitable place to start a description of the antibiotics and antibacterials. They are now of little importance in this area but are of great significance as a chemical group that provides a wide range of drugs used for disparate purposes such as diuretics, drugs for countering leprosy, and oral drugs for diabetics.

Bacterial diseases were a major cause of death until the discovery of sulfanilamide, the first sulfonamide drug, in 1935. Before that time, a significant cause of female mortality was puerperal or child-bed fever and it provides us with a useful measure of the prevalence of bacterial infections. Mortality from it is shown in Figure 1.2. In 1850 as many as 30% of the women who gave birth in hospitals died of puerperal fever caused by the hemolytic streptococcus. The Hungarian physician Semmelweis urged doctors to rinse their hands in dilute carbolic acid (phenol) before examining patients, particularly if they had been working in the dissecting room or carrying out autopsies. He was denounced as a charlatan. His ideas of antisepsis, unfortunately, were ahead of their time and he communicated them poorly. He died in 1865, himself ironically a victim of the hemolytic streptococcus. Eventually, of course, his ideas were universally accepted and Lister, who adopted many of them, was honored in his own lifetime. All through history there have been people with insights long before a body of scientific knowledge provided a basis for their opinions. Semmelweis, who is now regarded in all medical schools as a prophet, was one of these.

Seventy years had passed and bacterial diseases were well recognized before Domagk observed that Prontosil, a red azo dye, was an antibacterial agent particularly effective against streptococcal and staphylococcal infections.

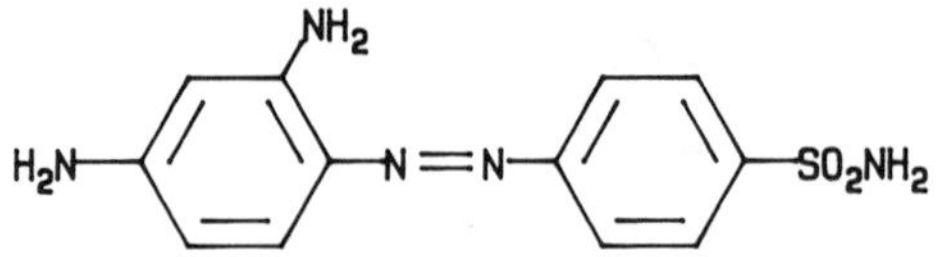

Prontosil

Ehrlich, who was mentioned in Chapter 4 as the founder of the receptor theory, felt that because certain dyes stained tissues selectively there might be a relation between the attraction of tissue for dye and therapeutic activity. It was this thinking that led to the discovery of Prontosil, but ironically Domagk was right for the wrong reasons. Prontosil was effective not because of its azo group, which enables it to stain tissues selectively, but because the azo compound underwent reductive cleavage in the gut to *p*-aminobenzenesulfonamide.

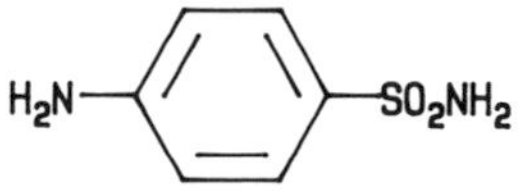

p-Aminobenzenesulfonamide

Once this was recognized, the metabolite replaced Prontosil and, under the name sulfanilamide, became the first sulfonamide drug. The second important sulfonamide was sulfapyridine, known as M & B 693 (the 693rd sulfonamide evaluated by the company, May and Baker), discovered in 1938:

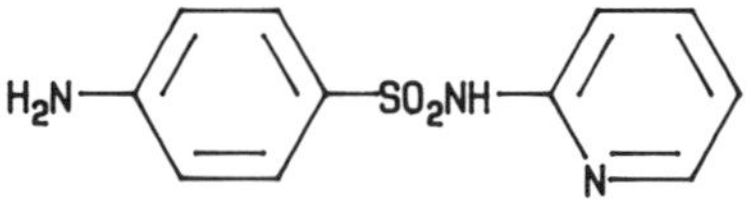

Sulfapyridine

The discovery of a single pharmacologically active compound motivates the synthesis and evaluation of numerous structural variations. Sulfanilamide was not an ideal drug. It was only sparingly soluble and, in the sort of doses that were required, it was liable to crystallize in the kidneys or the bladder. Derivatives of sulfanilamide were prepared in the hope that they would be more soluble, that their lipophilicity would take them to parts of the body that

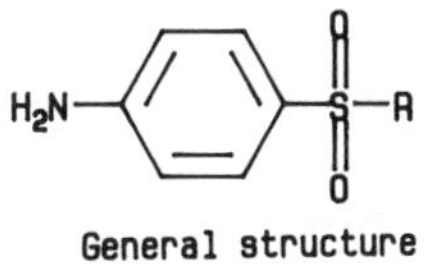

H_2N— —S(O)(O)—R

General structure

—R	Name	Comments
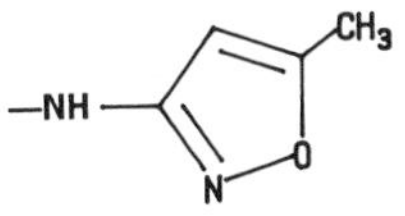 —NH, CH_3, N, O	Sulfamethoxazole (#25)	Used with trimethoprim in cotrimoxazole; active against bronchitis and acute respiratory tract infections.
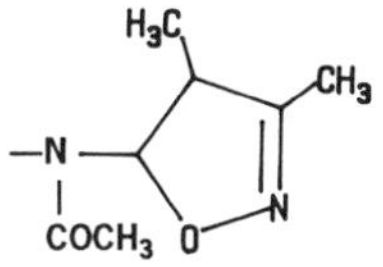 H_3C, CH_3, —N, $COCH_3$, O, N	Acetylsulfisoxazole (#104)	Used with erythromycin for the treatment of acute otitis media in children
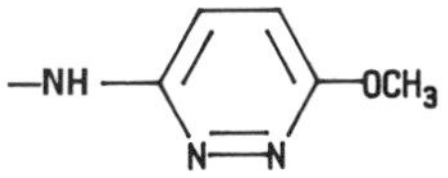 —NH, OCH_3, N—N	Sulfamethoxypyridazine	One of several "longer-acting" sulfonamides strongly bound to plasma proteins. Used in urinary tract infections where the advantage of less frequent administration outweighs possible toxic effects due to accumulation.
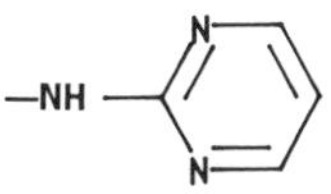 —NH, N, N	Sulfadiazine	Attains high levels in cerebrospinal fluid; hence a possible drug for treatment of meningococcal meningitis.

Figure 6.2. *Modern sulfonamides.*

sulfanilamide could not reach, and that a variant would be found active against a wider spectrum of bacteria and in particular against the staphylococci responsible for some stubborn infections.

Eventually, over 5000 sulfonamides were made and evaluated. Only about 40 of these are on the market and four of the most widely used are shown in Figure 6.2. Only sulfamethoxazole appears in the Top-100. Nonetheless, sulfonamides are still widely used in the developing world probably because they can be stored in adverse conditions, unlike penicillins and tetracyclines.

Sulfonamides that are readily absorbed in the gut may be used for treatment of infections of the respiratory tract. Sulfamethoxazole is in this category and is used in the combined antibacterial cotrimoxazole (#25), which will be discussed presently. Those that are scarcely absorbed in the gut, such as sulfamethoxypyridazine, are used in urinary tract infections because they concentrate in the urine. This is currently the main application for sulfonamides used alone. Sulfamethoxypyridazine is typical of several long-acting sulfonamides. The proportion of it that absorbs in the gut is strongly bound to plasma proteins and the complex acts as a depot of the material so that the drug may be administered less often. On the other hand, the danger of toxic effects due to accumulation is greater.

A mixture of erythromycin with the acetyl derivative of sulfisoxazole is marketed under the trade name Pediazole and appears at #113. It is recommended only for the treatment of acute otitis media (inflammation of the middle ear) in children.

Some sulfonamides of which sulfadiazine is typical pass readily into the cerebrospinal fluid and for this reason are useful in treatment of meningococcal septicemia and meningococcal meningitis.

Once sulfonamides had been discovered, their mode of action was soon elucidated. They interfere with the metabolic function of *p*-aminobenzoic acid. In modern jargon they are *p*-aminobenzoic acid antagonists. In normal metabolism, bacteria use *p*-aminobenzoic acid in a biosynthetic process, which is catalyzed by an enzyme. Sulfonamides have certain structural resemblances to *p*-aminobenzoic acid. They bind to the enzyme, keep off the *p*-aminobenzoic acid, and prevent the biosynthetic reaction from taking place.

Sulfonamides are usually manufactured by reaction of acetamidobenzenesulfonyl chloride with the amino derivative of a heterocyclic compound. Figure 6.3 shows the synthesis of sulfadiazine. Malonaldehyde (1) reacts with guanidine (2) to give the heterocyclic diazine, which, on treatment with phosphorus oxychloride, is aromatized to give 2-aminopyrimidine (4). The *p*-acetamidobenzenesulfonyl chloride (6) is made classically by reaction of acetamidobenzene ((5), from aniline and acetic anhydride) with chlorosulfonic acid. Reaction of (4) with (6) gives the acetyl derivative of sulfadiazine (7) and hydrolysis gives sulfadiazine (8) and acetic acid.

The importance of sulfonamides as antibacterials has diminished as a result of increasing bacterial resistance and their replacement by antibiotics which are more active and less toxic. Interest and sales revived a few years ago, however, when a British firm, Burroughs Wellcome, found that trimethoprim has a powerful synergistic action when mixed with sulfonamides. Figure 6.4 shows the structure and synthesis of trimethoprim. Trimethoxybenzaldehyde (1) is reduced to the alcohol (2) and converted to the halide (3) with phosphorus trichloride. A malonic ester synthesis gives the hydrocinnamic acid ester (4). Ethyl formate and a base give (5) and this can be cyclized with guanidine to give the pyrimidine (6). Phosphorus oxychloride replaces the hydroxyl with a chlorine, and ammonia in turn converts this to an amine to give trimethoprim. It is a folic acid antagonist. By itself, trimethoprim has antibiotic properties but is primarily an antimalarial agent.

The mixture with a sulfonamide, however, is of interest here. For the mixture to have optimum properties, a sulfonamide was required that was metabolized and excreted at the same rate as trimethoprim. Sulfamethoxazole came closest to this ideal. It turned out that the Swiss firm Hoffmann-La Roche held the patents on sulfamethoxazole and their price for the use of the compound by Burroughs Wellcome was that they should be able to market

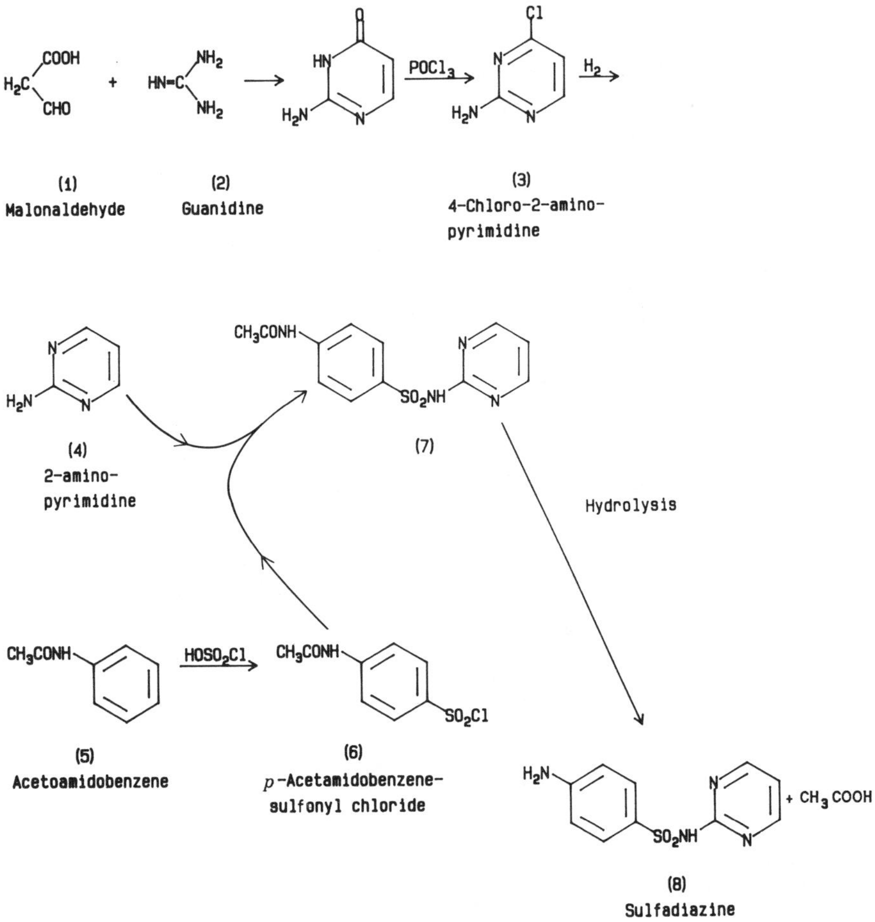

(1) Malonaldehyde

(2) Guanidine

(3) 4-Chloro-2-amino-pyrimidine

(4) 2-amino-pyrimidine

(7)

Hydrolysis

(5) Acetoamidobenzene

(6) *p*-Acetamidobenzene-sulfonyl chloride

(8) Sulfadiazine

Figure 6.3. *Sulfadiazine synthesis.*

an identical mixture. To the bafflement of the medical profession, Bactrim and Septrin were launched on the same day. The product, which has the generic name cotrimoxazole, has made a niche for itself for use with patients who are sensitive to the major antibiotics. Principal uses are in the treatment of urinary tract infections and bronchitis.

The research on sulfa drugs had certain interesting ramifications. It was noted, for example, that sulfa drugs inhibit an enzyme, carbonic anhydrase, that acidifies urine. This helped in the development of the carbonic anhydrase inhibitor diuretics that are used for heart disease (Section 7.4.1). Other investigators noted that some of the sulfa drugs lowered blood sugar levels and this led to the development of oral drugs for diabetes such as chlorpropamide and tolbutamide (Section 13.3). Sulfonamides and the related chemical group,

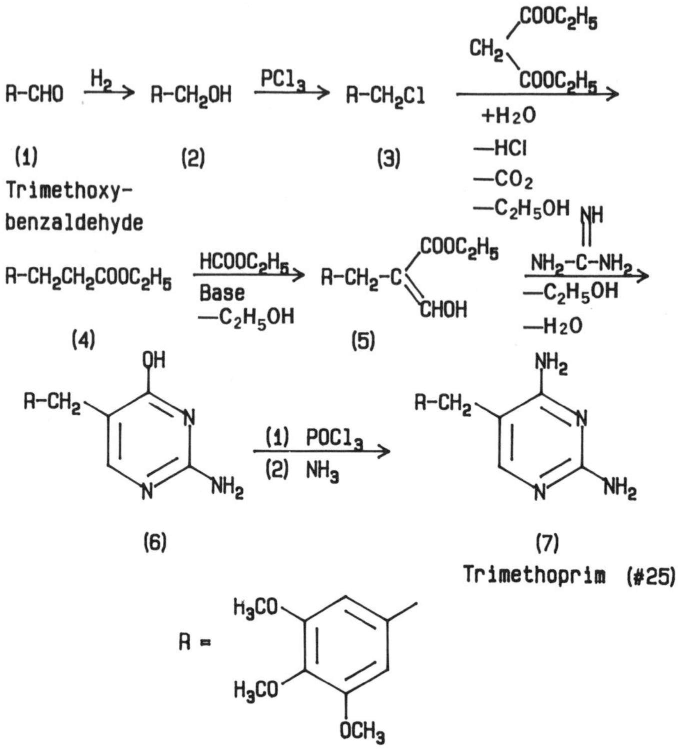

$$\underset{\substack{(1)\\ \text{Trimethoxy-}\\ \text{benzaldehyde}}}{R\text{-}CHO} \xrightarrow{H_2} \underset{(2)}{R\text{-}CH_2OH} \xrightarrow{PCl_3} \underset{(3)}{R\text{-}CH_2Cl} \xrightarrow[\substack{+H_2O\\ -HCl\\ -CO_2\\ -C_2H_5OH}]{CH_2(COOC_2H_5)_2}$$

$$\underset{(4)}{R\text{-}CH_2CH_2COOC_2H_5} \xrightarrow[\substack{\text{Base}\\ -C_2H_5OH}]{HCOOC_2H_5} \underset{(5)}{R\text{-}CH_2\text{-}C(COOC_2H_5)=CHOH} \xrightarrow[\substack{-C_2H_5OH\\ -H_2O}]{NH_2\text{-}C(=NH)\text{-}NH_2}$$

$$\underset{(6)}{\text{R-CH}_2\text{-(4-OH-2-NH}_2\text{-pyrimidine)}} \xrightarrow[(2)\ NH_3]{(1)\ POCl_3} \underset{\substack{(7)\\ \text{Trimethoprim (\#25)}}}{\text{R-CH}_2\text{-(2,4-diamino-pyrimidine)}}$$

R = 3,4,5-trimethoxyphenyl (H_3CO, H_3CO, OCH_3)

Figure 6.4. *Synthesis of trimethoprim.*

the sulfones, are also useful against malaria and leprosy (Section 18.4) and as anticonvulsants. Thus, the sulfonamides, like the steroids, are a group of drugs with wide application.

6.3 PENICILLINS

After the sulfonamides, the next breakthrough in the treatment of bacterial infections came with the discovery of penicillin, the first true antibiotic. In 1928 the Scottish bacteriologist Sir Alexander Fleming observed that a mold that accidentally contaminated one of his petri dishes was destroying a colony of staphylococci.

There is a popular story that a technician in Fleming's laboratory in St. Mary's Hospital, Paddington, London, left the window open on a Friday afternoon. When Fleming entered the laboratory on Monday morning, he observed to his disgust that contaminating material had come in through the window and that some of this mold had caused lysis or destruction of some of his prize cultures. Though Fleming reported his discovery, it is said that he did not recognize its significance and failed to follow it up. The story is charming but not true. For one thing, the windows in Fleming's lab were stuck tight and could not be opened. The full story is told by Kauffman and others (see bibliography). It is true that Fleming's discovery of penicillin was accidental but it depended on a complex series of unrelated events. Had only one of these failed to occur there would have been no discovery. Thus, there is often very little relation between science and discovery.

More to the point, Fleming certainly did realize that he had discovered an important material and he tried to isolate its active ingredient. He was unsuccessful, however, because the material was unstable and rapidly lost activity. In addition, purification was a tricky matter. It was difficult to eliminate foreign proteins that caused toxic reactions in the body. Thus, his preparation could only be used externally and could not compete with conventional antiseptics.

Eventually two workers in England, an Australian, Howard Florey, and a German Jewish refugee, Ernst Chain, obtained a $5000 grant from the Rockefeller Foundation to try to isolate the active ingredient. They hoped to use a new technique called lyophilization or freeze-drying to isolate the antibacterial component of Fleming's cultures. This is not the first time that a scientific advance has become possible because of an advance in equipment. Polyethylene is a case in point and much of modern biochemistry is possible only because of improvements in analytical chemistry.

The freeze-drier worked and Florey and Chain obtained a product a million times more active than Fleming's crude material. In 1939 they were treating infections in white mice and by 1942 the product had been used successfully on a human being. World War II was raging at the time and penicillin would obviously be valuable on the battlefields. An industrial process had to be developed.

Much of the scaling up was achieved in the United States by Coghill and his coworkers. Housewives all over the United States were asked to submit molds to an evaluation center and the low-yielding *Penicillium notatum* used by Fleming was replaced by *Penicillium chrysogenum*. This not only gave higher yields but permitted replacement of the expensive yeast extract substrate by the more readily available corn steep liquor, a waste product of wet corn milling. Surface cultures were replaced by deep fermentation when it was learned how to oxygenate the broth without contaminating it. By 1943 British and American factories were making penicillin on a large scale and by 1944 it became available for civilian use. In 1945 Fleming, Florey, and Chain received the Nobel Prize. Andre Maurois wrote of Fleming, "No man except Einstein in another field and before him Pasteur has had a more profound influence on the contemporary history of the human race."

In the same year the β-lactam–thiazolidone structure of benzylpenicillin was established:

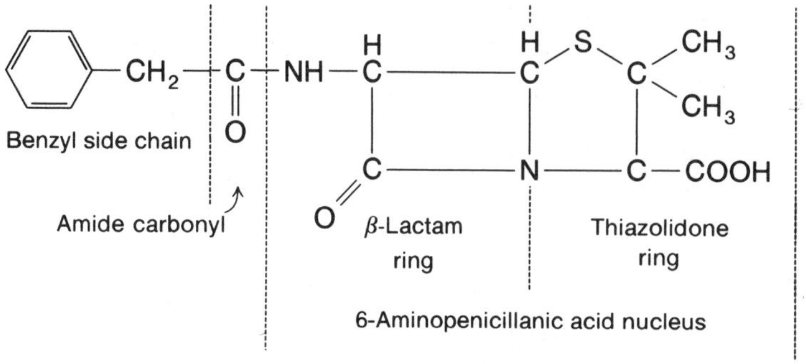

Benzyl side chain
Amide carbonyl
β-Lactam ring
Thiazolidone ring
6-Aminopenicillanic acid nucleus

Experimental work was difficult because the four-membered β-lactam ring is highly strained and thus readily hydrolyzed. The ring is not found in any other natural product and its existence had to be recognized before the structure of benzylpenicillin could be elucidated. It was done by Clarke and his coworkers in the United States and Dorothy Hodgkin in Britain (see bibliography).

Benzylpenicillin or penicillin G is the only naturally occurring penicillin of pharmaceutical value. Numerous modifications of penicillin G have been made either by people or by nature. They differ largely because of the way in which the amine group of the 6-aminopenicillanic acid nucleus, shown above, is amidified.

Penicillin functions by inactivating an enzyme necessary for the cross-linking of bacterial cell walls. The cell walls consist of peptidoglycan chains (Figure 6.5*a*) made up of *N*-acetylglucosamine and *N*-acetylmuramic acid residues. They are linked by rigid β (1–4)-glycosidic bonds of the kind that give the cellulose chain its rigidity. The walls are given further strength by the cross-linking of these chains by a pentapeptide moiety. The cross-linking is catalyzed by a transpeptidase. The transpeptidase forms an acyl intermediate with the penultimate D-alanine residue of the peptide as shown, a free D-alanine molecule also being produced (Figure 6.5*b*). This acyl-enzyme intermediate then reacts with the terminal glycine in another peptide to complete the cross-linking process.

Penicillin inhibits the transpeptidase by forming a covalent bond (amide linkage) with a serine residue at the active site of the enzyme (Figure 6.5*c*). This penicillinoyl–enzyme complex does not undergo deacylation; hence, the transpeptidase is irreversibly inhibited. Penicillin is uniquely effective in this reaction because the peptide bond in the β-lactam ring is highly strained and therefore reactive; also, the penicillin molecule is sterically similar to acyl-D-ala-D-ala, one of the substrates of the enzyme.

Because of the inhibition of the enzyme, the cell wall cannot form and the organism cannot reproduce. Figure 6.6 shows some *Escherichia coli* cells

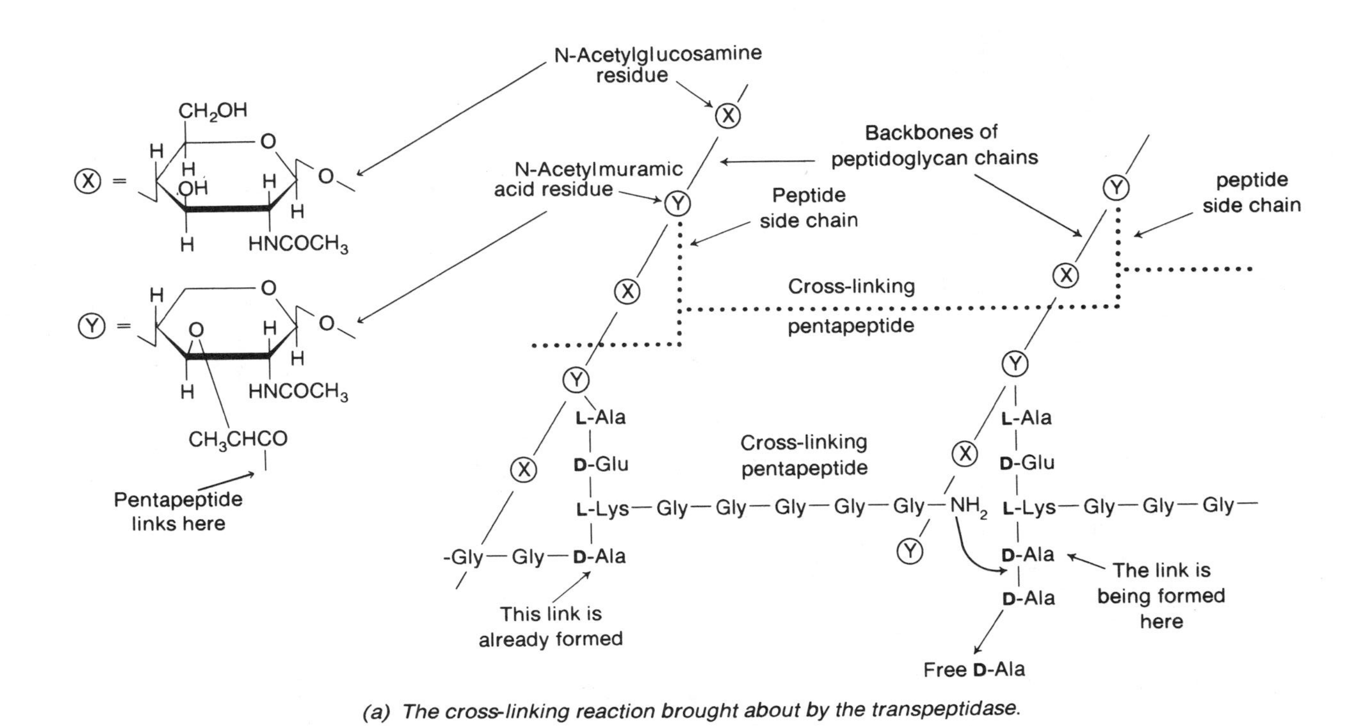

(a) *The cross-linking reaction brought about by the transpeptidase.*

Figure 6.5. *Mode of Action of Penicillin.*

$$R'—\text{D-Ala}—\overset{O}{\overset{\|}{C}}—\underset{H}{\underset{|}{N}}—\text{D-Ala}—COO^- \xrightarrow[]{\text{Enzyme} \quad \text{D-Ala}}$$

$$R'—\text{D-Ala}—\overset{O}{\overset{\|}{C}}—\boxed{\text{Enzyme}} \xrightarrow[]{R—Gly—NH_2 \quad \text{Enzyme}} R'—\text{D-Ala}—\overset{O}{\overset{\|}{C}}—\underset{H}{\underset{|}{N}}—Gly—R$$

Acyl – enzyme intermediate

(b) The role of the enzyme (R and R′ are peptidoglycan chains).

Penicillinoyl-enzyme complex (Enzymatically inactive)

(c) The blocking of the enzyme by penicillin.

Figure 6.5. *Continued.*

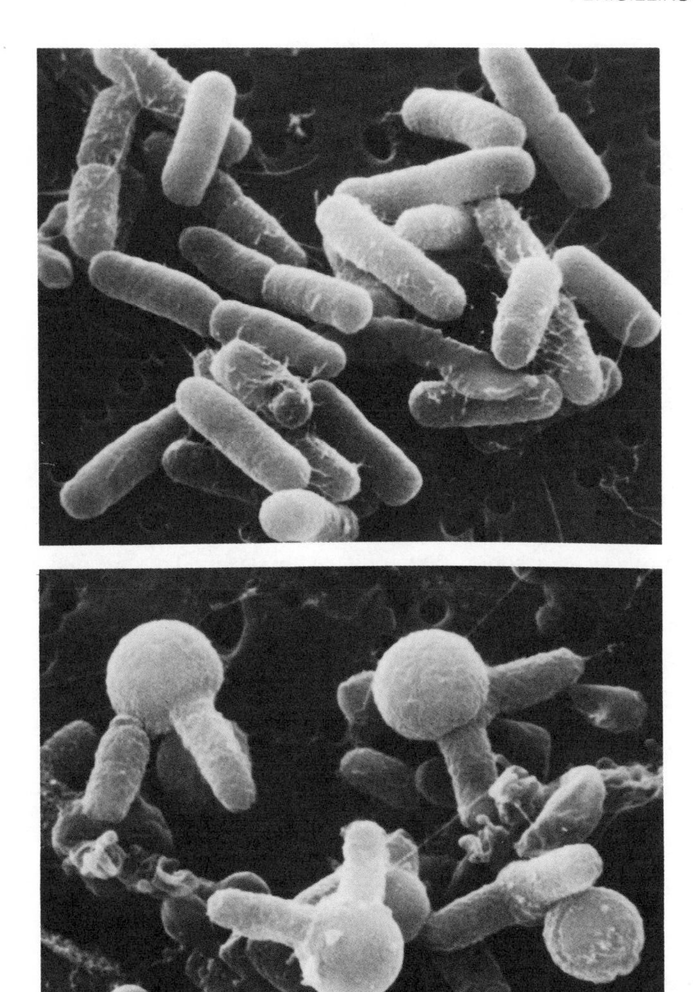

Figure 6.6. Effect of penicillin on Escherichia coli.
(a) Normal E. coli cells, with a dividing cell in the top left-hand corner.
(b) E. coli after contact with amoxycillin (10 μg / ml for 30 min). The cells become longer and show swellings.
Photographs used with permission of Beecham Pharmaceutical, Research Division, Betchworth, Surrey, England.

suffering from a bad attack of penicillin. Since humans do not make use of this cross-linking step, the penicillins are not toxic to them.

6.3.1 Penicillin Manufacture

Benzylpenicillin has been successfully synthesized by a complex series of reactions but with a low overall yield. Thus, fermentation is still the preferred route. A simplified flowsheet is shown in Figure 6.7. The fermentation substrate consists mainly of the sucrose found in corn steep liquor and lactose together with minerals and phenylacetic acid, the side chain precursor. The phenylacetic acid improves yields of benzylpenicillin and inhibits growth of other penicillins. The preferred mold is *Penicillium chrysogenum* and the first stage in the process is the transfer and development of the pure active mold culture from the slope (the test tube in which it is stored) through to a liquid culture in a flask, then to the seed tank and eventually to the fermenter. Nutrients and equipment are previously sterilized with steam at 120°C.

The process is aerobic, that is, it requires air. This created problems for chemical engineers who were familiar with only one other aerobic oxidation, the manufacture of vinegar. Eventually a way was found in which the viscous mixture was vigorously agitated, and aerated by a current of sterile air. The solution of this problem made possible all the fermentations used today to produce other antibiotics and other products of aerobic fermentation, such as amino acids and single cell proteins. Penicillin fermentation is carried out typically for 4 or 5 days at 25–27°C, by which time the reactive mixture contains about 0.5% penicillin. More recent strains of the mold are thought to give yields as high as 3%.

The problems are not over even when fermentation has taken place, for extraction of penicillin from the dilute solution is not easy. First the mycelium, that is the mold cells plus insoluble metabolites, is filtered off in a coated drum filter as shown in Figure 6.7. The filtrate or penicillin broth is adjusted to pH 2 with sulfuric acid. At this pH, the penicillin exists as an undissociated acid—its pK_a is about 2.8—and, consequently, it is soluble in organic solvents such as *n*-butyl acetate. After the extraction, the organic solution is decolorized and various impurities are removed by activated carbon, which is filtered off on a second drum filter. The penicillin is then precipitated by addition of a solution of potassium or sodium acetate in the organic solvent. It converts the undissociated penicillin to a negative ion and makes it insoluble in organic solvents. The final stages of the process are crystallization, filtration, washing, and drying of the crystals.

The dilemma is that the acid conditions that favor solvent extraction also favor acid hydrolysis of the penicillin. Figure 6.8 shows the reaction that gives benzylpenillic acid and benzylpenicillenic acid. Too much alkali also hydrolyzes the penicillin, this time to benzylpenicilloic acid. The labile nature of penicillin prevented Fleming from isolating it and presents many other problems. In this case the solvent extraction must be carried out rapidly in a special

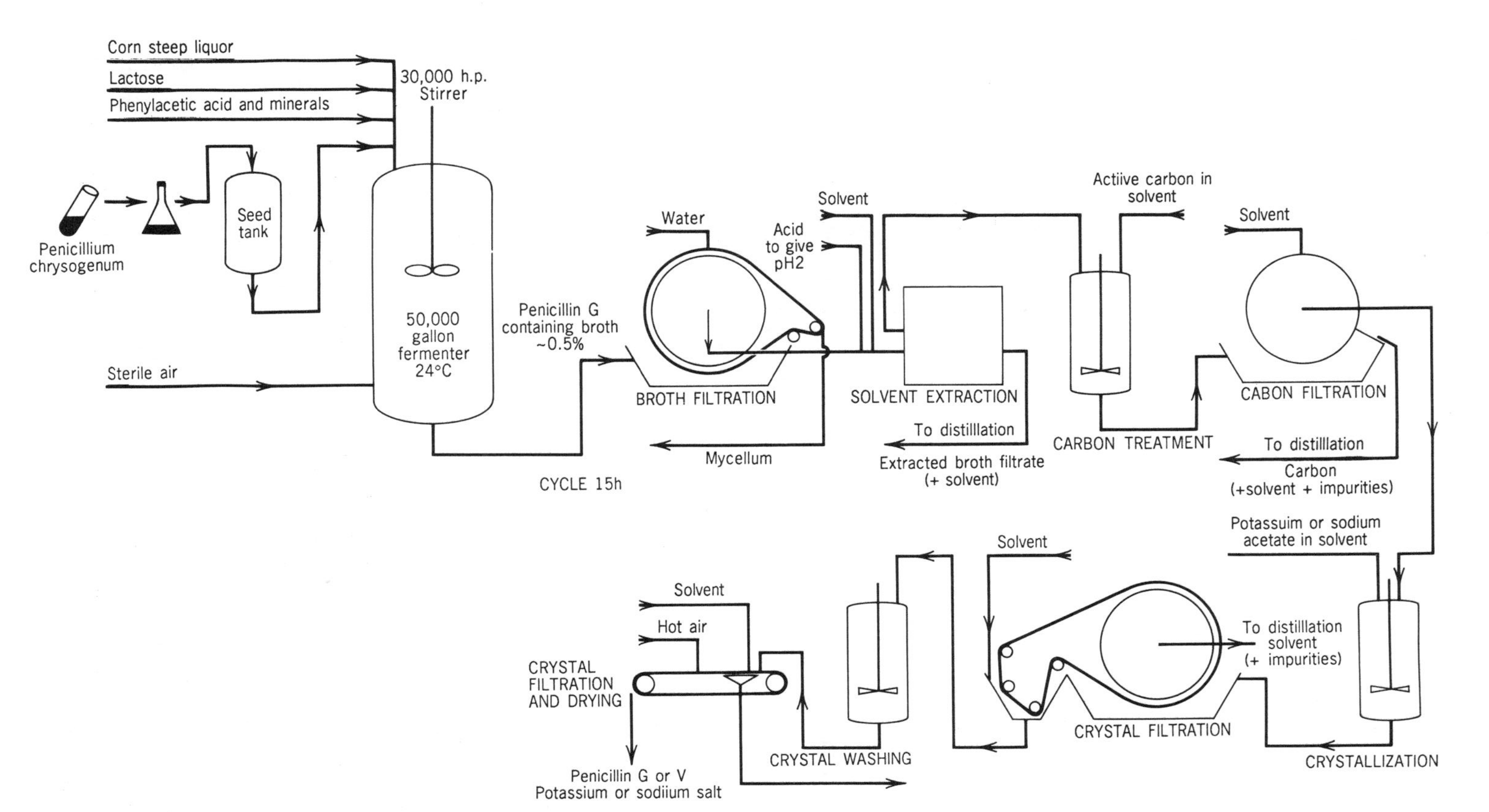

Figure 6.7. *Manufacture of penicillin: Penicillin purification process of Gist-Brocades. Reprinted from Ref 6 / 3 p. 106 by courtesy of Marcel Dekker Inc.*

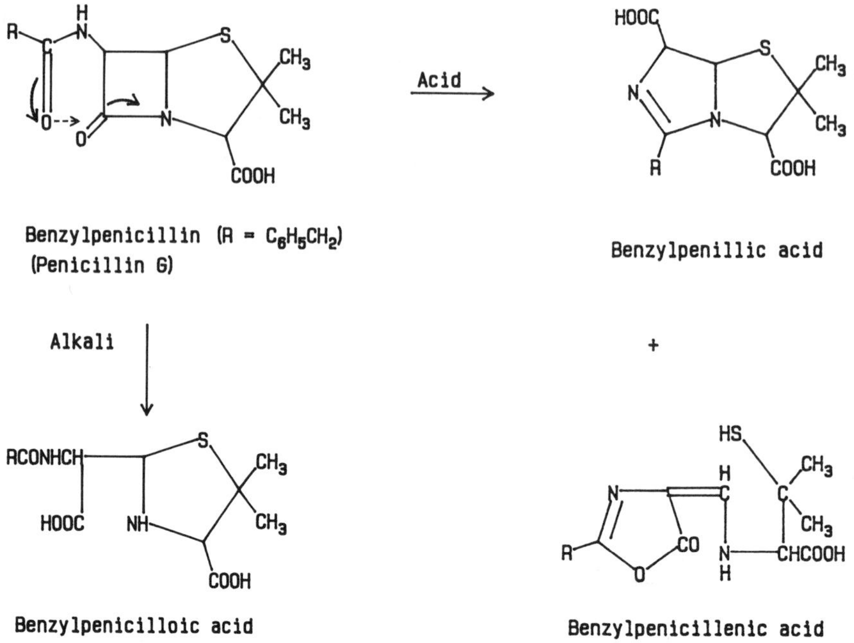

Benzylpenicillin ($R = C_6H_5CH_2$)
(Penicillin G)

Acid

Benzylpenillic acid

Alkali

+

Benzylpenicilloic acid

Benzylpenicillenic acid

Figure 6.8. *Acid and alkaline hydrolysis of penicillins.*

centrifugal mixer–settler to minimize acid hydrolysis. There is also an environmental problem involving the disposal of the mycelium and the mother liquors.

6.3.2 Improving Penicillins

When it appeared, benzylpenicillin was truly a miracle drug. Nonetheless, there are four problems associated with its use. First, it has a fairly narrow spectrum of activity and is active mainly against gram-positive bacteria. Second, it is hydrolyzed by the hydrochloric acid in the stomach because of its unstable β-lactam ring. Consequently, it has to be given by injection several times a day. Third, after penicillin had been in use for a few years, a strain of bacterium, *Staphylococcus aureus*, emerged. It produced an enzyme, β-lactamase or penicillinase, that attacked the reactive peptide bond in the penicillin and destroyed its activity. Fourth, about 1% of the population turned out to be allergic to penicillin. People are not born with allergies; they develop them as described in Section 9.1. No one reacts adversely to penicillin the first time he or she takes it. The penicillin becomes bound to body proteins and people become sensitized to this combination. On the second dose, there may be a reaction varying from a rash to a collapse of the circulation and fall in blood pressure known as anaphylactic shock. This can be fatal. The reaction is

R-NH S CH_3 CH_3 N O COOH

Penicillin nucleus

Structure of Side Chain R Attached to the 6-Amino-penicillanic Acid Nucleus	Name	Acid Resistance	Penicillinase Resistance	Comments
1. CH_2—C(=O)—	Benzyl-penicillin (penicillin G)	−	−	Naturally occurring penicillin. Must be injected. Mainly active against gram-positive bacteria.
2. —$COO^-HN^+(C_2H_5)_2CH_2CH_2OOC$—$NH_2$	Procaine penicillin G	−	−	Benzylpenicillin is converted to a salt with procaine [*2*-(diethylamino)ethyl-*p*-amino-benzoate].
3. O—CH_2—C(=O)—	Phenoxymethyl-penicillin (penicillin V)	+	−	Can be taken orally.
4. R—CH(NH_2)—C(=O)—	Ampicillin (R=H) Amoxycillin (R=OH)	+	−	Broad spectrum; active against gram-negative bacteria. Can be given orally. General purpose penicillins, widely used.
5. OCH_3, OCH_3, C(=O)—	Methicillin	−	+	Must be injected. Active against *Staphylococcus aureus.* (Now replaced by halogenated penicillins and nafcillin.)
6. R_1, R_2, N—O, CH_3, C(=O)—	R_1 R_2			All these penicillins are resistant to penicillinase but are much less active against other penicillin-sensitive bacteria.
	H H Oxacillin	+	+	
	Cl H Cloxacillin	+	+	
	Cl Cl Dicloxacillin	+	+	
	Cl F Flucloxacillin	+	+	
7. OC_2H_5, C(=O)—	Nafcillin	+	+	Ring system and steric hindrance make nafcillin penicillinase-resistant.
8. CH(COOH)—CO—	Carbenicillin	−	−	Must be injected. Gram-positive activity weak but gram-negative strong. Less active than ampicillin in most applications but particularly active against *Pseudomonas aeruginosa*.
9. S, CO—, HOOC, H	Ticarcillin	−	−	Similar to carbencillin. Active against *Pseudomonas aeruginosa*

Figure 6.9. *Important penicillins. Chart shows structure of the side chain R attached to the 6-aminopenicillanic acid nucleus for all compounds except compound 2. In compound 2, substitution is associated with the carboxyl group and occurs via salt formation. The side chain attached to the 6-aminopenicillanic acid nucleus in compound 2 is* $C_6H_5CH_2CO$.

unpredictable. At one time there was a scratch test for it but that turned out to be as dangerous as the drug itself. Because of the danger of sensitization, penicillin is never used in ointments.

Such problems are challenges to the chemist. Attempts to solve them centered on changing molecular structure. Over 2000 chemically modified penicillins have been synthesized and evaluated and the approach has solved the first three of the four problems. Figure 6.9 shows some important penicillins chosen either because of their historical significance or present day use.

The first improvement was the production of the procaine salt of benzylpenicillin. The structure of the salt-forming procaine moiety is structure 2 in Figure 6.9 and it combines with the carboxyl group of the penicillin, not the amine group as do the other structures shown. Procaine penicillin dissociates slowly in the body to benzylpenicillin and, consequently, need only be injected once a day. It is the first example in this section of a pro-drug (Section 4.7), that is, a drug that is deliberately administered in one form in the knowledge that it will be converted to another form within the body. Prontosil (section 6.2) was a sort of pro-drug in that it was converted to sulfanilamide in the body but that was not the intention and, as soon as sulfanilamide was recognized as the active principle, it was administered as such.

The next advance was the development of a penicillin that was stable to stomach acid and could therefore be taken orally. This was phenoxymethylpenicillin or penicillin V. The side chain is shown as structure 3, Figure 6.9. When benzylpenicillin is hydrolyzed by acid, it rearranges as shown in Figure 6.8. In phenoxymethylpenicillin there is a heteroatom, in this case an ether oxygen, on the α carbon of the side chain. This oxygen atom improves acid stability because it inhibits the participation of the side chain amide carbonyl in the electronic displacement that governs the rearrangement.

Phenoxymethylpenicillin is prepared by a method similar to the preparation of benzylpenicillin, the difference being that phenoxyacetic acid is used instead of phenylacetic acid in the substrate. Phenoxymethylpenicillin is usually marketed as its potassium salt. Several other penicillins were made by alteration of side chain precursors but they are of less importance than phenoxymethylpenicillin, which is still #6 in the Top-100.

6.3.3 Semisynthetic Penicillins

The new penicillins still had a narrow spectrum of activity and were still susceptible to hydrolysis by penicillinase. Indeed, there turned out to be a number of different penicillinases and β-lactamases. As penicillin-resistant bacteria multiplied, the continued usefulness of penicillin was in doubt. The major breakthrough was the discovery of the semisynthetic penicillins in 1959. The basic penicillin nucleus of 6-aminopenicillanic acid may have been synthesized first by Sheehan at MIT in 1958 but Beecham's, a British company better known at the time for proprietary medicines, toilet products, and soft drinks, obtained a pure crystalline material and won a subsequent court case. This work led to synthesis of a variety of penicillins by acylation of the amino group of 6-aminopenicillanic acid. The first successful drug to emerge was ampicillin. Its side chain is shown as the fourth structure in Figure 6.9. Ampicillin has been an important drug for over 20 years and is still #12 in the Top-100. The production of 6-aminopenicillanic acid and ampicillin will be described in the next section.

Ampicillin differs from benzylpenicillin only in the amino group attached to the α carbon, but its spectrum is much wider and it is active against some gram-negative bacteria. It is also resistant to stomach acid, although not to penicillinase. The key factor in acid resistance is the electron density in the side chain carbonyl group. In ampicillin, in acid solution, this is reduced by the adjacent $-NH_2$ group which becomes positive and therefore electron-attracting.

The synthesis of ampicillin, however, opened the door to many other semisynthetic penicillins. A closely related compound, whose side chain is also shown in Figure 6.9, is amoxycillin, and the most significant market trend in recent years has been the inroads this has made into the ampicillin market. Amoxycillin has a similar spectrum and stability but is better absorbed when taken orally so that higher blood levels are achieved with an equivalent dose. It is less toxic and produces better cell lysis. Furthermore, absorption is not affected by the presence of food in the stomach. Prescriptions for amoxycillin overtook those for ampicillin in 1980 in the United Kingdom and in 1982 in the United States.

Historically, the most significant penicillin to emerge after ampicillin was methicillin. It has side chain 5 in Figure 6.9 and was the first penicillinase-resistant penicillin to be marketed. Unfortunately, it was unstable to stomach acid and had to be injected. It was, however, active against *Staphylococcus aureus*, which was troublesome in many hospitals. Although a milestone in penicillin therapy, it has since been replaced by the five penicillins whose side chains are shown as 6 and 7 in Figure 6.9. Oxacillin, the three halogenated penicillins (cloxacillin, dicloxacillin and flucloxacillin), and nafcillin may all be taken by mouth and are resistant to penicillinase but they are less active than the conventional penicillins against bacteria that do not produce penicillinase. Their use should therefore be restricted to severe infections resistant to other penicillins and preferably only after bacteriological investigation.

The lipophilicity and, hence, rate of absorption of these penicillins increases with the number of halogen atoms and they are all strongly bound to serum protein. Flucloxacillin and cloxacillin are the preferred compounds at present in Britain and flucloxacillin is the favorite because it is absorbed twice as fast from the gut. In the United States, all the compounds except flucloxacillin are used and dicloxacillin is the favorite. In all these penicillins, a ring is attached directly to the side chain amide carbonyl group and there is 2,6-substitution in the benzene ring. It appears that penicillinase resistance is related to this sort of structure.

A number of penicillins are active against the difficult gram-negative bacteria. Carbenicillin (structure 8) was the first; the preferred compound at present appears to be ticarcillin, structure 9. They are especially active against *Pseudomonas aeruginosa*, but need to be injected.

The penicillin story provides the opportunity to point out that it is the chemist's job to improve on nature. Nature provided benzylpenicillin which is a poor product indeed in comparison with what is available today. But, lest the chemist take too much comfort from this statement, we must point out that nature always throws curves. The first two of the problems listed at the start of Section 6.3.2 have been overcome but there is still a problem with penicillin allergy. The problem of the penicillinases, the enzymes that inactivate penicillins, has been partly solved and many new drugs have been developed including the halogenated penicillins already mentioned. More recent developments will be described in Sections 6.3.7 and 6.3.8.

6.3.4 Production of 6-Aminopenicillanic Acid

The first stage in the manufacture of the semisynthetic penicillins is the cleavage of the existing side chain of penicillin G or V to give 6-aminopenicillanic acid (6-APA) followed by the addition of the appropriate side chain. 6-APA may be manufactured in three ways. A historically important route was the conventional fermentation route to penicillins carried out with careful exclusion of side chain precursors. The main product is 6-aminopenicillanic acid but yields are low and isolation is difficult. Therefore, better routes were sought.

A chemical route that is used commercially is shown in Figure 6.10. Benzylpenicillin (1) or phenoxymethylpenicillin is silylated to protect the carboxyl group. The silyl derivative (2) reacts with phosphorus pentachloride at $-59°C$ in an unusual reaction which leads to an iminochloride (3). The iminochloride is treated with methanol at $-70°C$ to give the imino ether (4). Addition of water detaches the side chain and hydrolyzes the silyl group to give 6-aminopenicillanic acid (5) which is precipitated. The chemical route reached the peak of its popularity in the late 1970s.

Most major manufacturers now use a third route which involves enzymatic hydrolysis. Beecham's had normally employed *E. coli* enzymes but the route became much more attractive with the availability of immobilized enzymes. The immobilized amidase cleaves benzylpenicillin to the desired 6-aminopenicillanic acid. The cleavage goes smoothly at 37°C and pH 7.8. The main difficulty is that, unless various pH control systems are installed, the phenylacetic acid that is liberated lowers the pH to 5 or 6, at which level of acidity the enzyme promotes the reverse reaction and regenerates benzylpenicillin.

The success of the semisynthetic penicillins is illustrated by the present-day classification of 6-aminopenicillanic acid as a bulk pharmaceutical intermediate with 1986 world production of 10,000 tonnes at $70/kg.

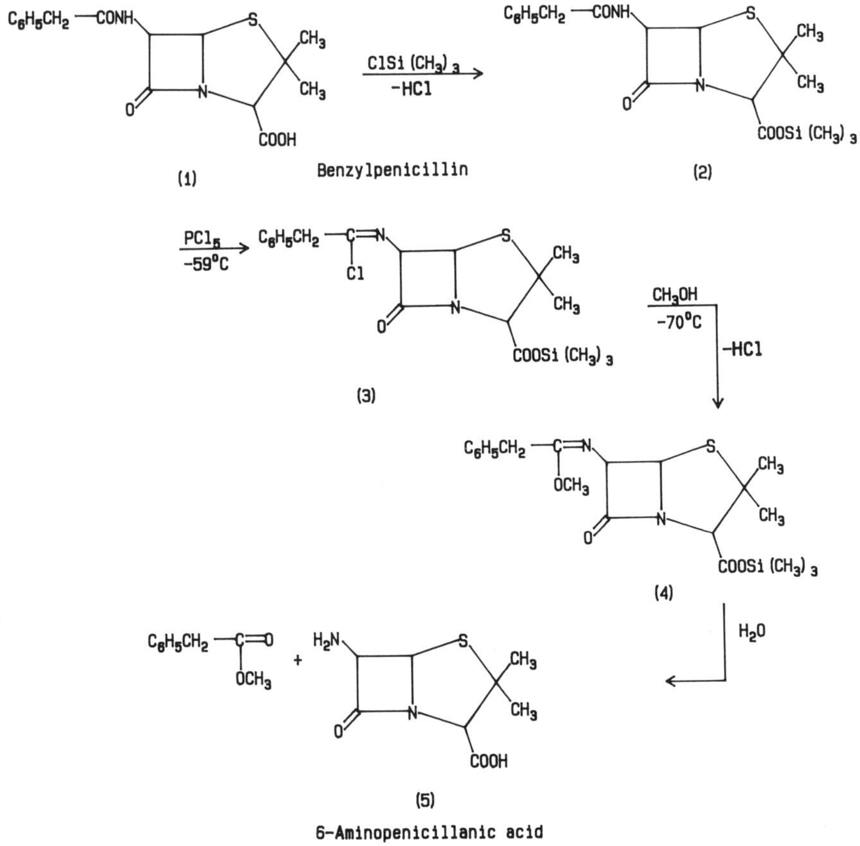

Figure 6.10. Chemical cleavage of benzylpenicillin to 6-aminopenicillanic acid.

6.3.5 Production of Ampicillin and Amoxycillin

The introduction of the side chain onto 6-APA may be illustrated by the synthesis of ampicillin shown in Figure 6.11. The amino group of phenylglycine (1) is first protected by reaction with carbobenzoxy chloride (2) to give (3). Reaction of (3) with ethyl chloroformate (4) in triethylamine gives a complex diester (5), which condenses with 6-aminopenicillanic acid (6) to give an N-substituted ampicillin (7). The protecting group is removed by hydrogenolysis and the ampicillin (8) is sold as the trihydrate.

Amoxycillin is manufactured in the same way, except that the side chain precursor is not phenylglycine but 4-hydroxyphenylglycine. There are two ingenious ways to manufacture this material. The first method is a modified Mannich reaction. Phenol in aqueous ammonia reacts with glyoxylic acid to give 4-hydroxyphenylglycine, presumably via the intermediate *p*-hydroxymandelic acid:

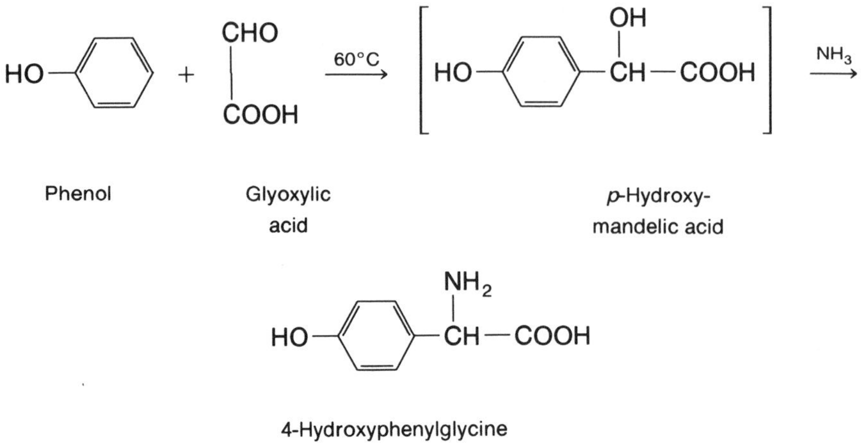

$$\text{HO}-C_6H_4 \;(\text{Phenol}) + \text{OHC}-\text{COOH} \;(\text{Glyoxylic acid}) \xrightarrow{60^\circ C} \left[\text{HO}-C_6H_4-\text{CH(OH)}-\text{COOH}\right] \;(p\text{-Hydroxymandelic acid}) \xrightarrow{NH_3}$$

$$\text{HO}-C_6H_4-\text{CH(NH}_2)-\text{COOH}$$

4-Hydroxyphenylglycine

In the second method, phenol, glyoxal, and urea react to give a hydroxyphenylhydantoin that is hydrolyzed by water to 4-hydroxyphenylglycine:

$$\text{HO}-C_6H_4 \;(\text{Phenol}) + \text{OHC}-\text{CHO} \;(\text{Glyoxal}) + H_2N-\text{CO}-NH_2 \;(\text{Urea}) \longrightarrow \text{HO}-C_6H_4-\text{CH}-\text{C(=O)}-\text{NH}-\text{C(=O)}-\text{NH} \text{ (ring)} \xrightarrow{H_2O}$$

$$\text{HO}-C_6H_4-\text{CH(NH}_2)-\text{COOH}$$

4-Hydroxyphenylglycine

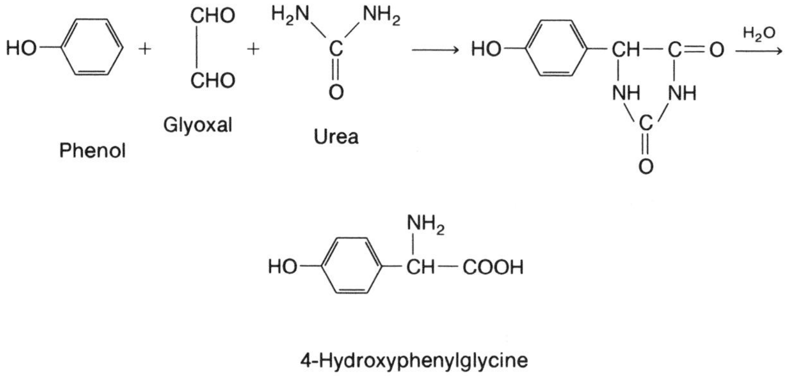

In 1986, 2800 tonnes of D-phenylglycine and 1450 tonnes of D-4-hydroxyphenylglycine were produced. Many other penicillin side chain precursors may also be made by these routes. They are subsequently resolved by classical methods to give the correct optical isomer.

6.3.6 Penicillin Esters

In addition to the problems with penicillins already mentioned, there is a minor one, which is that penicillin capsules have to be taken several times a

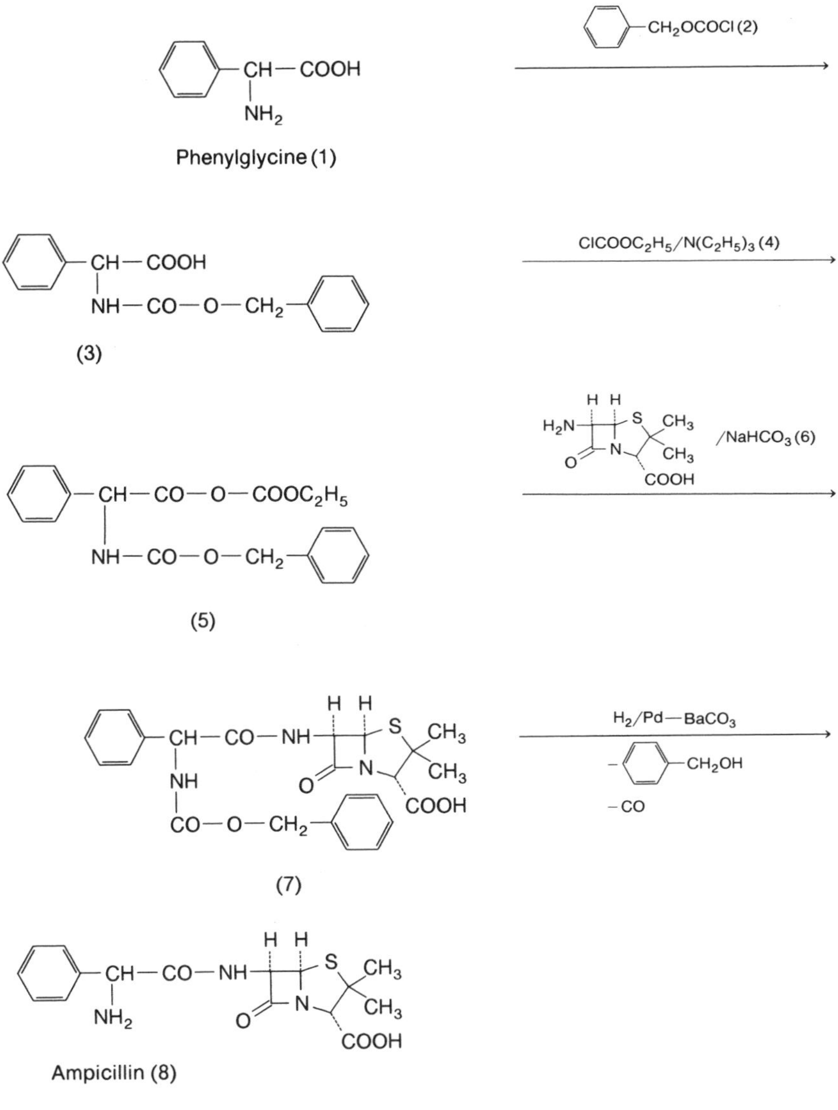

Phenylglycine (1)

$C_6H_5CH_2OCOCl$ (2)

(3)

$ClCOOC_2H_5/N(C_2H_5)_3$ (4)

(5)

$/NaHCO_3$ (6)

(7)

$H_2/Pd—BaCO_3$

$-C_6H_5CH_2OH$

$-CO$

Ampicillin (8)

Figure 6.11. *Synthesis of ampicillin from 6-aminopenicillanic acid.*

day. This led to the development of the ampicillin pro-drugs talampicillin, bacampicillin, and pivampicillin, shown in Figure 6.12. Whereas most penicillins are free acids or their sodium or potassium salts, the pro-drugs are esters of ampicillin. They are lipophilic and are rapidly absorbed in the gut. They hydrolyze to ampicillin in the intestinal cell wall and in the bloodstream. They can be used in smaller doses than ampicillin and are less irritating to the

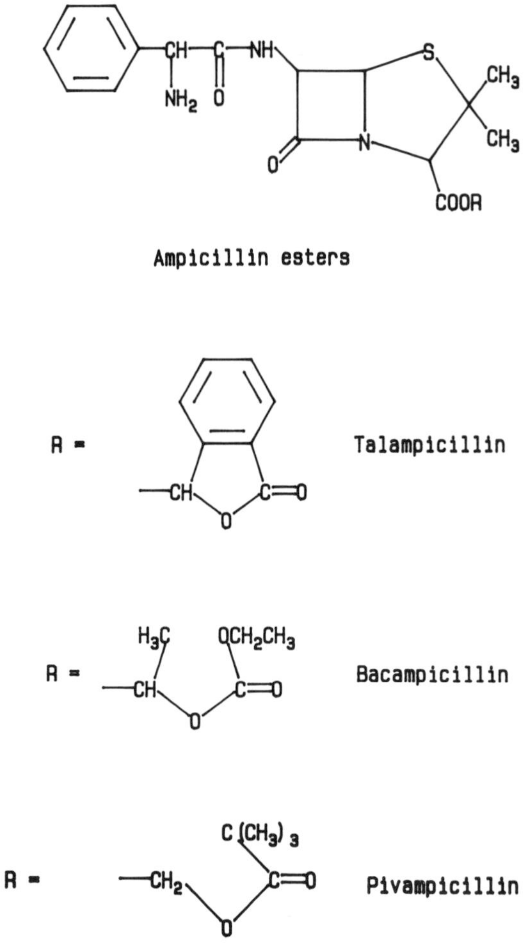

Ampicillin esters

R = Talampicillin

R = Bacampicillin

R = Pivampicillin

Figure 6.12. *Ampicillin pro-drugs.*

gut. The curious structures of the esterifying moieties arise because these labile acyloxymethyl esters are hydrolyzed in human tissues, unlike most esters of ampicillin.

Esterification of ampicillin presents difficulties. Straightforward acid catalysis is out of the question because of the instability of the β-lactam ring. The acyloxy compound would also be decomposed. Treatment of the sodium or potassium salt of ampicillin with an acyloxymethyl halide is attractive but the halide is insoluble in water while the ampicillin salt is insoluble in organic solvents. The difficulty is overcome by the use of phase transfer catalysis shown in Figure 6.13. Frame 1 shows the potassium salt of ampicillin dissolved in water, and the acyloxymethyl bromide in an organic solvent such as chloroform. There is no reaction. Frame 2 shows the addition of about 1% of

AMP =

NH_2 O CH_3 CH_3

(Ampicillin would thus be written AMP-COOH.)

TBA^+ = $(CH_3CH_2CH_2CH_2)_4N^+$

R =

(1)

Water AMP—COO^- K^+

$CHCl_3$ R-Br

The potassium salt of ampicillin and the acyloxymethyl bromide do not react because they reside in different phases.

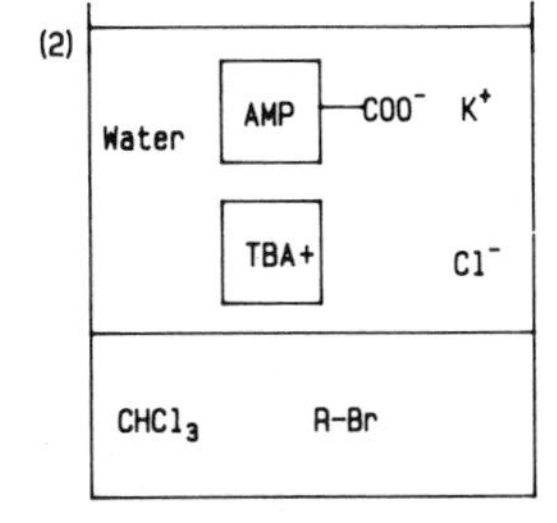

Addition of tetrabutylammonium chloride puts a lipophilic ion in the aqueous layer.

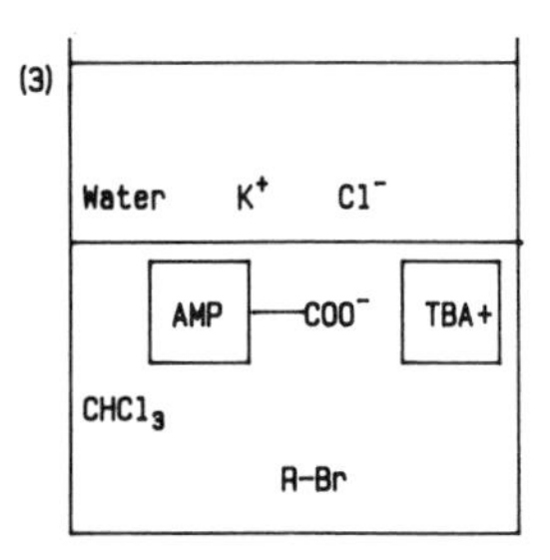

The tetrabutylammonium ion migrates to the organic layer carrying with it the ampicillin anion as an ion pair.

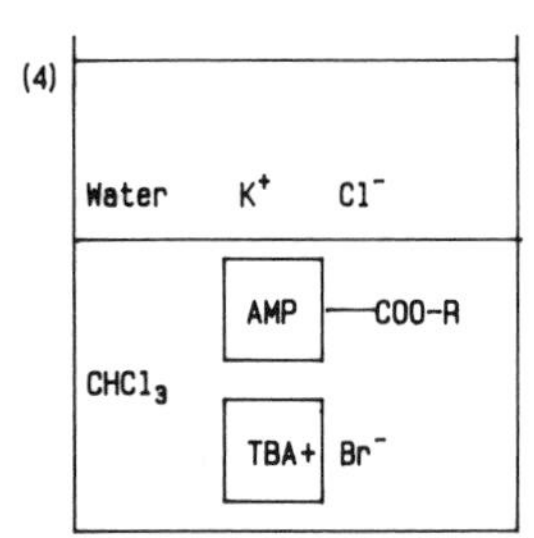

Reaction to give the ester is smooth in the organic layer. A bromide ion is also generated which is more hydrophilic than the ampicillin ion.

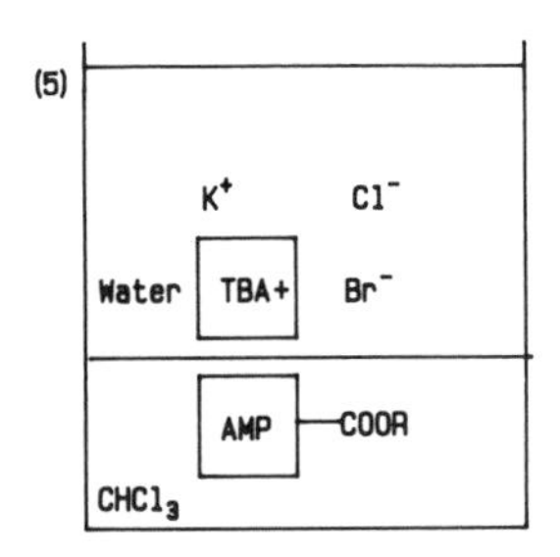

The ion pair migrates back to the aqueous layer and splits up, so the catalyst is regenerated and the ester remains in the $CHCl_3$ phase. The cycle is repeated.

Figure 6.13. *Esterification of ampicillin.*

tetrabutylammonium chloride which also dissolves in the aqueous layer. The tetrabutylammonium and ampicillin ions are both lipophilic and in frame 3 they both migrate to the organic layer as an ion pair. Frame 4 indicates that the ester is formed in the organic layer. A bromide ion is generated which is more hydrophilic than the ampicillin ion. Consequently, it carries the tetrabutylammonium ion back to the aqueous layer, so that the catalyst is regenerated and the cycle can be repeated.

Because of their expense, penicillin pro-drugs are not used widely.

6.3.7 β-Lactamase Inhibitors

Penicillin resistance arises in three ways. First, by transfer of the β-lactamase genes to and from bacterial chromosomes by plasmids and between chromosomes and plasmids by transposons (see bibliography). This interchange of genetic material not only is of importance to geneticists but is also of philosophical interest because it does not fit in with classical evolutionary theory. The second method is true bacterial evolution where a few resistant mutants in a large population survive and multiply. Third, nonpathogenic bacteria with penicillin resistance can produce β-lactamases and penicillinases that protect pathogenic bacteria in their vicinity.

The attempts to develop penicillins that resist the various penicillinases and β-lactamases resulted in methicillin, nafcillin, and the halogenated penicillins mentioned above. An alternative strategy is to develop a β-lactamase inhibitor, that is, a compound that bonds to the enzyme and prevents it from destroying a conventional penicillin. Clavulanic acid (#76) is the leading compound in the group and was brought onto the British market in 1981:

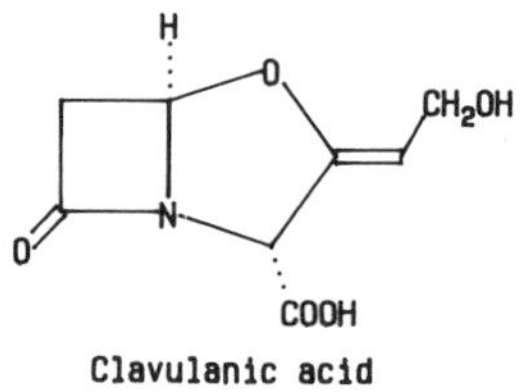

Clavulanic acid

Methicillin and the other penicillinase-resistant penicillins are in fact β-lactamase inhibitors. Compounds such as clavulanic acid, however, have only weak antibacterial action themselves but show powerful β-lactamase inhibition. Clavulanic acid, made by fermentation, is used in conjunction with conventional penicillins (amoxycillin and ticarcillin) to protect them against β-lactamase. Like all β-lactamase inhibitors it has a simple β-lactam structure and this is undoubtedly the basis for its action.

A competing β-lactamase inhibitor is sulbactam. This can be used like clavulanic acid in a mixture with another penicillin but in one new drug it is offered covalently bonded to the antibiotic. Sultamicillin is a mixed acylal of sulbactam and ampicillin made by combining the two drugs with formaldehyde.

Ampicillin + Formaldehyde + Sulbactam $\xrightarrow{-2H_2O}$

Sultamicillin

6.3.8 The Future of Penicillins

Penicillins have been declared to be no longer worth further research every year since 1951 when Ernst Chain persuaded the Weizmann Institute in Israel to abandon their projects. In spite of this, the search for new ones will continue and the problem of penicillin allergy has still to be solved. An encouraging factor is that new strains of penicillin-resistant organisms are appearing relatively slowly. Admittedly, penicillin-resistant gonorrhea—the so-called Asian gonorrhea—is reaching alarming levels in the United States, and syphilis too is having a new lease of life. Procaine penicillin is listed in Table 6.3 as the antibiotic of first choice in these diseases and there is a question as to how long this will continue to be true. Nonetheless, the spectrum of ampicillin remains broadly what it was when the drug was introduced.

There are, however, people who seem no longer to respond to antibiotics. There are two ways in which this can occur. First, they might acquire an antibiotic-resistant bacterium from another carrier, and, second, they might harbor dormant antibiotic-resistant microorganisms on dead tissue within themselves. Occasionally this will flare up into a full-scale infection. Resistant organisms, of course, must be treated with other penicillins or with other antibiotics, such as the cephalosporins, which are described in the next section.

There is always the hope that a broader spectrum penicillin with fewer side effects will be discovered. While this remains a goal, research continues into more specialized penicillin or β-lactam antibiotics. Indeed, there has been a change of philosophy in recent years. There is concern that widespread use of broad-spectrum antibiotics leads to the more rapid development of resistant bacteria and that perhaps, wherever possible, an antibiotic should be chosen that counters only the specific bacteria responsible for a given infection. Many of the compounds in this section have a narrow spectrum of activity.

The highly strained β-lactam ring of penicillin is central to its biological activity. If so, then simpler drugs containing this structure should be effective, and many of them are. One such structure has been given the name of penem. Penemcarboxylic acid is the simplest known active β-lactam and hydroxyethylpenem is more active:

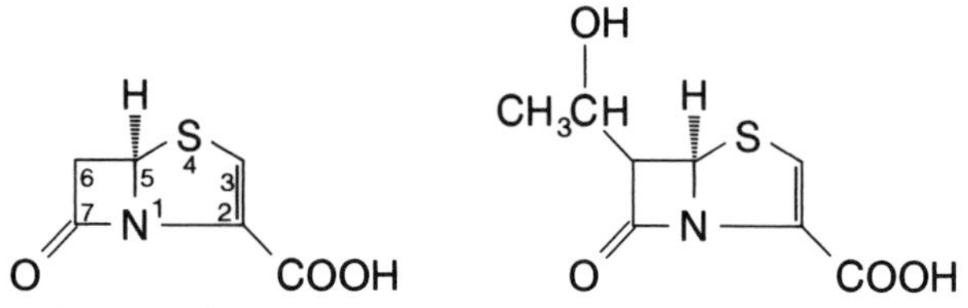

Penemcarboxylic acid

Hydroxyethylpenem

The above penems have an *R* stereo configuration at carbon 5. Penems with the unnatural 5*S* configuration are biologically inactive. It has been shown that the *R* stereo configuration is the only essential stereochemical requirement to achieve activity in a β-lactam antibiotic. Since the penems are effective without the side chain, it can be concluded that β-lactam antibiotics are effective, first of all, because of the strained β-lactam ring and, second, because of the *R* stereo configuration on carbon 5.

The penems are active against β-lactamase-producing bacteria but, unfortunately, they do not seem to be useful clinically. Animal trials have shown that they are excreted much more rapidly than conventional penicillins and too rapidly to be effective antibiotics. Furthermore, they have relatively poor acylating power. Acylation is the mechanism by which penicillin attaches itself to enzymes in bacteria (Section 6.3). In the penem structure, the strained β-lactam ring puckers the fused rings and increases the height of the ring nitrogen over the pyramid of carbon atoms to which it is attached. Its height, as measured by x-ray diffraction, is greater in penems than it is in conventional penicillins. This is believed to make penems less reactive acylating agents. More recent work has suggested that penems and carbapenems (see below) kill dormant and growing bacteria by inactivating a previously unknown enzyme that bacteria use while dormant to keep cell walls intact.

Another group of penicillinase-resistant β-lactam antibiotics is the carbapenems. They are penems to which sulfur-containing groups are attached. A typical compound is imipenem, the formimidoyl derivative of thienamycin:

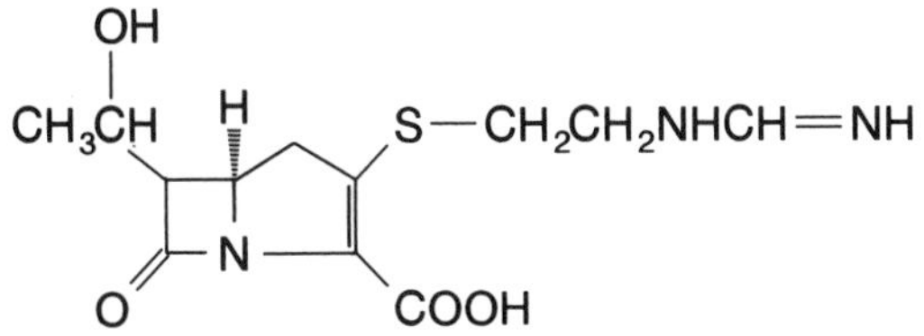

Imipenem (formimidoyl derivative of thienamycin)

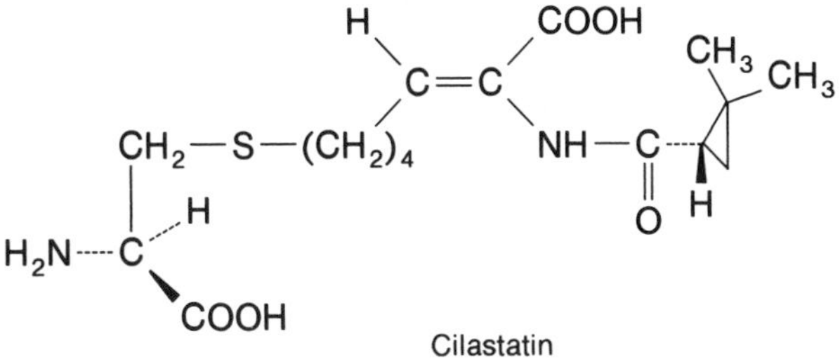

Cilastatin

Its structure is the same as that of hydroxyethylpenem shown above except that there is a "tail" attached at carbon 6 and the sulfur atom is now outside the ring. Thienamycin is too unstable to be useful but the formimidoyl group lends sufficient stability for it to be stored. In the body, it is destroyed by the kidney enzyme, dihydropeptase-1. It is therefore marketed as a mixture with cilastatin, a dehydropeptidase-1 inhibitor (see above). It is said to be the widest spectrum antibiotic known.

The amide group in the penicillin side chain has also been modified to provide amdinocillin:

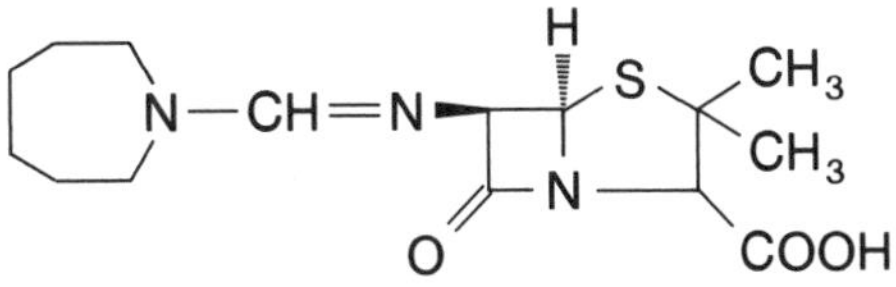

Amdinocillin

This compound bonds to a different site from conventional penicillins. It is β-lactamase-resistant and active against gram-negative bacteria. It appears to restore activity to other penicillins when used together with them.

The final group of β-lactams is the monobactams. Aztreonam is the first totally synthetic monocyclic β-lactam. The fused ring has proved unnecessary!

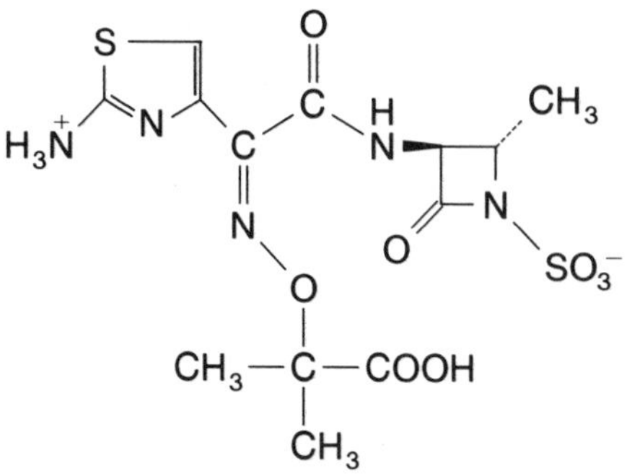

Aztreonam

It must be injected and is active against aerobic gram-negative bacteria. It does not appear to induce β-lactamase production and is expected to be valuable in hospital-acquired infections and to replace some aminoglycosides which carry greater toxicity hazards.

The penicillins are of great commercial and historical importance and in addition they are of intrinsic interest and provide a fine example of the way in which development of more effective drugs proceeds. Fleming's initial experiments failed to produce a clinically useful material. Florey and Chain produced a drug that had to be injected. Coghill and his coworkers in the United States scaled up the process successfully, and a few years later the Distillers' Company in Britain developed an oral penicillin. The appearance of penicillin-resistant bacteria threatened the continuing usefulness of penicillins and this was overcome by the development of the semisynthetic penicillins. Since then a range of penicillins with specialized applications has become available and the future of these drugs looks bright. There is still the problem of reducing allergic response in a small percentage of the population but even this in time may be solved by a combination of chemistry and physiological understanding.

6.4 CEPHALOSPORINS

The cephalosporins are a group of β-lactam antibiotics closely related to the penicillins. The difference is that most of the cephalosporins have a sulfur-containing ring with six rather than five members. Some of the important ones are shown in Figure 6.14. Cephalexin is #22 in the Top-100, cefaclor is #42, and cefadroxil is #91. Cephalexin and cefaclor have side chains identical with that of ampicillin while the cefadroxil side chain is like that of amoxycillin.

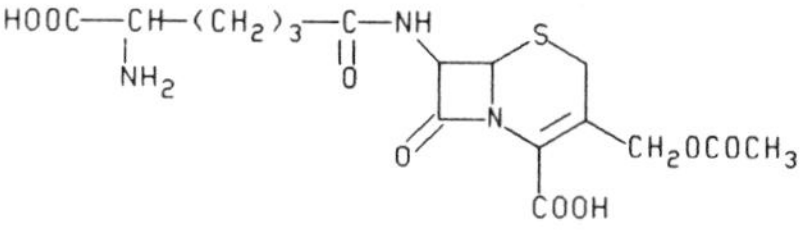

HOOC—CH—(CH2)3—C—NH
NH2 O S
O N CH2OCOCH3
COOH

1. Cephalosporin C (fungal metabolite)

First generation

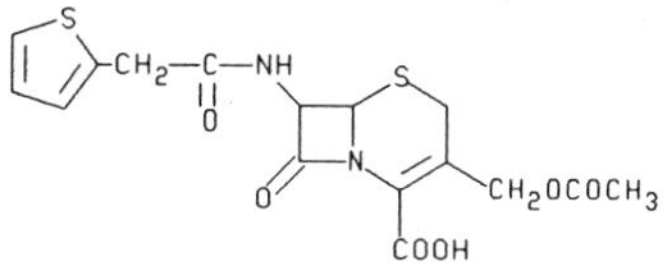

S CH2—C—NH S
O
O N CH2OCOCH3
COOH

2. Cephalothin

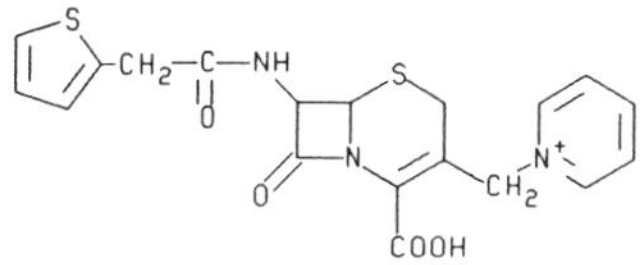

S CH2—C—NH S
O
O N CH2 N+
COOH

3. Cephaloridine

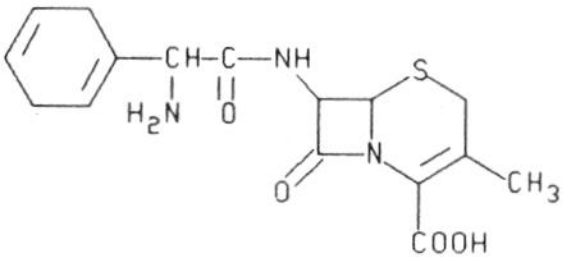

CH—C—NH S
H2N O
O N CH3
COOH

4. Cephradine (O)

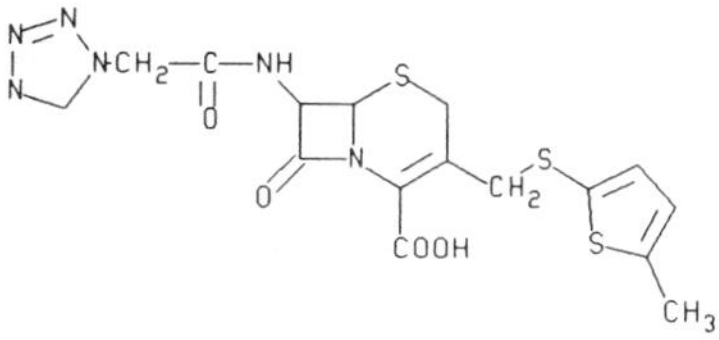

N=N
N N CH2—C—NH S
O
O N CH2 S S
COOH CH3

5. Cephazolin

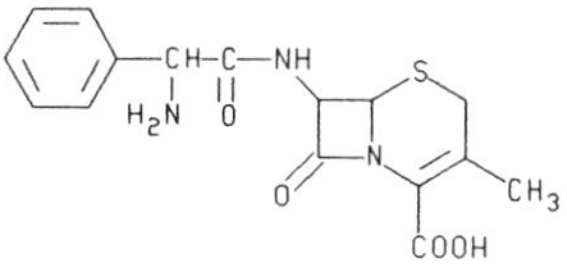

CH—C—NH S
H2N O
O N CH3
COOH

6. Cephalexin (O)

HO CH—C—NH S
H2N O
O N CH3
COOH

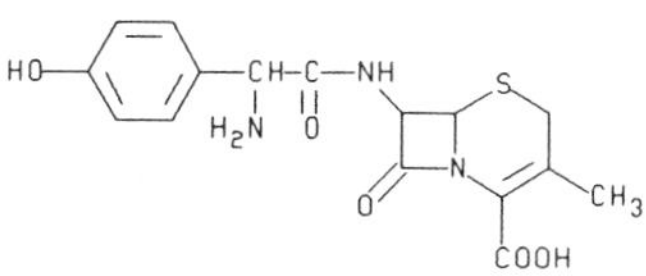

7. Cefadroxil (O)

Second generation

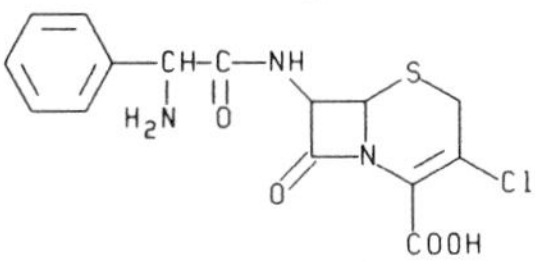

CH—C—NH S
H2N O
O N Cl
COOH

8. Cefaclor (O)

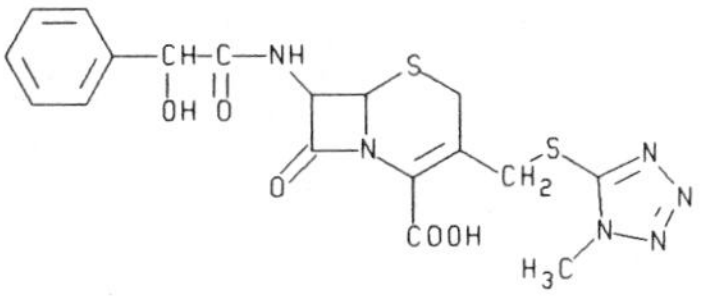

CH—C—NH S
OH O
O N CH2 S N N N N
COOH H3C

9. Cefamandole

O C—C—NH S
O
N O N CH2 OCOCH3
O
CH3 COOH

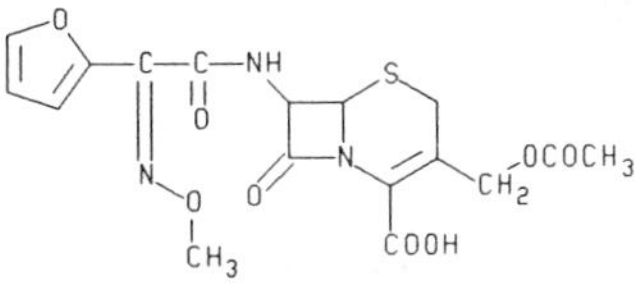

10. Cefuroxime

Third generation

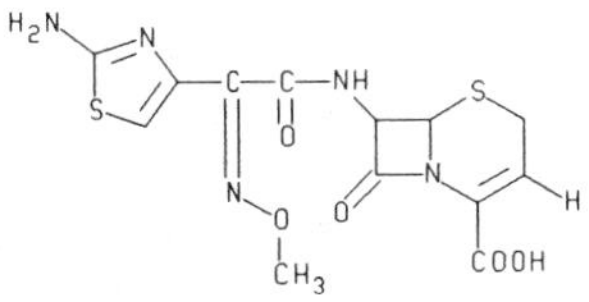

H2N N
S C—C—NH S
O
N O N H
O
CH3 COOH

11. Cefotaxime

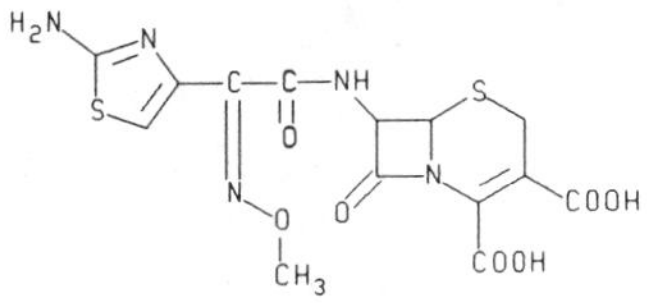

H2N N
S C—C—NH S
O
N O N COOH
O
CH3 COOH

12. Ceftizoxime

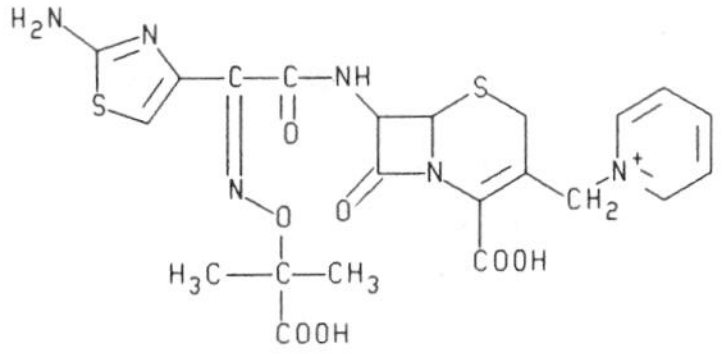

H2N N
S C—C—NH S
O
N O N CH2 N+
O
H3C—C—CH3 COOH
COOH

13. Ceftazidime

CH3CH2N N—C—NH—CH—C—NH S
O O O O
O N CH2 S N N N N
OH COOH CH3

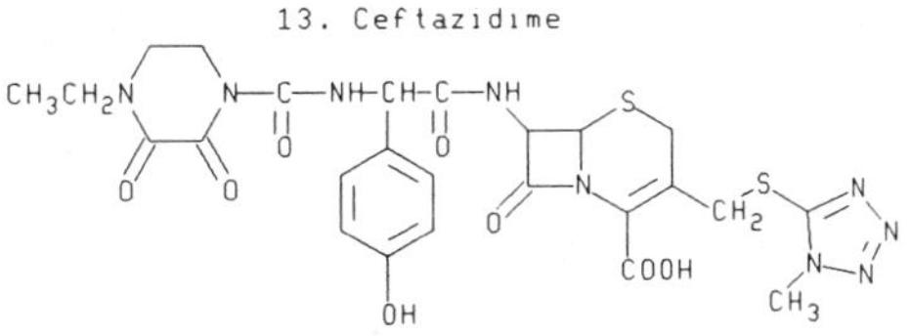

14. Cefoperazone

Figure 6.14. *Cephalosporins (O = Oral cephalosporins; all others must be injected. Cephradine may be administered either way).*

The cephalosporins are produced by chemical modification of the fungal metabolite cephalosporin C (structure 1, Figure 6.14). The fungus is *Cephalosporium acrimonium* and it was first isolated from Sardinian sewage. Two other compounds were isolated from the same fungus before cephalosporin C—one a steroid and the other a penicillin. Cephalosporin C has only weak antibiotic activity. Indeed, useful cephalosporins cannot be obtained by fermentation. Instead, most of them are produced by cleavage of cephalosporin C to 7-aminocephalosporanic acid (7-ACA, Figure 6.15), followed by addition of a suitable side chain.

The first cephalosporin on the market was cephalothin (structure 2, Figure 6.14). It had useful activity against many bacteria but the 3-acetoxymethyl side chain $—CH_2OCOCH_3$ could be cleaved by esterases present in mammals to leave a $—CH_2OH$ side chain. This compound was less active. It turned out that the labile acetoxy group could be displaced by various nucleophiles. Reaction of cephalothin with pyridine gave cephaloridine (3). Cephaloridine retained the activity of cephalothin and was metabolically stable but occasionally caused renal tubular necrosis. It was also susceptible to β-lactamases and both compounds had to be injected, which limited their scope. Neither was very active against gram-negative bacteria.

Structural modification led to cephradine (4), cephazolin (5), and cephalexin (6), still the most popular cephalosporin. Most cephalosporins are poorly absorbed from the gut and many are unstable at the pH of the stomach. Unlike the others, cephalexin (#22) and cephradine can be taken orally and the former still appears in the Top-100. Cephradine is the only cephalosporin on the market that can be taken either by mouth or by injection. Cephalexin is not sufficiently soluble to be injected.

These were the so-called first-generation cephalosporins, characterized by their weak gram-negative activity. A recent addition to this group is cefadroxil (#91, structure 7), which may be taken orally and has a longer plasma half-life than most cephalosporins, although it is less active against *Hemophilus influenzae*. The second-generation cephalosporins, for example, cefaclor (8) and cefamandole (9), are generally more active against gram-negative bacteria. Cefaclor can be taken orally and ranks #42 in the Top-100. Third-generation cephalosporins are characterized by even stronger gram-negative activity and resistance to β-lactamases produced by gram-negative bacteria. Selected second- and third-generation drugs are shown in Figure 6.14.

Two general solutions were found to the β-lactamase problem with cephalosporins and these were chemical in their approach. One was based on the 7-oximinoacyl cephalosporins and the other on cephamycin C. They will be discussed in Sections 6.4.2 and 6.4.3, respectively.

The mode of action of the cephalosporins is similar to that of the penicillins. The principal side effect is hypersensitivity and about 10% of penicillin-sensitive patients will also be allergic to cephalosporins. The first- and many of the second-generation cephalosporins were usually not regarded as drugs of first choice because they had a spectrum of activity close to that of ampicillin, and various equally effective and less expensive drugs were available. The third-generation cephalosporins, however, have spectra that differ between themselves and are becoming primary drugs for various types of meningitis and gonorrhea.

6.4.1 Manufacture of Cephalosporins

Figure 6.15 shows the first stage in the manufacture of cephalosporins, that is, the production of 7-ACA. In the original route, shown on the left, cephalosporin C (1) is treated with nitrosyl chloride in formic acid. A diazotization reaction takes place and the amide nitrogen enolizes to an imino group attached to a pyran ring (2) via the intermediates shown. The pyran hydrolyzes to the desired product (3). Nitrosyl chloride and formic acid are unusual diazotizing agents and are used in place of nitrous acid because otherwise the 7-amino group would be diazotized and lost as soon as it was formed.

The more modern route, which we think is now generally preferred, is shown on the right of Figure 6.15. It is analogous to the production of 6-aminopenicillanic acid (Figure 6.10) involving imino chloride and imino ether formation.

Figure 6.16 shows the acylation of the amino group on the four-membered ring of 7-ACA to give cephalexin. The amino group of phenylglycine (1) is first protected by reaction with *t*-butoxychloroformate (2) to give the carbamate (3). Reaction of (3) with isobutoxychloroformate (4) gives a substituted anhydride (5). Condensation of (5) with 7-aminocephalosporanic acid (6) gives an intermediate (7), which, on treatment with trifluoroacetic or formic acid, gives an amine, cephaloglycin (8). Hydrogenolysis of the allylic acetoxy group gives cephalexin (9).

Cephalosporins lacking the acetoxy group in the 3 position can sometimes be made from penicillins, which are much cheaper, by a curious ring expansion shown in Figure 6.17. The principal deacetoxycephalosporins are cephalexin and cephradine. The first stage is the production of the synthon, 7-aminodeacetoxycephalosporanic acid (7-ADCA). The carboxyl group of penicillin V (1) is protected as the trimethylsilyl or trichloroformyl ester. The sulfur atom is then converted to a sulfoxide (2) with sodium metaperiodate or an organic peracid. Ring expansion may be brought about by a range of reagents, including trimethyl phosphite in benzene. A sulfenic acid intermediate (3) is formed which can either cyclize back to the sulfoxide or ring-expand to the intermediate (4), which loses water to give a deacetoxycephalosporanic acid amidified in the 7-position (5). Conventional cleavage of the side chain, as illustrated in Figures 6.10 and 6.15, gives 7-ADCA, compound 6. This may be converted to cephradine by conventional chemistry. Cephalexin is said to be made more easily from the intermediate 2,2,2-trichloroformyl ester of 7-ADCA (6).

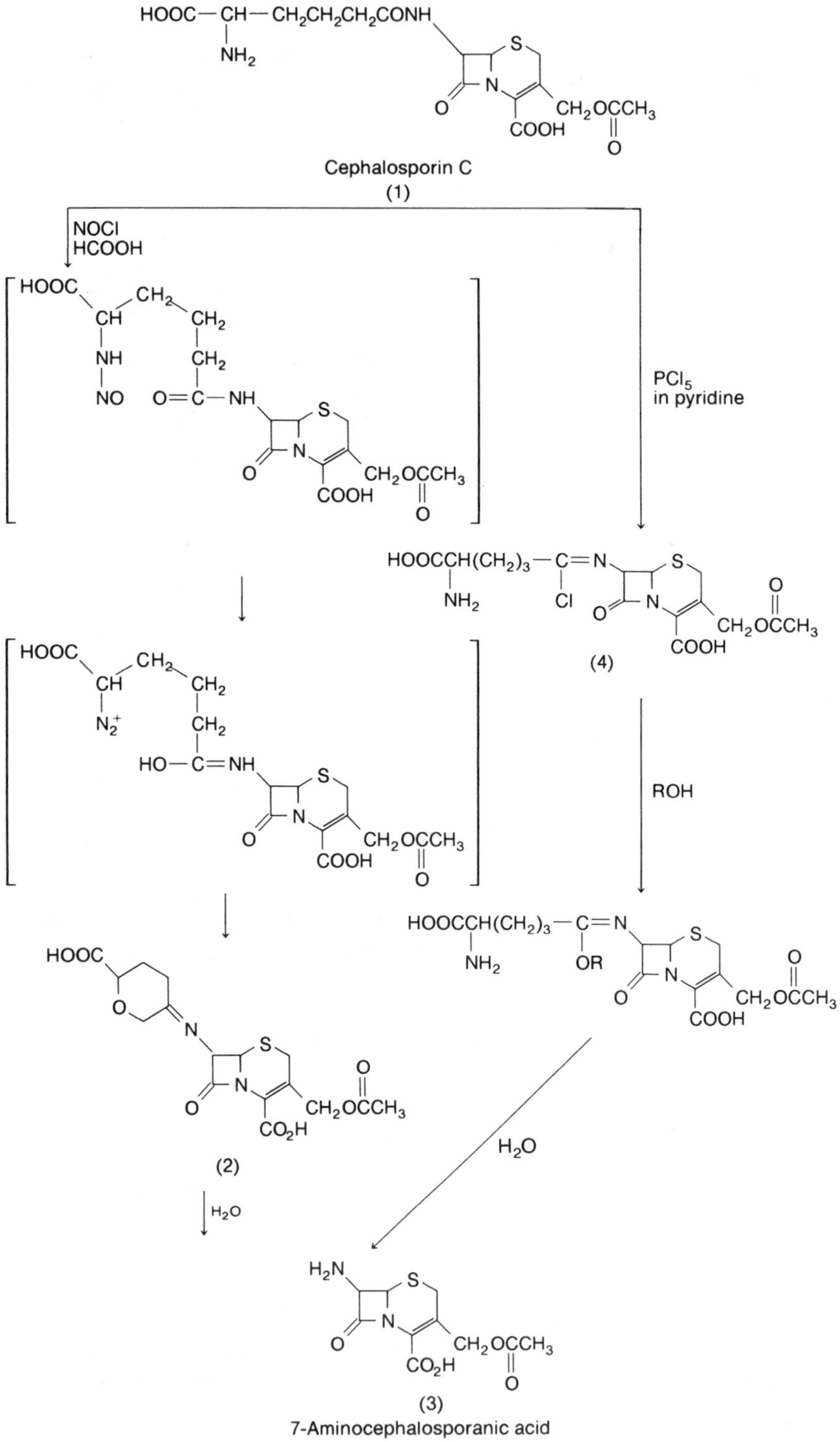

HOOC—CH—CH$_2$CH$_2$CH$_2$CONH
NH$_2$
S
N
O
CH$_2$OCCH$_3$
COOH
O

Cephalosporin C
(1)

NOCl
HCOOH

PCl$_5$
in pyridine

HOOC CH$_2$ CH CH$_2$ NH CH$_2$ NO O=C—NH S N O CH$_2$OCCH$_3$ COOH O

HOOCCH(CH$_2$)$_3$—C=N S
NH$_2$ Cl O N O CH$_2$OCCH$_3$ COOH
(4)

HOOC CH$_2$ CH CH$_2$ N$_2^+$ CH$_2$ HO—C=NH S N O CH$_2$OCCH$_3$ COOH O

ROH

HOOCCH(CH$_2$)$_3$—C=N S
NH$_2$ OR O N O CH$_2$OCCH$_3$ COOH

HOOC O N S N O O CH$_2$OCCH$_3$ CO$_2$H
(2)

H$_2$O

H$_2$O

H$_2$N S N O CH$_2$OCCH$_3$ CO$_2$H O
(3)
7-Aminocephalosporanic acid

Figure 6.15. *7-Aminocephalosporanic acid synthesis.*

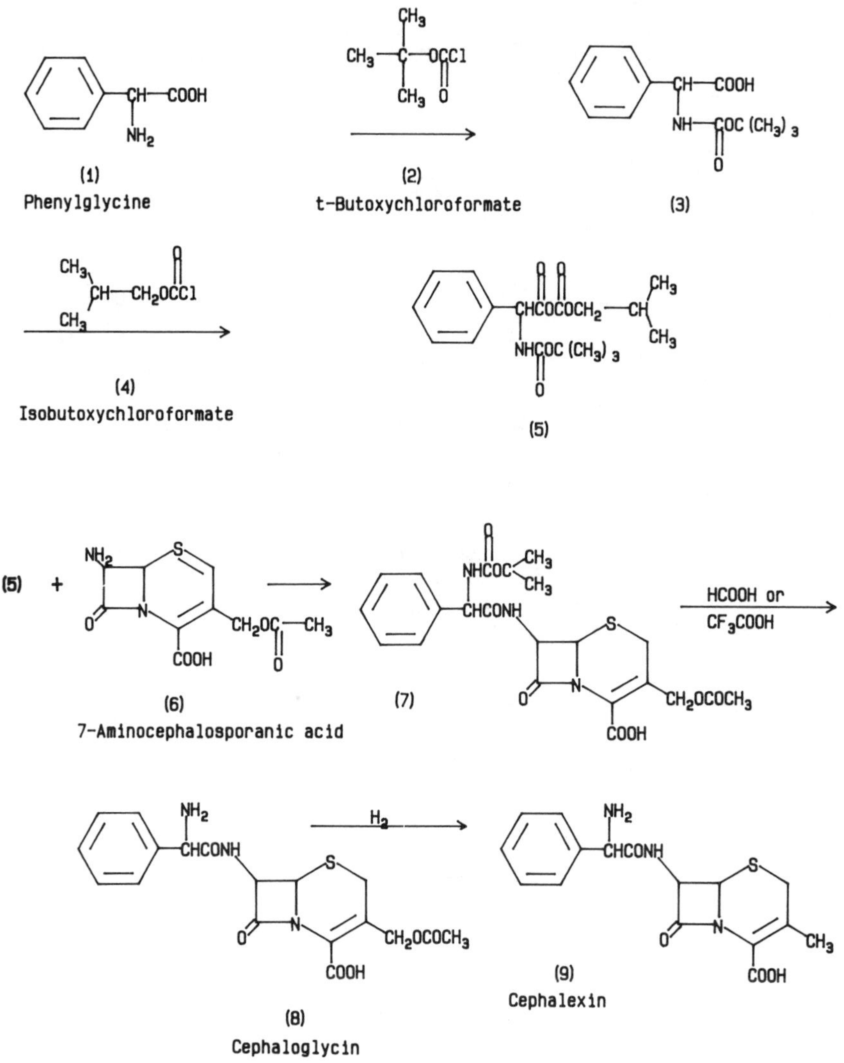

(1) Phenylglycine

(2) t-Butoxychloroformate

(3)

(4) Isobutoxychloroformate

(5)

(6) 7-Aminocephalosporanic acid

(7)

HCOOH or CF_3COOH

H_2

(8) Cephaloglycin

(9) Cephalexin

Figure 6.16. *Synthesis of cephalexin.*

Some other cephalosporins are of commercial importance and hundreds have been synthesized. The side chains frequently have a chiral center. The main problem is to protect the functional group on the side chain during the acylation reaction with a group that can subsequently be removed without damage to the β-lactam ring. Amine groups can be protected as chloroformates or Schiff's bases. Hydroxyl groups can be protected by esterification with formic acid and dichloroacetic acid, whereas carboxyl groups can also be protected as labile esters; silyl esters are preferred.

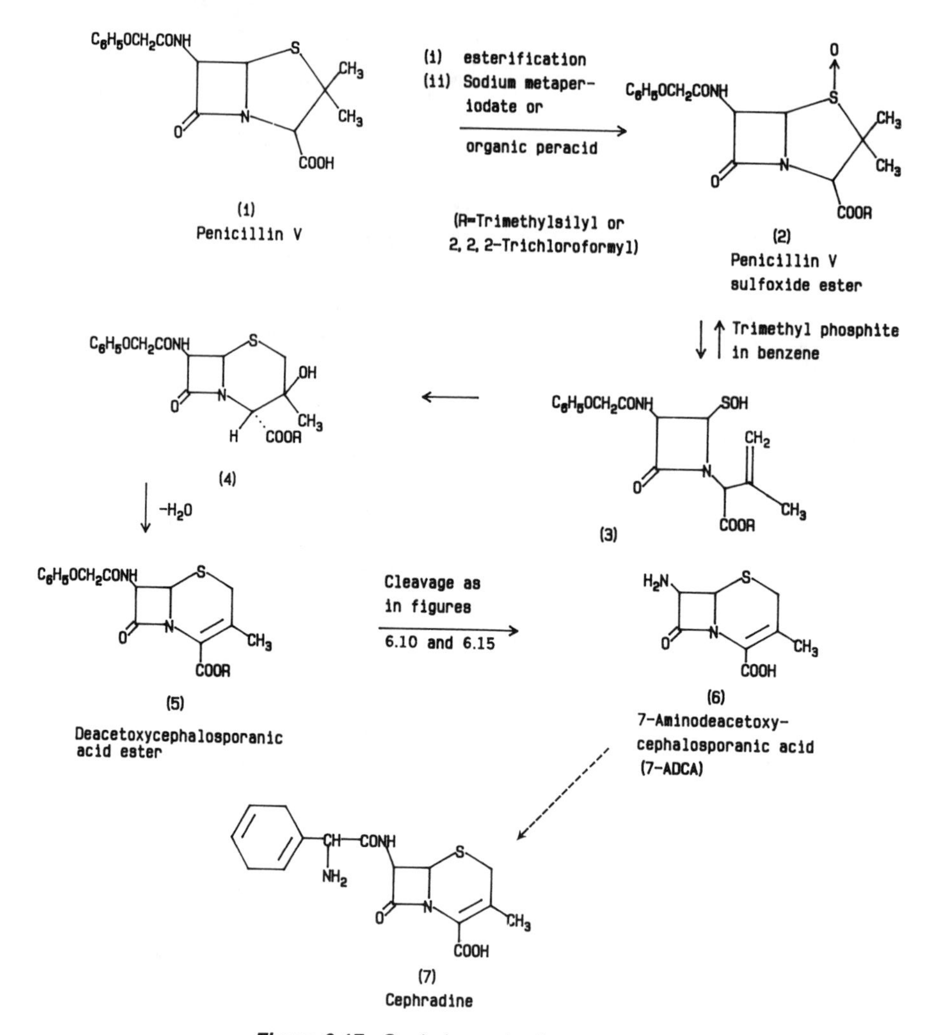

Figure 6.17. *Cephalosporins from penicillins.*

6.4.2 7-Oximinoacylcephalosporins

The 7-oximinoacylcephalosporins were found to be relatively stable to β-lactamases. The first commercial example was cefuroxime (structure 10, Figure 6.14). In addition to the 7-oximinoacyl group, it has a 3-carbamoyloxymethyl group which is stable to mammalian esterases. It has a broader spectrum of activity than cephaloridine but still lacks activity against the more difficult pathogens such as *Pseudomonas*.

The stereochemistry of the oximes turned out to be important, the syn configuration being more active than the anti isomer. An ingenious way of obtaining the syn side chain was based on the reaction of furanglyoxylic acid with hydroxylamine at pH 9, room temperature, and in the presence of magnesium hydroxide. About 90% syn isomer results. Alkylation gives the side chain precursor for cefuroxime.

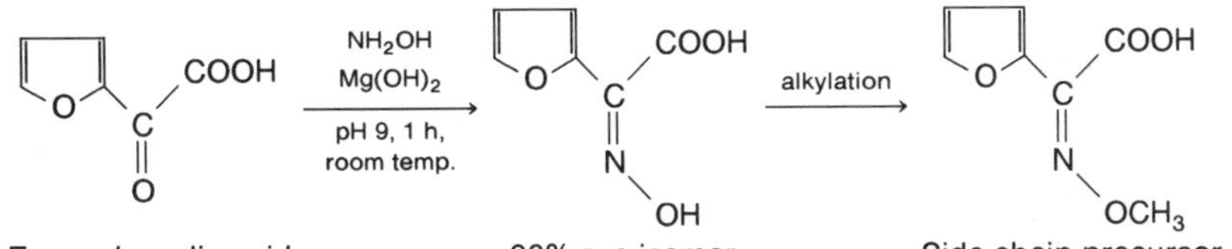

NH_2OH, $Mg(OH)_2$; pH 9, 1 h, room temp. → ; alkylation →

Furanglyoxylic acid — 90% syn isomer — Side chain precursor

Cefotaxime (structure 11, Figure 6.14) was an early third-generation cephalosporin. It possesses the syn oxime grouping of cefuroxime but has an aminothiazole ring attached to it. It has resistance to β-lactamases but the 3-acetoxymethyl group (as in cephalothin) is susceptible to mammalian esterases. Several aminothiazole cephalosporins with metabolically stable 3-substituents have since been developed, for example, ceftizoxime (12). They all have similar antibacterial activity.

It was then found that the methyl substitutent on the oxime could be replaced by other substituents and that, if this replacement was combined with a quaternary ammonium substituent in the 3-position, antibiotics emerged with high gram-negative activity. Ceftazidime (13, Figure 6.14) is the most important compound in the group. It is synthesized as shown in Figure 6.18. The starting material is cephaloridine (1) (also 3, Figure 6.14), obtainable in high purity from the action of pyridine on cephalothin. It is deacylated in the 7-position by a route similar to that used for penicillin in Figure 6.10. The first two stages (treatment with trimethylsilyl chloride then phosphorus pentachloride) are the same, then 1,4-butanediol is substituted for methanol and aqueous hydrochloric acid for water in stages 3 and 4. 7-Amino-3-pyridinium-methylcephalosporanic acid (2) results. It is an interesting point that quaternary nitrogen compounds such as dodecylbenzyldimethylammonium chloride (benzalkonium chloride) and certain pyridinium compounds have been used as germicides for decades.

The side chain precursor is made from acetoacetic ester (3). Treatment with sodium nitrite in acetic acid gives an oxime, and sulfuryl chloride then halogenates the methyl group to give the haloketone (4). To minimize formation of the unwanted anti isomer, this is cyclized with thiourea in the presence of an acid-binding agent such as triethylamine or propene oxide to give (5). Treatment with triphenylmethyl (trityl) chloride protects the amine group as in (6). The oxime group is protected by a group derived from the *tert*-butyl ester of 2-bromoisobutyric acid to give (7). Selective saponification gives the side chain precursor (8).

The acid (8) is coupled with the amine (2) with the aid of phosphorus oxychloride in dimethylformamide to give a ceftazidime precursor from which the trityl and *tert*-butyl groups are removed by hydrogen chloride in formic acid to leave ceftazidime (9).

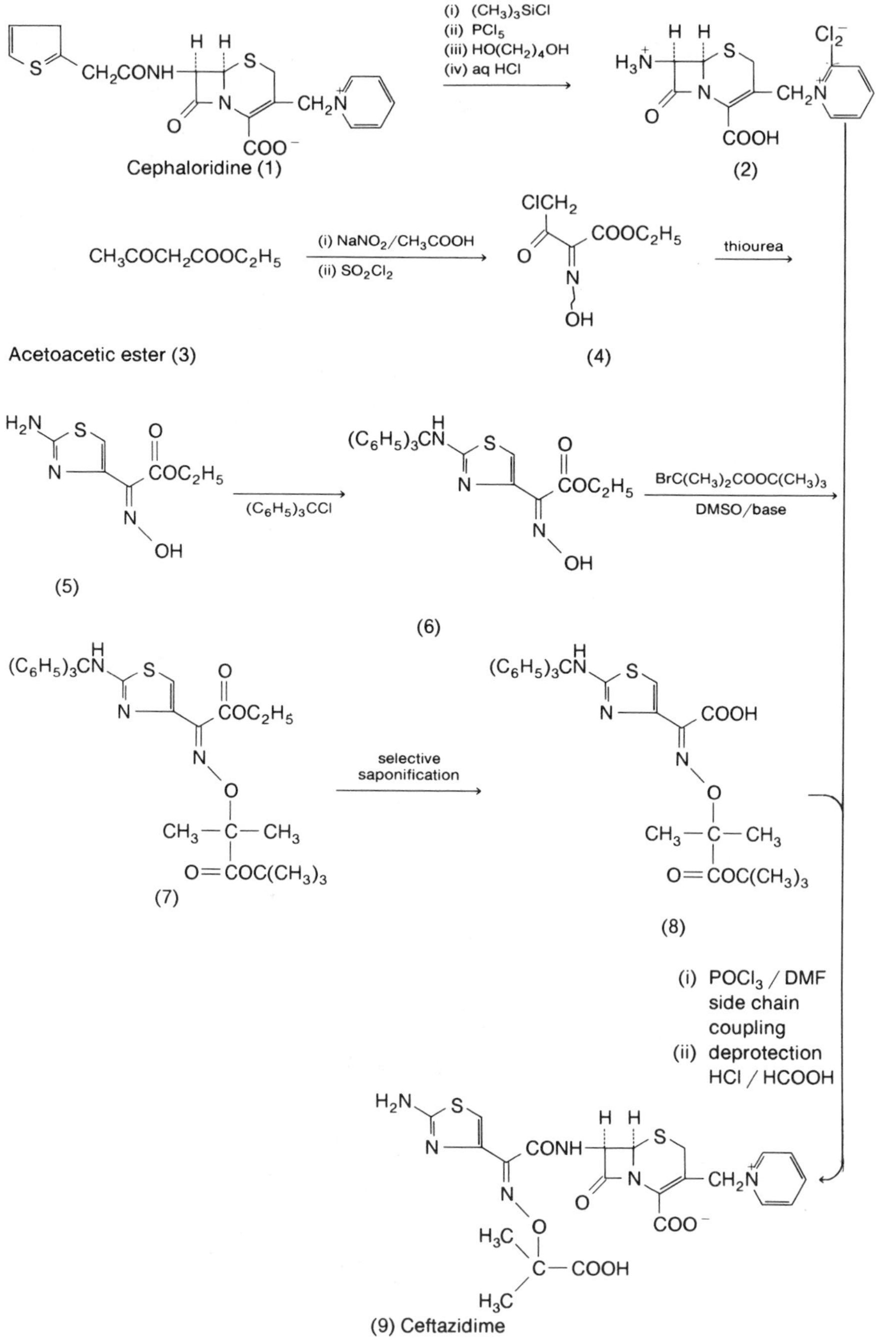

S CH_2CONH H H S N O $CH_2\overset{+}{N}$ COO^-

Cephaloridine (1)

(i) $(CH_3)_3SiCl$
(ii) PCl_5
(iii) $HO(CH_2)_4OH$
(iv) aq HCl

$H_3\overset{+}{N}$ H H S N O $CH_2\overset{+}{N}$ Cl_2^- COOH

(2)

$CH_3COCH_2COOC_2H_5$

(i) $NaNO_2/CH_3COOH$
(ii) SO_2Cl_2

$ClCH_2$ O $COOC_2H_5$ N OH

thiourea

Acetoacetic ester (3)

(4)

H_2N S N O COC_2H_5 N OH

(5)

$(C_6H_5)_3CCl$

$(C_6H_5)_3C\overset{H}{N}$ S N O COC_2H_5 N OH

$BrC(CH_3)_2COOC(CH_3)_3$
DMSO/base

(6)

$(C_6H_5)_3C\overset{H}{N}$ S N O COC_2H_5 N O CH_3-C-CH_3 $O=COC(CH_3)_3$

(7)

selective saponification

$(C_6H_5)_3C\overset{H}{N}$ S N COOH N O CH_3-C-CH_3 $O=COC(CH_3)_3$

(8)

(i) $POCl_3$ / DMF side chain coupling
(ii) deprotection HCl / HCOOH

H_2N S N CONH H H S N O $CH_2\overset{+}{N}$ COO^- N O H_3C C—COOH H_3C

(9) Ceftazidime

Figure 6.18. *Synthesis of ceftazidime.*

6.4.3 Cephamycin C Derivatives

The second solution to the β-lactamase problem depended on cephamycin C, a metabolite from a strain of *Streptomyces clavuligerus*:

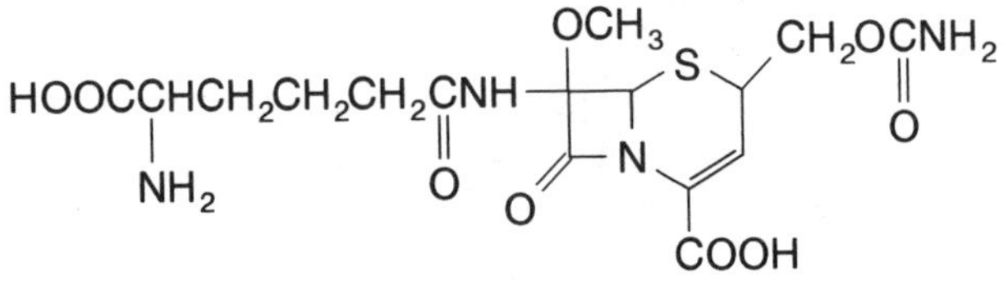

Cephamycin C

It has the same ring system and 7-side chain as cephalosporin C but has a 3-carbamoyloxymethyl substituent. More significantly, it has a second substituent—a methoxy group—in the 7-position. The steric hindrance from this group apparently confers resistance to β-lactamases. This is analogous to the shielding of the amide group of penicillin by halogen atoms (Section 6.3.3) to inhibit attack by β-lactamases. The first cephamycin antibiotic was cefoxitin:

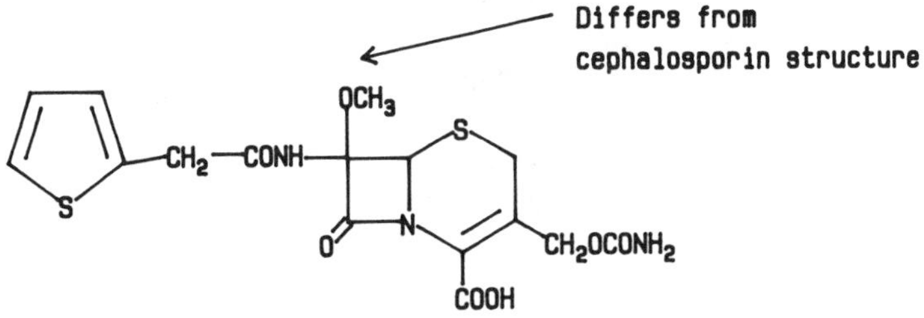

Differs from cephalosporin structure

Cefoxitin

It is moderately active (second generation) against gram-negative bacteria, especially *Bacteroides fragilis*, and is recommended for cases of abdominal sepsis such as peritonitis.

Moxalactam has a 7-methoxy group like cefoxitin and also has the striking feature that the hetero atom in the six-membered ring is not sulfur but oxygen:

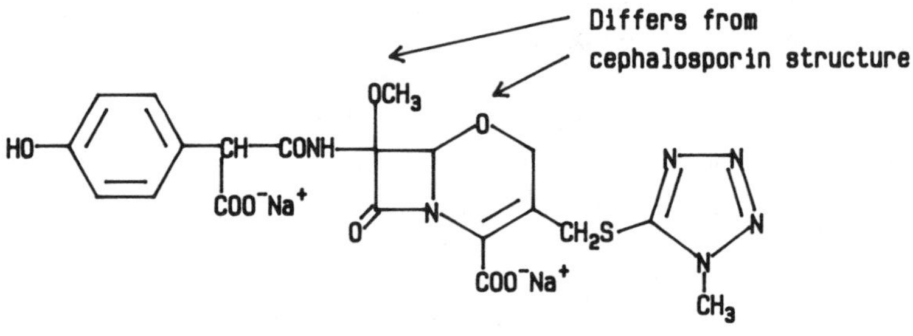

Differs from cephalosporin structure

Moxalactam

In this it resembles clavulanic acid (Section 6.3.7), where oxygen replaced sulfur in the thiazolidone ring of penicillin. This structural feature gives moxalactam enhanced resistance to β-lactamases. It is highly active against gram-negative bacteria. Like all third generation cephalosporins, it must be injected but a single injection per day is sufficient. Unfortunately it has recently been shown to cause prolonged bleeding time in some patients. It is therefore administered with vitamin K to reduce the problem but sometimes the effect is not reversible. Hence use of moxalactam is declining.

6.4.4 Cephalosporins in Hospitals

Cephalosporins often need to be injected and they make up at least half of all injectable antibiotics consumed. Because of this, they are mainly used in hospitals and do not figure prominently in the Top-100. On the other hand, as indicated in Table 6.2, they head the world market for antibiotic sales and are financially the most important hospital drugs. Here is a case where our classification on the basis of numbers of prescriptions is misleading.

It is expensive to stay in a hospital in the United States and any drug that will limit such a stay is welcomed. Antibiotics are relatively cheap to develop because the patient takes only a single course of them, and long-term toxicity tests are unnecessary. Hence there is a steady stream of novel cephalosporin antibiotics, each of which is said to have that slightly broader spectrum that will get the average patient out of hospital more quickly. The drugs are expensive in comparison with other drugs, but not when compared with hospital fees. This aspect of marketing was demonstrated recently by Pfizer's "Cefobid guarantee." Medicare's Diagnosis Related Group (DRG) Prospective Payment System covers a certain length of hospital stay. Pfizer will reimburse the patient for any amount of Cefobid (cefoperazone 13, Figure 6.14) required for treatment beyond the length of stay covered by DRGs.

6.5 TETRACYCLINES

The tetracycline antibiotics, some of which are shown in Figure 6.19, have conventionally been regarded as second in importance only to the β-lactams, but their position has weakened recently. The Top-100 contains only one of them, tetracycline at #20, although doxycycline comes at #103 and minocycline at #115. Tetracyclines are fermentation products like penicillins, and the first two to be discovered were chlortetracycline produced by *Streptomyces rimosus* and introduced in 1948 and oxytetracycline produced by *Streptomyces aureofaciens* and introduced in 1950. The parent compound, tetracycline itself, was not discovered until 1952 and its introduction made chlortetracycline obsolete. A new family of tetracyclines was developed in the late 1950s, characterized by the absence of the methyl group from position R_2 in Figure 6.19. The first member of the group to be introduced clinically was demethyl-

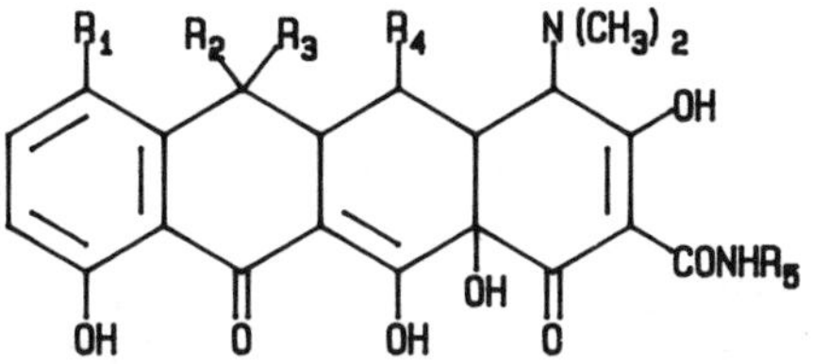

	R_1	R_2	R_3	R_4	R_5
Tetracycline (#20)	H	CH_3	OH	H	H
Oxytetracycline	H	CH_3	OH	OH	H
Chlortetracycline	Cl	CH_3	OH	H	H
Rolitetracycline	H	CH_3	OH	H	$-CH_2N$ 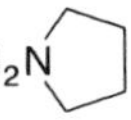 (pyrrolidine ring)
Methacycline	H	$=CH_2$		OH	H
Doxycycline (#103)	H	CH_3	H	OH	H
Demeclocycline	Cl	H	OH	H	H
Sancycline	H	H	H	H	H
Lymecycline	H	CH_3	OH	H	$-CH_2NHCH(CH_2)_4NH_2$ with COOH on the CH
Clomocycline	Cl	CH_3	OH	H	$-CH_2OH$
Minocycline (#115)	$-N(CH_3)_2$	H	H	H	H

Figure 6.19. *Tetracyclines.*

chlortetracycline, known as demeclocycline, and it was produced by a mutant strain of the mold which had given chlortetracycline.

Tetracyclines had the broadest spectrum of antibiotic activity to have been discovered and, consequently, were widely used. This usefulness has diminished because of increasing bacterial resistance and they have slipped in rank.

The tetracyclines may be either bacteriostatic or bactericidal. They are active against *Hemophilus influenzae* and thus are useful in exacerbations of chronic bronchitis. They are the only drugs effective against rickettsial infections, a group of rare but usually fatal diseases including Rocky Mountain spotted fever and certain forms of typhus. They are remarkably safe but like many other antibiotics can produce diarrhea. They deposit in growing teeth and in bones. Consequently, if children use them regularly, they may develop yellow-stained teeth. Tetracyclines can also cross the placenta and affect the teeth of a fetus, so they are not given to pregnant women. Only minocycline and doxycycline should be given to patients with kidney disease.

Like penicillins, tetracyclines may produce allergic reactions and even anaphylactic shock (Section 6.3.2), a condition that can prove fatal. They also have a profound effect on normal bacterial populations and the elimination of these may permit so-called superinfection, a condition where the disappearance of harmless bacteria opens the door to infection by pathogenic ones.

All tetracyclines except minocycline have a similar spectrum of activity and they differ only in marginal properties related to stability, absorption, and excretion rates. The converse of this is that the tetracycline molecule can be modified dramatically by alteration of the various groups attached to the rings and the antibiotic power is retained.

Tetracyclines are believed to inhibit protein synthesis by sensitive bacteria. They do this by interfering with biochemical reactions occurring in the ribosomes. A few strains of resistant bacteria have emerged, but unfortunately a strain resistant to one tetracycline is resistant to all of them; that is, absolute cross-resistance exists among all members of the group.

6.5.1 Manufacture of Tetracyclines

Tetracycline is obtained from chlortetracycline by catalytic hydrogenolysis, which removes the aromatic chlorine atom:

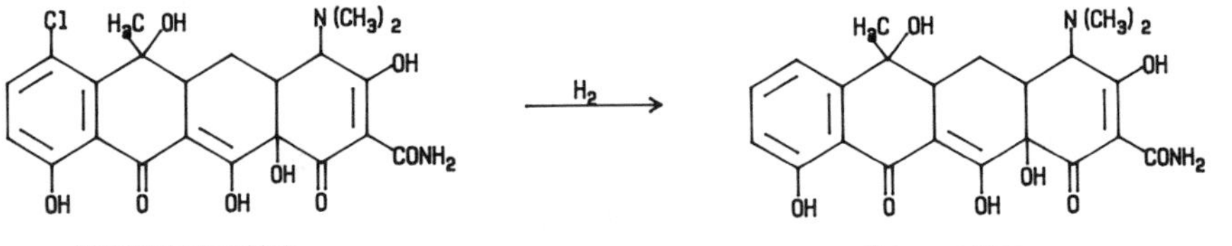

Chlortetracycline $\xrightarrow{H_2}$ Tetracycline

The same process applied to demethylchlortetracycline removes both the chlorine atom at R_1 and the labile benzylic hydroxyl at R_3 to give 6-demethyl-6-deoxytetracycline:

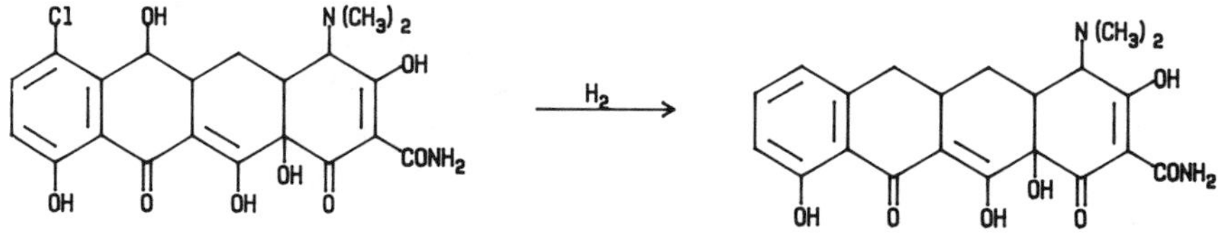

Demethylchlortetracycline (Demeclocycline) $\xrightarrow{H_2}$ Sancycline (6-Demethyl-6-deoxytetracycline)

This is known as sancycline and is the simplest tetracycline with full antibiotic properties. It has not been used clinically, but, like 6-APA, is a valuable synthon. The problem with the above tetracyclines is that they are all liable to dehydration and epimerization, and this deactivates them. Indeed, the breakdown products are toxic and tetracycline containers are always marked with an expiration date after which they should not be used. Sancycline, on the other hand, is stable even in concentrated sulfuric acid and can thus be used as the raw material for a range of semisynthetic tetracyclines.

Minocycline is a typical semisynthetic tetracycline based on sancycline. Nitration of sancycline (1) with potassium nitrate and sulfuric acid yields a mixture of the isomers (2) and (3). The two isomers must be separated, for only the *para*-nitro compound is useful. It is reductively methylated with hydrogen, formaldehyde, and a hydrogenation catalyst to give minocycline (4). Minocycline, demeclocycline, and sancycline, shown in Figure 6.19, are all examples of tetracyclines that have been demethylated in the R_2 position. Minocycline has a wider spectrum of activity than the other tetracyclines.

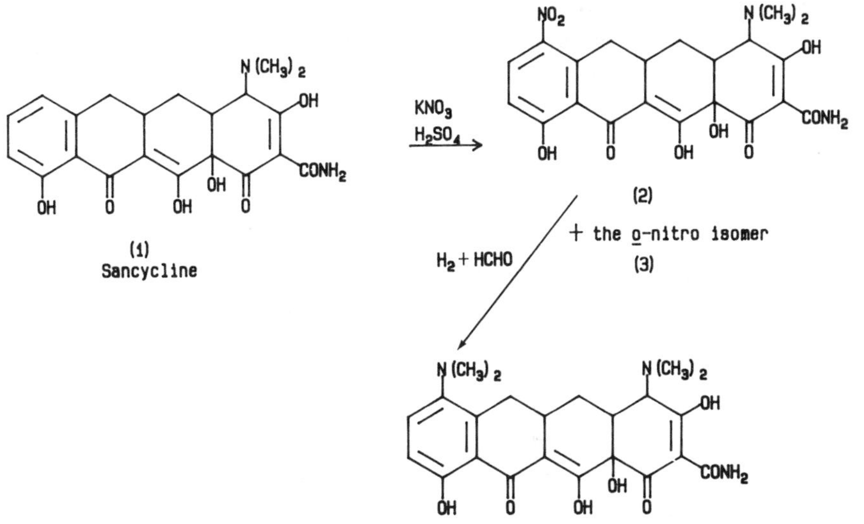

(1)
Sancycline

KNO_3, H_2SO_4 →

(2)
+ the o-nitro isomer
(3)

$H_2 + HCHO$

Minocycline (4)
(7-Dimethylamino-6-demethyl-6-deoxytetracycline)

Doxycycline is one of the newer tetracyclines and is prepared from oxytetracycline by catalytic hydrogenolysis of the R_3 hydroxyl:

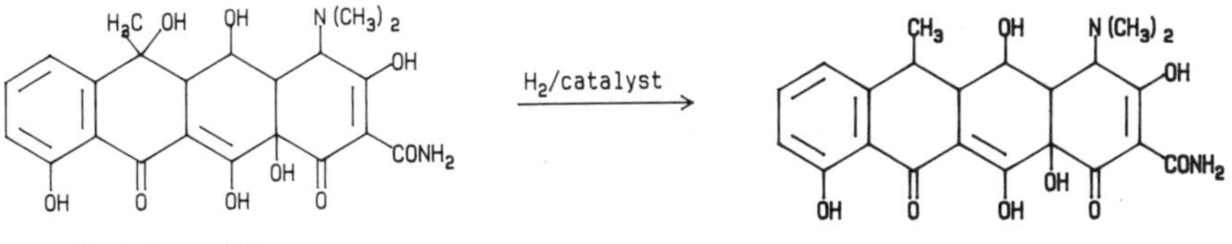

Oxytetracycline →(H_2/catalyst) Doxycycline

The loss of a hydroxyl group makes the compound more lipophilic and high blood levels can be maintained with only a single dose per day. Nevertheless, in severe infections, several doses per day are required as with other antibiotics. Doxycycline is well absorbed even when taken with food and this is an advantage. Other tetracyclines, apart from minocycline, chelate with calcium, iron, and magnesium salts and interact with milk, alkalis, and antacids such as aluminum hydroxide. This reduces their rate of absorption to a point where they are of no therapeutic value.

6.5.2 Rolitetracycline

Tetracyclines are not soluble enough to be injected and rolitetracycline (Figure 6.19) is a pro-drug designed to overcome this. It is made from tetracycline, formaldehyde, and pyrrolidine by an unusual Mannich reaction:

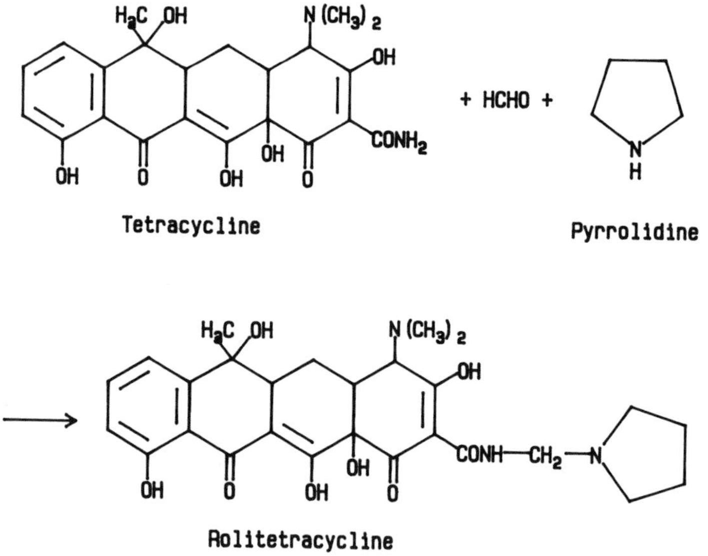

Tetracycline + HCHO + Pyrrolidine → Rolitetracycline

Amides are not normally nucleophilic enough to enter into the Mannich reaction but in this case the amide is made so by the ring system to which it is attached. Rolitetracycline has enhanced water solubility and hydrolyzes to tetracycline at the pH of the blood.

6.6 MACROLIDE ANTIBIOTICS

The macrolide antibiotics are represented in the Top-100 by erythromycin at #5. The structure of erythromycin A, the most important compound, is

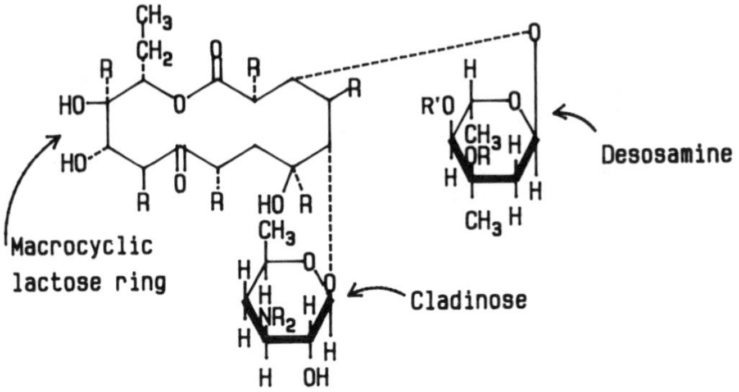

$R = CH_3$ **Erythromycin A (R' = H)**

Erythromycin ethylsuccinate (R' = $-COCH_2CH_2COOCH_2CH_3$)

The group derives its name from a macrocyclic lactone ring containing 12–16 carbon atoms to which one or more sugar units are attached. Here it is attached to units of the sugars L-cladinose and D-desosamine. Erythromycin is proliferated by *Streptomyces erythreus*. It is active against gram-positive bacteria and fungal infections, and is widely prescribed especially against penicillin-resistant staphylococci. It is useful for patients who are allergic to penicillin, and particularly if they are children whose teeth might be discolored by tetracyclines.

Erythromycin functions by inhibiting protein synthesis in microorganisms by bonding to ribosomes. This is similar to the tetracycline and streptomycin mechanisms (see below). Bacteria are said to develop resistance to it fairly readily.

Erythromycin is poorly absorbed in the body, is unstable to the acid in the stomach, and has an unpleasant taste. It is usually used as a more palatable derivative. It can be given in enteric coated tablets, but dosage with an ester is an alternative and the most widely prescribed ester is the ethylsuccinate. This is the ester of erythromycin with $HOOCCH_2CH_2COOC_2H_5$.

Most of the macrolide antibiotics are bacteriostatic but it appears that high doses of erythromycin may be bactericidal to certain organisms. Adverse reactions are rare and take the form of abdominal pains and skin rashes.

6.7 AMINOGLYCOSIDES AND ANTITUBERCULOSIS DRUGS

The aminoglycosides are a group of broad-spectrum antibiotics. Streptomycin is typical. It is characterized by a sugar or aminosugar unit bound by a glycoside linkage to a cyclitol unit, which is also a sugar. This portion of the molecule is called streptobiosamine. It is bound to an aglycone, streptidine, which is an aminosugar containing two guanidine groups:

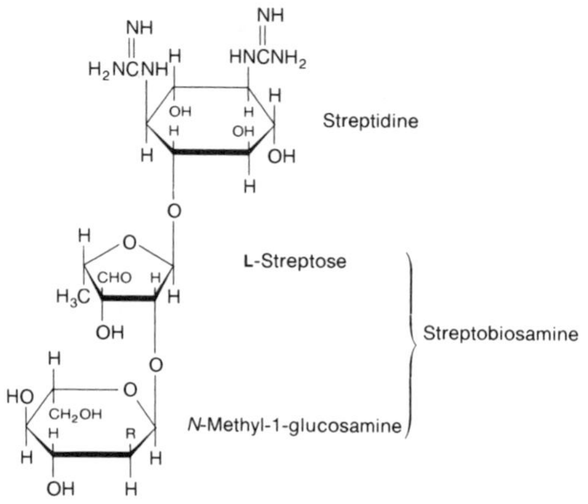

Streptidine

L-Streptose

N-Methyl-1-glucosamine

Streptobiosamine

Streptomycin (R = CH_3NH)

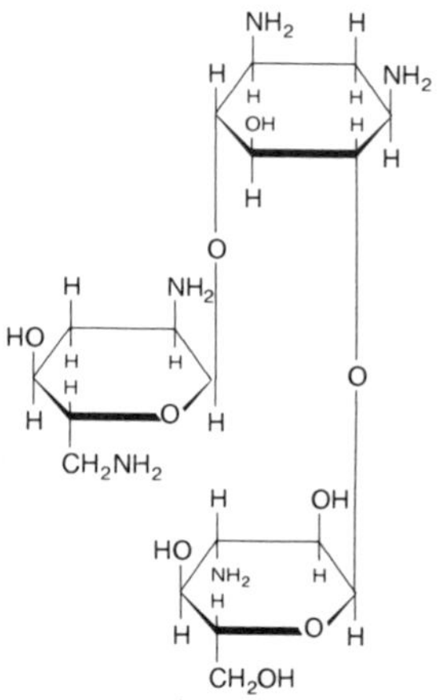

Tobramycin

The most widely used aminoglycoside in the developed world is tobramycin, an injectable antibiotic active against *Staphylococcus aureus* and *Pseudomonas aeruginosa*. The aminoglycoside family also includes neomycins, gentamycins, and kanamycins.

Streptomycin was the first of the aminoglycoside antibiotics. It was discovered by Abraham Waksman, a Russian immigrant to the United States. He was interested in microorganisms living in the soil and his study took a sudden new direction in 1939 when Rene Dubos, one of his students, discovered a bactericidal agent produced by a soil microorganism. This discovery gave an impetus to the work on penicillin and it was Waksman who coined the term antibiotic for such materials. In 1940 he isolated actinomycin, a peptide antibiotic that we shall mention later, and in 1943 he obtained streptomycin from the species *Streptomyces griseus*. It was active against the difficult gram-negative bacteria that were not affected by penicillin and was first used clinically in 1945.

Why did Waksman and so many subsequent workers look in the soil for antibiotics? The reason is that pathogenic bacteria are continually being discharged into the soil, yet these are seldom detected when soil is cultured. It was logically argued, therefore, that the soil contains natural products that inhibit the growth of bacteria.

When discovered, streptomycin was hailed as the broadest spectrum and least toxic antibiotic known. It acts on the ribosome concerned with protein biosynthesis in the bacterium and, hence, inhibits the polymerization of amino acids to give protein. It does not have the same effect on the mammalian enzyme. Unfortunately, bacteria rapidly develop resistance to streptomycin. In some patients it damages the eighth cranial nerve, the inner ear, leading to deafness and loss of balance. Accordingly, streptomycin is now used only for treatment of tuberculosis and this gives us a chance to discuss tuberculosis therapy.

6.7.1 Tuberculosis Therapy

Because of improved living conditions in the Western world and Japan, tuberculosis is no longer the dread killer disease it once was, and it persists only among the aged, who suffered poor nutrition as youngsters, and the poor. In third world countries, however, it is still a threat and is particularly prevalent in overcrowded areas such as the slums of Hong Kong and areas of rural poverty in China, Africa, India, and the Amazon jungle.

Because of the emergence of resistant strains of bacteria and the necessity for long-term therapy, tuberculosis is treated with several antibiotics simultaneously. This is known as concurrent therapy and is not possible with all antibiotics. For example, tetracyclines and chloramphenicol antagonize penicillin activity and must not be used with it. The idea of concurrent therapy is that if one bacterium in 10^9 is resistant to one antibiotic, only one in 10^{27} will be resistant to three.

Drugs against tuberculosis are shown in Figure 6.20. In the past the preferred combination was streptomycin, isoniazid, and *p*-aminosalicylic acid as its sodium salt. The simple structures of the last two compounds are noteworthy. They are not used more widely against infections because they are scarcely active by themselves, but they prevent the development of resistant tuberculosis strains. The direction of therapy changed with the introduction of rifampin; the drugs of first choice are now rifampin, streptomycin, isoniazid, and ethambutol (Figure 6.20). Streptomycin and rifampin are both bacteriostatic at low concentrations and bactericidal at concentrations not much higher.

Drugs of second choice are pyrizinamide, kanamycin, ethionamide, and *p*-aminosalicylic acid, shown in the lower section of Figure 6.20. The most important is rifampin because its introduction in 1967 revolutionized antituberculosis therapy.

The synthesis of rifampin is shown in Figure 6.21. It starts with rifamycin B, a product of the microorganism *Streptomyces mediterranei*. Five rifamycins were in fact isolated and rifamycin B was the most stable. Initial studies showed it was active against tuberculosis but it had to be injected, which

1. *Preferred combination in the past:*

CONHNH$_2$ COOH OH

Streptomycin + N + NH$_2$

Isoniazid p-Aminosalicylic acid

2. *Drugs of first choice:*

CH_3 CH_3 CH_3 CH_3 CH_3O O OH OH OC CH_3 CO CH_3 OH OH CH_3 NH O CH=N–N N–CH_3 O OH CH_3 O

Rifampin

+ Streptomycin

+ Isoniazid

CH_2OH CH_2OH

$CH_3CH_2CHNHCH_2CH_2NHCHCH_2CH_3$

+ Ethambutol

3. *Drugs of second choice:*

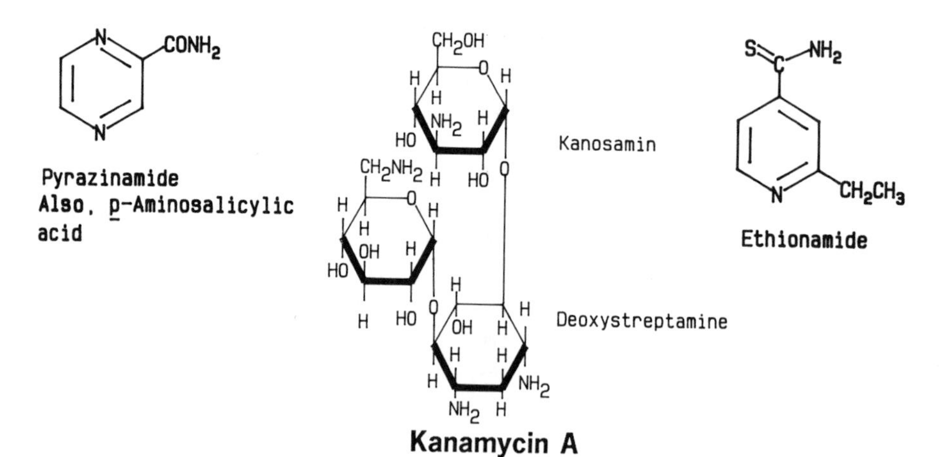

Figure 6.20. *Drugs against tuberculosis.*

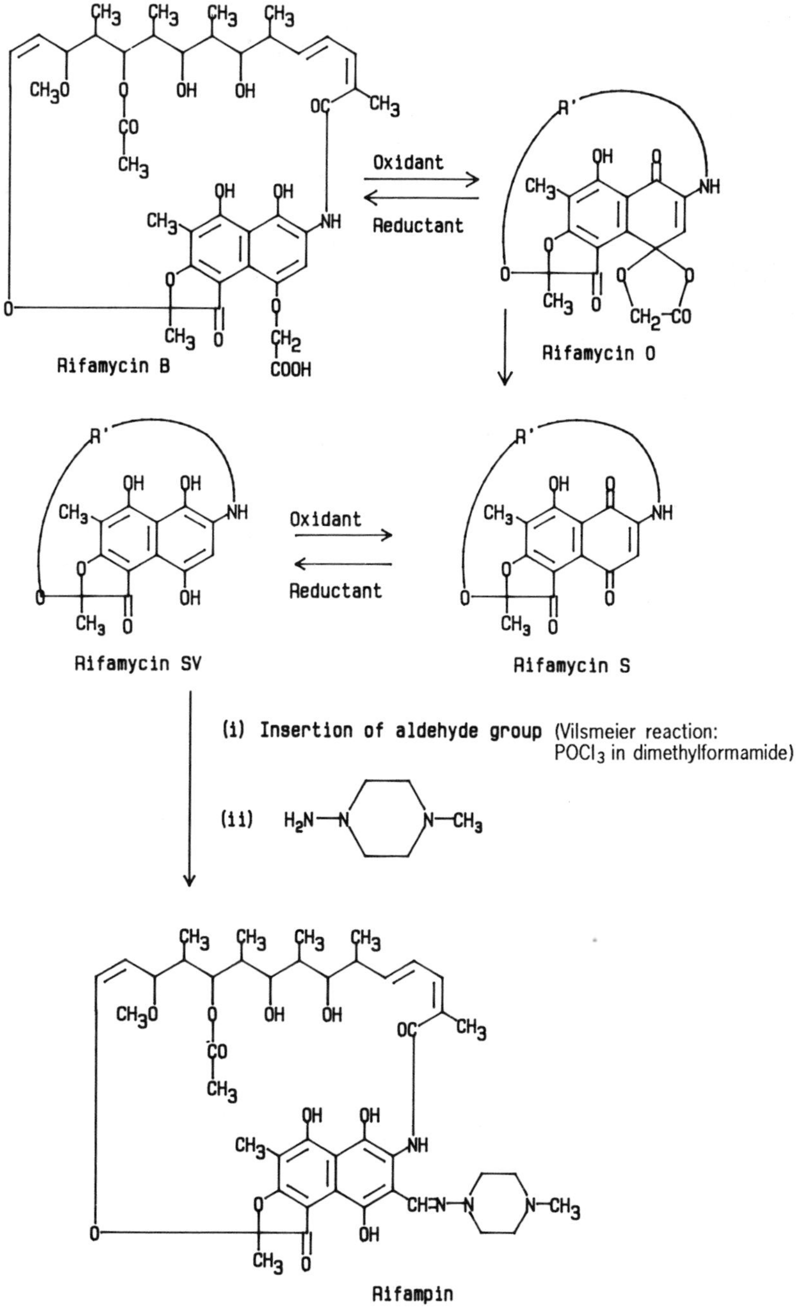

Rifamycin B ⇌ (Oxidant / Reductant) Rifamycin O

↓

Rifamycin SV ⇌ (Oxidant / Reductant) Rifamycin S

(i) Insertion of aldehyde group (Vilsmeier reaction: $POCl_3$ in dimethylformamide)

(ii) H_2N–N(piperazine)N–CH_3

↓

Rifampin

Figure 6.21. *Synthesis of rifampin.*

diminished its value. Rifamycin B is a strongly acid material and is easily isolated from the culture medium in which it is formed. Oxidation, as shown in the same figure, leads to rifamycin O. The reaction is carried out rapidly and takes place in aqueous solution. If the solution is allowed to stand for three days, rifamycin S results. This has a high antibiotic activity but is also more toxic than its precursors. Reduction of rifamycin S gives rifamycin SV. Insertion of an aldehyde group in the 3-position of rifamycin SV followed by formation of a Schiff's base with 1-amino-4-methylpiperazine yields rifampin.

Rifampin may be taken orally. In addition to its antituberculosis activity, it is also useful against leprosy and is active against some viruses. The basis for its action is that it inhibits an RNA polymerase and for the same reason is used in molecular biology.

Neither streptomycin nor any of the other anti-tuberculosis drugs appears in the Top-100 because of the low incidence of tuberculosis in the United States.

6.8 PEPTIDE ANTIBIOTICS

Dubos' discovery of tyrothricin in 1939 and Waksman's discovery of actinomycin in 1940 were mentioned above. These were found to be mixtures of polypeptides, their molecules consisting, like proteins, of chains of amino acids, but the chains were relatively short. The structure of actinomycin D is

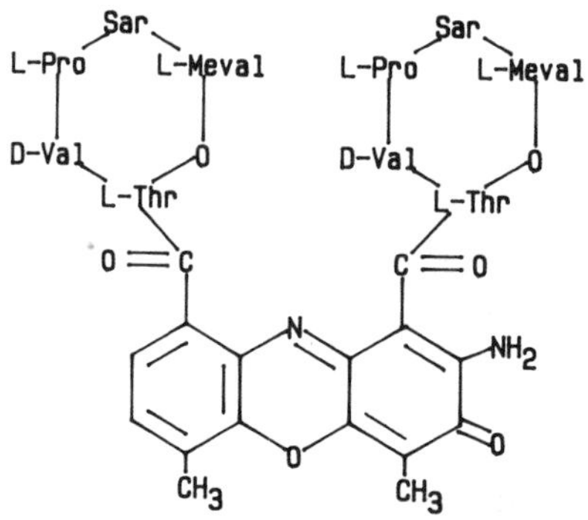

Sar
L-Pro L-Meval
D-Val O
L-Thr
O=C
Sar
L-Pro L-Meval
D-Val O
L-Thr
C=O
N
NH2
O
O
CH3 CH3

Actinomycin D

Sar = Sarcosine, Meval = Methylvaline, Pro = Proline,
Thr = Threonine, Val = Valine

Although over 200 peptide antibiotics have been described, structures have been proposed for only about 40 and proof of structure and synthesis achieved for about 10. They contain a range of unusual amino acids and have unique antibiotic properties. The drawback that has inhibited extensive work on them is that most of them are too toxic, especially to the kidneys, to be taken by injection or by mouth.

The peptide antibiotic in the Top-100 is polymyxin at #96:

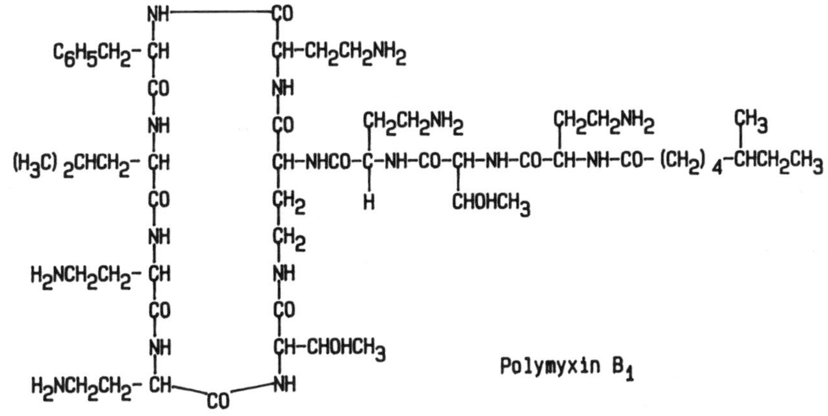

```
               NH-------------CO
               |              |
  C6H5CH2-     CH             CH-CH2CH2NH2
               |              |
               CO             NH
               |              |
               NH             CO        CH2CH2NH2            CH2CH2NH2             CH3
               |              |         |                    |                     |
(H3C)2CHCH2-   CH             CH-NHCO-C-NH-CO-CH-NH-CO-CH-NH-CO-(CH2)4-CHCH2CH3
               |              |         |       |
               CO             CH2       H       CHOHCH3
               |              |
               NH             CH2
               |              |
 H2NCH2CH2-    CH             NH
               |              |
               CO             CO
               |              |
               NH             CH-CHOHCH3
               |              |
 H2NCH2CH2-    CH------CO-----NH
```

Polymyxin B_1

The polymyxins are a complex group of closely related substances. They are proliferated by various strains of *Bacillus polymyxa*. Polymyxin B_1 is a cationic surface-active agent and has a molecular weight of about 1000. It is highly active against gram-negative bacteria, but is poorly absorbed from the gut and even with injections it is difficult to reach a satisfactory blood level without the production of side effects. It is mainly used as an ointment. Its mode of action is interesting. Its surface activity facilitates its reaction with both the polar and fatty portions of the phospholipid molecules that protect the cell wall of the microorganism. This makes it easier for lysozyme to attack the cell wall and destroy the bacterium.

Polypeptide antibiotics are of particular interest at present because they are anticancer agents for specific tumors such as the so-called Wilms' tumor and trophoplastic tumors. This activity seems unrelated to their antibiotic properties but could be a function of their surface activity.

6.9 QUINOLONE ANTIBACTERIALS

The first quinolone antibacterial to be discovered was nalidixic acid:

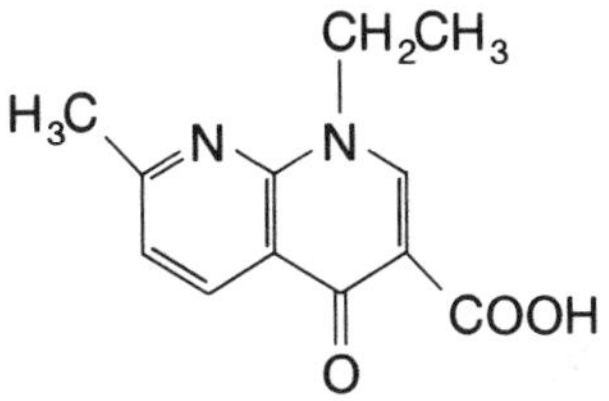

CH_2CH_3 H_3C N N COOH O

Nalidixic acid

It can be taken orally and may be used for treatment of uncomplicated lower urinary tract infections. Other antibiotics, however, have usually been preferred.

Most of the early quinolones caused gastrointestinal disturbance and other side effects. Since 1980, however, they have received renewed attention because of the development of the 6-fluoro analogues, which have fewer side effects, require smaller dosages, and have greater bioavailability and reduced development of bacterial resistance.

The new quinolones may be made by chemical synthesis and taken orally. They act by inhibiting DNA gyrase. This enzyme mediates the uncoiling of DNA during replication of circular bacterial chromosomes and the separation of the two circular chromosomes afterwards. Inhibition of this process is lethal for the bacterium. The new quinolones are expected to be licensed in the United States in the near future. The leading compounds are

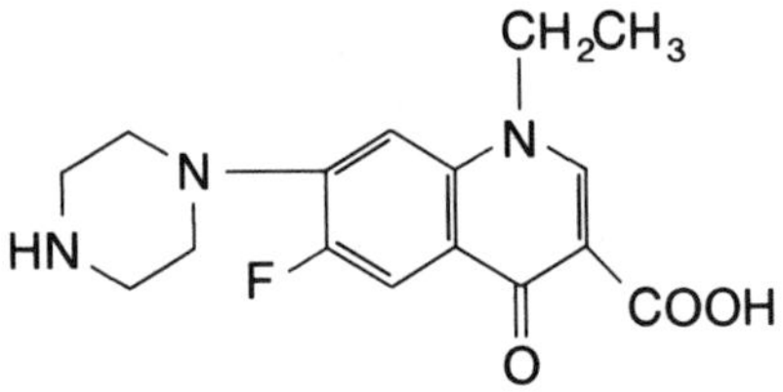

Norfloxacin

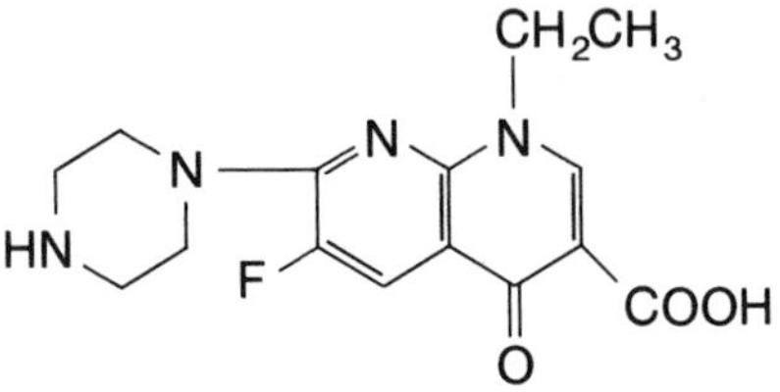

Enoxacin

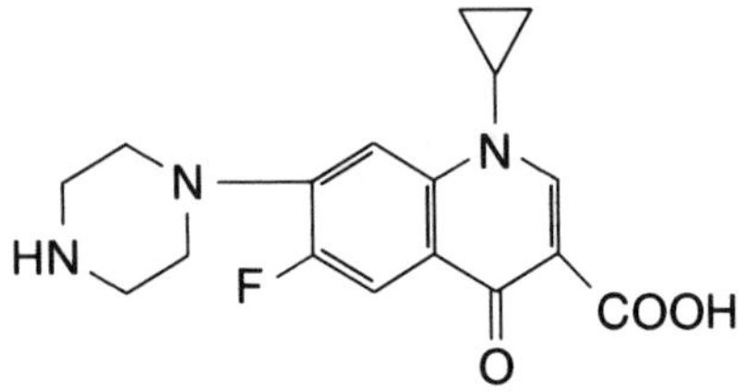

Ciprofloxacin

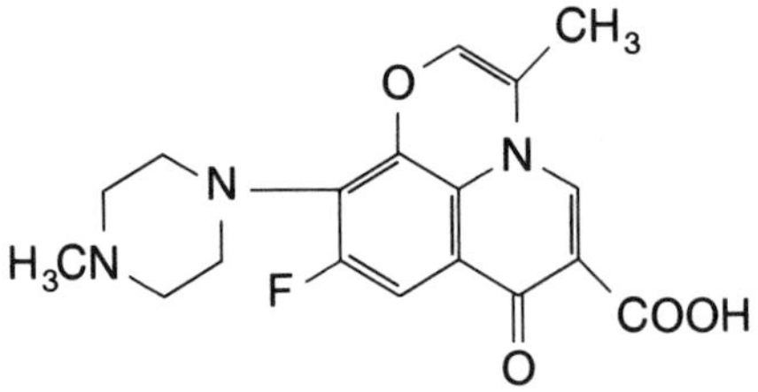

Ofloxacin

6.10 MISCELLANEOUS ANTIBACTERIALS

6.10.1 Chloramphenicol and Nitrofurantoin

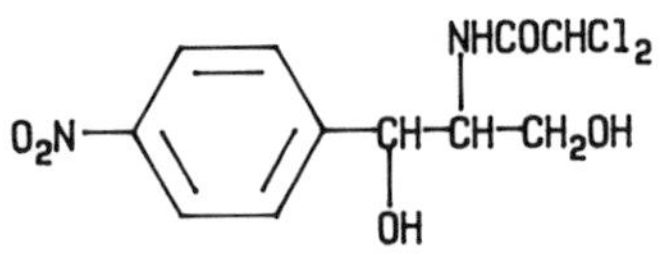

Chloramphenicol

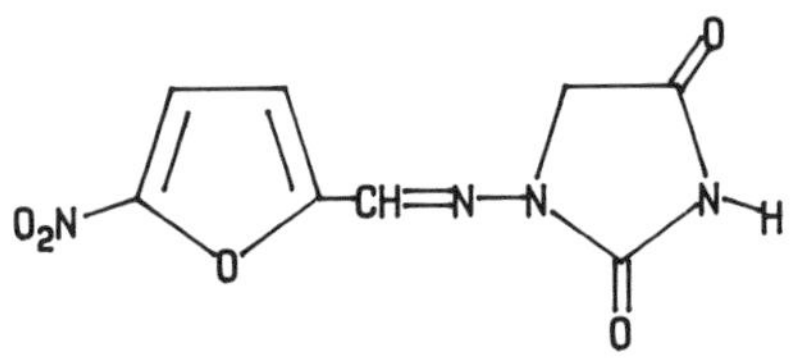

Nitrofurantoin (#62)

Chloramphenicol is of interest because for many years it was the only important antibiotic to be made by chemical synthesis and not at least partly by fermentation. It and nitrofurantoin are unusual in that they are among the few pharmaceuticals to contain a nitro group. Indeed chloramphenicol was the first such material to be discovered in nature. The reason for the scarcity of nitro groups in pharmaceuticals is that they have the potential for being reduced to nitrosamines in living systems. Nitrosamines are believed to cause cancer and are at the heart of the controversy about the use of nitrates and nitrites in food preservation.

Chloramphenicol was found in soil from Venezuela in 1947 and was the first broad-spectrum antibiotic to be isolated. It does not appear in the Top-100 but is important because it is the only effective agent against typhoid, paratyphoid, which is a milder form of typhoid, and some forms of meningitis. Its drawback is that in about one case in 50,000 it produces aplastic anemia, an irreversible and fatal disease of the bone marrow. In many developing countries, the occasional antibiotic-induced death is regarded as acceptable, and chloramphenicol is used for routine infections. In most developed countries, however, it is reserved for the infections mentioned.

The synthesis of chloramphenicol illustrates a common problem in the manufacture of natural products, the achievement of the proper stereochemistry. The first industrial route is shown at the top of Figure 6.22 and involved the condensation of benzaldehyde (1) with β-nitroethanol (2) to give 2-nitro-1-phenyl-1,3-propanediol (3). This unfortunately had an erythro:threo ratio of about 2 : 1, while the most active product is pure ($-$)-threo. The valueless erythro compound must be separated and discarded, which makes the process expensive. The nitro group of the threo compound is reduced to an amine (4) and the D,L mixture is resolved to give the desired D compound. The amine group is then protected either by acetylation or formylation, after which the aromatic ring is nitrated. Next the acyl group is removed to give (5) and the amine dichloroacetylated to give chloramphenicol (6).

Boehringer Mannheim in Germany is said to have developed a direct though tedious route, shown in the lower half of the diagram, which gives only the D,L-threo compound. In this route, *p*-nitroacetophenone (7) is brominated to give the α-bromoketone (8), which in turn is quaternized with hexamethylene tetramine to give (9). Alcoholic mineral acid hydrolysis gives the mineral acid salt of the aminoketone (10). The amine is then acetylated and the

1. *First industrial route:*

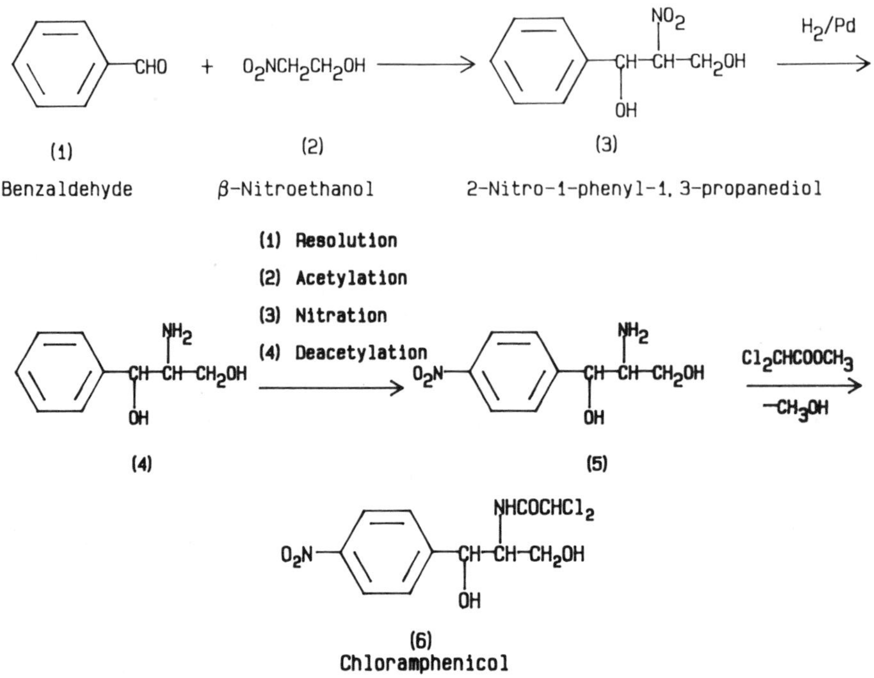

(1) Benzaldehyde + (2) β-Nitroethanol ($O_2NCH_2CH_2OH$) → (3) 2-Nitro-1-phenyl-1, 3-propanediol; H_2/Pd →

(4) → (1) Resolution (2) Acetylation (3) Nitration (4) Deacetylation → (5); $Cl_2CHCOOCH_3$, $-CH_3OH$ →

(6) Chloramphenicol

2. *Boehringer Mannheim approach:*

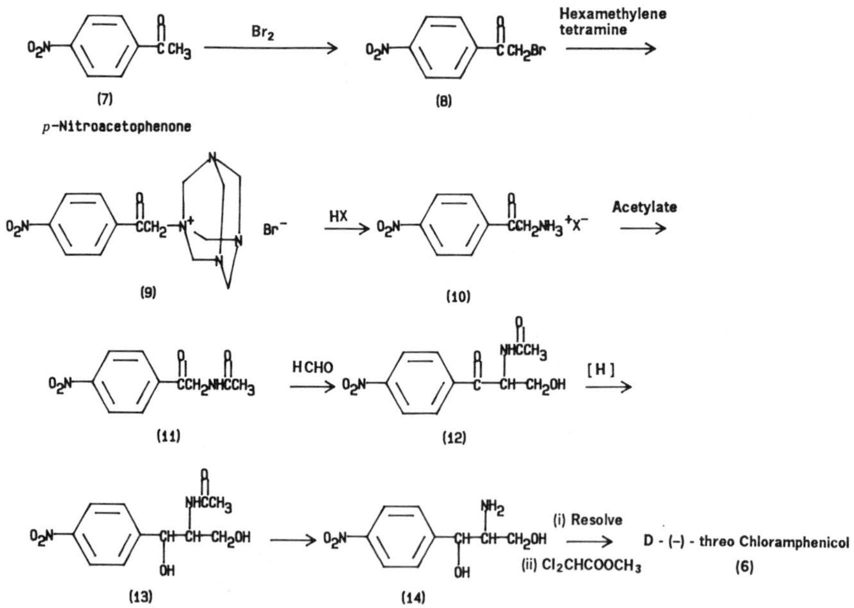

(7) *p*-Nitroacetophenone; Br_2 → (8); Hexamethylene tetramine →

(9) Br^-; HX → (10) $O_2N-C_6H_4-CCH_2NH_3^+X^-$; Acetylate →

(11) $O_2N-C_6H_4-CCH_2NHCCH_3$; HCHO → (12); [H] →

(13) → (14); (i) Resolve (ii) $Cl_2CHCOOCH_3$ → D - (-) - threo Chloramphenicol (6)

Figure 6.22. *Syntheses of chloramphenicol.*

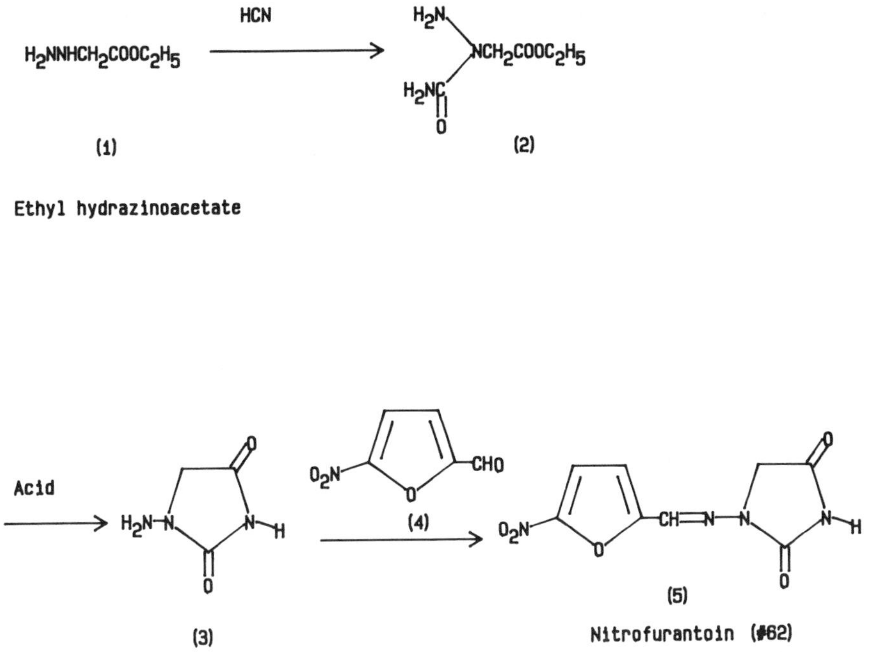

$H_2NNHCH_2COOC_2H_5$ (1) —HCN→ (2)

Ethyl hydrazinoacetate

—Acid→ (3) —(4)→ (5)

Nitrofurantoin (#62)

Figure 6.23. Synthesis of nitrofurantoin.

resulting compound (11) treated with formaldehyde to insert a methylol group. A Meerwein–Ponndorf reduction of the ketone (12) gives the acetylated amino alcohol (13) and deacetylation gives the D,L-threo racemate (14), which can be converted to chloramphenicol by the steps mentioned in the earlier route. The overall yield of the threo compound is said to be 41%.

Nitrofurantoin is an antibacterial, not an antibiotic, and is used to counter prolonged infections of the urinary tract. It is #86 in the Top-100. Its synthesis is shown in Figure 6.23. Ethyl hydrazinoacetate (1) is treated with hydrogen cyanide to give a carbamate (2). Cyclization with a strong acid yields an aminohydantoin (3). Compound (3) will condense with the synthon, 5-nitrofurfural (4), to give nitrofurantoin (5). Chapter 13, Volume 1, of Lednicer and Mitscher (see Bibliography) lists the various other processes in which compound 4 is involved.

6.10.2 Lincomycin and Clindamycin

Lincomycin and clindamycin are antibiotics used primarily in hospitals because of toxicity problems. Oral administration is possible but injection is usually preferred. They are active against the difficult *Staphylococcus aureus*.

Lincomycin is an octose sugar combined with an amino acid:

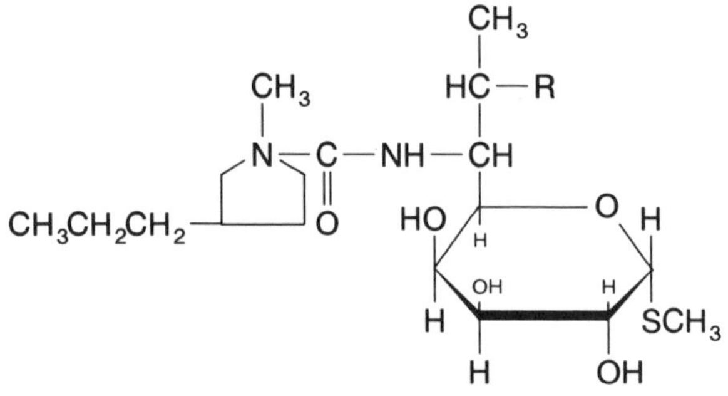

Lincomycin (R = OH)
Clindamycin (R = Cl)

It is proliferated by *Streptomyces lincolnensis*. Chemical replacement of the hydroxyl by chlorine gives clindamycin, a more potent antibiotic, which has generally replaced lincomycin.

6.10.3 Vancomycin

Vancomycin is another somewhat toxic antibiotic for hospital use and may be given orally or intravenously. It was widely used in the late 1950s against penicillin-resistant staphylococci but was replaced by the second and third generation β-lactams. It has re-emerged in the past few years because of the increase in methicillin-resistant staphylococci and recognition of certain dangers from *Clostridium difficile*. It is a glycopeptide proliferated by *Streptomyces orientalis*.

6.11 ANTIFUNGALS

There are many agents causing infection other than bacteria, notably viruses, fungi, insects, protozoa, and worms. Viruses will be discussed in Chapter 20 and the others in Chapter 18; both chapters deal with drugs outside the Top-100. Two antifungals are within the Top-100, however, so they will be considered here. They are miconazole at #61 and clotrimazole at #72. Both are imidazoles.

The more important is miconazole and it is used specifically against vaginal fungi. Its synthesis is shown in Figure 6.24. 2,4-Dichloroacetophenone (1) is brominated to give (2) and this is used to alkylate imidazole to give the *N*-substituted imidazole (3). The carbonyl group is reduced to a hydroxyl group with sodium borohydride. The resulting alcohol (4) is converted to an alkoxide and alkylated with 2,4-dichlorobenzyl chloride to give miconazole (5).

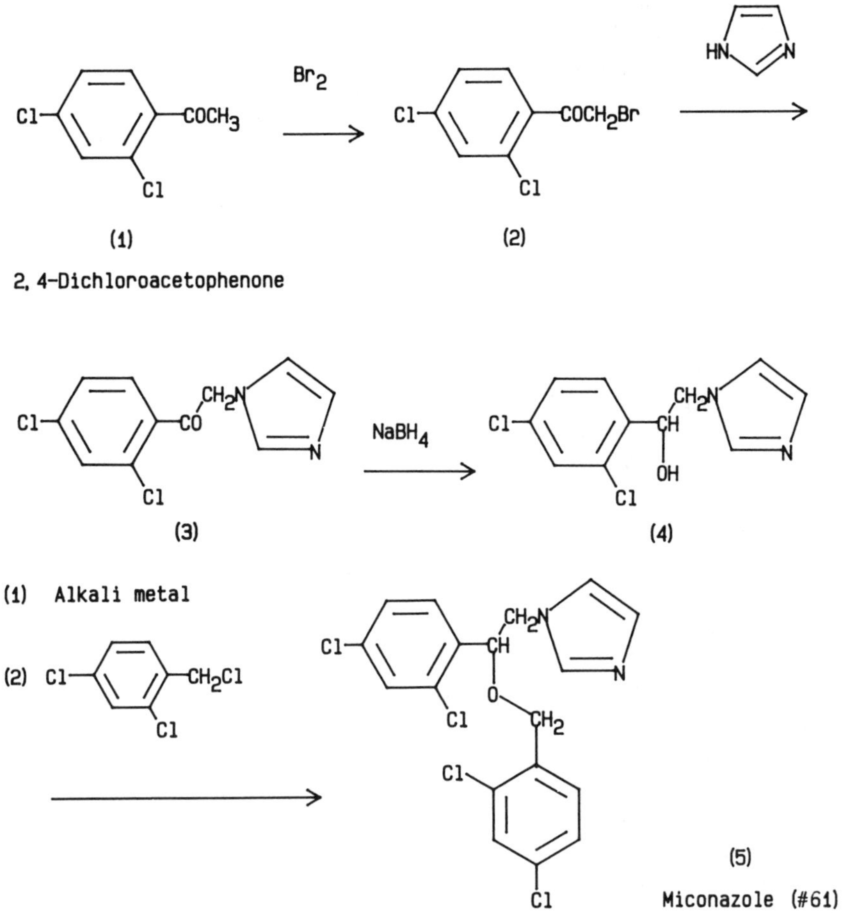

Figure 6.24. Synthesis of miconazole.

The synthesis of clotrimazole is similar, involving the alkylation of imidazole with *o*-chlorophenyldiphenylmethyl chloride.

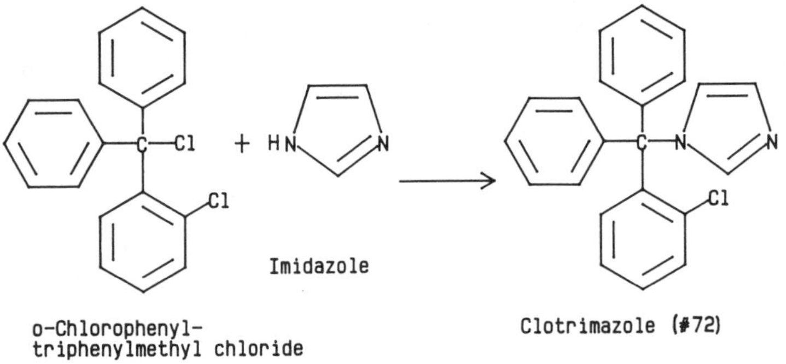

Clotrimazole may be used against vaginal fungi but is also a general fungicide and is used for conditions such as athlete's foot. The other preparation used against athlete's foot in the United States is zinc undecylenate:

$Zn^{++}[CH_2{=}CH(CH_2)_8COO]^-_2$

Zinc undecyclenate

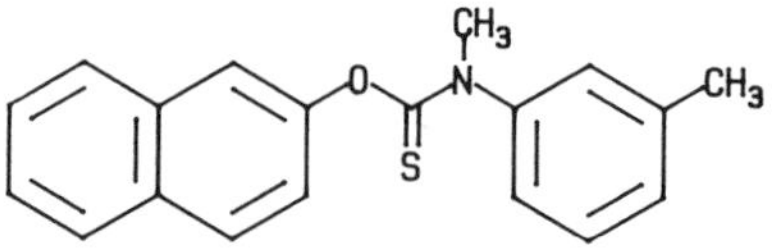

Tolnaftate

Another important antifungal agent, shown above, is tolnaftate. The β-naphthol ester is so potent that only 1 or 2 drops of a 1% solution in polyethylene glycol is adequate for application to areas as large as the hand. It should be noted, however, that β-naphthol derivatives are suspect because some of them have been shown to be carcinogenic.

6.12 THE FUTURE

There are many other antibiotics in the pharmacopoeia but we have covered a large number and have strayed at times a long way from the Top-100. Table 6.3 is a summary of which antibiotics are used for what diseases. This should provide a feel for the large number of infections that need to be cured and the range of chemicals available for curing them.

Antibiotics are a success story. In the 30 years after 1935, bacterial diseases were effectively conquered. Certainly, there are still parts of the world where there is a high death rate from them, but that is due to a lack of medical infrastructure and not to a failure of medical science. Antibiotics are cheap and effective, and a range of materials is available if the antibiotic of first choice does not work or if there are side effects. Contrary to the pessimistic views expressed in the 1950s, bacteria have not in general evolved strains that are immune to the major antibiotics. That is not to say that bacteria have not evolved resistant strains and we have mentioned these repeatedly in our discussion and have shown how new antibiotics have been developed to cope with them (Sections 6.3.3, 6.3.7, and 6.3.8). The point is that the chemists seem to be keeping ahead of the microorganisms and, for the foreseeable future, we can discount major epidemics of the nastier bacterial infections.

In the future there will undoubtedly be new antibiotics which are either specialized or without the side effects of some of the present ones. No major new group of antibiotics has appeared for over 20 years but present emphasis on the quinolones, because of the emergence of more effective compounds, is noteworthy.

TABLE 6.3. Summary of Antibacterial Therapy

Infection	Suggested Antibacterial[a]	Comment
	1 Gastro-intestinal System	
Gastro-enteritis	Antibiotic not usually indicated	Frequently nonbacterial etiology
Bacillary dysentery	Antibiotic usually not indicated	Co-trimoxazole in severe illness
Campylobacter enteritis	Erythromycin	
Invasive salmonellosis	Co-trimoxazole or ampicillin	
Typhoid fever	Chloramphenicol or co-trimoxazole or amoxycillin	
Biliary-tract infection	Gentamicin or a cephalosporin	
Peritonitis	Gentamicin + metronidazole (or clindamycin)	
	2 Cardiovascular System	
Endocarditis caused by:		
Staphylococcus aureus	Flucloxacillin + fusidic acid (or gentamicin)	Treat for at least 4 weeks
Streptococci with reduced sensitivity to penicillin, e.g., *Streptococcus faecalis*	Benzylpenicillin + low-dose gentamicin (i.e., 60 – 80 mg twice daily)	Treat for at least 4 weeks
Penicillin-sensitive streptococci (e.g., viridans streptococci)	Benzylpenicillin (+ low-dose gentamicin, i.e., 60 – 80 mg twice daily)	Treat for 4 weeks; stop gentamicin after 2 weeks if organism fully sensitive to penicillin. Oral amoxycillin may be substituted for benzylpenicillin after 2 weeks
	3 Respiratory System	
Haemophilus epiglottis	Chloramphenicol	Give intravenously
Exacerbations of chronic bronchitis	Tetracycline or ampicillin or ampicillin derivative or cotrimoxazole (or trimethoprim)	Note that 20% of pneumococci and some *Haemophilus influenzae* strains are tetracycline-resistant
Pneumonia:		
Previously healthy chest	Benzylpenicillin or ampicillin or ampicillin derivative	Add flucloxacillin if *Staphylococcus* suspected, e.g., in influenza or measles; use erythromycin if *Legionella* infection is suspected

TABLE 6.3. Continued

Infection	Suggested Antibacterial[a]	Comment
Previously unhealthy chest	Flucloxacillin + ampicillin (or co-trimoxazole or erythromycin)	Substitute erythromycin for flucloxacillin if *Legionella* infection is suspected
	4 Central Nervous System	
Meningitis caused by:		
Meningococcus	Benzylpenicillin	Intrathecal therapy not necessary
Pneumococcus	Benzylpenicillin	
Haemophilus influenzae	Chloramphenicol	
	7 Genital System	
Syphilis	Procaine penicillin (or tetracycline or erythromycin if penicillin-allergic)	Treat for 10 – 21 days
Gonorrhoea	Procaine penicillin with probenecid or ampicillin with probenecid (or co-trimoxazole, spectinomycin or cefuroxime if penicillin-allergic)	Single-dose treatment
Nongonococcal-urethritis	Tetracycline	Treat for 10 – 21 days
	8 Urinary Tract	
Acute pyelonephritis or prostatitis	Cotrimoxazole (or trimethoprim) or gentamicin or cephalosporin	Do not give cotrimoxazole (or trimethoprim) in pregnancy. Treat prostatitis with cotrimoxazole (or trimethoprim) for 4 weeks
"Lower" UTI	Trimethoprim or ampicillin or nitrofurantoin or oral cephalosporin	
	9 Blood	
Septicaemia		
Initial "blind" therapy	Aminoglycoside + a penicillin or a cephalosporin alone	Choice of agents depends on local bacterial resistance patterns and clinical presentation

TABLE 6.3. Continued

Infection	Suggested Antibacterial[a]	Comment
	10 Musculoskeletal System	
Osteomyelitis and septic arthritis	Clindamycin or flucloxacillin + fusidic acid. If *Haemophilus influenzae*, give ampicillin or cotrimoxazole	Under 5 years of age may be *H. influenzae*. Treat acute disease for at least 6 weeks and chronic infection for at least 12 weeks
	11 Eye	
Purulent conjunctivitis	Chloramphenicol or gentamicin eye-drops	
	12 Ear, Nose, and Oropharynx	
Dental infections	Amoxycillin (or ampicillin ester)	Metronidazole for anaerobic infections or those not responding to amoxycillin
Sinusitis	Erythromycin or cotrimoxazole	
Otitis media	Benzylpenicillin Phenoxymethylpenicillin	Initial i / m therapy (if possible) with benzylpenicillin, then oral therapy with phenoxymethylpenicillin
	Amoxycillin (or ampicillin ester) if under 5 years (or erythromycin if penicillin-allergic)	Under 5 years of age may be *Haemophilus influenzae*
Tonsillitis	Benzylpenicillin Phenoxymethylpenicillin	Initial i / m therapy (in severe infection) with benzylpenicillin, then oral therapy with phenoxymethylpenicillin. Most infections are caused by viruses
	13 Skin	
Impetigo	Topical chlortetracycline or oral flucloxacillin if systemic toxicity	
Erysipelas	Benzylpenicillin Phenoxymethylpenicillin	Initial i / m therapy (if possible) with benzylpenicillin, then oral ltherapy with phenoxymethylpenicillin
Cellulitis	Flucloxacillin (or erythromycin if penicillin-allergic)	
Acne	Tetracycline	Treat for at least 3 – 4 months

[a]Where ampicillin is suggested in the table amoxycillin or an ester of ampicillin may be used, and where flucloxacillin is suggested cloxacillin may be used.

Source: British National Formulary, **13**, (1987).

7

Cardiovascular Drugs

Heart disease is America's number one killer. It is also the disease that motivated some of the most interesting pharmaceutical advances of the 1970s. In 1971, the four most widely prescribed drugs for heart disease were digoxin, nitroglycerol, quinidine, and reserpine (structures shown in Sections 7.2, 7.5.1, 7.3, and 7.6.1, respectively). Reserpine was an early tranquilizer that also reduced blood pressure. Three of the four were alkaloids from plants, and the fourth was a nitrate already known as an explosive. Eight years later, by 1979, a revolution had occurred. Of the above four drugs, digoxin and nitroglycerol remained important, but they had been joined in the Top-100 by another 19 cardiovascular drugs. By 1986 the Top-100 contained 24 cardiovascular drugs. This is the largest number of drugs in any category and it includes 6 of the first 10. They are listed in Table 7.1 under several headings because there are different kinds of heart disease, which are treated in different ways. In addition to drugs in the Top-100, Table 7.1 includes some modern drugs that had not found their way into the Top-100 by 1986, a few drugs that are not widely used but exemplify important points, and others, such as reserpine, that are of historical importance.

7.1 THE FUNCTIONING OF THE HEART

The mammalian heart is illustrated in Figure 7.1. It is a muscle that acts like two pumps circulating blood through the body. The two upper chambers are called the atria and the two lower chambers are called the ventricles. There are valves between the atria and the ventricles called the tricuspid and mitral

TABLE 7.1. Cardiovascular Drugs[a]

Cardiac Glycosides
- Digoxin (#7)

Antiarrhythmics
- Quinidine
- Procainamide (#122)
- Flecainide acetate
- Encainide

Diuretics
- Acetazolamide (carbonic anhydrase inhibitor)
- Hydrochlorothiazide (#1), Chlorthalidone (#74), Chlorothiazide } Low ceiling diuretics
- Furosemide (#9), Bumetanide (#123) } Loop diuretics
- Triamterene (#4), Amiloride (#70), Spironolactone } Potassium-sparing diuretics

Potassium Replenisher
- Potassium chloride (#8)

Coronary Vasodilators
- Nitroglycerol (#21)
- Amyl nitrite
- Isosorbide dinitrate (#54)
- Nifedipine (#37), Diltiazem (#45), Verapamil (#93) } Calcium antagonists
- Pentoxifylline (#100) (Hemorheologic agent)

β-Adrenergic Blocking Agents
- Pronethalol
- Practolol
- Propranolol (#10)
- Atenolol (#17)
- Metoprolol (#32)
- Timolol (#55)
- Nadolol (#64)
- Labetalol (#132) (α and β blocker)

Antihypertensive Drugs
- Reserpine, Methyldopa (#29), Clonidine (#56) } Centrally acting antihypertensives
- Prazosin (vasodilator)
- Guanethidine (adrenergic neuron blocker)
- Captopril (#47), Enalapril (#101), Cilazapril } Angiotensin converting enzyme inhibitors

Anticoagulants
- Warfarin (#68)
- Heparin
- Dipyridamole (#36)

Hypolipemics
- Clofibrate
- Cholestyramine
- Colestipol
- Gemfibrizol
- Lovastatin
- Niacin

Thrombolytic Agents
- Tissue-type plasminogen activators "Eminase" and "Activase"

[a]Some drugs have several uses but have been classified under a single heading.

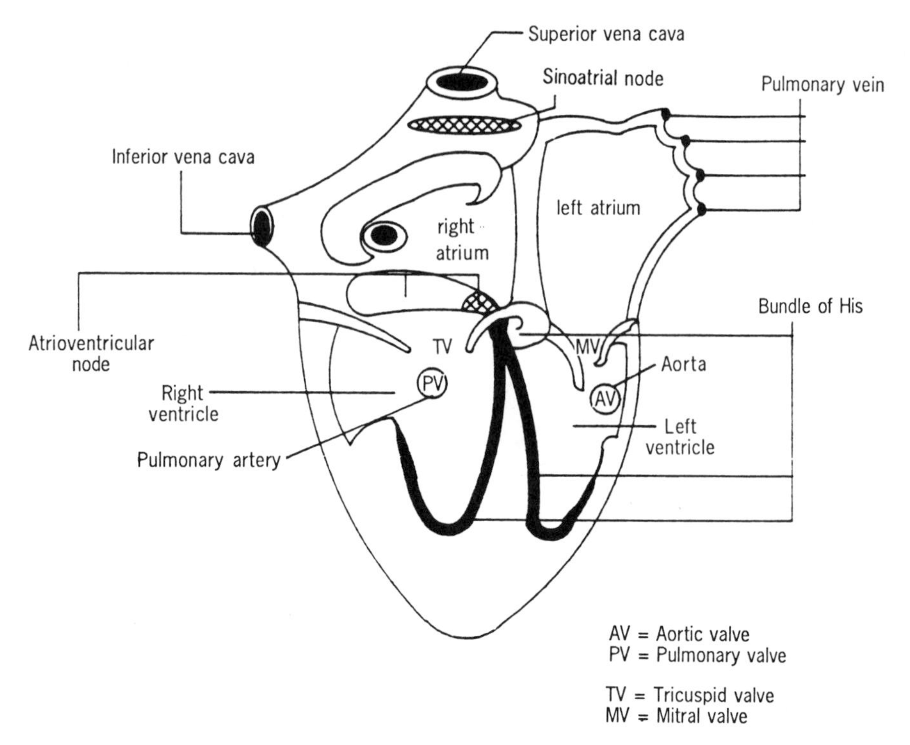

Figure 7.1. *Mammalian heart. The black lines running down through the heart wall separating the ventricles (the intraventricular septum) represent specialized muscle cells which conduct nervous impulses from the atrioventricular node. The pulmonary artery and aorta are added to the diagram to show the chambers from which they arise, but would not appear in a section of the heart along this plane.*

valves and there are also valves at the openings to the pulmonary artery and the aorta.

The cardiac cycle begins with the atria full of blood. An electrical impulse from special pacemaker cells in the wall of the right atrium at the sinoatrial node causes both atria to contract. The tricuspid and mitral valves open and the pulmonary and aortic valves close. Blood is forced into the ventricles. This is called the period of contraction.

There is a second control center in the heart called the atrioventricular node. When the signals from the sinoatrial node reach it, it sends out further signals via the bundle of His that allow the atria to relax and cause the ventricles to contract. The tricuspid and mitral valves close and the pulmonary and aortic valves open. Blood from the left ventricle is forced into the aorta and carries oxygen, nutrients, ions, hormones, and heat around the body to wherever they are needed. Particularly important are three major coronary arteries that supply the heart itself with blood. In the body tissues, the oxygen is, in effect, burned to provide the body with energy. The waste carbon dioxide and deoxygenated blood return to the right atrium via the inferior and superior venae cavae. Meanwhile, the blood from the right ventricle is forced

TABLE 7.2. Diseases of the Heart

Disease	Characteristics	Examples of Drugs
Congestive heart failure	Heart muscle deteriorates; edema	Cardiac glycosides, diuretics
Arrhythmia	Disorder of electrical impulse formation or conduction; tachy- and bradycardia	Digoxin, β blockers, quinidine, verapamil, flecainide
Coronary artery disease	Atheroma; narrowing of coronary arteries, ischemia, angina, myocardial infarction, coronary thrombosis	Coronary vasodilators, β blockers, calcium, antagonists, nitrites, nitrates, (bypass surgery)
High blood pressure (hypertension)	May rupture blood vessel (aneurysm) or cause heart or kidney failure	α and β blockers, peripheral vasodilators, ACE inhibitors, centrally acting antihypertensives
Atherosclerosis	Thickening / roughening of artery walls by fatty deposits	Hypolipemics
Thromboembolism (embolism)	Blocking of artery by blood clot that has broken loose	Anticoagulants, thrombolytic agents

through the pulmonary artery to the blood capillaries in the lungs. There the blood picks up oxygen by an iron-catalyzed gas exchange mechanism and gives up carbon dioxide. It returns to the left atrium via the pulmonary vein and the cycle is repeated.

The nodes that control the cycle are themselves regulated by the autonomic nervous system (Section 7.5.4), but they can contract independently of nervous stimulation.

7.1.1 Diseases of the Heart

There are a number of ways in which this system can malfunction and the most important are listed in Table 7.2 together with some of the drugs used to treat them. The first two, congestive heart failure and arrhythmia, are a consequence of the deterioration of the heart muscle so that it pumps inefficiently. It cannot maintain adequate blood circulation and the tissues do not receive sufficient oxygen and nutrients. Inadequate pumping of blood out of the heart can lead to back pressure. Hence, there is congestion in the lungs and excess fluid in the tissues. This was earlier known as dropsy and is now called edema. Arrhythmia is a disorder of electrical impulse formation or conduction in the heart's pacemaker system.

Coronary artery disease is responsible for over half the deaths from heart disease and is usually due to a narrowing of the coronary arteries as a result of

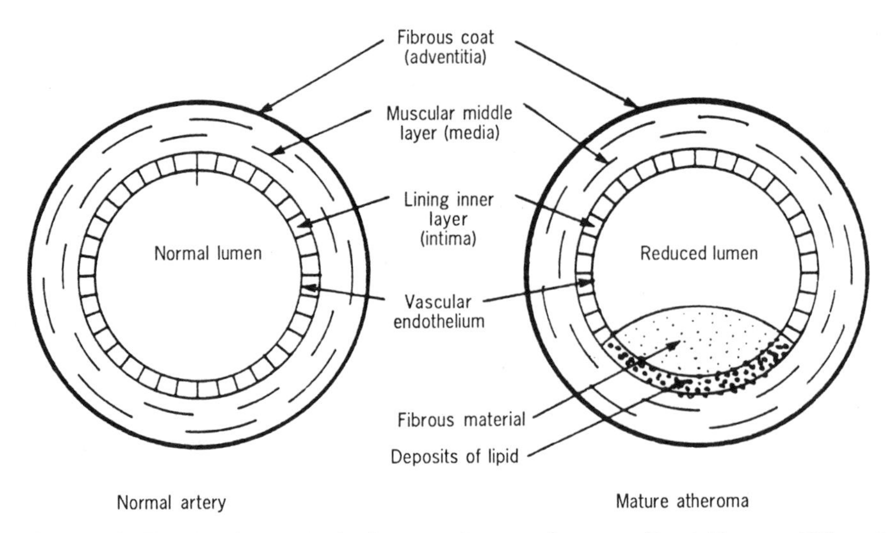

Figure 7.2. *The development of atheroma. Source: Coronary Heart Disease, Office of Health Economics, London, 1982.*

atheroma. The development of atheroma is shown in Figure 7.2. Atheroma starts with some damage to the inside of an artery, the so-called vascular endothelium. Damage occurs most readily at points of stress or turbulence. The normal response of the body to damage is for a blood clot to form from platelets in the blood at the point of damage. This covers and protects the injury until it is repaired. The process is self-limiting and the clot is absorbed when it is no longer needed.

In atheroma, the process for some reason is not self-limiting. Aided by high hydrostatic pressure in the artery, platelets and lipids in the blood invade the next layer of the artery below the endothelium called the intima. They release chemicals that cause migration of muscle cells to the endothelium. Sometimes new endothelium grows over this layer and the process halts at the expense of the narrowing and loss of elasticity of the artery.

More often, the process continues and may be worsened by an intravascular thrombosis where a blood clot gets caught in the narrowed section of the artery causing further obstruction. In the end, fibrous materials in the form of a scar deposit inside the artery. These not only reduce the diameter of the artery but also reduce its elasticity so it can no longer dilate.

Atheroma in the coronary arteries leads to a deficiency of blood to the heart muscle called ischemia. A short, mild ischemia produces a violent pain across the chest called angina pectoris. A prolonged attack leads to permanent damage and scarring called myocardial infarction. If it results from a blood clot it is called coronary thrombosis.

High blood pressure or hypertension may kill by rupturing a blood vessel in a vital organism (aneurysm) or by causing the heart or kidneys to fail. Its cause is uncertain although there may be a genetic factor and it may also be due to a narrowing and hardening of the peripheral arteries, known as arteriosclerosis.

Atherosclerosis is a thickening and roughening of the walls of arteries, not coronary arteries in this case, caused by fatty deposits, that is, by atheroma. Thromboembolism is the blocking of an artery by a blood clot that has broken loose from its site of formation. If the lung is affected, death may be instantaneous.

In addition to the above diseases of the heart, strokes are also classified under the cardiovascular heading. In 1986, strokes accounted for 12% of U.S. mortality. In a stroke, there is a sudden interruption of blood supply to the brain. Since the brain has no oxygen reserve, part of it dies and, if the patient survives, he or she may do so with impaired speech, balance, vision, touch, sensation, and movement. Strokes are related to high blood pressure, cigarette smoking, excessive alcohol consumption, high blood viscosity, and diabetes.

7.2 CARDIAC GLYCOSIDES

Congestive heart failure is usually treated with digoxin, one of a group of drugs known as the cardiac glycosides. Digoxin is obtained from the leaves of the white foxglove, *Digitalis lanata*. Digoxin is a glycoside made up of one molecule of digoxigenin, which has a steroid-like structure, and three molecules of a sugar, digitoxose:

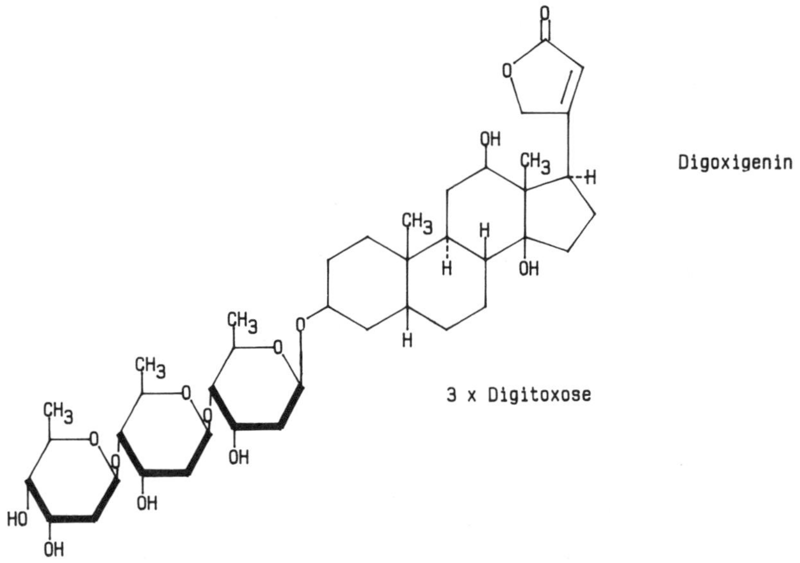

Digoxigenin

3 x Digitoxose

Digoxin (#7)

Digoxin stimulates both the nerves that control the heart and the heart muscle itself. It increases the force of myocardial contraction. That means that it increases the force of contraction of the heart and the heart works harder without demanding more oxygen. Digoxin has a number of other uses and is widely effective except in cases of coronary thrombosis and angina pectoris.

Digoxin in the form of digitalis is easily the oldest drug in the Top-100. A Welsh physician noted its action in 1250 and it was used clinically by the British physician William Withering in 1785. Bioavailability problems were mentioned in Section 4.9.

A problem associated with the production of digoxin is that it occurs as a mixture with other cardiac glycosides which are wasted. Boehringer Mannheim in Germany has developed a method of converting one of them, digitoxin, to digoxin by means of plant cell culture. In the living plant, digitoxin is the precursor of digoxin and undergoes 12β-hydroxylation in certain cells to give digoxin. Boehringer Mannheim has identified the cells in *Digitalis lanata* that contain the appropriate enzyme and carry out the conversion with immobilized cultures of them.

7.3 ANTIARRHYTHMICS

Diuretics, too, are used in congestive heart failure but, because digoxin is also used for arrhythmia of the heart, treatment of arrhythmia will be mentioned before diuretics. In arrhythmia or atrial fibrillation, the heart beats irregularly rather than pumping inadequately. Causes of too fast a heartbeat, that is, tachycardia, include exercise, fever, anxiety, drinking too much coffee, tea, or alcohol, and taking certain drugs, but it can also result from heart disorders.

Arrhythmia can have two causes. Sometimes the muscles of the atria cease to respond to the electrical signals provided by the sinoatrial node, perhaps because of problems with their conduction, and sometimes the node itself may cease to provide signals.

Digoxin usually has a favorable effect on tachycardia because it stimulates the ventricles—the pumping muscles—independently of the sinoatrial node. On the other hand, it can also induce tachycardias in certain cases. Other important antiarrhythmic drugs include the β blockers and the calcium antagonists, which will be discussed in Sections 7.5.5 and 7.5.2 as angina remedies, procainamide (#122, see below), and quinidine, which is an alkaloid.

Three new drugs introduced in the mid 1980s are encainide, amiodarone, and flecainide acetate:

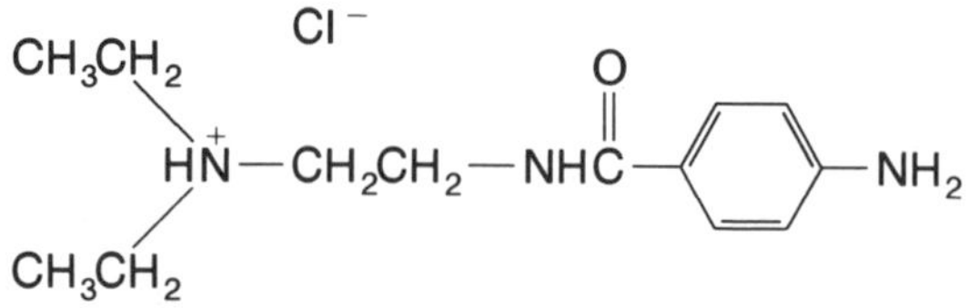

$$(CH_3CH_2)_2\overset{+}{N}H-CH_2CH_2-NHC(=O)-C_6H_4-NH_2 \quad Cl^-$$

Procainamide hydrochloride

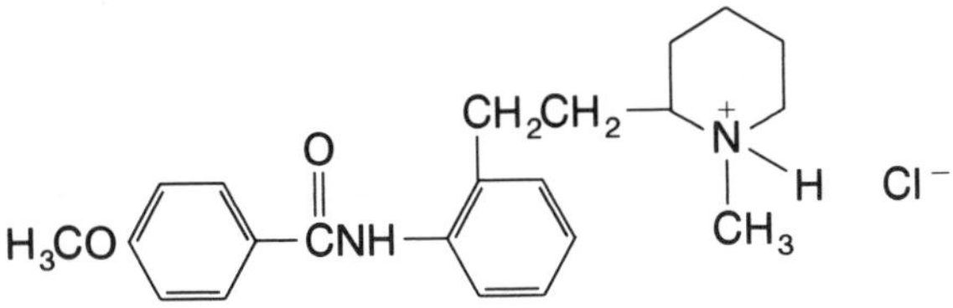

Encainide hydrochloride

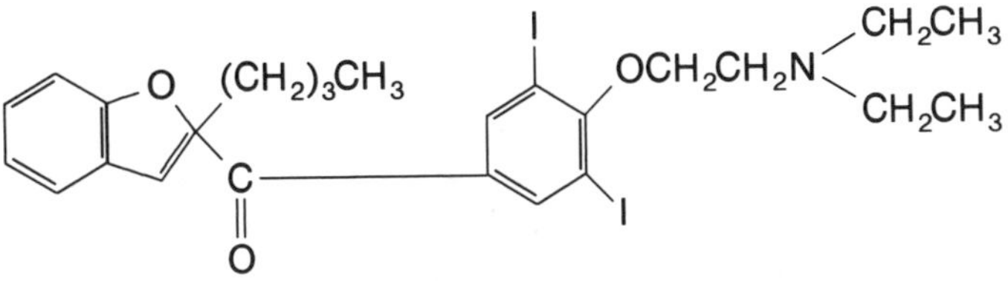

Amiodarone

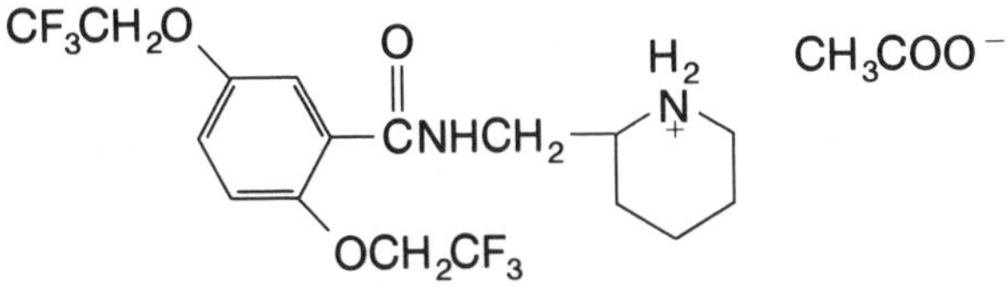

Flecainide acetate

All the above materials are used to treat tachycardias. They have slightly different applications and side effects. Bradycardias are treated with atropine (Section 14.1) or isoproterenol (Section 15.1). In severe long-term cases, pacemakers may be implanted.

All the antiarrhythmics depress the excitability of the heart muscle and therefore suppress extra beats. The ones whose structures are shown (plus quinidine) also reduce the rate at which electrical impulses travel through the heart and depress the ease with which it contracts. Some, but not all, lengthen the period between heart contractions by prolonging the action of the slowing nerve of the heart called the vagus nerve.

Quinidine is a stereoisomer of quinine (Section 18.2.1), the classical drug for the treatment of malaria:

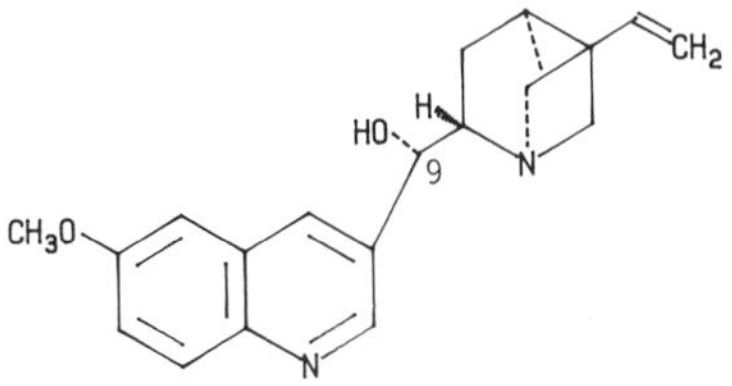

Quinidine

It is still an adequate arrhythmic but since the advent of pacemakers and the newer drugs, especially procainamide, which has a similar therapeutic pattern, it has been used less and has dropped out of the Top-100.

Quinine and quinidine are nonetheless a rare example of two diastereoisomers, each of which is a drug in its own right. The molecule has two adjacent asymmetric centers which lead to two erythro and two threo forms. The erythro forms are quinine and quinidine. It appears that, for the molecule to have the antimalarial activity associated with quinine, the hydroxyl group in position 9 must be in juxtaposition with the nonaromatic nitrogen atom of the quinuclidine nucleus. This is the case only with the erythro forms. If the rigid quinuclidine nucleus is replaced with a more flexible piperidyl group, which can place itself in juxtaposition with the hydroxyl group, slight activity is observed.

Quinine was synthesized by Woodward and Doering in 1944. It was the first of a series of brilliant syntheses by Woodward, which included cholesterol, cortisone, strychnine, reserpine, chlorophyll, and a tetracycline antibiotic. He won a Nobel prize in 1965. A shorter synthesis has since been devised that is not sterically selective and gives equal quantities of quinine and quinidine. The references are listed in our bibliography, but neither synthesis is a practical source of the drugs and both are still obtained from cinchona bark. Most of the quinidine used today is obtained by isomerization of quinine.

7.4 DIURETICS

Digoxin provides one way of treating heart failure, which is by stimulating the heart. An alternative is to alleviate the edema. Edema can also result from other causes such as kidney or liver failure. The tissues swell because of excessive retention of fluids and the typical patient has swollen feet and ankles.

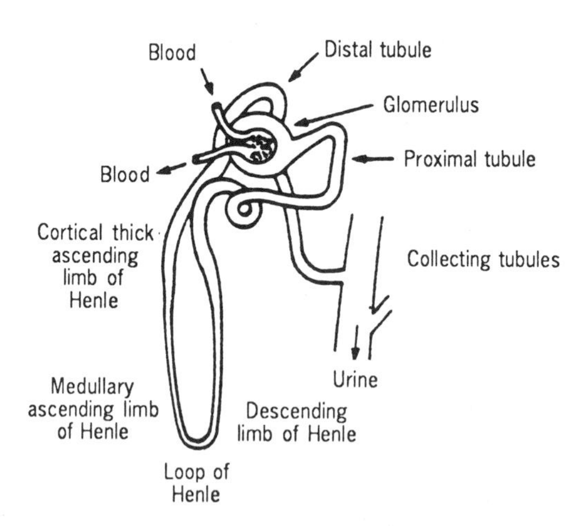

Figure 7.3. Simplified diagram of a kidney nephron.

Drugs that counter fluid retention are called diuretics. In general, they act on the kidneys to increase the output of sodium ions and water in the urine. Figure 7.3 is a much simplified sketch of a single kidney nephron of which there are 2–2.5 million in a kidney. The nephron is made up of a glomerulus and a long tubule consisting of a proximal tubule, the loop of Henle, and a distal tubule. Blood enters a complex of capillaries in the glomerulus and, because the pressure is higher inside than out, water, various molecules, and electrolytes are filtered through into the glomerulus. This fluid passes through the tubule where 99% of it is reabsorbed and 1% passes into the collecting tubule and is excreted as urine. The reabsorption, however, is selective and by this means the kidneys control the delicate balance between salt and water in the body.

Salt in the tubule may be reabsorbed into the bloodstream or remain in the urine. If urine is strongly saline, body water will osmose into it and be excreted. If it is dilute, this will happen to a lesser extent.

The system is subject to a complicated set of hormonal and enzyme controls, two of which are particularly significant. Aldosterone is a hormone produced by the adrenal gland:

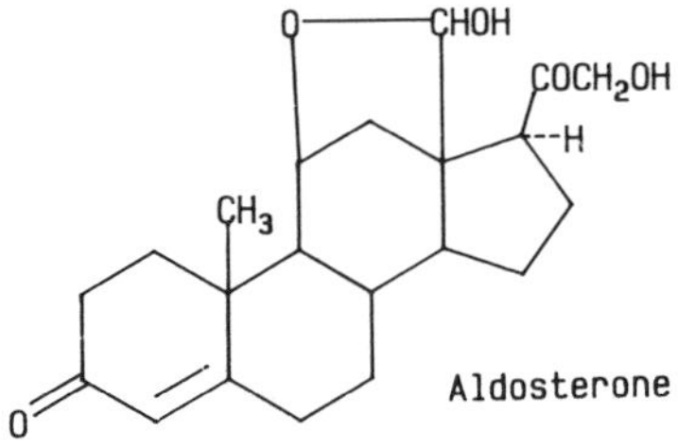

CHOH
$COCH_2OH$
H
CH_3
O
O
Aldosterone

It increases the reabsorption of sodium ions in the tubule; hence, more water is retained and the volume of urine is decreased. The enzyme carbonic anhydrase speeds up the achievement of the carbonic acid equilibrium

$$CO_2 + H_2O \rightleftharpoons H_2CO_3 \rightleftharpoons H^+ + HCO_3^-$$

Because sodium ions exchange with hydrogen ions in the tubule, rapid equilibration increases transport of ions and again more water and salt are retained.

The main requirements of a diuretic are that it is of low toxicity, that it does not change the pH of body fluid, and that it causes the excretion of sodium and potassium ions in ratios comparable with edemas. The third requirement is the most difficult and diuretic users usually take potassium supplements to replace the excess ions lost in the urine.

There are four main classes of diuretics and the three still in use are shown in Figure 7.4. They will be discussed in turn.

7.4.1 Carbonic Anhydrase Inhibitors

The earliest diuretics were tea and coffee, which were found to contain theophylline (Section 15.2). Another early group was the organomercury compounds, which, not unexpectedly, showed toxic side effects. An advance came in 1936 when sulfonamide antibacterials, for example, *p*-aminobenzenesulfonamide (Section 6.2), were found to have diuretic action. Unfortunately, the diuretic action was accompanied by a pH change in body fluids and urine from 5.5 to nearly 8.5. The sulfonamides were found to block carbonic anhydrase; hence, bicarbonate does not revert to carbon dioxide and water and is eliminated in the urine with a consequent disturbance of pH.

The carbonic anhydrase blockers were far less toxic than the mercurials but the pH change limited their use. They have been replaced by the low-ceiling diuretics. Acetazolamide, which was the prototypal carbonic anhydrase inhibitor, is still used in treating glaucoma to reduce fluid pressure in the eye. In this case, loss of bicarbonate ion is beneficial.

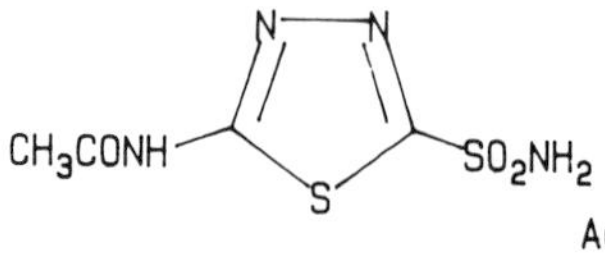

CH_3CONH — (N—N, S ring) — SO_2NH_2

Acetazolamide

7.4.2 Low-Ceiling Diuretics

A breakthrough in therapy occurred when a group of sulfonamides was discovered that acted as diuretics but showed only weak carbonic anhydrase inhibition. They were thiazides or related compounds and the most widely prescribed members of the class are shown in Figure 7.4. Because of their weak carbonic anhydrase inhibition, thiazides promote excretion of sodium and bicarbonate ions. They also promote excretion of approximately equal amounts of sodium and chloride ions and this is of benefit. In addition to inhibiting carbonic anhydrase, they inhibit transfer of chloride ions and the attendant sodium ions in the cortical thick ascending limb of the loop of Henle and the distal tubule. Thiazides cause loss of potassium and, unless taken in very small doses, are accompanied by a potassium replenisher, usually potassium chloride.

The first diuretic in this class was chlorothiazide. It was introduced in 1953 and has only recently dropped out of the Top-100. Its onset of action is mild and the time of action is 6–10 hours. It presents no pH problems and the introduction of the ortho chlorine atom and the second sulfonamido group have markedly changed the pharmacological properties of the original sulfonamide antibacterials. Once a certain dose of chlorothiazide has been taken, further doses have no additional effect. It and others in its class are therefore known as low-ceiling diuretics.

1. Low-ceiling diuretics (thiazides and related compounds):

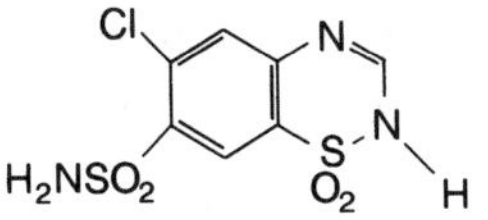

Cl
H_2NSO_2
N
S
N
O_2
H

Chlorothiazide

H
Cl
N
N
H_2NSO_2
S
O_2
H

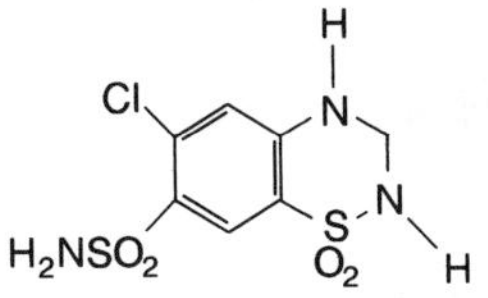

Hydrochlorothiazide (#1)

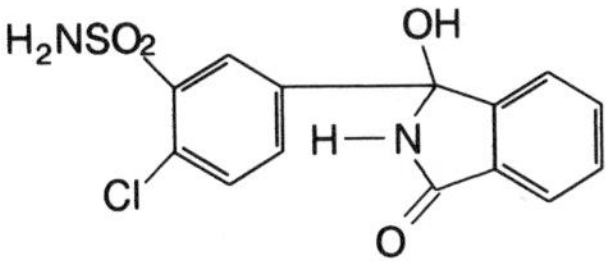

H_2NSO_2
OH
H—N
Cl
O

Chlorthalidone (#74)

2. High-ceiling or loop diuretics:

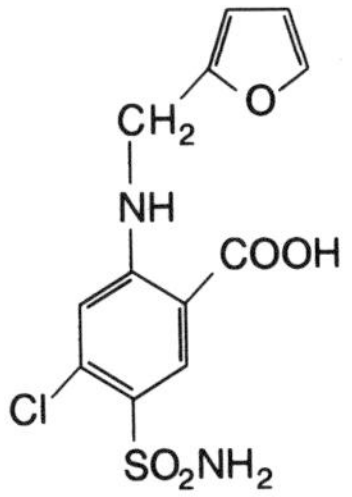

O
CH_2
NH
COOH
Cl
SO_2NH_2

Furosemide (#9)

HOOC
$NHCH_2CH_2CH_2CH_3$
O
SO_2NH_2

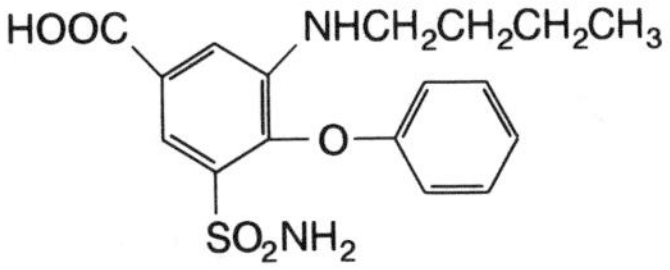

Bumetamide (#123)

3. Potassium-sparing diuretics (including aldosterone antagonists):

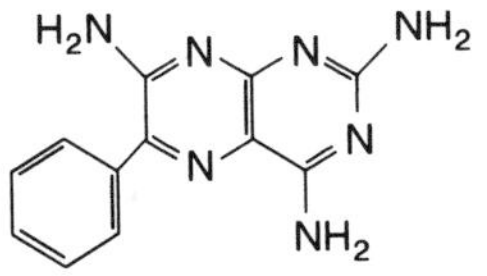

H_2N
N
N
NH_2
N
N
NH_2

Triamterene (#2)

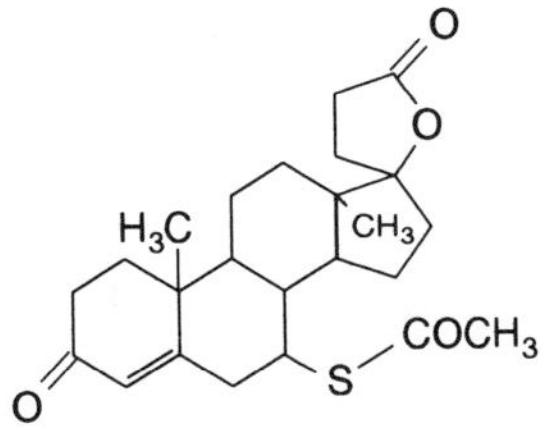

O
O
H_3C
CH_3
$COCH_3$
S
O

Spironolactone

O
NH_2
Cl
N
CN=C
NH_2
H_2N
N
NH_2

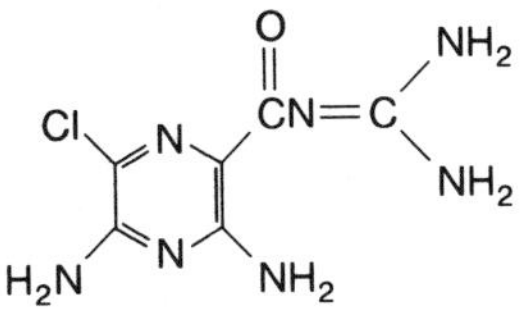

Amiloride (#70)

***Figure 7.4**. Diuretics.*

At the number one position in the Top-100 is hydrochlorothiazide, a drug that differs from chlorothiazide only by two hydrogens in the 3 and 4 positions. Here is another example of what continues to amaze chemists. Tiny structural changes in a molecule will affect physiological activity greatly. The reduction of the 3, 4 double bond of the thiadiazine ring leads to a tenfold increase in diuretic activity and a tenfold decrease in carbonic anhydrase inhibition. Certainly one would expect the metabolic properties of the two compounds to be similar and, indeed, they are both diuretics. But the two extra hydrogens mean less loss of potassium and bicarbonate ions and there is less tendency to produce an electrolyte inbalance. Also, hydrochlorothiazide is better tolerated by patients and produces fewer side effects such as headache, nausea, restlessness, constipation, and anorexia.

The preeminent position of hydrochlorothiazide in the Top-100 is due partly to its role as a constituent of many compound drugs. Mixed with the potassium-sparing diuretic, triamterene, it is the most widely prescribed branded drug; it is also prescribed on its own and mixed with amiloride (#70) and spironolactone, the other potassium-sparing diuretics.

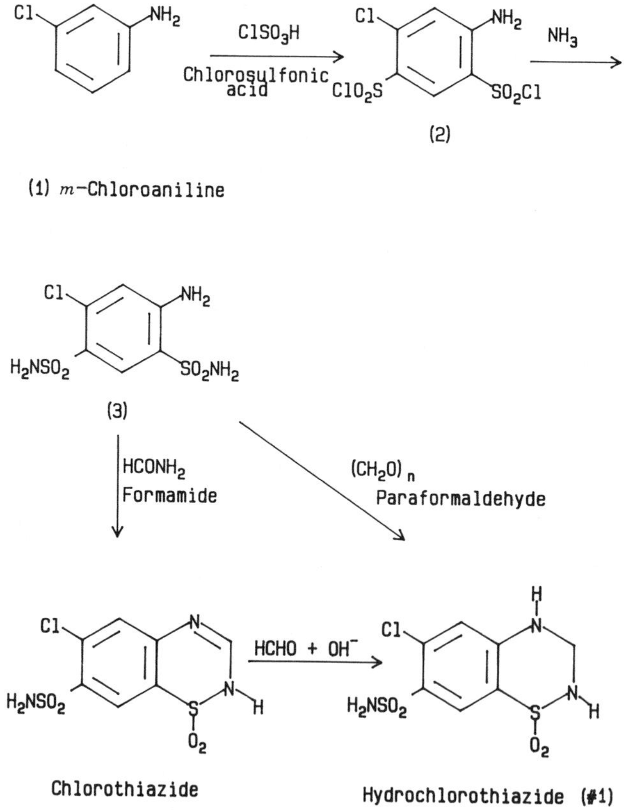
Cl
NH2
ClSO3H
Chlorosulfonic acid
ClO2S
SO2Cl
NH3
(2)
(1) m-Chloroaniline
H2NSO2
SO2NH2
(3)
HCONH2
Formamide
(CH2O)n
Paraformaldehyde
HCHO + OH⁻
N
H
S
O2
Chlorothiazide
Hydrochlorothiazide (#1)

Figure 7.5. Synthesis of chlorothiazide and hydrochlorothiazide.

Chlorothiazide and hydrochlorothiazide are made as shown in Figure 7.5 from a common intermediate (3). This is made by high temperature chlorosulfonation of *m*-chloroaniline (1) to give the bissulfonyl chloride (2) which on ammonolysis yields (3). If (3) is treated with formamide, chlorothiazide results. If it is treated with paraformaldehyde, hydrochlorothiazide is the product. Chlorothiazide itself may be reduced to hydrochlorothiazide with alkaline formaldehyde.

Chlorthalidone, the second diuretic in the thiazide group, is not a true thiazide. It is 70 times as active a carbonic anhydrase inhibitor as hydrochlorothiazide and leads to increased output of potassium and bicarbonate ions. On the other hand, it is unusually long acting—up to 72 hours after a single dose—and may be given daily or on alternate days. The synthesis of chlorthalidone starts with the substituted benzophenone (1). It is transformed to a sulfonamide (2) by diazotization with nitrous acid, followed by a Sandmeyer reaction with sulfur dioxide and cuprous chloride in acetic acid. (2) Is treated with thionyl chloride and ammonia in aqueous ethanol to give chlorthalidone (3).

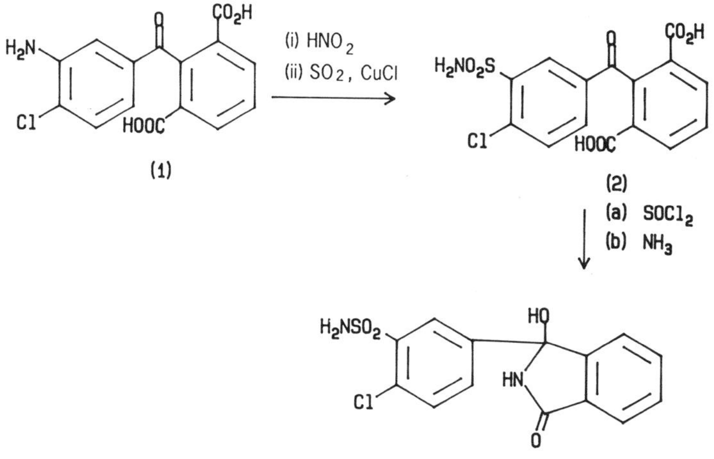

H_2N, Cl, HOOC, CO_2H (1)

(i) HNO_2
(ii) SO_2, CuCl

H_2NO_2S, Cl, HOOC, CO_2H (2)

(a) $SOCl_2$
(b) NH_3

H_2NSO_2, Cl, HO, HN, O

(3)
Chlorthalidone

7.4.3 High-Ceiling (Loop) Diuretics

The newest and most potent group of diuretics used clinically is the loop diuretics. The loop diuretics have a more rapid and violent onset of action than the thiazides. They are shorter acting and additional doses have an additional effect; that is, they are "high ceiling." Furosemide (Figure 7.4) is the most widely prescribed member of the group.

Loop diuretics act by inhibiting transfer of sodium ions in the loop of Henle, especially in the medullary ascending limb and cortical thick ascending limb. Inhibition is also believed to occur in the proximal tubule.

Furosemide is especially useful for patients with pulmonary edema (retention of fluid in the lungs) because of deterioration of the left ventricle and also for patients who do not respond to the thiazides. Furosemide causes about three times as much excretion of salt as do the thiazides. It also causes loss of potassium ions from the body and is usually given with a potassium replenisher.

Furosemide is synthesized from 2,4-dichlorobenzoic acid by reaction with chlorosulfonic acid followed by amidification with ammonia. Reaction of the active chlorine atom in the 2-position with furfurylamine leads to furosemide.

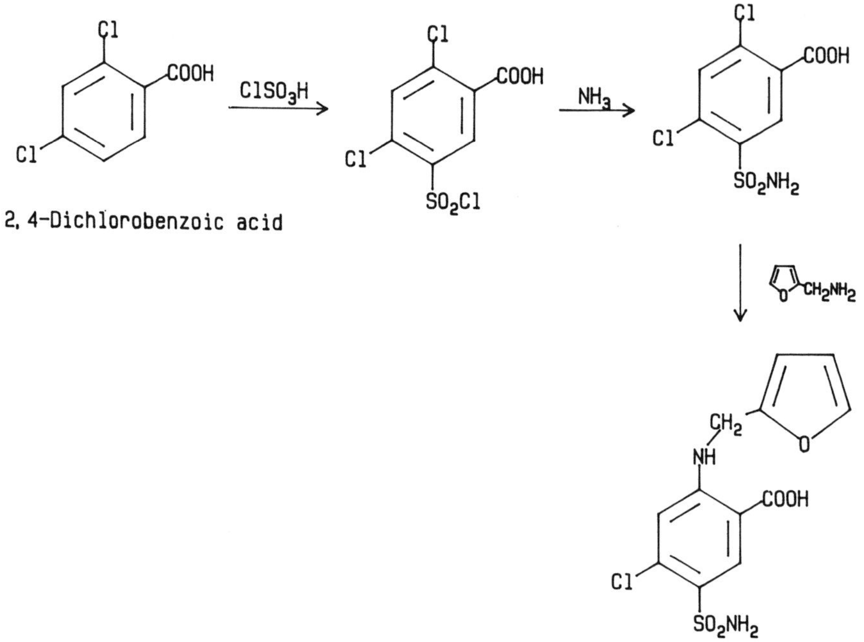

Cl COOH Cl $ClSO_3H$ → Cl COOH Cl SO_2Cl NH_3 → Cl COOH Cl SO_2NH_2

2,4-Dichlorobenzoic acid

↓ CH_2NH_2 (furfurylamine)

CH_2 NH O COOH Cl SO_2NH_2

Furosemide (#9)

Bumetanide (#123) is similar in action to furosemide but is 40 times as potent on a weight for weight basis.

7.4.4 Potassium-Sparing Diuretics

The final group of diuretics is the potassium-sparing diuretics and the most widely prescribed member of the group is triamterene (#2) shown in Figure 7.4. Triamterene is a mild diuretic, which does not lead to excretion of potassium. In some way or another it permits the reabsorption of potassium but not sodium ions in the tubules. Neither does it cause serious uric acid retention, which is a problem with some of the other diuretics. It is usually

(1)

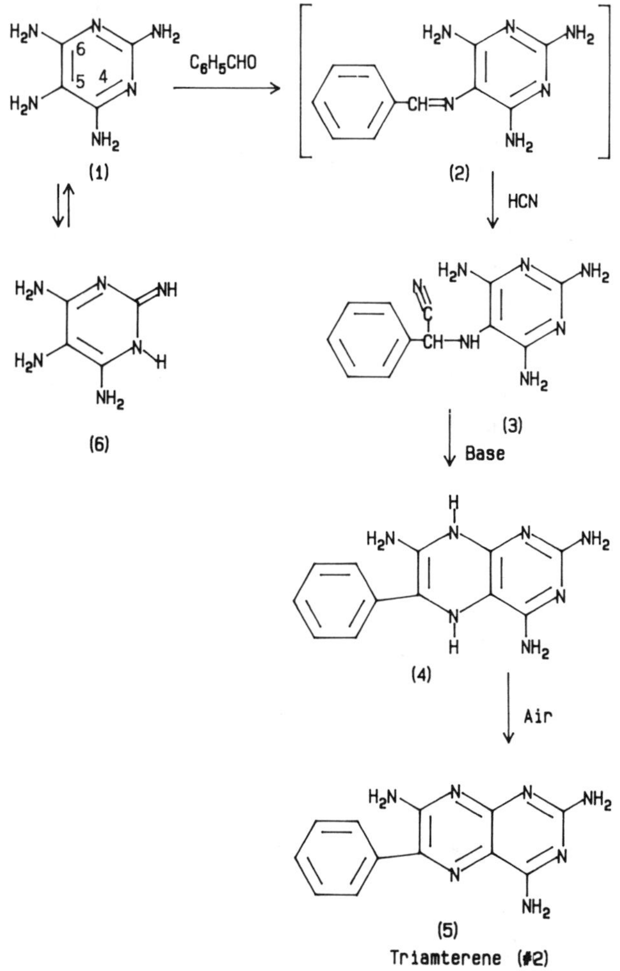

(1)

(2)

HCN

(3)

Base

(4)

Air

(5)

Triamterene (#2)

(6)

(2)

Guanidine nitrate

1. CH_3ONa
2. $NC—CH_2—CN$

2,4,6-Triamino-pyrimidine

$NaNO_2/CH_3COOH$

5-Nitroso-2,4,6-triaminopyrimidine

$CH_2—CN/NaOCH_3$

Benzyl cyanide

Triamterene

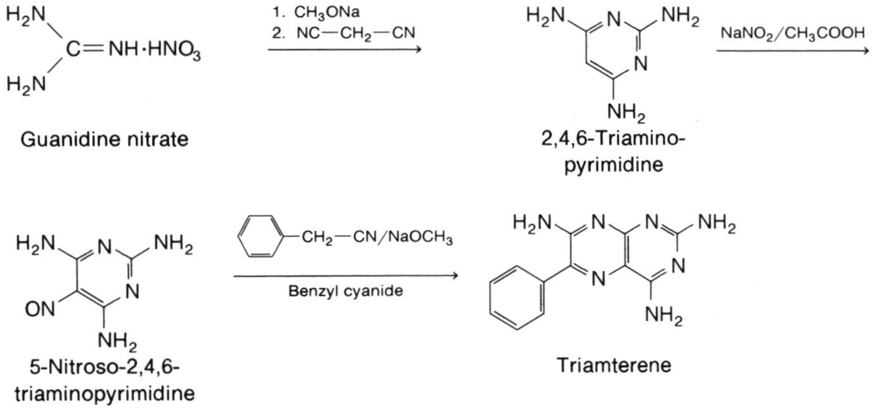

Figure 7.6. *Syntheses of triamterene.*

used in combination with a thiazide so that a potassium replenisher is unnecessary. It should not be used, however, in the presence of renal disease or hepatitis.

One synthesis of triamterene starts with the tetraaminopyrimidine, Figure 7.6, structure (1). Interestingly, the amino at the 5 position is more basic than the other amino groups because of the existence of a tautomeric form (6). Condensation with benzaldehyde yields the Schiff's base (2), which is not isolated but is reacted with HCN to provide the α-aminonitrile (3). This is an interesting example of addition of HCN to a double bond. Treatment of (3) with base yields (4) which oxidizes spontaneously in air to triamterene (5).

A second synthesis of triamterene starting with 2,4,6-triamino-5-nitroso-pyrimidine and benzyl cyanide is shown in the lower part of Figure 7.6.

The second potassium-sparing diuretic is amiloride, which has a similar pharmacological profile to triamterene and is preferred in some countries. The synthesis is shown in Figure 7.7. *o*-Phenylene diamine (1) plus glyoxal (2) give a dicarboxylic acid (3). Amidification gives (4), and a single equivalent of hypobromous acid gives the Hoffmann rearrangement of only one of the amide groups. Methanolysis of the intermediate carbamate gives the amino ester (5). Exposure of this ester to sulfuryl chloride gives (6). Ammonia

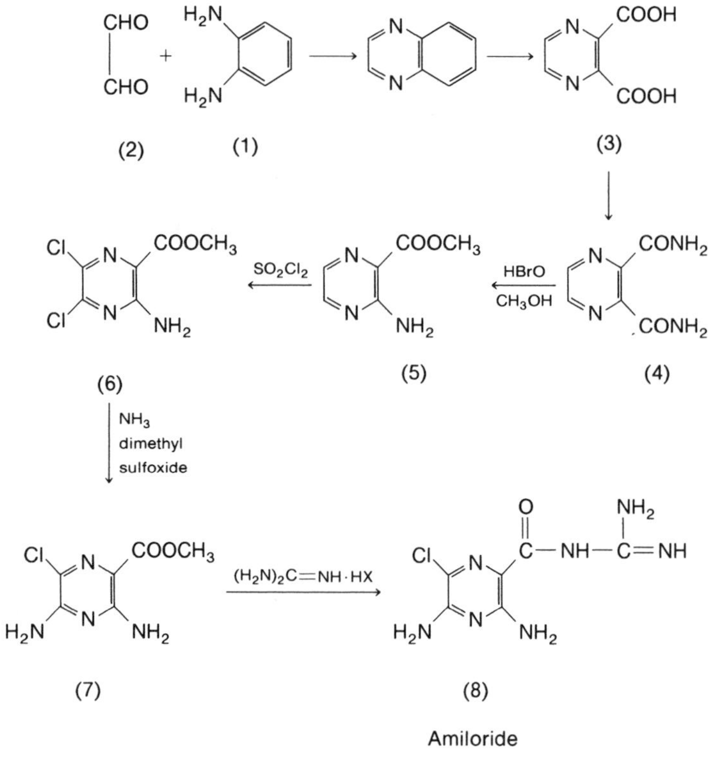

(2) (1) (3) (4) (5) (6) (7) (8)

SO_2Cl_2 — HBrO, CH_3OH — NH_3 dimethyl sulfoxide — $(H_2N)_2C{=}NH \cdot HX$

Amiloride

Figure 7.7. *Synthesis of amiloride.*

displaces the para chlorine, which is activated by the carboxyl to give (7). Heating a salt of guanidine with (7) gives amiloride (8).

The third potassium-sparing diuretic is spironolactone, shown in Figure 7.4. Potassium replenishers need not be given with it. It is a mild diuretic and is normally used with a loop or thiazide diuretic to avoid the need for replenishers and to potentiate their action.

Spironolactone has a steroid structure. It has been shown to be an antagonist to aldosterone, also a steroid. Aldosterone (Section 7.4) aids retention of salt and water. Having a similar structure, spironolactone competes with it for

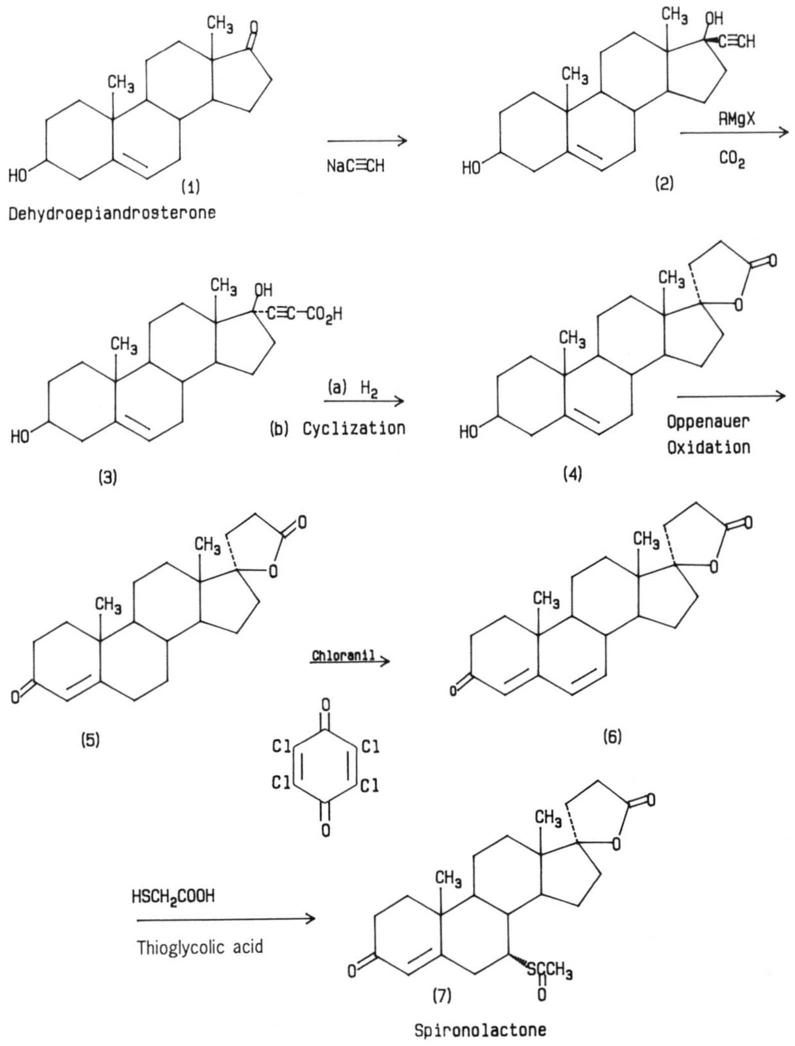

Figure 7.8. Synthesis of spironolactone.

receptor sites and is the victor. When the aldosterone is blocked, diuresis occurs.

In certain cases, excess levels of aldosterone in the body lead to the retention of sodium chloride and water and this in turn leads to a kind of hypertension known as hyperaldosteronism. Spironolactone relieves this hypertension. In addition, however, it has been shown to have some diuretic effect even where there is no excess of aldosterone in the body.

The synthesis of spironolactone is shown in Figure 7.8. It starts with the steroid dehydroepiandrosterone (1), which is also an important intermediate for the synthesis of oral contraceptive components. It is made from diosgenin and will be discussed in Section 11.6 on steroid drugs. Reaction with sodium acetylide yields (2). This is a reaction well known in steroid chemistry. Compound (2) in turn reacts with a large excess of organomagnesium halide in a Grignard reaction to give an acetylide salt which can be carbonated to give the carboxylic acid (3). Hydrogenation followed by cyclization gives the spirolactone (4) which undergoes the Oppenauer oxidation to give the 3-keto compound (5). (5) Is dehydrogenated to the conjugated structure (6) by treatment with chloranil. Addition of thiolacetic acid to the diene yields spironolactone (7).

7.4.5 Potassium Replenishers

Thiazide and loop diuretics lead to potassium loss. The effect varies from diuretic to diuretic, as already indicated, but is usually a problem. Lack of potassium may cause muscle weakness, constipation, and loss of appetite, and, most seriously, it may affect the heart. Addition of potassium salts to diuretic tablets has no noticeable effect, but potassium levels may be replenished if salts are taken on different days from diuretics. Potassium chloride is the form in which potassium replenishers are usually taken. Bicarbonate and citrate are occasionally preferred. Potassium replenishers occupy eighth place in the Top-100.

7.5 CORONARY VASODILATORS AND β-ADRENERGIC AGENTS

Coronary heart disease is due to narrowing of the coronary arteries because of atheroma or thrombosis. It can lead to angina, myocardial infarction, and sudden death. Cases of uncontrollable angina are often treated by coronary artery bypass grafting. In this technique, a length of vein taken from the patient's leg is grafted to replace the blocked coronary artery. It has become possible because of the development of the heart–lung machine and the technique of coronary arteriography but also because of modern anesthetics. In the United States in 1987, over 200,000 bypass operations were performed at a cost of over $4 billion. The U.K. figure was 11,800 in 1985. Allowing for population differences, this is about a quarter of the U.S. rate and there is a debate as to whether Americans are too ready or British too reluctant to operate. Other surgical techniques are also used for coronary heart disease including the dilatation of a blocked vessel by a "balloon" and the melting of fatty layers with the aid of lasers.

Drug therapy is useful to relieve or prevent angina and to prevent recurrence of myocardial infarction. The drugs fall into two groups, the vasodilators and the β-adrenergic agents or β blockers. The vasodilators may be subdivided further into nitrites and nitrates on the one hand and calcium antagonists on the other. For convenience, we have also included pentoxifylline, a hemorheologic agent (Section 7.5.3), in this group.

7.5.1 Nitrites and Nitrates

The nitrites and nitrates of importance are nitroglycerol, isoamyl nitrite, and isosorbide dinitrate:

$$\begin{array}{l} CH_2-O-NO_2 \\ | \\ CH-O-NO_2 \\ | \\ CH_2-O-NO_2 \end{array} \qquad \begin{array}{l} CH_3CHCH_2CH_2ONO \\ \quad | \\ \quad CH_3 \end{array}$$

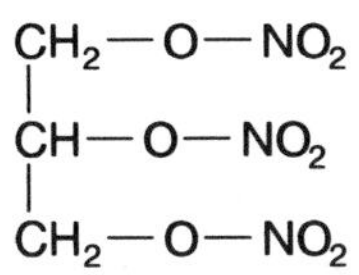

Nitroglycerol (#21)

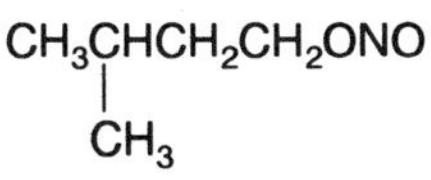

Isoamyl nitrite ("amyl nitrite")

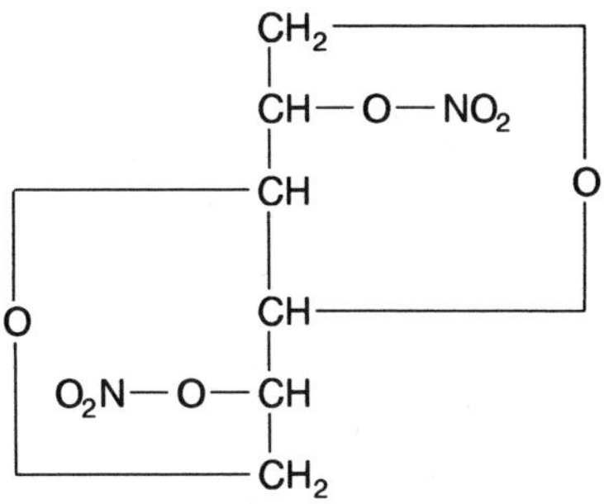

Isosorbide dinitrate (#54)

In general they relax the involuntary muscles especially those in the walls of the blood vessels near the heart. The blood vessels expand; the work load on the left ventricle is reduced and so is the heart's demand for oxygen. The mode of action of these compounds is largely unknown.

Nitroglycerol is the cornerstone of acute pain control. It is taken as a tablet under the tongue but its effect lasts for only 20–30 minutes. Repeated doses increase the chance of side effects. Sustained release tablets are available with this as with the other nitrates but there are doubts as to their effectiveness. Thus β blockers and calcium antagonists are often used instead. A new dosage form consists of an adhesive patch attached to the chest, which releases nitroglycerol, which in turn is absorbed through the skin (Figure 4.10).

Nitroglycerol has been used for over a century and amyl nitrite and other organic nitrates have also been used for decades. The choice among them is unimportant. Amyl nitrite has to be sniffed. It is a mixture of compounds, the chief of which is isoamyl nitrite. It appears to be used rarely in modern practice although it turns up as a street drug. Nitroglycerol is made by esterification of glycerol with nitric acid.

Isosorbide dinitrate is taken under the tongue but can also be swallowed. It is said to be the most effective of the nitrates and nitrites for decreasing blood pressure and relieving angina. It is made by the simultaneous nitration and dehydration of sorbitol.

7.5.2 Calcium Antagonists

The second group of vasodilators is the calcium antagonists. One of them, nifedipine, has been known for over 20 years, but the understanding of the mode of action of these drugs has led to an increased interest in them and three have entered the Top-100.

Variation of levels of free calcium ions within muscle cells is an important control mechanism in cell biochemistry and is related to the ability of calcium ions to form bridges (salts) between anionic groups. Muscle cells contain filaments of the so-called contractile proteins, myosin and actin, and these have anionic groups. When calcium ions enter the filaments, they cross-link these groups and, hence, cause the protein filaments to contract. When calcium ions leave the protein filaments, they relax again.

Calcium ions gain access to muscle cells through calcium channels or pores in the cell's outer membrane. In addition, the cell itself stores calcium ions whose release can be triggered by entering ions.

The calcium antagonists inhibit the inward flow of calcium across the membranes of heart cells and also the smooth muscle cells in the arteries. This relaxes the muscle cells, decreases the demand of the heart for oxygen, and allows the arteries to dilate. As explained in Section 7.5.4, this is what the β blockers also do, but in the case of the calcium antagonists the β receptors remain unblocked. This is an advantage, for example, in patients with bronchial problems where β blockers might cause asthma (Section 15.1).

The three most widely prescribed calcium antagonists are nifedipine (#37), verapamil (#93), and diltiazem (#45). Nifedipine is more potent than verapamil for angina but verapamil also counters arrhythmia. Nifedipine is useful as an adjunct to β blockers or for patients who react adversely to them. Many other calcium blockers have recently been licensed or are in the development stage and the area is one of great research interest.

Nifedipine is a 1,4-dihydropyridine. It is synthesized by condensation of 2 moles of ethyl acetoacetate with ammonia and *o*-nitrobenzaldehyde. The reaction is a step in the classical Hantzsch pyridine synthesis, a subsequent stage being the oxidation of the dihydropyridine to the pyridine.

NO_2, CHO, CH_3CH_2OOC, CH_2, H_3C, C=O, $COOCH_2CH_3$, CH_2, O=C, CH_3, NH_3 ⟶ NO_2, H, CH_3CH_2OOC, $COOCH_2CH_3$, H_3C, N–H, CH_3

Nifedipine

Diltiazem is shown in Figure 7.9. It is structurally similar to the benzodiazepine anxiolytics (Section 8.3) and, like chlorpromazine (Section 8.2), has a sulfur and nitrogen-containing ring with a dimethylaminoalkyl side chain. The first stage in the diltiazem synthesis is production of the epoxide (2) by epoxidation of the unsaturated precursor methyl *p*-methoxycinnamate (1). Reaction of (2) with the anion of 2-nitrothiophenol (3) opens the epoxide ring to give (4). The ester is saponified and the product resolved with cinchonidine to give the more active optical isomer. This is reduced to the amine (5), which

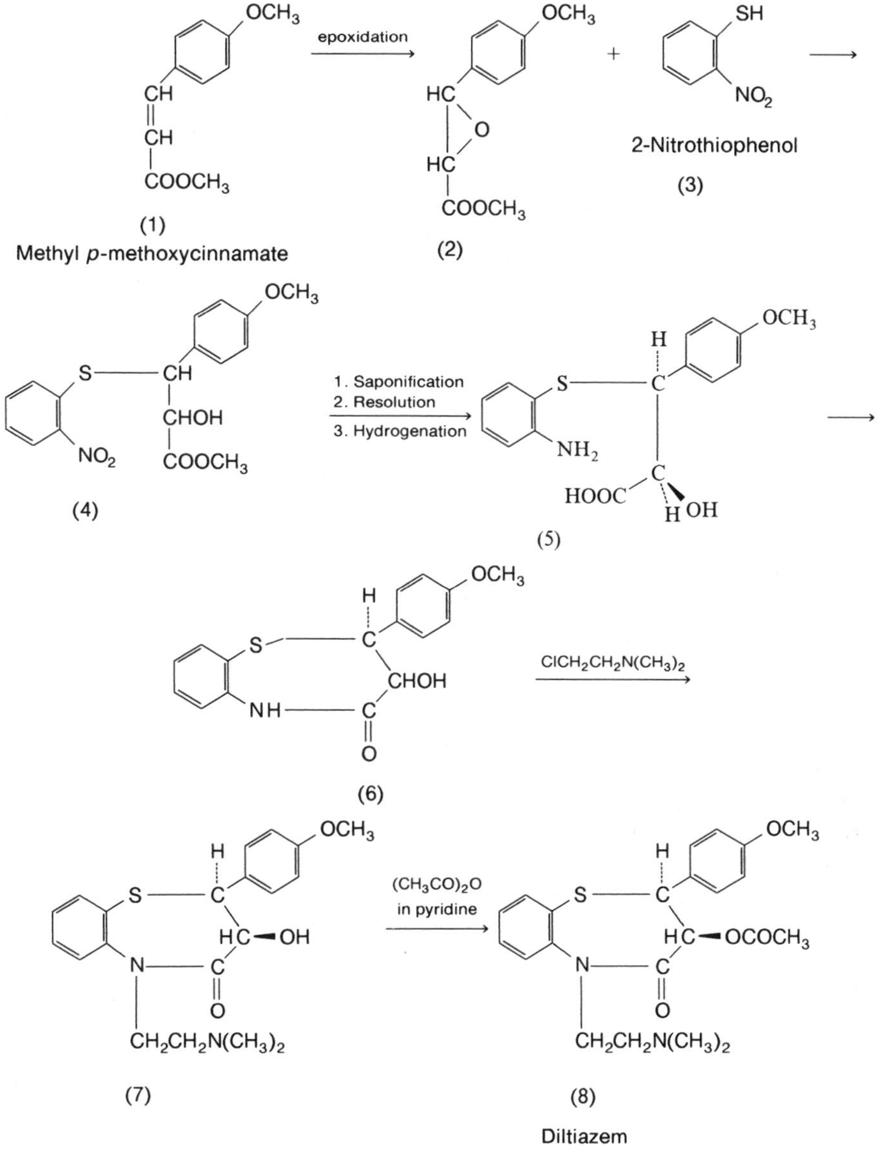

epoxidation

(1)
Methyl *p*-methoxycinnamate

(2)

2-Nitrothiophenol
(3)

1. Saponification
2. Resolution
3. Hydrogenation

(4)

(5)

$ClCH_2CH_2N(CH_3)_2$

(6)

$(CH_3CO)_2O$ in pyridine

(7)

(8)
Diltiazem

Figure 7.9. *Synthesis of diltiazem.*

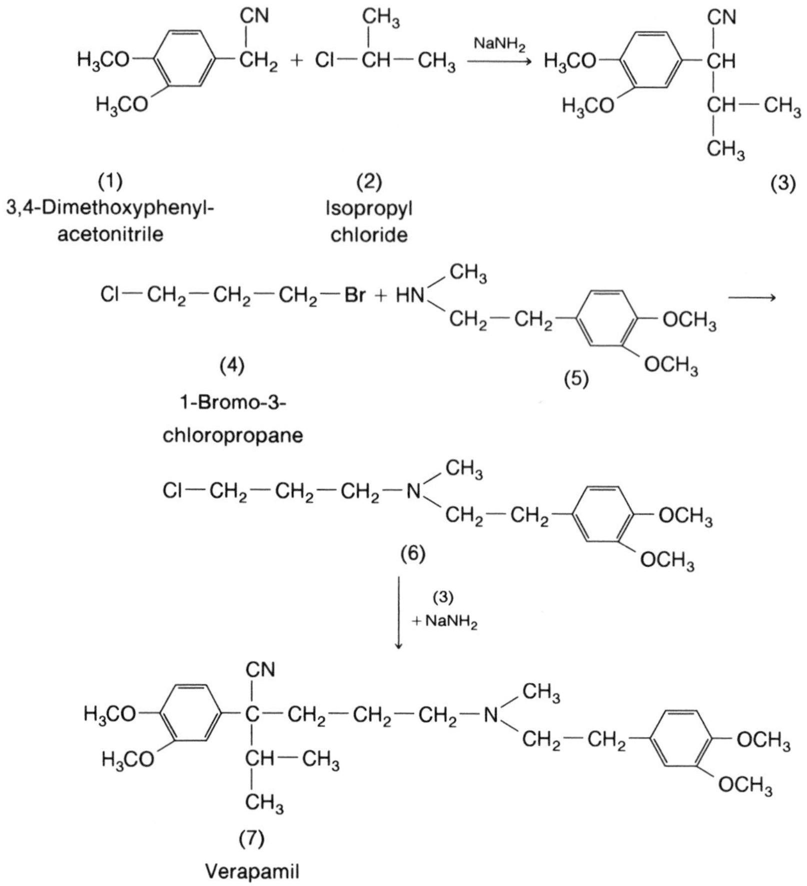

(1) 3,4-Dimethoxyphenyl-acetonitrile

(2) Isopropyl chloride

(3)

(4) 1-Bromo-3-chloropropane

(5)

(6)

(7) Verapamil

Figure 7.10. *Synthesis of verapamil.*

is cyclodehydrated to a 7-membered lactam ring (6). Alkylation of the amine with 2-chloroethyldimethylamine in the presence of dimethylsulfinyl sodium gives (7), and acetylation with acetic anhydride in pyridine gives diltiazem (8).

A synthesis of verapamil is shown in Figure 7.10. 3,4-Dimethoxyphenyl-acetonitrile (1) reacts with isopropyl chloride (2) in the presence of sodium amide to give the intermediate (3). Two other starting materials, 1-bromo-3-chloropropane (4) and the secondary amine (5), react to give another intermediate (6). (3) and (6) Combine in the presence of sodium amide to give verapamil (7).

7.5.3 Hemorheologic Agent

Pentoxifylline (#100) is a drug that reduces the effects of peripheral arterial disease by decreasing the viscosity of the blood; hence, it is classified as a hemorheologic agent. It improves the oxygenation of the ischemic tissues. Like

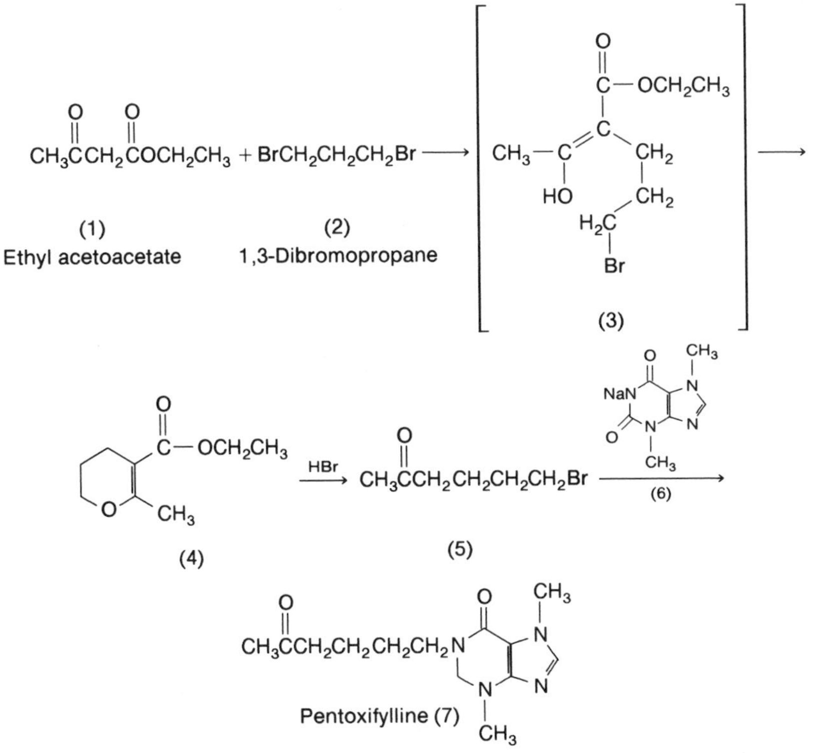

$CH_3COCH_2COOCH_2CH_3$ + $BrCH_2CH_2CH_2Br$ ⟶ [(3)] ⟶

(1) Ethyl acetoacetate (2) 1,3-Dibromopropane

(4) $\xrightarrow{HBr}$ $CH_3COCH_2CH_2CH_2CH_2Br$ (5) $\xrightarrow{(6)}$

Pentoxifylline (7)

Figure 7.11. Synthesis of Pentoxifylline.

caffeine (Section 8.5.2) and theophylline (Section 15.2), it is a xanthine derivative.

The side chain is prepared as shown in Figure 7.11. Ethyl acetoacetate (1) reacts with 1,3-dibromopropane (2) to give a bromoketone (3), which spontaneously cyclizes to the dihydropyran (4). Hydrogen bromide opens the ring again to give the bromoketone (5). Presumably a ketoester intermediate is formed which loses the carboxyethyl group. The sodium salt of theobromine (6) is treated with (5) and pentoxifylline (7) results. Theobromine itself has long been known as a mild diuretic, vasodilator, and smooth muscle relaxant.

7.5.4 Mode of Action of β-Adrenergic Agents

The discovery of the *β* blockers or *β*-adrenergic agents was, in many respects, the most important advance in heart therapy in the past 25 years. It was based on an understanding of body chemistry; hence, an appreciation of the significance of the *β* blockers requires an understanding of their mode of action. We will therefore digress to describe the nervous system (Figure 7.12), especially

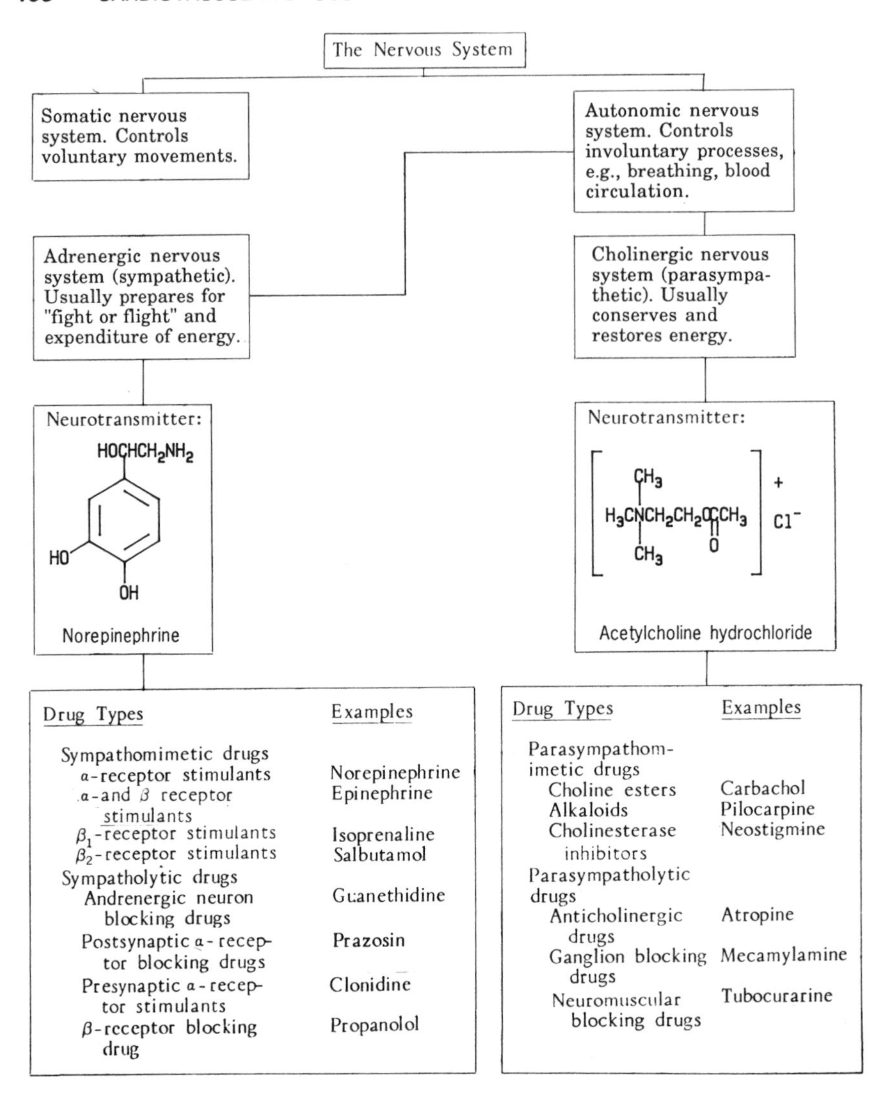

Figure 7.12. *Drugs affecting the nervous system.*

the autonomic nervous system. This will also be important for Chapter 8, which deals with drugs affecting the central nervous system.

The functions of the body are brought about by stimuli originating in the brain. Some of these functions are voluntary. A person, unless incapacitated, can raise an arm or move a hand at will and such actions are controlled by the so-called somatic nerves. Such functions as breathing and the circulation of the blood, however, are involuntary and occur without a conscious decision. They are controlled by the autonomic nervous system. The autonomic nervous system was formerly classified into so-called sympathetic and parasympathetic sections, but this classification has now been replaced by one based on the chemicals released at the nerve synapses. Some synapses release nor-

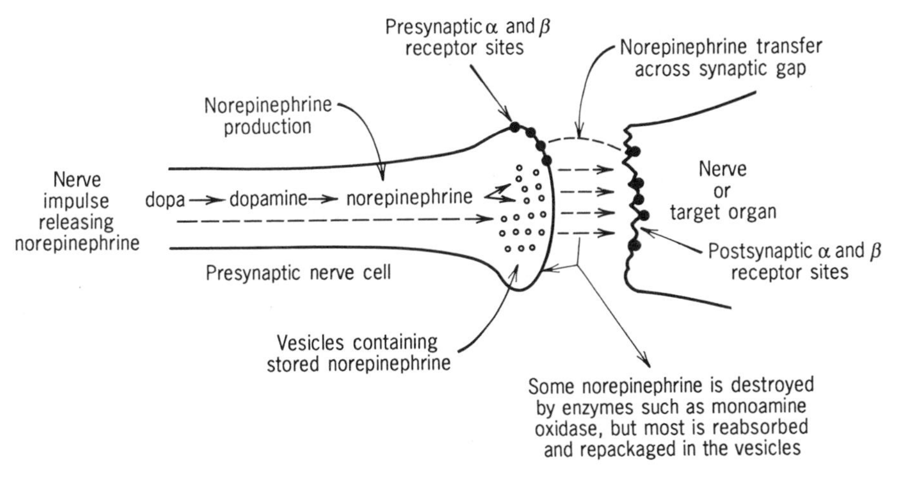

Figure 7.13. *Schematic representation of an adrenergic synapse.*

epinephrine, otherwise known as noradrenaline. They are called adrenergic nerves. Usually, but not always, they prepare the body for "fight or flight," that is, the expenditure of energy. Other synapses release acetylcholine, shown in the figure as the chloride, and these are known as cholinergic nerves. They are usually concerned with conserving and restoring energy. Often both sets of nerves impinge on the same organ with opposite effects. Because they help transmit messages along nerves, chemicals such as norepinephrine and acetylcholine are known as chemical messengers or neurotransmitters.

Drugs that stimulate the adrenergic nervous system are called sympathomimetic drugs and those that depress it are called sympatholytic. Drugs that stimulate the cholinergic nervous system are called parasympathomimetic and those that depress it are called parasympatholytic. Some examples of such drugs are given in Figure 7.12 and they will be discussed later.

Figure 7.13 illustrates what happens at a synapse in the adrenergic nervous system. On the left is a nerve cell which is a very long thin specialized cell and on the right is another nerve cell on the target organ which is controlled by the nerve. The gap between them is the synapse. L-Norepinephrine is biosynthesized in the nerve cells from L-dopa:

$$\text{L-Dopa} \xrightarrow[\text{decarboxylase}]{\text{L-Dopa}} \text{Dopamine} \xrightarrow[\beta\text{- hydroxylase}]{\text{Dopamine}} \text{L-Norepinephrine}$$

L-Dopa: NH_2, $CH_2CHCOOH$, HO, OH — Dopamine: $CH_2CH_2NH_2$, HO, OH — L-Norepinephrine: $HOCHCH_2NH_2$, HO, OH

L-Dopa is first decarboxylated to dopamine and then hydroxylated to norepinephrine by two enzymes. The norepinephrine is then stored in vesicles in the nerve ending.

When stimulated by an electrical impulse along the nerve, the vesicles release norepinephrine, which diffuses across the synaptic gap and interacts with receptors on the other side of the gap. This propagates another electrical impulse at an intermediate receptor or causes the appropriate response at an organ.

Once released by the nerve synapses, the neurotransmitters cannot be allowed to accumulate or the adjoining receptors would be permanently occupied. In the case of cholinergic nerves, the acetylcholine is destroyed by an enzyme, acetylcholinesterase. In the case of norepinephrine, there are enzymes that destroy it such as monoamine oxidase. But most of it is reabsorbed into the presynaptic nerve ending and repackaged in the vesicles for further use.

If the synapse mechanism operated in only this way, there would be a strictly unidirectional flow of information along the nerve. This is not so; there is also a feedback loop. The presynaptic nerve ending also possesses receptors that respond to norepinephrine and, if local concentrations of norepinephrine are high, they inhibit its further release.

Once it reaches the receptors, norepinephrine can have a variety of effects. In the past it was believed to be a stimulant always, but it is now known that it depresses certain kinds of bodily functions. This difference in action arises because of the existence of at least two kinds of adrenergic receptors. Those receptors responsible for excitatory action are called α receptors and cause muscles in the cells of which they occur to contract. Stimulation of α receptors constricts the blood vessels in the skin, raises blood pressure, and dilates the pupils of the eye.

Receptors with a mainly inhibitory action are called β receptors and allow muscles to relax. The ratio of α to β receptors in an organ decides how it will respond to a rise in norepinephrine in the body. The β receptors are subdivided again into β_1 receptors, which are found in the heart, and β_2 receptors, which are found in the bronchi. Although β receptors are usually inhibitory, those found in the heart are excitatory. Stimulation of β_1 receptors increases the heart rate and the output of blood from the heart. Stimulation of β_2 receptors relaxes the bronchial muscles and the uterus and dilates the blood vessels. Sympathomimetic drugs are drugs that stimulate these receptors and sympatholytic drugs block them.

7.5.5 β-Adrenergic Drugs

Figure 7.14 shows the structures and uses of the eight drugs mentioned at the bottom of Figure 7.12 as acting on the adrenergic nervous system. Norepinephrine, epinephrine, isoprenaline, and salbutamol are all agonists. They have structures close to that of norepinephrine and clearly interact with

α STIMULANT:

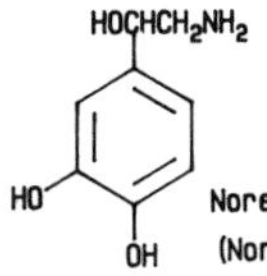

HOCHCH$_2$NH$_2$ HO OH Norepinephrine (Noradrenaline)

(Used for acute hypotension and cardiac arrest.)

α & β STIMULANT:

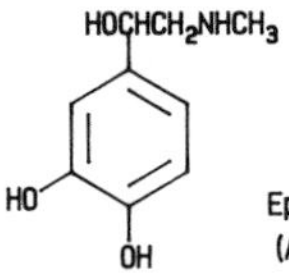

HOCHCH$_2$NHCH$_3$ HO OH Epinephrine (Adrenaline)

(Used for acute allergic and anaphylactic reactions.)

β(1) STIMULANT:

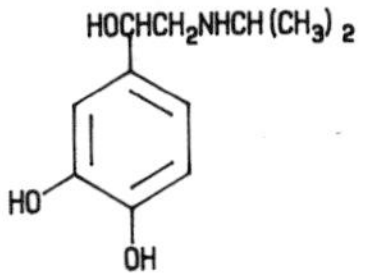

HOCHCH$_2$NHCH(CH$_3$)$_2$ HO OH Isoprenaline

(Used for bradycardia (slow heartbeat) and bronchospasm.)

β(2) STIMULANT:

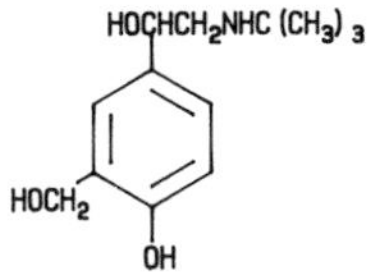

HOCHCH$_2$NHC(CH$_3$)$_3$ HOCH$_2$ OH Salbutamol

(Used for asthma.)

ADRENERGIC NEURON BLOCKER:

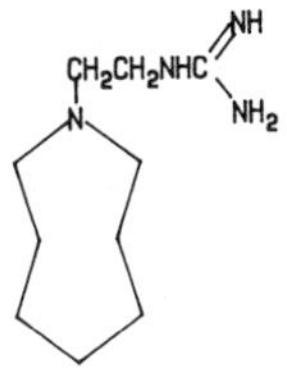

CH$_2$CH$_2$NHC(=NH)NH$_2$ N Guanethidine

(Used for hypertension.)

POST-SYNAPTIC α BLOCKER:

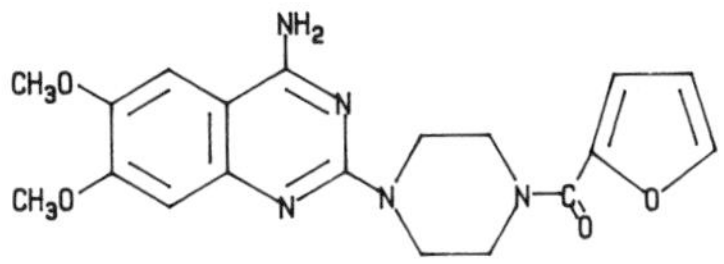

NH$_2$ CH$_3$O CH$_3$O N N N N–C=O O Prazosin

(Used for hypertension.)

PRE-SYNAPTIC α STIMULANT:

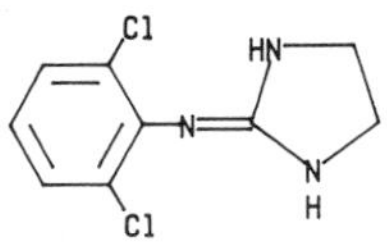

Cl Cl HN N= N H Clonidine

(Used for hypertension.)

β BLOCKER:

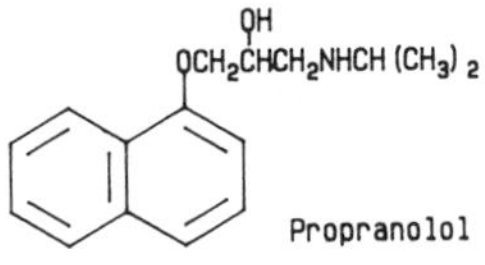

OH OCH$_2$CHCH$_2$NHCH(CH$_3$)$_2$ Propranolol

(Used for angina.)

***Figure 7.14**. Sympathomimetic and sympatholytic drugs.*

receptors in the same way. The first three compounds are used to stimulate the heart in one way or another, while the fourth, salbutamol, is used to relax the bronchi in cases of asthma.

The four sympatholytic drugs or blockers are shown in the lower half of the diagram. Of the four, only propranolol in the lower right corner has a structure resembling norepinephrine. Even propranolol differs from it more than the agonists in the top half of the diagram and clearly it attaches to the same receptor as norepinephrine and blocks the β receptor. Prazosin blocks the postsynaptic α receptors in the same way in spite of bearing no obvious structural resemblance to norepinephrine. Guanethidine inhibits the take-up of norepinephrine by the storage vesicles in the synapse so that supplies become depleted. Clonidine behaves as a postsynaptic α blocker. It is in fact an agonist for the presynaptic α receptors and by stimulating them it inhibits the output of norepinephrine from the vesicles. These four drugs are all used for heart disease and will be discussed later in this chapter.

After that digression, we return to the use of β-receptor-blocking drugs in the treatment of angina. Angina was mentioned briefly in Sections 7.1 and 7.5 but, now that the adrenergic nervous system has been described, it can be portrayed in more detail. The heart contains β_1 receptors and these are stimulated by norepinephrine. They are also stimulated by epinephrine, shown in Figure 7.14. Epinephrine is not thought to be a neurotransmitter. It is a hormone produced along with norepinephrine by the adrenal medulla, a gland located near the kidneys. It is the body's activator and is released in response to exercise, fear or anxiety.

In primitive humans, the system worked successfully. When frightened, their options were either to stand their ground and fight, or to run away. Both responses would require extra supplies of blood and oxygen to the muscles. Fright would trigger the adrenal gland, which would pump epinephrine into the bloodstream. The heart would beat faster and the additional blood and oxygen supply would be provided.

Today, fighting or fleeing is usually an inappropriate response to stress. Nonetheless, the adrenal gland continues to do its job and, in a person with an occluded coronary artery, this leads to angina and myocardial infarction.

If the patient takes β-blocking drugs, however, they occupy the β_1 receptor sites in the heart which would otherwise be stimulated by epinephrine. Consequently, the stimulus to the heart is reduced and it needs do less work. The total volume of blood pumped in a given time lessens and less oxygen is required. In addition, certain peripheral blood vessels relax and the blood pressure drops.

One theory of this antihypertensive action is that the epinephrine released under stress not only stimulates the β receptors of the heart but also the presynaptic β receptors on sympathetic nerves (Figure 7.13). This enhances norepinephrine release. The norepinephrine overstimulates postsynaptic α receptors, causing vasoconstriction and increasing hypertension. This mechanism explains the rapid antihypertensive effect of α-blocking drugs such as prazosin (Figure 7.17). Equally, blockade of the presynaptic β receptors reduces the release of norepinephrine at the synapse but does not reduce the nerve-stimulated release. Hence, there is an antihypertensive action. The problem with this theory is that the presynaptic β receptors are β_2, while selective β_1 blockers are nonetheless antihypertensive. Further work remains to be done and, clearly, the complete mechanism by which β blockers work is not fully understood.

7.5.6 β-Adrenergic Blocking Drugs

The search for a β blocker was started by J. W. Black in England in 1959 working at the time for ICI. There were some clues to the sort of structure that might work. In particular, it should resemble norepinephrine, but not too closely; otherwise it would be an agonist.

Figure 7.15 shows historically important β blockers and selected compounds currently on the market. The first compound to attach strongly to β receptors was dichloroisoproterenol. It was discovered in the United States in 1958, but it was a modified agonist not a blocker and stimulated the heart excessively. The first compound seriously considered as a drug was pronethalol, but it turned out to cause tumors in mice and was therefore unusable. Nonetheless, it underwent some tests on humans and the idea of β blocking was shown to work. Pronethalol differs from norepinephrine and propranolol in having β substitution on the naphthalene ring. Other β-substituted naphthalenes, particularly β-naphthol derivatives, are carcinogenic and it may be that this is the structural feature that accounts for the carcinogenicity of pronethalol. It would be pleasing to record that the switch from β- to α-substituted naphthols in the ICI research program came about because of these considerations, but the chemist concerned has since recorded that he switched because he could not find the bottle of β-naphthol.

The change to α-naphthols led rapidly to propranolol. It was a huge success. Introduced by ICI in the mid 1960s, it was still the third most widely prescribed drug in the United States in 1984. In addition to the antianginal effect it was expected to have, it turned out also to lower the blood pressure of patients with hypertension.

Research on β blockers did not cease, however, with the discovery of propranolol. First of all, ICI's competitors hoped to discover compounds equally good or better and thus gain a share of the market. Second, although propranolol was effective and had few side effects, there were still possible improvements and variations. The various β blockers differed, for example, in their cardioselectivity, agonist activity, and quinidine-like activity (Figure 7.15).

First, there is the question of cardioselectivity. Propranolol blocks β_2 receptors as well as β_1 receptors. This could be dangerous for patients who are liable to asthma because it could depress their already inhibited breathing. ICI developed practolol, which was much more selective than propranolol, and brought it onto the market in 1970. Unfortunately, it turned out to cause eye damage in a small proportion of cases and was withdrawn from general use in 1974. Since then metoprolol has been introduced by Ciba-Geigy and atenolol by ICI. Both of these affect β_1 receptors much more strongly than β_2

	Cardio-selectivity	Agonist Activity	Quinidine-Like Activity	Starting material for synthesis
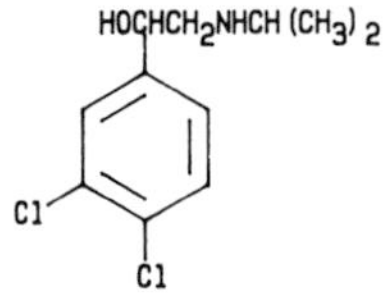 Dichloroisoproterenol	No	Yes	?	
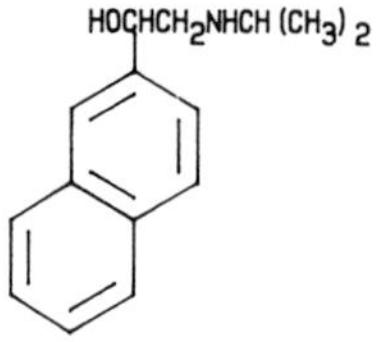 Pronethalol	?	?	?	
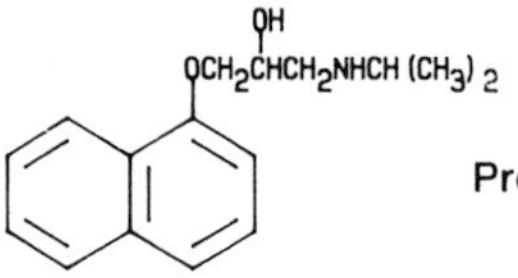 Propranolol (#10)	No	No	Yes	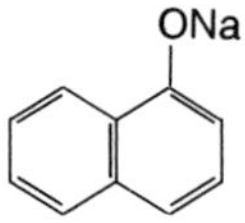 ONa
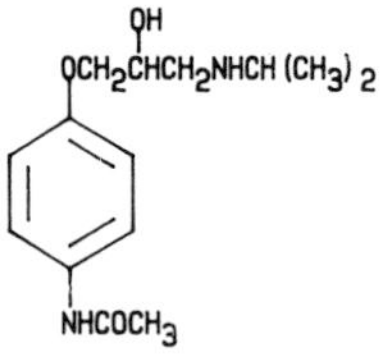 Practolol	Yes	Yes	?	
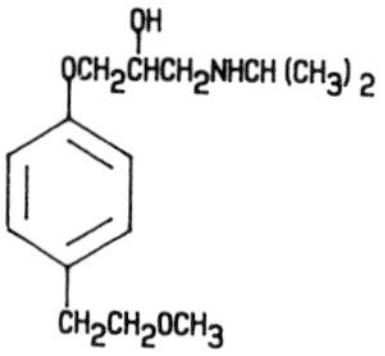 Metoprolol (#32)	Yes	No	No	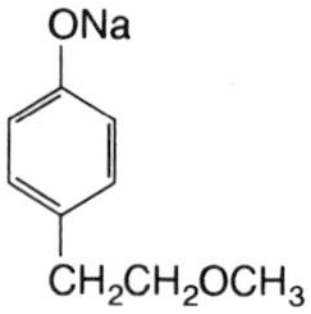 ONa, $CH_2CH_2OCH_3$
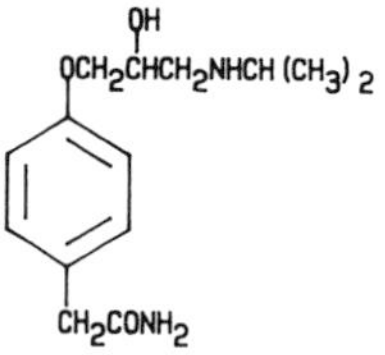 Atenolol (#17)	Yes	No	No	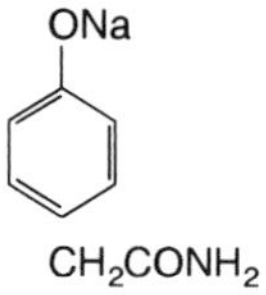 ONa, CH_2CONH_2
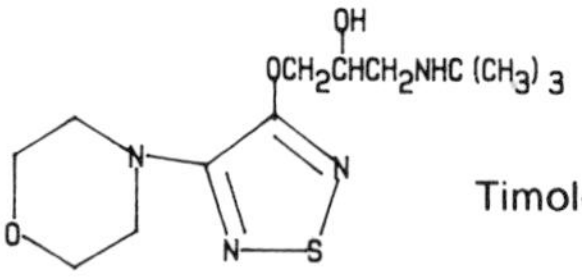 Timolol (#55)	No	No	Yes	
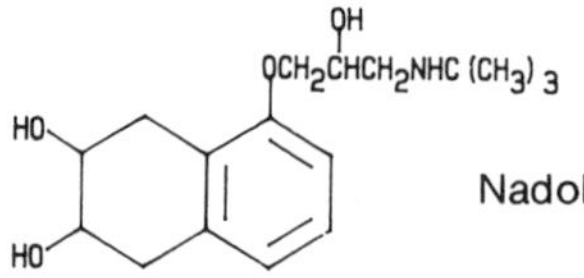 Nadolol (#64)	No	No	Yes	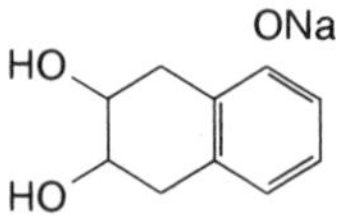 HO, HO, ONa

Figure 7.15. *β Blockers.*

receptors. They are a help to asthmatic patients but carry with them an enhanced risk of congestive heart failure and they are also less effective at reducing blood pressure.

A second possible drawback with propranolol was that it occasionally depressed the heart rate and lowered the blood pressure abnormally. Certain of the β blockers, although they blocked adrenaline, were themselves modified agonists; that is, they stimulated the heart mildly. This is indicated in Figure 7.15 in the column headed agonist activity. Practolol was an example. It prevented excessive slowing of the heart but this carried with it the corresponding risk that in some patients it might raise the blood pressure.

Third, propranolol acted like the alkaloid quinidine to stabilize the rhythm of the heart when it was disturbed, that is, it was an antiarrhythmic. This is indicated in the third column in Figure 7.15. A side effect was that it

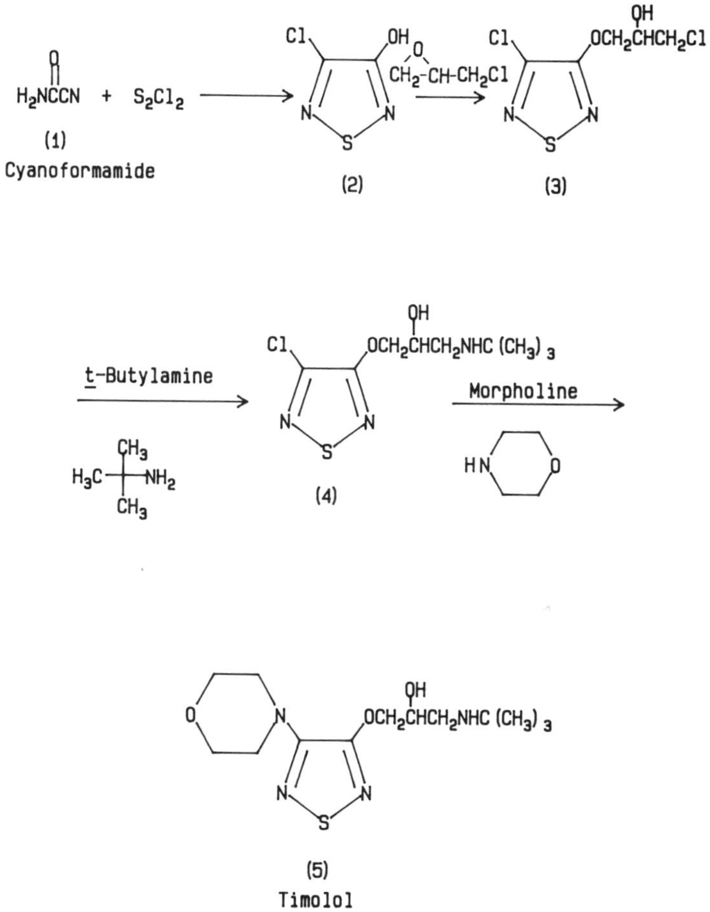

Figure 7.16. *Synthesis of timolol.*

sometimes led to bad dreams and lethargy. This quinidine-like activity was shared with timolol but absent from metoprolol.

A wide range of β blockers, some of which are shown in Figure 7.15, is now available to fit the needs of patients who do not react to propranolol in an optimum manner. Indeed, several so-called third-generation β blockers are undergoing clinical trials. Although commercial competition is one motive for their development, many of them are genuine advances and not just "me-too" drugs. The physician has to weigh, for example, the advantages of cardioselectivity against the drawbacks of possible congestive heart failure. Atenolol need only be taken once-daily. The range of drugs available makes it possible to make the best possible decision for each individual patient.

The β blockers turned out to have a number of other uses apart from treatment of heart disease. Most of them have a drying effect on the eye and this can be made use of in a condition called glaucoma. Timolol is the drug of choice for this application and has cut down the use of the alkaloid pilocarpine.

There are five β blockers currently in the Top-100 and they have similar side chains. Propranolol, metoprolol, and atenolol are synthesized in the same way. The sodium salt of the phenolic compound shown in the final column of Figure 7.15 is reacted with epichlorohydrin. The resulting ether is either reacted directly with isopropylamine or reacted first with hydrochloric acid to give the chlorohydrin, which is then reacted with isopropylamine:

$$\text{—ONa} + \text{ClCH}_2\text{CH}\overset{\text{O}}{\text{——}}\text{CH}_2 \longrightarrow$$

$$\text{—OCH}_2\text{CH}\overset{\text{O}}{\text{——}}\text{CH}_2 \xrightarrow{(\text{CH}_3)_2\text{CHNH}_2}$$

$$\text{—OCH}_2\underset{}{\overset{\text{OH}}{\text{CH}}}\text{CH}_2\text{NHCH}(\text{CH}_3)_2$$

Nadolol follows the same pattern except that isopropylamine is replaced by *tert*-butylamine. Timolol has the same side chain as nadolol but the synthesis is different in several ways including the reaction with epichlorohydrin. It is shown in Figure 7.16. It starts with cyanoformamide (1) which is reacted with sulfur chloride to give the thiadiazole (2). *O*-Alkylation of (2) with epichlorohydrin in the presence of catalytic amounts of piperidine yields the chlorohydrin (3). Reaction of (3) with *tert*-butylamine gives (4). The final step involves reaction of (4) with morpholine to yield timolol (5).

The β-blocking activity of the above compounds rests with one enantiomer which has the (S) absolute configuration. Conventional resolution techniques via optically active acids are tedious and time-consuming and a general asymmetric synthesis has been sought (see bibliography).

7.5.7 α- and β-Adrenergic Drug

A single drug, labetalol (#132) blocks α and β receptors. Its structure is

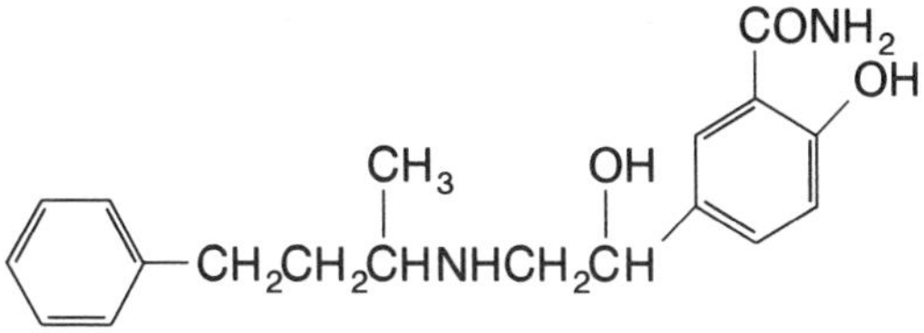

$CONH_2$
OH
CH_3 OH
$C_6H_5-CH_2CH_2CHNHCH_2CH$

Labetalol

It is the preferred drug for patients with simultaneous angina and hypertension and may save them from having to take both β blockers and antihypertensives.

Labetalol exists as four stereoisomers and it turns out that two are α blockers and two β blockers. The α blocker is due to be launched as a drug in its own right.

7.6 ANTIHYPERTENSIVE AGENTS

The next disease of the heart for which drugs are listed in Table 7.1 is high blood pressure or hypertension. Two forms of hypertension are recognized. Essential or ideopathic hypertension may be inherited or it may be due to arteriosclerosis, that is, the constriction of the peripheral arteries so that greater pressure is required to circulate blood through them. Secondary hypertension, on the other hand, is the result of a known disorder such as kidney disease. Mild hypertension is sometimes treated with β blockers (Section 7.5.5) or thiazide diuretics (Section 7.4.2). More serious hypertension is treated with special drugs, the most important of which are shown in Figure 7.17, together with reserpine which is of historical importance. Four of them appear in the Top-100.

7.6.1 Centrally Acting Antihypertensives

The first group of antihypertensives is those acting on the central nervous system. The oldest member of the group, indeed the first drug against hypertension, is reserpine, which was discovered in the early 1950s. It is an alkaloid and was isolated from the plant *Rauwolfia serpentina*. Its structure is

1. Acting on the central nervous system:

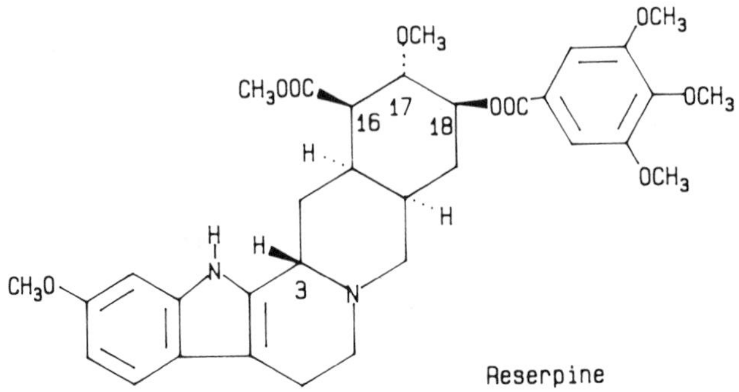

Reserpine

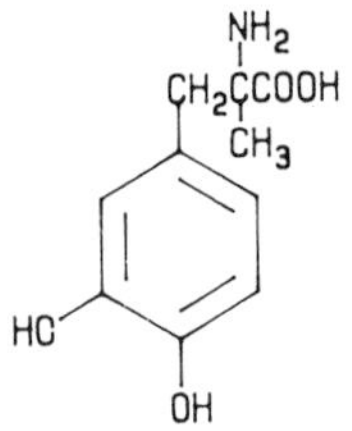

Methyldopa (#29)

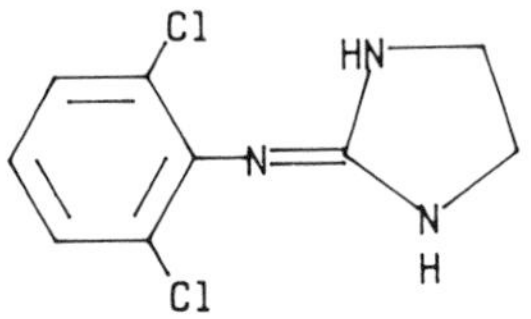

Clonidine (#56)

2. Vasodilators:

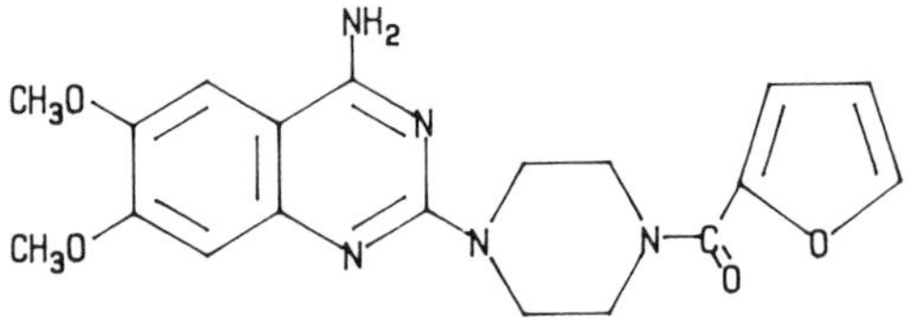

Prazosin (#50)

Figure 7.17. *Antihypertensive drugs.*

shown in the figure and the stereochemistry is crucial. Reserpine is one of 64 possible stereoisomers and none of the others show pharmacological activity. The β configuration at positions 3, 16, and 18 plus a 17-α configuration are essential.

An important discovery resulting from early work with reserpine was its ability to relieve the symptoms of certain psychotic disorders. Thus, reserpine was one of the first tranquilizers and opened up the science of psychopharmacology. Better tranquilizers are now available and will be discussed in Chapter 8.

Reserpine and many of the other antihypertensive drugs are sympatholytic drugs, that is, blockers of the adrenergic nervous system (Figure 7.12). The mode of action of reserpine is not clear. It seems to block α receptors to some extent, but its action is mainly due to inhibition of the norepinephrine storage

3. Adrenergic neuron blocker:

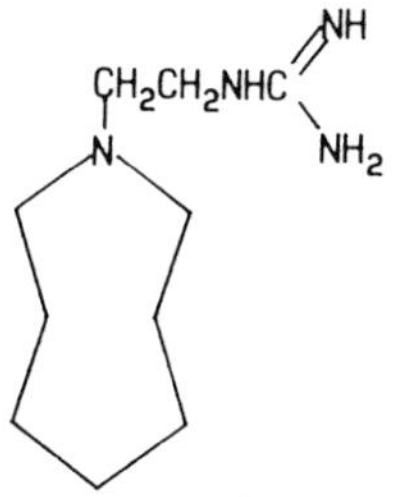

Guanethidine

4. Angiotensin converting enzyme inhibitors

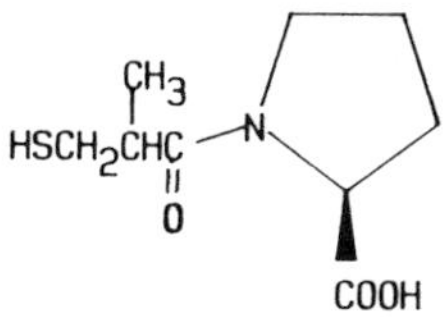

Captopril (#47)

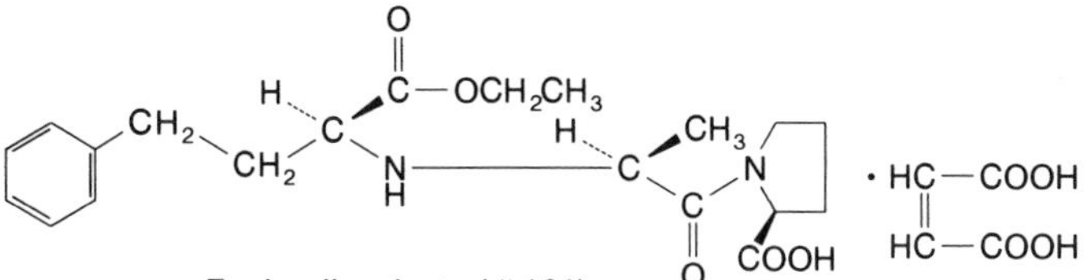

Enalapril maleate (#101)

Figure 7.17. *Continued.*

mechanism and the reentry of norepinephrine into the vesicles. Stores of norepinephrine and related neurotransmitters in the brain and various other organs are depleted.

When reserpine is given by mouth, the onset of action is slow and it has no effect for three to six days. It may cause mental depression and on occasions has led to suicide. This and other side effects have led to reserpine's replacement by more recently discovered antihypertensives, and it has dropped out of the Top-100.

Methyldopa (#29) is currently the most widely prescribed antihypertensive, meriting its position because of its lack of side effects. It is an example of a drug whose discovery was based on theoretical considerations. In Section 7.5.4 the conversion in the body of L-dopa to norepinephrine is shown. It is brought about by two enzymes. The equivalent processes for methyldopa are

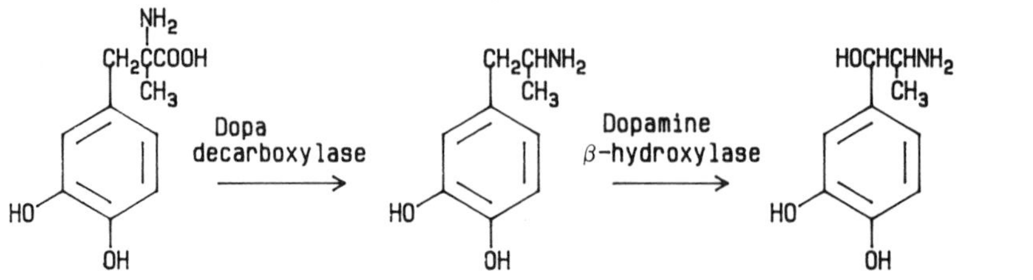

Methyldopa, taken as a drug, competes with dopa for supplies of dopa decarboxylase, so less dopamine is produced in the body. Instead, methyldopa gives methyldopamine but it gives it more slowly because the methyl group next to the carboxyl slows the decarboxylation. The methyldopamine in turn gives methylnorepinephrine. The latter is a weaker α receptor stimulant than norepinephrine and is produced in smaller amounts. Consequently, the α receptors receive less stimulus and the blood pressure drops. Modified agonists of this kind are sometimes said to be false transmitters. Methyldopa is also converted to methylnorepinephrine in the brain where its effect is slightly different but adds to the peripheral effect.

The synthesis of methyldopa starts with the ketone (1) which reacts with ammonium chloride and potassium cyanide to give an α-aminonitrile (2). The L isomer is separated from the racemate by a conventional resolution with camphorsulfonate. The unwanted D isomer is racemized back to the starting ketone which is recycled. The L isomer is treated with concentrated sulfuric acid which hydrolyzes the nitrile to a carboxyl and removes the ether methyl group to give L-methyldopa:

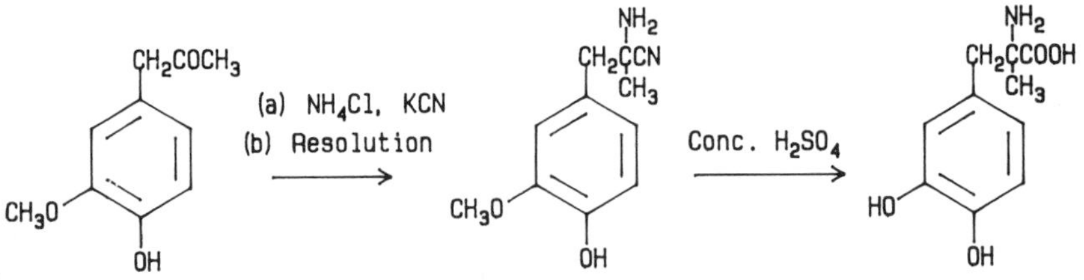

Clonidine is the third centrally acting antihypertensive and is a presynaptic α blocker. Its main side effects are sedation and lack of salivation. Also, blood pressure may rise sharply when therapy ceases. Clonidine is an imidazoline

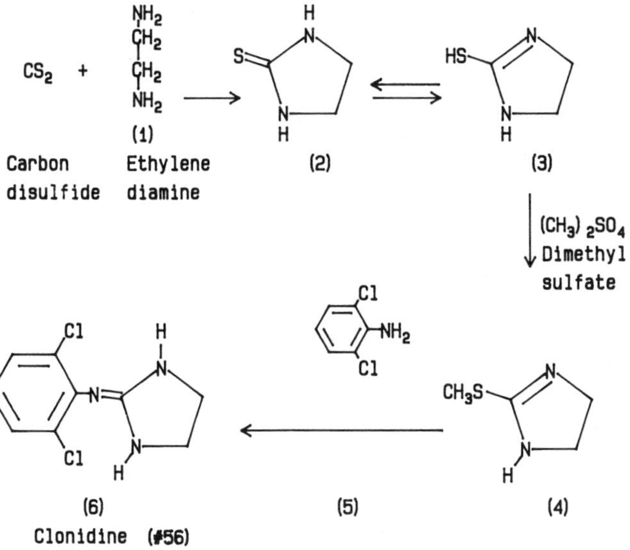

Figure 7.18. Synthesis of clonidine.

and this class of compounds generally has a similar biological effect to epinephrine and norepinephrine.

The synthesis of clonidine is shown in Figure 7.18. Reaction of ethylene diamine (1) with carbon disulfide gives (2) which is tautomeric with a sulfhydryl (3). Methylation gives the *S*-methylimidazoline (4). This is reacted with 2,6-dichloroaniline (5) to give clonidine (6).

7.6.2 Vasodilators

Vasodilators act by allowing the smooth muscles in the peripheral blood vessels to relax so that the blood can pass through them more easily. They are often used together with β blockers.

Prazosin (#50) is a postsynaptic α blocker. Its synthesis, shown in Figure 7.19, requires nine steps. It starts with 2,3-dimethoxybenzaldehyde (1). Nitration is followed by oxidation of the aldehyde to yield the acid (2). The carboxyl group in turn is converted to an amide. Catalytic reduction of the nitro group gives an amine (3). Reaction of this amine–amide with urea contributes a heterocyclic ring (4). Reaction of (4) with phosphorus oxychloride gives a dichloride (5). The two chlorines have differing reactivities and the more reactive, which is allylic, is replaced by an amine group on treatment with ammonia in tetrahydrofuran to give (6). The second chlorine is then reacted with excess piperazine to give (7). Acylation of the secondary amine group with furoyl chloride gives prazosin (8). The secondary amine reacts preferentially to the primary amine which is much more aromatic in character.

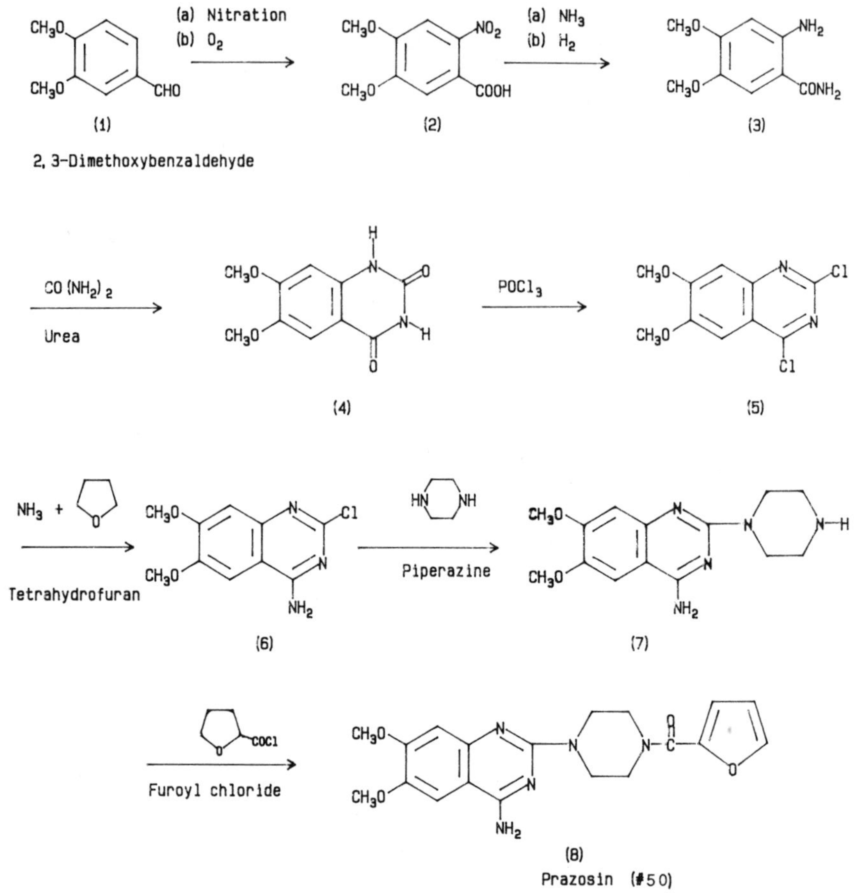

(a) Nitration (b) O_2 → (a) NH_3 (b) H_2 →

(1) 2, 3-Dimethoxybenzaldehyde (2) (3)

$CO(NH_2)_2$ Urea → $POCl_3$ →

(4) (5)

NH_3 + Tetrahydrofuran → Piperazine →

(6) (7)

Furoyl chloride →

(8) Prazosin (#50)

Figure 7.19. Prazosin.

7.6.3 Adrenergic Neuron Blockers

Guanethidine is an adrenergic neuron blocker. It inhibits storage of norepinephrine in the nerve vesicles (Section 7.5.4). It is reserved for patients with severe hypertension who do not respond to other drugs, but it is not commonly used today because of problems with regulation of dosage, and the availability of the ACE inhibitors (Section 7.6.4).

7.6.4 Angiotensin-Converting Enzyme Inhibitors

The final group of antihypertensives is the angiotensin-converting enzyme (ACE) inhibitors. They are used in severe hypertension that does not respond to other drugs.

Angiotensinogen is a protein in the body and is converted to angiotensin I by renin, an enzyme released by the kidney. Angiotensin I is a decapeptide and is biologically inactive but it is converted by the angiotensin-converting enzyme to angiotensin II. Angiotensin II is an octapeptide and is the most potent material known for raising blood pressure. It does so by causing the smooth muscle of the arteries to contract and also acts on the adrenal cortex to release aldosterone, the salt and fluid retaining hormone (Section 7.4). It is destroyed rapidly in the body by angiotensinase A to give various inactive metabolites.

There are various stages at which this process could be modified and the field is one of great interest for researchers. Drugs have been investigated that block the action of renin on angiotensinogen, of angiotensin-converting enzyme on angiotensin I, and of angiotensin II on arterial smooth muscle. Two compounds, captopril (#47) and enalapril maleate (#101), were in use at the beginning of 1987 and many more were in the pipeline including lisinopril, a lysine derivative of enalapril (see below). Captopril and enalapril share the same ring structure (Figure 7.17) but have different side chains. Enalapril is a pro-drug and the ethyl group hydrolyzes in the body to give the true drug, enalaprilat, which has a free carboxyl in the center of the molecule rather than an ester grouping.

The enzyme that converts angiotensin I to angiotensin II (and that is blocked by ACE inhibitors) had not been studied by x-ray diffraction at the time that captopril was discovered but it was known to contain zinc. The thiol group of captopril was designed to "anchor" to the zinc atom since the alanine–proline residues were known to bind to the active site. Enalapril lacks the thiol group but has a powerful zinc-binding facility at the carboxyl group which results from hydrolysis of the ester group.

Figure 7.20 shows the supposed binding of captopril to the enzyme with its carboxyl interacting with the basic side chain of an amino acid residue and its carbonyl hydrogen bonding with a secondary amine group. A molecular graphics representation of this structure suggested that a drug would be more potent if the methyl side chain were anchored in position and made part of a ring system. The outcome of this was the experimental drug cilazapril, which has the same side chain as enalapril but a rigid conformation:

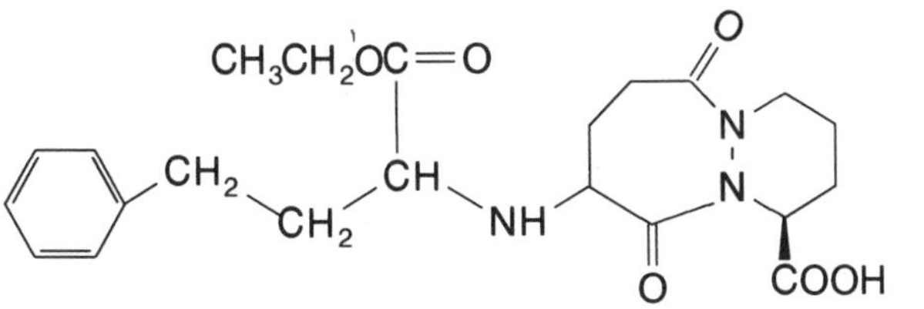

Cilazapril

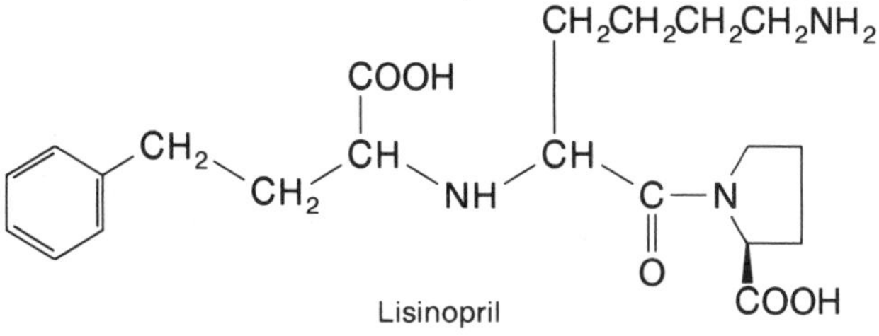

Lisinopril

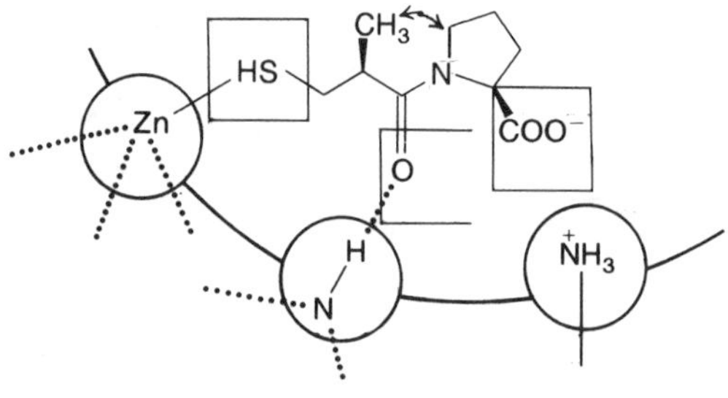

Figure 7.20. A schematic proposal for the binding of captopril to the angiotensin-converting enzyme. The binding functions of captopril (dashed squares) interact with the corresponding functions of the angiotensin-converting enzyme (circles). The dashed arrow shows the basis for preparation of more rigid bicyclic analogues such as cilazapril. Reproduced from reference 7/12 by permission of Roche Products.

Captopril may be made by a number of routes, one of which is shown in Figure 7.21. The amino acid L-proline (1) is treated with carbobenzoxy chloride to give (2). Isobutylene plus acid gives (3) in which the carboxy group is protected by a *tert*-butyl group. Hydrogenolysis gives the *tert*-butyl ester of L-proline (4). A second reagent (8) is prepared by reaction of methacrylic acid (5) with thioacetic acid (6) to give (7), which, on treatment with thionyl chloride, gives (8). (4) And (8) react in the presence of base to give (9), and the protective *tert*-butyl group is removed with trifluoroacetic acid to give (10). Resolution of the dicyclohexylamine salt in ethyl acetate yields the D isomer (11) and treatment with ammonia in methanol removes the acetyl group from the side chain to give captopril.

7.7 ANTICOAGULANTS

Warfarin (#68) is the preferred oral anticoagulant. It functions by reducing the concentration of prothrombin in the blood. Prothrombin is the agent responsible for blood clotting and its presence also gives blood a higher viscosity. Administration of warfarin reduces blood viscosity and, hence, the workload on the heart and also reduces the chance of thrombosis. The corresponding disadvantage is an increase in the possibility of hemorrhage, and patients taking warfarin must be carefully monitored.

In addition to its role in humans, warfarin is also used to kill rats and mice. After taking the bait, they are liable to bleed to death from some trivial injury, preferably a long way away from where the bait was set.

Warfarin is a coumarin derivative and is made as shown in Figure 7.22 by condensation of *o*-hydroxyacetophenone (1) with diethyl carbonate. Either O- or C-alkylation may occur. Formula (2) shows C-alkylation but it is of no significance. Either the O- or C-alkylated compound will cyclize to yield the coumarin (3). This compound in turn will undergo a Michael condensation with benzalacetone (4) to give warfarin (5).

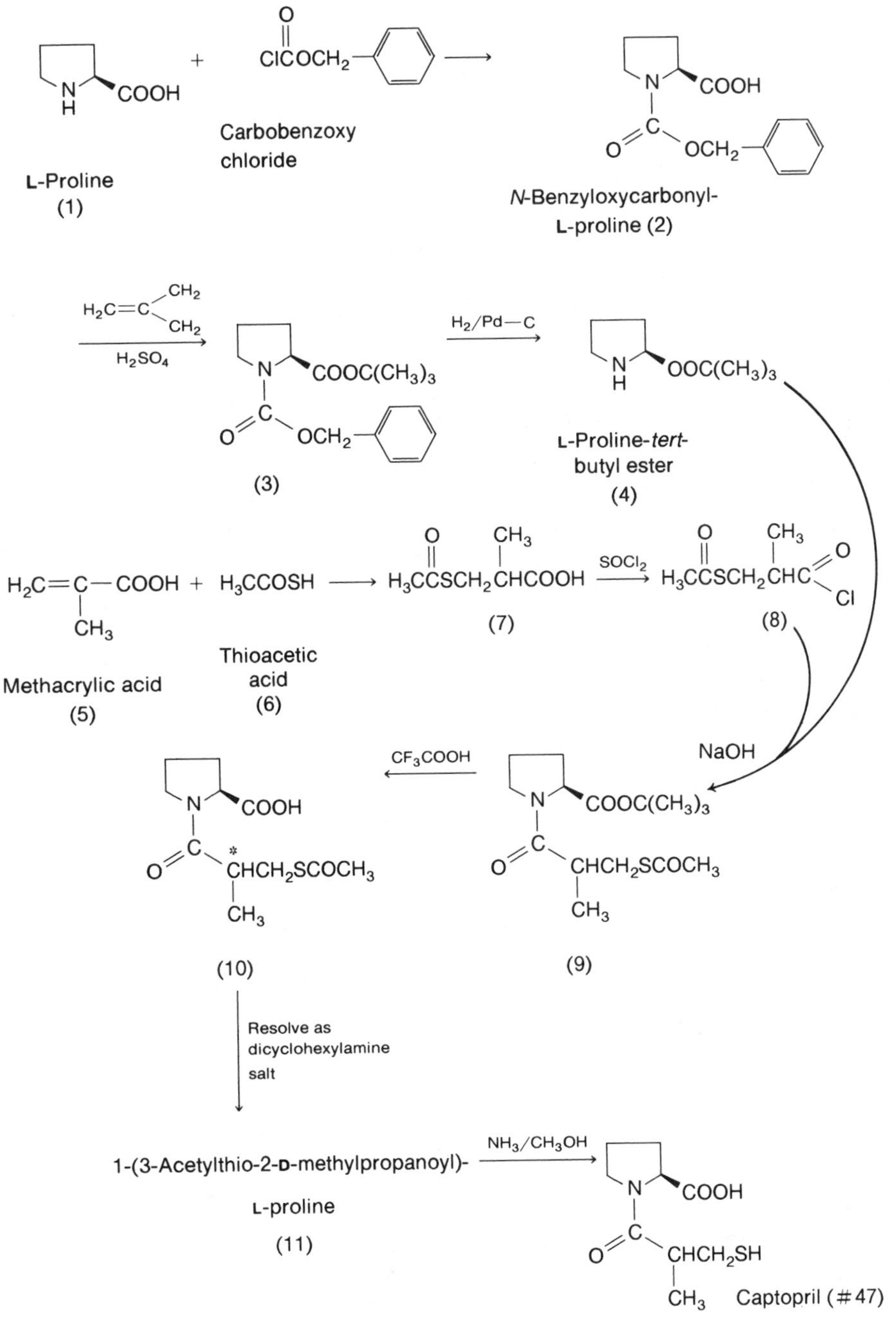

L-Proline (1) + Carbobenzoxy chloride ($ClCOCH_2C_6H_5$) ⟶ *N*-Benzyloxycarbonyl-L-proline (2)

(2) $\xrightarrow[H_2SO_4]{H_2C{=}C(CH_3)_2}$ (3) $\xrightarrow{H_2/Pd{-}C}$ L-Proline-*tert*-butyl ester (4)

$H_2C{=}C(CH_3){-}COOH$ + H_3CCOSH ⟶ $H_3CCOSCH_2CH(CH_3)COOH$ (7) $\xrightarrow{SOCl_2}$ $H_3CCOSCH_2CH(CH_3)COCl$ (8)

Methacrylic acid (5)

Thioacetic acid (6)

(4) + (8) $\xrightarrow{NaOH}$ (9) $\xrightarrow{CF_3COOH}$ (10)

(9): *N*-($COCH(CH_3)CH_2SCOCH_3$)-proline $COOC(CH_3)_3$

(10): *N*-($CO\overset{*}{C}H(CH_3)CH_2SCOCH_3$)-proline COOH

(10) ⟶ Resolve as dicyclohexylamine salt ⟶ 1-(3-Acetylthio-2-D-methylpropanoyl)-L-proline (11)

(11) $\xrightarrow{NH_3/CH_3OH}$ *N*-($COCH(CH_3)CH_2SH$)-proline COOH — Captopril (#47)

Figure 7.21. *Synthesis of captopril.*

The most effective anticoagulant is heparin. It was discovered in 1916 but still has not been completely synthesized. It is a polysaccharide and the present view of its structure is

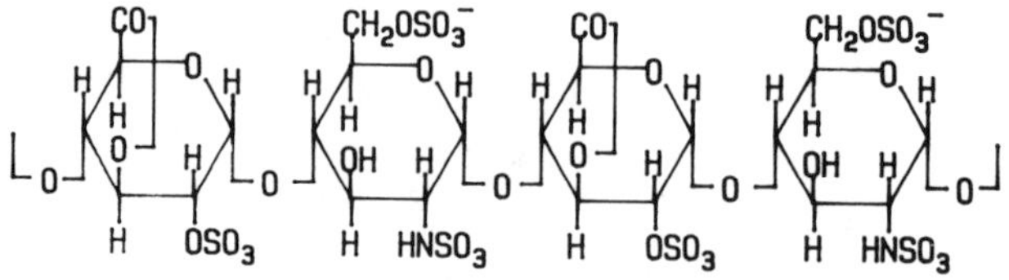

Heparin for therapeutic use is extracted from bovine lung tissue and the intestinal mucosa of pigs and cattle. It is one of the drugs still obtained from animal sources. In 1973, 35 million pounds of mucosa were processed to give about 3000 pounds of heparin. Heparin does not appear in the Top-100 because it is not pH stable and must be injected. It is used in the treatment of thrombosis and is especially useful before and after surgery to prevent thrombotic complications.

An unusual heart drug with effects similar to warfarin is ankrod, the venom from the Malayan pit viper, which contains an enzyme that reduces a material called fibrogen in the blood and, hence, its viscosity. It thus helps people with narrowed arteries. Unfortunately, people taking ankrod develop antibodies that negate its effect so that long-term therapy is impossible at present. It is

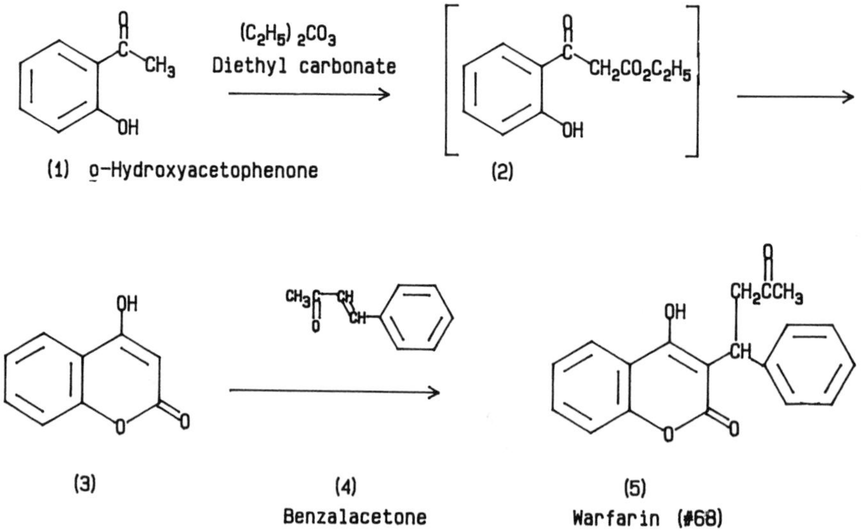

(1) o-Hydroxyacetophenone

$(C_2H_5)_2CO_3$
Diethyl carbonate

(2)

(3)

(4)
Benzalacetone

(5)
Warfarin (#68)

Figure 7.22. Synthesis of warfarin.

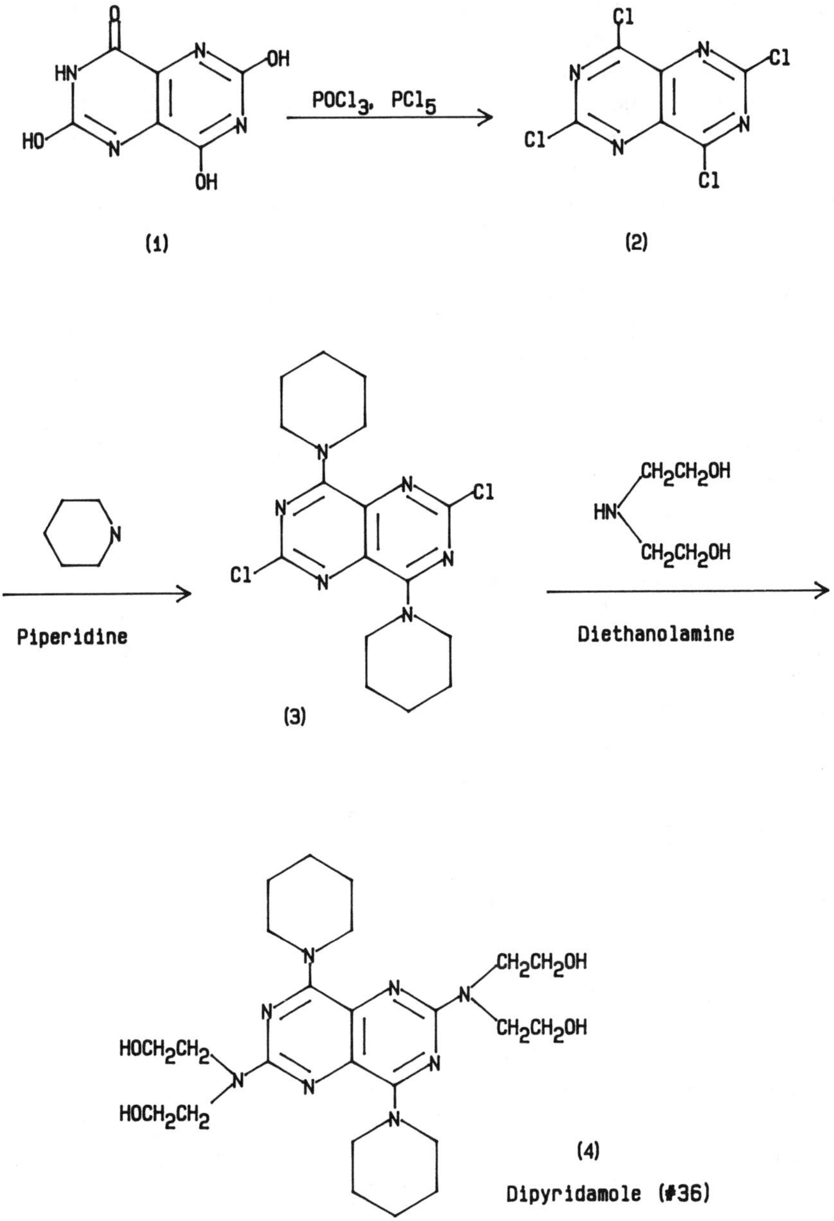

$POCl_3$, PCl_5

(1) (2)

Piperidine

(3)

Diethanolamine

(4)

Dipyridamole (#36)

Figure 7.23. Synthesis of dipyridamole.

reasonable to believe that chemical modification could alter this undesirable property.

The above anticoagulants are effective in preventing blood clotting in the veins but have little effect on the arteries. On the arterial side of the circulation, clotting occurs by platelet aggregation. Blood cells are subdivided into red cells, white cells, and platelets. Platelets are minute, granular, disk-shaped bodies, more numerous than white cells but less numerous than red cells. Their aggregation is a complicated process and is mediated by the prostacyclin (PGI_2)–thromboxane A_2 balance (Sections 19.1.2, 19.1.3).

Many drugs interfere with platelet aggregation but aspirin and dipyridamole are the most widely used. Aspirin inhibits the cyclooxygenase that leads to both prostacyclin and thromboxane.

Dipyridamole (Figure 7.23) inhibits an enzyme, phosphodiesterase, which degrades cyclic adenosine-3′,5′-monophosphate (cAMP). Hence, levels of cAMP increase. This in turn stimulates prostacyclin formation and inhibits the aggregation of platelets.

Dipyridamole was originally marketed as a coronary vasodilator (Section 7.5). It is chemically different from the other vasodilators and contains a pyrimido-pyrimidine ring system. It was #64 in the Top-100 in 1979 but is no longer recommended for angina in the United Kingdom and declared only "possibly effective" in the United States. Its subsequently recognized role as an antiplatelet drug, however, sent it to #36 in 1986.

Figure 7.23 shows the synthesis of dipyridamole, one that is more straightforward than most. It is made from a pyrimido-pyrimidine (1) closely related to triamterene (Figure 7.6). When treated with phosphorus oxychloride and phosphorus pentachloride, compound (1) gives a tetrachloro compound (2). Two of the chlorine atoms are more reactive than the others and will react with piperidine at room temperature to give (3). Further reaction with diethanolamine gives dipyridamole (4).

7.8 THROMBOLYTIC AGENTS

Thrombolytic agents are materials that cause blood clots to dissolve. In acute myocardial infarction, the coronary artery is blocked by a blood clot and the heart muscle dies from lack of oxygen. This is not necessarily a sudden event and, if a thrombolytic agent could be given within a few hours of the heart attack, the blood supply might be restored in time to save the victim.

The area is one of great research interest. Two products have been licensed in some markets at the time of writing: Activase and Eminase. Activase is a recombinant human tissue-type plasminogen activator (t-PA). t-PA is an enzyme produced naturally by the human body but the drug is made by a cell-cloning biotechnological technique. Eminase is a *p*-anisoylated (human) lys–plasminogen–streptokinase activator complex.

Both materials act by converting the inert proenzyme plasminogen into its

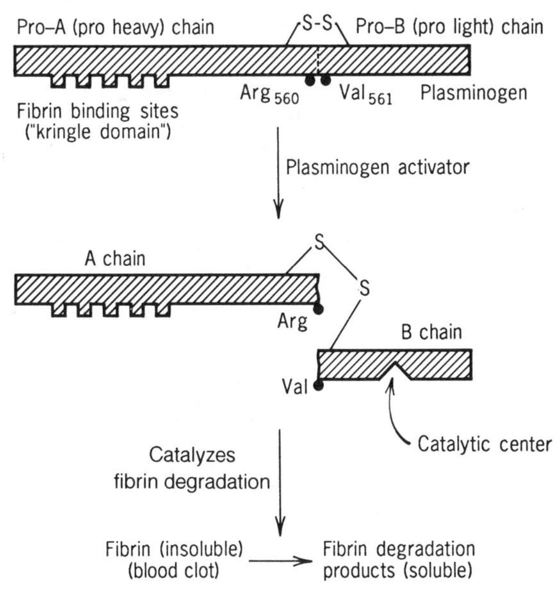

Figure 7.24. *Mode of action of plasminogen activators.*

proteolytically active form, plasmin (Figure 7.24). They do this by hydrolyzing a single chemical bond, arginyl(560) to valine(561), which links together the A (heavy) and B (light) chains of plasminogen. The newly formed B chain changes its conformation and a catalytic center is created. This catalytic center is capable of digesting the insoluble fibrin network that makes up the blood clot into smaller soluble degradation products.

A difference between the products is that Eminase, because of the anisoyl modification to the natural enzyme, is longer-lived in the body. Hence, it may be given by injection into a vein, whereas Activase is administered as a drip.

The appearance of Eminase and Activase revived interest in streptokinase, a protein from *Streptomyces haemolyticus*, which was an earlier tissue plasminogen activator. It was suggested that this had been incorrectly assessed when it first appeared and that, given with aspirin, it is as effective as the newer materials.

7.9 HYPOLIPEMICS

Hypolipemics are drugs that reduce the level of lipids, especially cholesterol, in the blood. A high level of cholesterol in the blood is linked, although in a complicated way, with atherosclerosis, that is, with fatty deposits on the artery walls. Levels above 200 mg cholesterol/100 mL blood have been associated with early death from heart disease.

Cholesterol is made by body cells especially in the liver and is also obtained from animal fats in the diet. It is not soluble in the blood and is converted by the intestine and liver into protein–lipid compounds (lipoproteins) in which form it is transported around the body. Low-density lipoproteins (LDL) carry cholesterol from the liver to cells throughout the body, including those lining the arteries near the heart, and this leads to atherosclerosis. High-density lipoproteins (HDL) perform a protective function by transporting cholesterol from cells to the liver, where it is broken down to give bile acids.

An early hypolipemic, administered in massive doses and widely used today, was niacin or nicotinic acid, a member of the vitamin B complex:

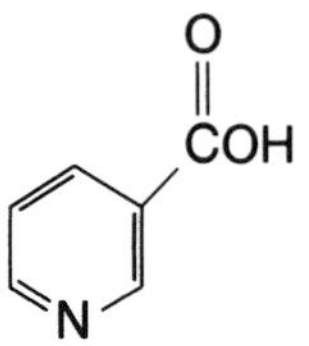

O ‖ COH — N

Niacin

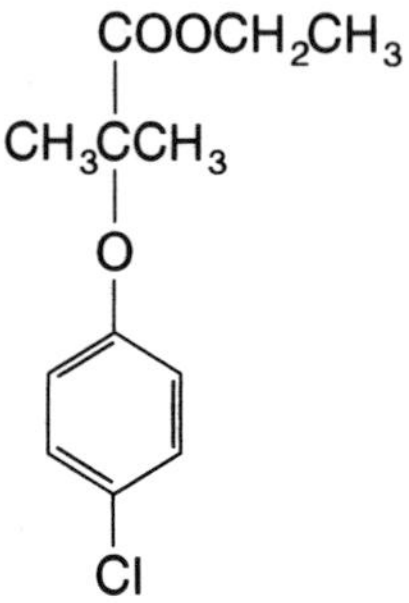

$COOCH_2CH_3$ — CH_3CCH_3 — O — Cl

Clofibrate

It is available over the counter and, hence, does not feature in the Top-100. It reduces the rate of synthesis of LDL. It has been shown to decrease significantly the incidence of recurrent nonfatal myocardial infarction. Also believed to be effective for the lowering of cholesterol levels are the complex carbohydrates in oats, bran, and other high-roughage food. This provides an example of how diet can be used therapeutically.

Another early hypolipemic was clofibrate, which only recently dropped from the Top-100. Its structure (see above) resembles a plant hormone more than a drug. How it reduces cholesterol is not completely understood. Indeed, it is known to be more effective in reducing triglycerides that occur naturally in the body. Probably, it inhibits synthesis of triglycerides and cholesterol in the liver. It is known that, when the drug is taken, there is an increase in the rate of synthesis of α-glycerophosphate dehydrogenase in the liver. This enzyme destroys α-glycerophosphate, which is the enzyme required for triglyceride synthesis.

Since 1982, various questions have arisen as to possible side effects of clofibrate. A huge 12-year testing project involving thousands of subjects showed that clofibrate certainly did lower cholesterol and that this was accompanied by a reduction of nonfatal heart attacks. Unfortunately, more of the treated group died than of those who were given a placebo, although the deaths were from seemingly unrelated conditions such as cancer. More recent studies have shown a statistical, but not necessarily causal, link between low cholesterol and cancer. Thus, it is possible that raised cholesterol in the blood may protect against certain diseases. While there is a case, therefore, for healthier eating and the use of cholesterol-lowering drugs for patients with especially high levels, there is uncertainty as to whether a general lowering of cholesterol levels is desirable.

Meanwhile, the development of hypolipemics is currently one of the most active areas of pharmaceutical research. Two drugs already on the market–cholestyramine and colestipol–are ion exchange resins that absorb bile acids in the intestine.

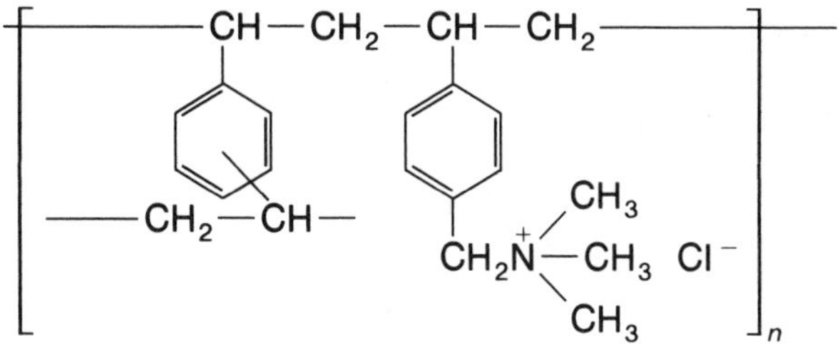

Cholestyramine resin

They form a resin–bile complex that is excreted so that the bile acids cannot be returned to the liver. To replace them, the liver increases its oxidation of cholesterol and this in turn leads to a drop in blood cholesterol. Drawbacks of the drugs include the large amounts of resin that have to be swallowed, which many patients say is like eating wet sand.

Two other hypolipemics are probucol and gemfibrizol:

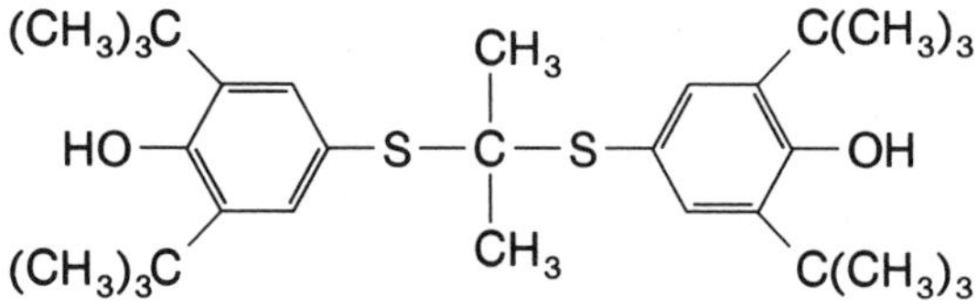

Probucol

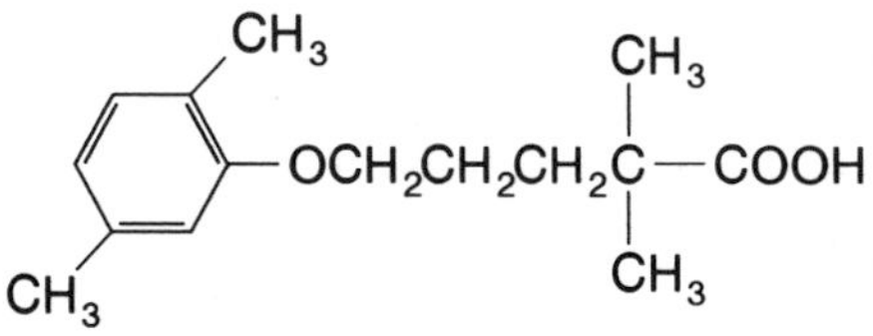

Gemfibrizol

Probucol lowers both LDL and HDL levels, but HDL is affected more, so there is an unfavorable HDL : LDL ratio. The drug is therefore of value only in certain types of excess blood lipids diseases. Gemfibrizol is chemically related to clofibrate. It is thought to increase the activity of an enzyme, lipoprotein lipase, which breaks down various lipoproteins and possibly even gives high-density lipoproteins as degradation products.

Finally, there is a new approach based on enzyme inhibitors, and a number of compounds are in the pipeline. Lovastatin was approved in 1987:

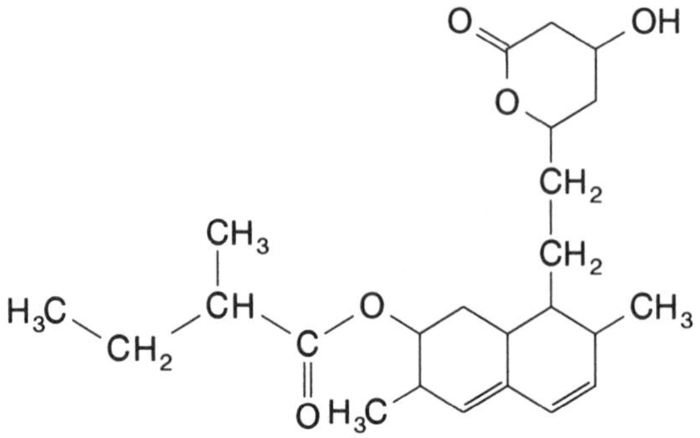

Lovastatin

It inhibits the key enzyme in cholesterol synthesis, 3-hydroxy-3-methylglutaryl-CoA reductase and thus reduces cellular synthesis of cholesterol. Cells therefore increase uptake of blood cholesterol to meet their needs and blood cholesterol levels drop.

7.10 CONCLUSION

Heart drugs were the pharmaceutical growth market of the 1970s and 1980s. By 1986 they amounted to 18.2% ($18 billion) of the total world pharmaceutical market. The shares of the cardiovascular market held by the different drug groups are shown in Figure 7.25. Since 1980, calcium antagonists and ACE inhibitors have shown the most dramatic growth, while diuretics, synthetic hypotensives, and coronary vasodilators have declined. Beta blockers went through a peak in 1985 but then lost ground to calcium antagonists.

That heart drugs have had considerable success is indicated by the marked decrease in mortality from heart disease in the United States. A rising trend flattened out in Europe. Undoubtedly this is partly due to changes in diet and life-style but some of the credit must go to pharmaceuticals. Curiously enough, the drop in mortality rate in the United States, where the new drugs have been introduced only slowly, has been greater than in Britain, where they were introduced more quickly. Among middle-aged men in 1968 the chance of a coronary heart disease death of an American was 40% higher than that of a Briton, while by 1976 the American risk had declined below that of the British. The British have been reluctant to change their life-style and appear apathetic about its risks.

Meanwhile, it is much clearer now than it was 20 years ago what steps a person should take to avoid a heart attack. Cigarette smoking, high blood

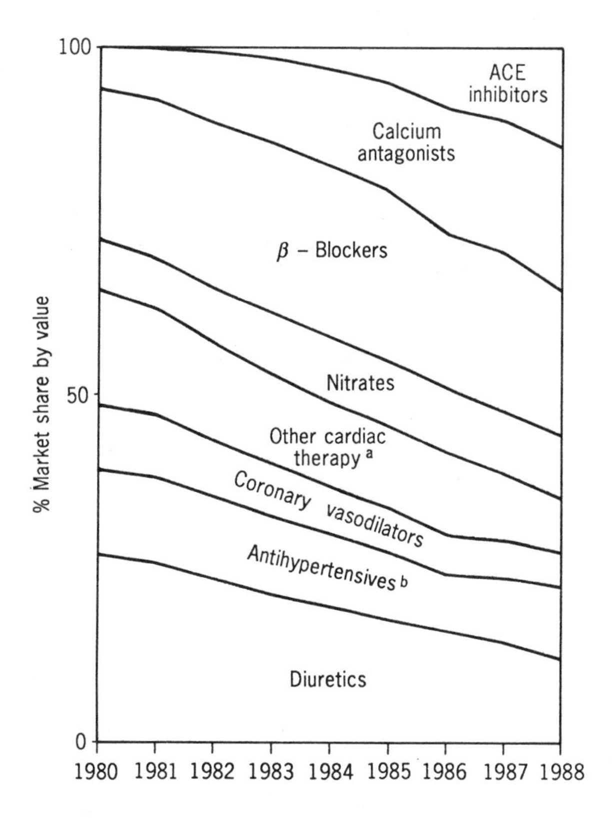

Figure 7.25. *World cardiovascular market.*

[a] *Hypolipemics, included under this heading, were only $304 million of the total $5430 million market in 1988.*

[b] *Drugs used only for hypertension, principally α blockers.*

pressure, and high blood cholesterol are the major risk factors and diabetes and family history are contributory. Obesity, stress, personality type, lack of exercise, and hardness of tap water may all be significant, but the evidence is inconclusive. For those afflicted by high blood pressure or high cholesterol, the range of drugs available has grown enormously. There is no longer any reason why a person should knowingly walk about with high blood pressure.

8

Drugs Affecting the Central Nervous System

In the Top-100 drugs there are 18 that affect the central nervous system, that is, the mind. They are also known as psychotropic drugs. In number they are second only to heart drugs. They include drugs to cheer people up or calm them down, to put them to sleep or wake them up. Although the illnesses that psychotropic drugs are intended to alleviate account for only about 1% of deaths in the United States, they are responsible for 10% of all visits to general practitioners and 25% of all prescriptions. One estimate suggests that in the United States one person in three has a raised level of anxiety, although not serious enough to warrant treatment. One in ten has mental and emotional problems requiring treatment and one in a hundred has a true psychotic illness. The same estimate suggests about one in two hundred people with cancer, but, of course, the death rate from cancer is much higher than that from emotional or mental illness.

This is so even if we classify suicides as deaths from mental illness. The U.S. suicide rate is 12.5 and the British 8.2 per 100,000 population. About 93% of suicides are associated with depressive illness, alcoholism, schizophrenia, neurosis and personality disorder, and drug addiction. Depressive illness is the leading cause (64% of cases), followed by alcoholism (15%).

The high incidence of mental illness has been attributed by some to the "future shock"-laden civilization of the developed world. It would indeed be interesting to know how the incidence of mental ailments today compares with that of 50 or 100 years ago when recognition and diagnosis were much less advanced, and life expectancy was much less.

TABLE 8.1. Drugs Affecting the Central Nervous System

Drug	Class
Hypnotics and Sedatives	
Triazolam (#41)	Benzodiazepine
Butalbital (#53)	Barbiturate
Flurazepam (#58)	Benzodiazepine
Temazepam (#66)	Benzodiazepine
Nitrazepam	Benzodiazepine
Methaqualone	Benzothiadiazine
Chloral hydrate	
Anticonvulsants	
Phenytoin (#40)	Hydantoin
Phenobarbital (#44)	Barbiturate
Carbamazepine (#85)	Dibenzazepine
Valproic acid	Dialkylalkanoic acid
Major Tranquilizers (Neuroleptics)	
Reserpine	Rauwolfia alkaloid
Chlorpromazine	Phenothiazine
Haloperidol (#75)	Butyrophenone
Thioridazine (#87)	Phenothiazine
Prochlorperazine (#106)	Phenothiazine
Flupenthixol	Thioxanthene
Pimozide	*N*-(Diphenylbutyl)piperidine
Clozapine	Dibenzothiazepine
Anxiolytics (Minor Tranquilizers)	
Diazepam (#15)	Benzodiazepine
Alprazolam (#19)	Benzodiazepine
Lorazepam (#49)	Benzodiazepine
Clorazepate (#52)	Benzodiazepine
Chlordiazepoxide (#79)	Benzodiazepine
Hydroxyzine (#108)	Antihistamine
Oxazepam (#117)	Benzodiazepine
Prazepam (#134)	Benzodiazepine
Buspirone	Azaspirodecane
Meprobamate	Carbamate
Antidepressants	
Iproniazid	MAO inhibitor
Isocarboxazid	MAO inhibitor
Phenelzine	MAO inhibitor
Tranylcypromine	MAO inhibitor
Imipramine	Tricyclic
Amitriptyline (#34)	Tricyclic
Doxepin (#63)	Dibenzoxepin
Trazodone (#83)	Triazolopyridine

TABLE 8.1. Continued.

Antidepressants	
Maprotiline	Bicyclooctadiene
Mianserin	Tetracyclic
Fluoxetine	
Lithium salts	
Stimulants and Anorectics	
Amphetamine	Phenylpropylamine
Fenfluramine	Phenylpropylamine
Phentermine	Phenylpropylamine
Diethylpropion	Phenylpropanolamine
Caffeine (#43)	Xanthine
Cocaine	Alkaloid
Psychotomimetics	
Mescaline	Phenylethylamine
Dimethyltryptamine	Indoleamine
Psilocybin	Indoleamine
Lysergic acid diethylamide	Indoleamine
Tetrahydrocannabinol	Dibenzopyran

Types of mental illness vary from the anxiety that we all know through mild depression, mania or manic depression, severe depression, hysteria, phobias, obsessions, and schizophrenia. These often respond well to drug treatment. The two most prevalent and serious mental disorders are schizophrenia and manic–depressive illness. Half the persons confined to mental institutions are schizophrenic. The pioneer psychiatrist, Kraepalin, identified four categories of schizophrenia: paranoid, involving delusions of persecution; catatonic, involving periods of total withdrawal and sometimes also of overexcitement; hebephrenic, characterized by bizarre emotion and thought; and simple, where there is a gradual progression toward abnormal behavior. Major tranquilizers can be used to treat schizophrenia (Section 8.2), but they are not wholly effective.

Schizophrenia is more narrowly defined in the United Kingdom than in the United States and the category of simple schizophrenia has largely disappeared. There is an increasing reluctance to prescribe major tranquilizers, especially to children. It must be stressed that schizophrenia is not the Dr. Jekyll–Mr. Hyde split personality of popular fiction.

The other serious and widespread mental disorder is manic–depressive illness. The manic phase is characterized by hyperactivity, which eventually gives way to disorganization of thought, followed by periods of acute depression. The depression does not respond to the tricyclic antidepressants discussed in Section 8.4. The drug of choice is lithium carbonate which, taken during the manic phase, helps alleviate the depressive phase. There are few drugs as specific as lithium carbonate but again a more effective drug would be welcome.

Drugs affecting the central nervous system fall into six main categories, shown in Table 8.1. These are hypnotics and sedatives, major tranquilizers, anxiolytics, antidepressants, stimulants and anorectics, and psychotomimetics. Anticonvulsants have been added to the table and to this chapter because they are similar in some respects to the barbiturates that are listed under hypnotics and sedatives. Anesthetics and drugs for Parkinson's disease have been excluded because they are not usually classified with psychotropics. Anesthetics lie outside the scope of this book and anti-Parkinsonism drugs will be dealt with separately in Chapter 16. The table lists the compounds in the Top-100 and the chemical class to which they belong, together with other compounds of historical or general interest.

8.1 HYPNOTICS AND SEDATIVES

The oldest group of psychotropic drugs is the soporifics or hypnotic sedatives. They induce sleep but are neither so potent nor so dangerous as anesthetics. The oldest chemical entity is chloral hydrate, first described by von Liebig in 1832.

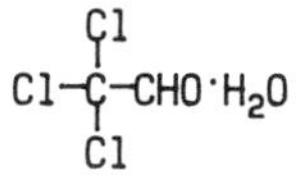

$$Cl_3C{-}CHO \cdot H_2O$$

Chloral hydrate

It tastes horrible and irritates the stomach. In combination with alcohol, it constitutes a Mickey Finn, famous in dime novels. It was popularly believed that many young men were shanghaied onto boats to become unwilling sailors with the aid of Mickey Finns. Chloral hydrate is rarely used now but is of value for children and old people who may easily get confused or suffer from hangovers after more conventional sleeping medications.

8.1.1 Barbiturates

Another long-established group of soporifics is the barbiturates shown in Figure 8.1. Barbituric acid was discovered by Bayer in 1863 and the first

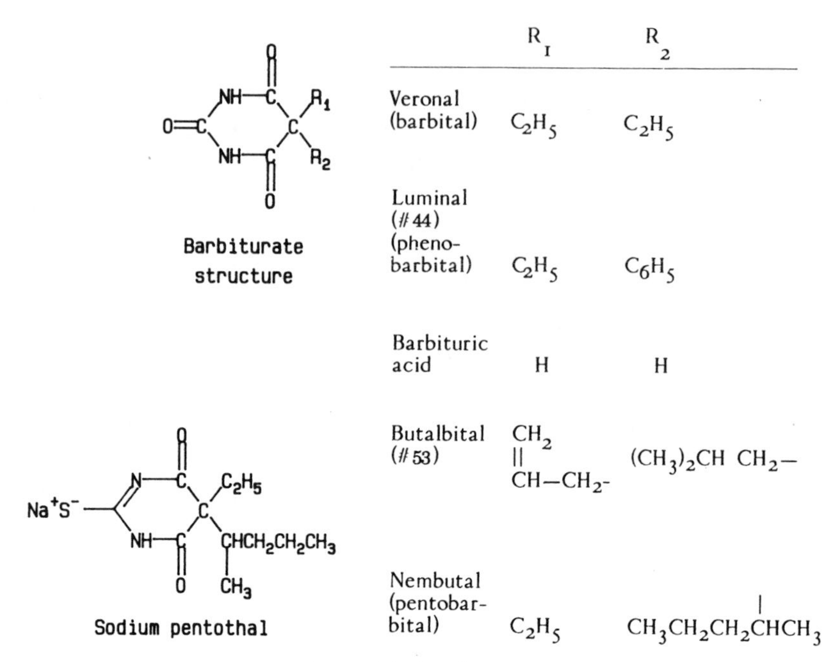

	R_1	R_2
Veronal (barbital)	C_2H_5	C_2H_5
Luminal (#44) (phenobarbital)	C_2H_5	C_6H_5
Barbituric acid	H	H
Butalbital (#53)	$CH_2{=}CH{-}CH_2{-}$	$(CH_3)_2CH\ CH_2{-}$
Nembutal (pentobarbital)	C_2H_5	$CH_3CH_2CH_2CHCH_3$

Figure 8.1. *Barbiturates.*

barbiturate soporific was barbital, discovered by Fischer and von Mering in Germany in 1903 and known as veronal. The second was phenobarbital, discovered in 1911 and named luminal. Also shown in Figure 8.1 is the parent compound barbituric acid which is pharmacologically inactive. The two barbiturates in the Top-100 are phenobarbital (#44) and butalbital (#53). Barbiturates are subdivided clinically according to the time taken for their action to begin and the length of time it lasts. Phenobarbital is long acting. Time of onset is an hour and the effect lasts for 6 to 12 hours. It is used in various mixtures as a sedative in cases of anxiety, and in various neuroses and menopausal and menstrual disorders. It is mainly used as an anticonvulsant in certain types of epilepsy.

Although phenobarbital induces sleep in patients troubled by anxiety, the true soporifics are the intermediate acting barbiturates, which have a 30-minute onset and last for 5 to 6 hours. Butalbital falls into this class. Pentobarbitone, known as Nembutal, is a short-acting barbiturate and is sometimes used to induce sleepiness before anesthesia. Thiopental sodium, known as Pentothal, is shown separately in Figure 8.1. It is the fastest-acting barbiturate and is also used for preanesthetic sedation. The factors affecting the length of action and degree of activity of barbiturates are well understood and are discussed in Burger, Chapter 54 (see bibliography).

Barbiturates present problems because patients develop tolerance for them and ever higher doses are required. They are habit-forming and their therapeutic index is low, that is, the fatal dose is not far removed from the therapeutic dose. Thus, they are truly dangerous drugs and are widely used by suicides. They also present the danger of automatism. This means that a patient may take a dose, leave the bottle on the bedside table, and then, when half asleep, forget that a dose has been taken, and take some more. This is one of the theories for the death of Marilyn Monroe.

Barbiturates are especially dangerous when taken with alcohol. Not only do both materials depress the central nervous system, but they are both metabolized by the liver and the joint effect of the two may lead to liver failure.

As safer soporifics have become available, (mainly the benzodiazepines, Section 8.3.1) prescribing of barbiturates has declined and so have barbiturate-induced suicides. In England and Wales, barbiturate prescriptions halved between 1964 and 1974 and barbiturate suicides dropped from 1500 to 1000 annually.

Barbiturates are synthesized by condensation of urea with a substituted malonic ester in a classical barbituric acid synthesis:

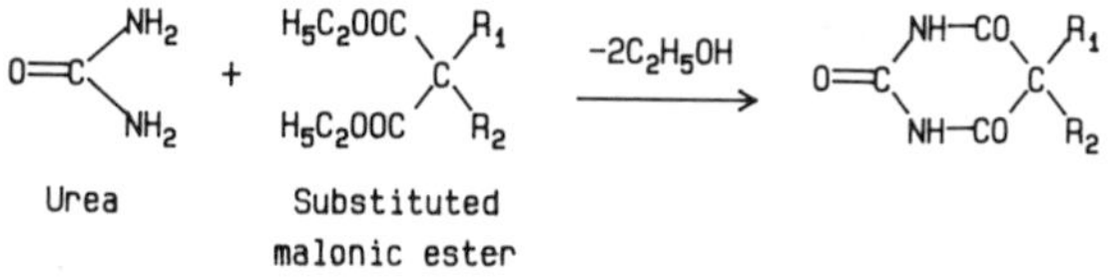

$$O{=}C(NH_2)_2 + (H_5C_2OOC)_2CR_1R_2 \xrightarrow{-2C_2H_5OH} O{=}C(NH{-}CO)_2CR_1R_2$$

Urea

Substituted malonic ester

The synthesis of some of the substituted malonic esters, however, may require ingenuity.

8.1.2 Nonbarbiturate Hypnotics and Sedatives

It was the dangers of barbiturates that led to the early success of thalidomide:

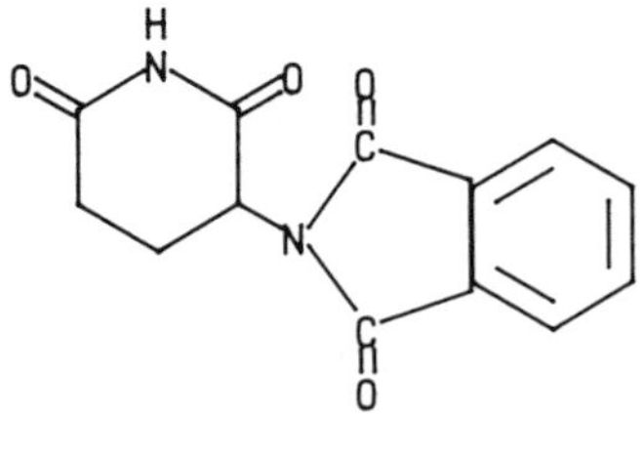

Thalidomide

Thalidomide was supposed to be a uniquely safe soporific, but, when taken during early pregnancy, turned out to cause fetal abnormalities involving short or nonexistent arms and legs. Worldwide, about 8000 children with birth deformities survived and perhaps twice that number died at birth. In the United States, Dr. Frances Kelsey of the FDA fought a successful delaying battle against approval of the drug and only experimental tablets were distributed.

One nonbarbiturate soporific, which appeared shortly after thalidomide, was methaqualone. In one formulation, it was combined with diphenhydramine, an antihistamine, and sold under the name Mandrax:

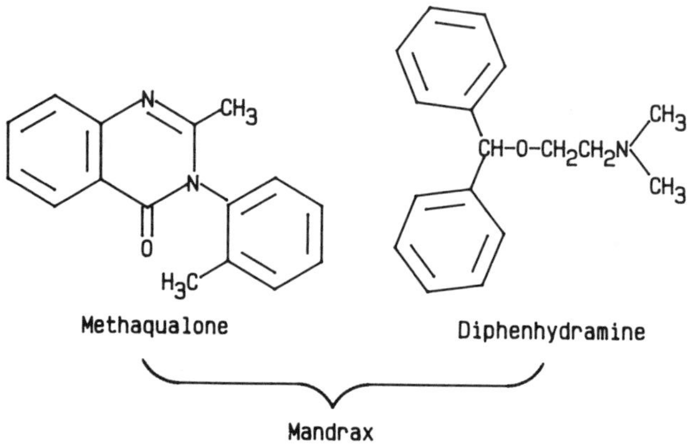

Methaqualone Diphenhydramine

Mandrax

Although less dangerous than barbiturates, it turned out to be habit-forming and liable to abuse. Users said it produced a relaxed state and heightened sexual responsiveness. A celebrated case in the United States was that of an entertainer named Freddie Prince, who apparently needed methaqualone to help him handle the problems that fame and fortune had heaped upon him at an early age. Unfortunately, he took an overdose that killed him.

In 1981 the New York Department of Health reported a huge increase in methaqualone prescribing, under the trade name of Quaalude, and said that it was connected with the proliferation of so-called "stress" and "insomnia" centers. A Federal Grand Jury investigated whether the centers were an ingenious way of legally exploiting the demand for a fashionable drug. Methaqualone manufacture was discontinued in 1984.

The discovery of the benzodiazepine tranquilizers (Section 8.3) provided a better nonbarbiturate soporific. All the benzodiazepine tranquilizers in high enough doses will produce sleep, but two are particularly suitable. Flurazepam, known as Dalmane, is #58 in the Top-100 and nitrazepam, known as Mogadon, is widely used in the United Kingdom, but not in the United States. Both induce restful sleep with normal dreaming as characterized by rapid eye movements. Their chemistry will be discussed along with the other benzodiazepines (Section 8.3.1).

A truly satisfactory soporific is a goal of current research. Scientists at Harvard have shown that a chemical substance, factor S, is produced in the brains of animals when they have been deprived of sleep for a long time. This chemical can be extracted and the extract inserted into the brains of rabbits. The rabbits become sleepy and their brains give off the waves associated with tranquil sleep. The isolation, identification, and synthesis of factor S is still a long way off. The development of a drug based on it that would cross the blood–brain barrier is even more remote. Nonetheless, the steps by which this natural "sleeping pill" might be produced are clear if difficult, and it might well be accomplished in the next 20 years.

8.1.3 Anticonvulsants

Anticonvulsants are included with hypnotics and sedatives because phenobarbital serves both purposes. The other anticonvulsants in the Top-100 are phenytoin (#40) and carbamazepine (#85):

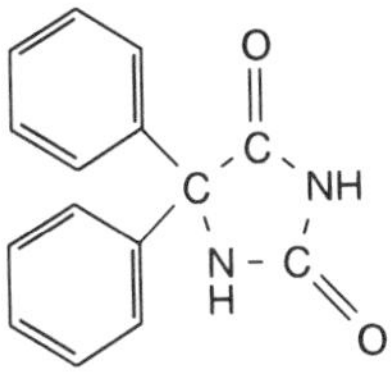

Phenytoin (#40)

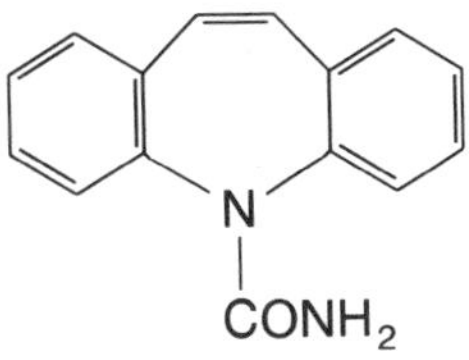

Carbamazepine (#85)

Phenytoin, diphenylhydantoin, is used as its sodium salt to achieve solubility. It is made by a classical Bucherer hydantoin synthesis (Figure 8.2). This starts with benzophenone (1), which reacts with a mixture of potassium cyanide and ammonium carbonate. Presumably the first step is the formation of an α-aminonitrile (2). Cyclization gives the amino–amidine (3), which hydrolyzes to phenytoin (4). Phenytoin has partly displaced phenobarbital as an anticonvulsant, but is sometimes used together with it to reduce the amount of barbiturate required.

Carbamazepine is the drug of choice in partial seizures. It is closely related to the tricyclic antidepressant imipramine (Section 8.4.2), and its synthesis, shown in Figure 8.3, starts with the synthon (5) from the imipramine synthesis (Figure 8.23). This is N-acetylated and brominated with *N*-bromosuccinimide to give (1). Dehydrobromination by heating in collidine introduces a double bond in the 10–11 position. Deacetylation with potassium hydroxide in ethanol gives the dibenzazepine (3). Treatment with phosgene gives the carbamoyl chloride (4) and ammonia plus heat gives carbamazepine (5).

Another anticonvulsant, valproic acid, has been used in Europe for a long time but has only been marketed in the United States for the past few years. It is ineffective in partial seizures but is finding increasing application in a type

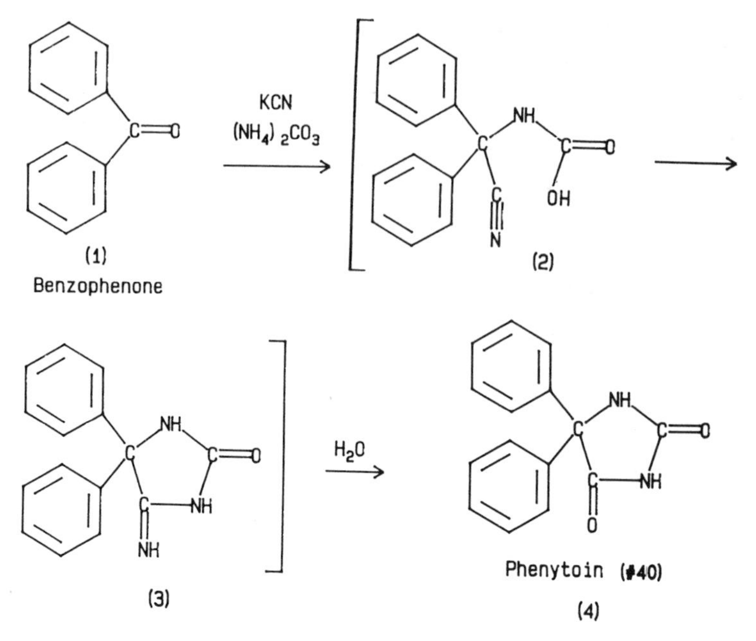

Figure 8.2. *Synthesis of phenytoin.*

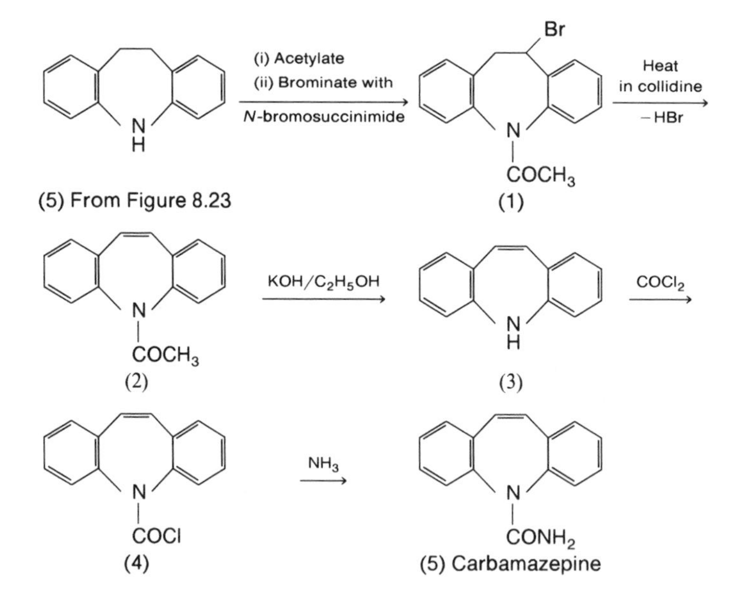

Figure 8.3. *Syntheses of carbamazepine.*

of general seizure called "tonic–clonic", characterized by initial contraction of muscles, leading to rigidity, followed by relaxation interrupted by further spasms.

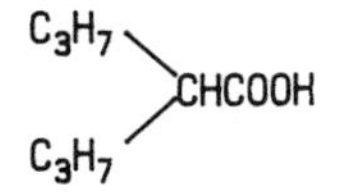

$(C_3H_7)_2CHCOOH$

Valproic acid

8.2 MAJOR TRANQUILIZERS (NEUROLEPTICS)

Major tranquilizers, or neuroleptics, produce an effect on people who are seriously mentally disturbed. They are of value in controlling the manic phase of manic–depressive illness and some can stimulate withdrawn schizophrenics into more normal contact with the world. The first two neuroleptics were discovered in the early 1950s. Reserpine has already been discussed in Section 7.6.1 because it is antihypertensive as well as tranquilizing.

In 1952, at about the same time that reserpine was discovered, French workers noted the tranquilizing effects of chlorpromazine, a phenothiazine

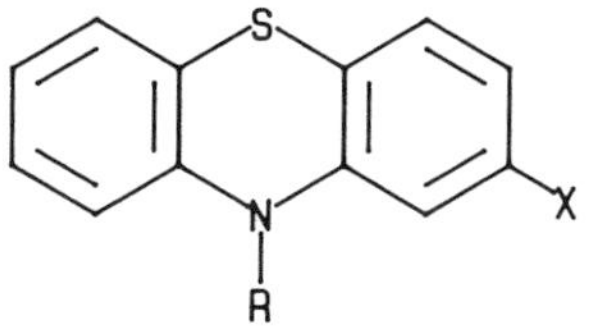

Basic structure

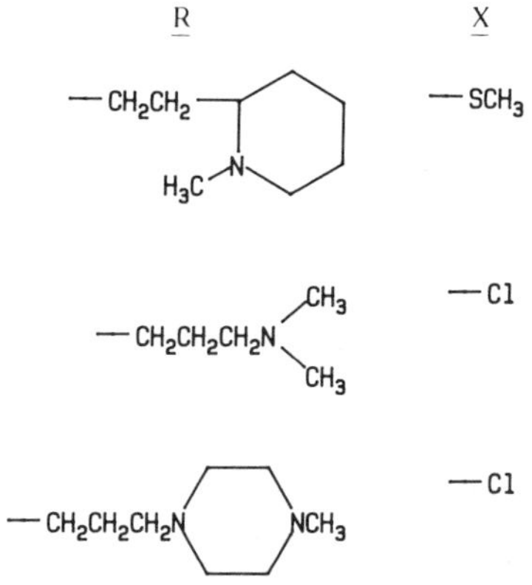

		R	X
Thioridazine	(#71)	$-CH_2CH_2-$(N-methylpiperidin-2-yl)	$-SCH_3$
Chlorpromazine		$-CH_2CH_2CH_2N(CH_3)_2$	$-Cl$
Prochlorperazine	(#106)	$-CH_2CH_2CH_2N$(piperazine)NCH_3	$-Cl$

Figure 8.4. *Phenothiazine neuroleptics.*

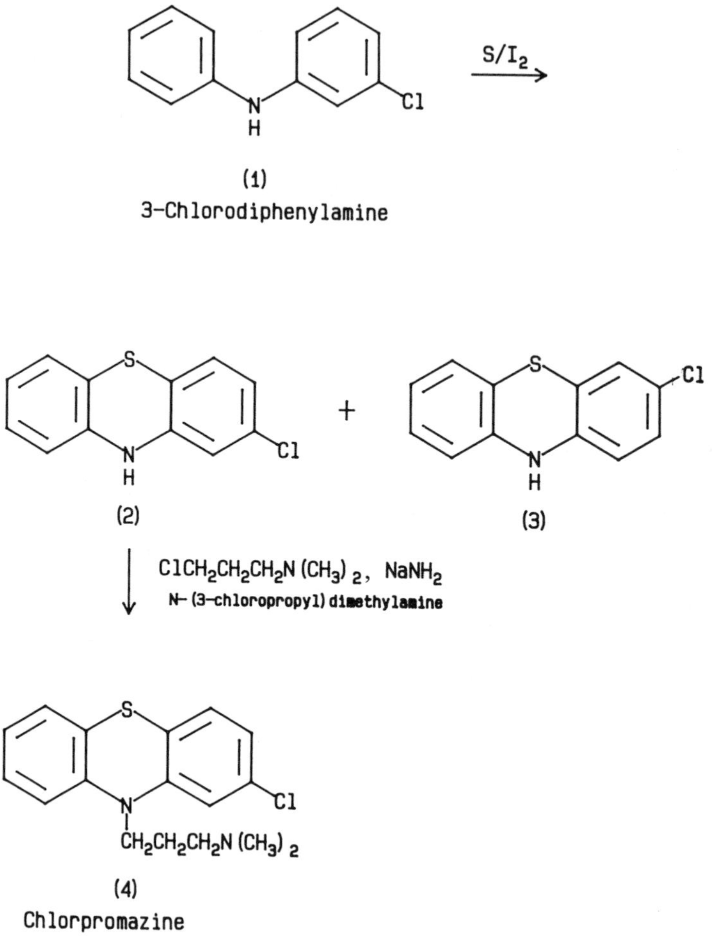

S/I_2

(1)
3-Chlorodiphenylamine

(2) + (3)

$ClCH_2CH_2CH_2N(CH_3)_2$, $NaNH_2$
N-(3-chloropropyl)dimethylamine

$CH_2CH_2CH_2N(CH_3)_2$

(4)
Chlorpromazine

Figure 8.5. *Synthesis of chlorpromazine.*

whose discovery led to the development of a range of other phenothiazine tranquilizers. Five of these appeared in the Top-100 in 1979 but only one remains. A second, prochlorperazine, is at #106. Figure 8.4 shows their structures. Mental hospitals are, of course, major users of neuroleptics but their consumption does not appear in prescription statistics.

The syntheses of most phenothiazines depend on similar intermediates. The chlorpromazine synthesis in Figure 8.5 is straightforward and typical. 3-Chlorodiphenylamine (1) is heated with sulfur and an iodine catalyst to give 3-chlorophenothiazine (2). Some 4-substituted phenothiazine (3) is also produced, which is of no value but, because of steric effects, the proportion is relatively small. Further treatment of (2) with *N*-(3-chloropropyl)dimethylamine adds a side chain to give chlorpromazine (4). This side chain occurs widely in pharmaceuticals.

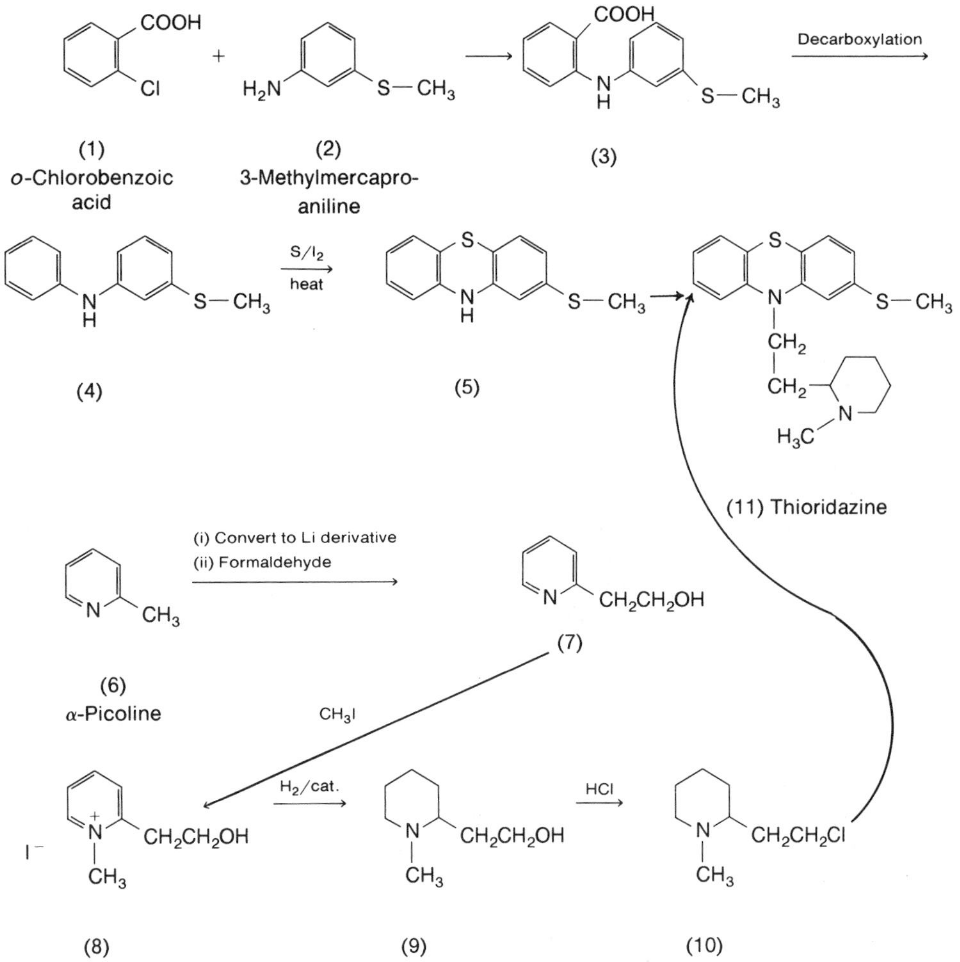

Figure 8.6. *Synthesis of thioridazine.*

The synthesis of thioridazine is shown in more detail in Figure 8.6. The diphenylamine derivative (4) corresponding to (1) in Figure 8.5 is made by reaction of *o*-chlorobenzoic acid (1) with the substituted aniline (2) to give (3), which is then decarboxylated to (4) to remove the activating carboxyl group. As with the chlorpromazine synthesis, ring closure with sulfur and iodine gives a phenothiazine (5). The reagent that leads to the side chain is made from α-picoline (6), which is converted to a lithium derivative and treated with formaldehyde to give (7). Quaternization with methyl iodide gives the pyridinium salt (8). Catalytic reduction gives the piperidine (9). Reaction with HCl converts the hydroxyethyl group to chloroethyl as in (10). Reaction of (10) with (5) yields thioridazine (11).

1. Butyrophenones:

Haloperidol (#75)

2. Thioxanthenes:

Flupenthixol

3. N - (Diphenylbutyl) piperidines

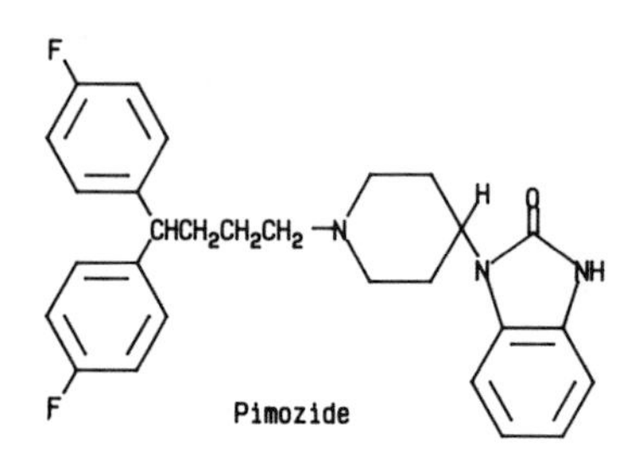

4. Dibenzothiazepines:

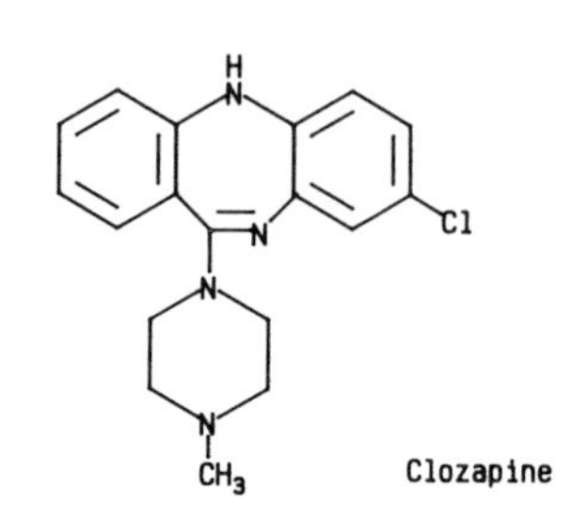

Figure 8.7. *Other neuroleptics.*

Other groups of neuroleptics are represented in Figure 8.7 and include butyrophenones, thioxanthenes, *N*-(diphenylbutyl)piperidines, and dibenzothiazepines. An example is given of each class. The only one appearing in the Top-100 is the butyrophenone, haloperidol (#75). The synthesis of haloperidol is shown in Figure 8.8 and involves the condensation of 4-chlorophenyl-4-hydroxypiperidine (1) with *p*-fluorophenyl-3-chloropropyl ketone (2) to give haloperidol (3). The intermediate (1) results from the interaction of *p*-chloro-α-methylstyrene (4) with formaldehyde and ammonium chloride. This is a Mannich reaction with inorganic ammonium chloride rather than an organic amine. A proposed route for this curious reaction is given by Lednicer and Mitscher (see bibliography). The initial reaction gives (5), which on heating with acid rearranges to (6), which in turn adds HBr to provide the 4-arylbromopiperidine (7). Alkaline hydrolysis gives the desired hydroxypiperidine (1).

The chemistry that leads to the ketone (2) is more conventional and involves the acylation of fluorobenzene (8) with 4-chlorobutyryl chloride (9) under Friedel Crafts conditions.

The above synthesis of haloperidol is described in the patent literature, but an apparently simpler one (although it involves a Grignard reaction) is shown at the bottom of the figure. The ketone group of a substituted piperidinone (10) will react with the Grignard reagent, 4-chlorophenylmagnesium bromide,

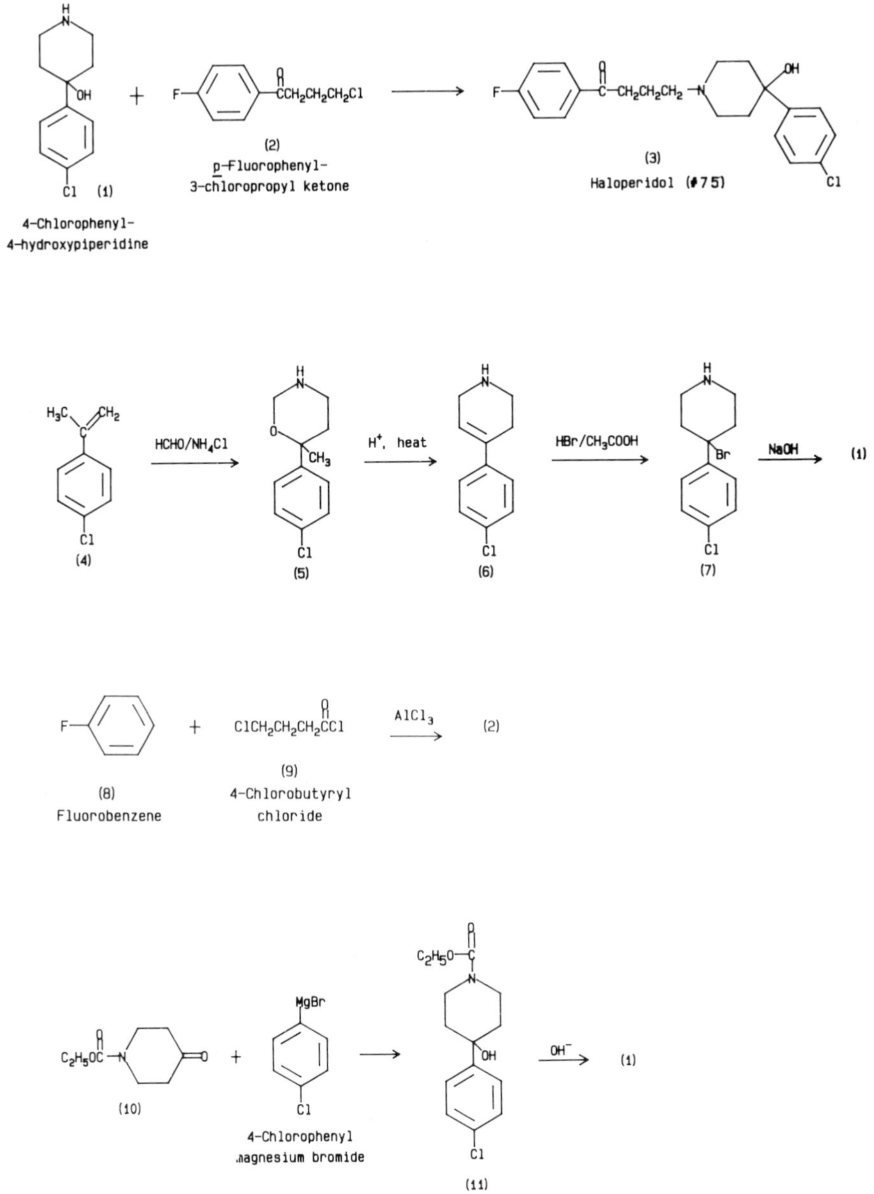

H N OH Cl (1) + F $\overset{O}{\|}$ $CCH_2CH_2CH_2Cl$ ⟶ F $\overset{O}{\|}$ $C-CH_2CH_2CH_2-N$ OH Cl

(1) 4-Chlorophenyl-4-hydroxypiperidine

(2) p-Fluorophenyl-3-chloropropyl ketone

(3) Haloperidol (#75)

H_3C CH_2 C Cl (4) $\xrightarrow{HCHO/NH_4Cl}$ H N O CH_3 Cl (5) $\xrightarrow{H^+,\ heat}$ H N Cl (6) $\xrightarrow{HBr/CH_3COOH}$ H N Br Cl (7) $\xrightarrow{NaOH}$ (1)

F + $ClCH_2CH_2CH_2\overset{O}{\overset{\|}{C}}Cl$ $\xrightarrow{AlCl_3}$ (2)

(8) Fluorobenzene

(9) 4-Chlorobutyryl chloride

$C_2H_5O\overset{O}{\overset{\|}{C}}-N$ =O (10) + MgBr Cl ⟶ $C_2H_5O-\overset{O}{\overset{\|}{C}}$ N OH Cl (11) $\xrightarrow{OH^-}$ (1)

4-Chlorophenyl magnesium bromide

Figure 8.8. *Syntheses of haloperidol.*

prepared from 1-chloro-4-bromobenzene. The product is a tertiary alcohol (11) from which the protecting group on the nitrogen atom can readily be removed to give the desired compound (1).

The neuroleptics differ from the hypnotic sedatives for, although they produce considerable sedation, the patient is easily roused. They impair sustained attention but they also reduce spontaneous aggression. An interesting offshoot of this property is that neuroleptics can be used to tame wild animals.

The various neuroleptics shown in Figures 8.4 and 8.7 differ in their relative sedative and stimulating properties and in their side effects. For example, the butyrophenones and the phenothiazines with piperazine side chains are stimulants. Flupenthixol and pimozide share this property and are valuable for the treatment of withdrawn, apathetic patients. On the other hand, phenothiazines with an aminopropyl side chain, such as chlorpromazine, have more marked sedative properties.

Low doses of phenothiazines, especially chlorpromazine and prochlorperazine, are useful for control of nausea, vomiting, dizziness, and vertigo.

Certain long-acting neuroleptics are available, known as "depot" neuroleptics. They are given by injection and are useful for the treatment of schizophrenics who might not comply with doctor's orders. An example is haloperidol decanoate, which need be administered only once per month.

The most serious side effect of the neuroleptics is that they produce extrapyramidal symptoms, that is, symptoms like those of Parkinson's disease, together with confusion, dizziness and changes in muscle tone and blood pressure. Thioridazine is relatively free from this and is consequently used for elderly patients. A drug that is claimed to be even better is clozapine (Figure 8.7). It is not a classical neuroleptic because it enhances dopamine levels and, indeed, has some antidepressant activity. It is hoped that it will be the first of a new series of neuroleptics whose effectiveness will reach a level unattainable by the phenothiazines.

The social implications of tranquilizer consumption will be discussed in Section 8.7 but it is worthwhile noting now that the phenothiazines have revolutionized the treatment of serious mental illness and have enabled some patients to be treated in the community rather than in institutions. Phenothiazines are not curative but they relieve thought disturbance, paranoia, hallucinations, delusions, loss of self-care, and social withdrawal. Many schizophrenics are able to stop medication after a year or two although relapses may occur two to eight weeks after the drugs are stopped.

8.3 ANXIOLYTICS (MINOR TRANQUILIZERS)

By the end of the 1950s, great strides had been made in treatment of mental illness. A range of hypnotics and sedatives was available together with various neuroleptics and antidepressants (Section 8.4). The use of lithium came later,

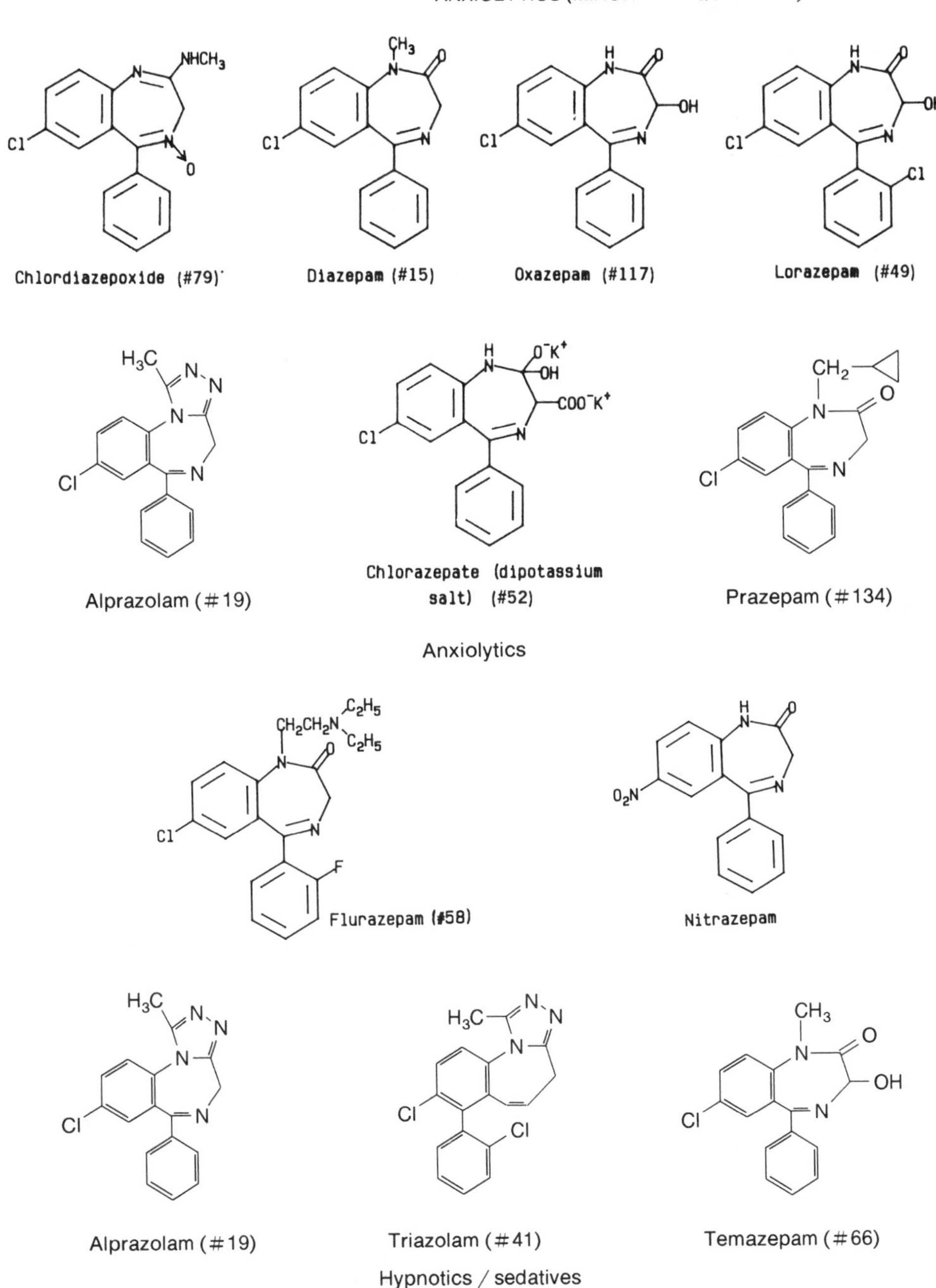

Figure 8.9. *Anxiolytics and benzodiazepine / hypnotics / sedatives.*

but the most widely discussed innovation in psychopharmacology since 1960 has been the introduction of anxiolytics (minor tranquilizers) and in particular the benzodiazepines. Chlordiazepoxide was introduced in 1960. There are five anxiolytics in the Top-100 plus three others at #108, #117, and #134. Seven are benzodiazepines, namely chlordiazepoxide, diazepam, clorazepate, oxazepam, alprazolam, lorazepam, and prazepam, and they are shown in Figure 8.9. The sedatives flurazepam, nitrazepam, temazepam, and triazolam are also benzodiazepines. The remaining anxiolytic, hydroxyzine, is described in Section 8.3.3.

Anxiolytics are frequently described as minor tranquilizers. The usage is somewhat misleading. These drugs differ markedly from the major tranquilizers and their use is anything but minor. Because they allay anxiety, they are better called anxiolytics, and that is the term we shall use.

8.3.1 Benzodiazepines

As noted above, the action of anxiolytics is to reduce anxiety. What is anxiety? One definition is that it is what everyone is taking Valium for. At best it is a loose assembly of poorly defined symptoms including insomnia, vague unfocused fears, loss of security, nervous mannerisms, headache, stomach pains, wheezing, tachycardia, and sweating. In the past, such symptoms were borne stoically and no one died of them. On the other hand, psychosomatic symptoms are just as keenly felt by the patient as symptoms with a physical basis, and it would be a harsh physician who denied relief to a genuine sufferer.

In addition to relieving anxiety, anxiolytics are sedative and anticonvulsant. They are also used for preanesthetic medication and alcohol withdrawal. For relief of chronic anxiety, chlordiazepoxide, diazepam, clorazepate, and prazepam are the preferred compounds because they have a long half-life. Alprazolam, oxazepam, and lorazepam are more appropriate for acute anxiety and attacks of panic, since they are shorter acting. Clorazepate is also used for symptoms of acute alcohol withdrawal.

Flurazepam and nitrazepam are the preferred benzodiazepines for use as hypnotics and sedatives because they are very long-acting. They may give rise to residual hangovers and, with repeated dosage, these effects may be cumulative. In such cases, the medium-acting temazepam and triazolam may be better, but they are correspondingly less useful for patients who suffer from early awakening.

Benzodiazepines are also used for anticonvulsant therapy and for muscle spasm, and diazepam is the preferred compound. The latter use for diazepam is not widely known, so athletes may react angrily to its prescription, thinking that their physician is dismissing their muscle spasm as a psychiatric disorder.

Under normal circumstances, the benzodiazepines are very safe drugs and meprobamate (Section 8.3.3) is only slightly more toxic, although none of them should be taken with alcohol. We cannot find a record of any of them being successfully used for suicide, which makes them much safer than the barbiturates. Rare meprobamate and benzodiazepine poisonings have been recorded. Occasional allergies to anxiolytics are also known.

All the anxiolytics are nonetheless addictive and they can lead to compulsive use and physical dependence. Cessation of large doses can lead to severe withdrawal symptoms. There is some evidence that the brain produces a benzodiazepine-like compound of its own, and that production is much reduced when benzodiazepine drugs are taken. Hence, withdrawal leads initially to acute anxiety and the patient concludes he or she is still ill and rushes back to the physician for a further prescription.

Computerized axial tomographic scans, so-called CAT scans, have been performed on the brains of patients who have taken diazepam three times a day for over 7 years. Preliminary findings suggest changes in the brain similar to those occurring in acute alcoholism. This has not yet been firmly established; what is worrisome is the number of patients who appear to have been allowed to take diazepam over such a period.

The British National Formulary (see bibliography) declares that, "Anxiolytic treatment should be limited to short periods because tolerance to its effects develops within four months of continuous use and because of the danger of insidious development of dependence." This advice has been widely disregarded in the past. It is some reassurance that physicians are now becoming more aware of the problem and anxiolytic prescriptions in the United States have dropped by a third from their peak as indicated in Figure 8.10. The same figure shows the smaller but increasing usage of benzodiazepines as sedatives. UK figures show a 45% drop in anxiolytic prescriptions between 1979 and 1985.

The benzodiazepines have been a huge commercial and, in spite of problems with overprescribing, a therapeutic success. In the 20 years from 1965 to 1985, 1.5 billion benzodiazepine prescriptions were written in the United States, with diazepam accounting for half of these and chlordiazepoxide for

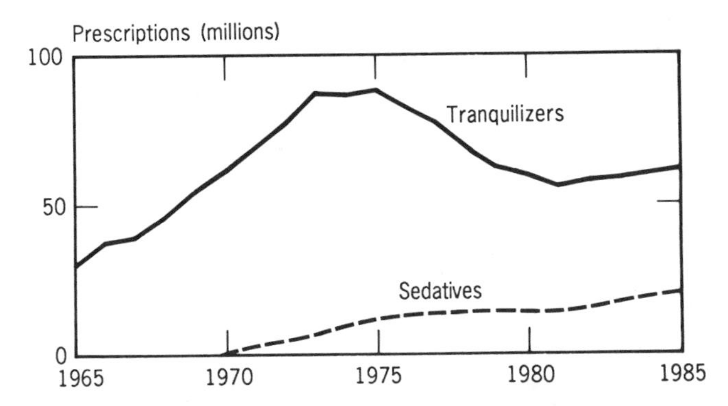

Figure 8.10. *Benzodiazepine prescriptions 1965–1985. Source: U.S. Food and Drugs Administration. Reference 8 / 12.*

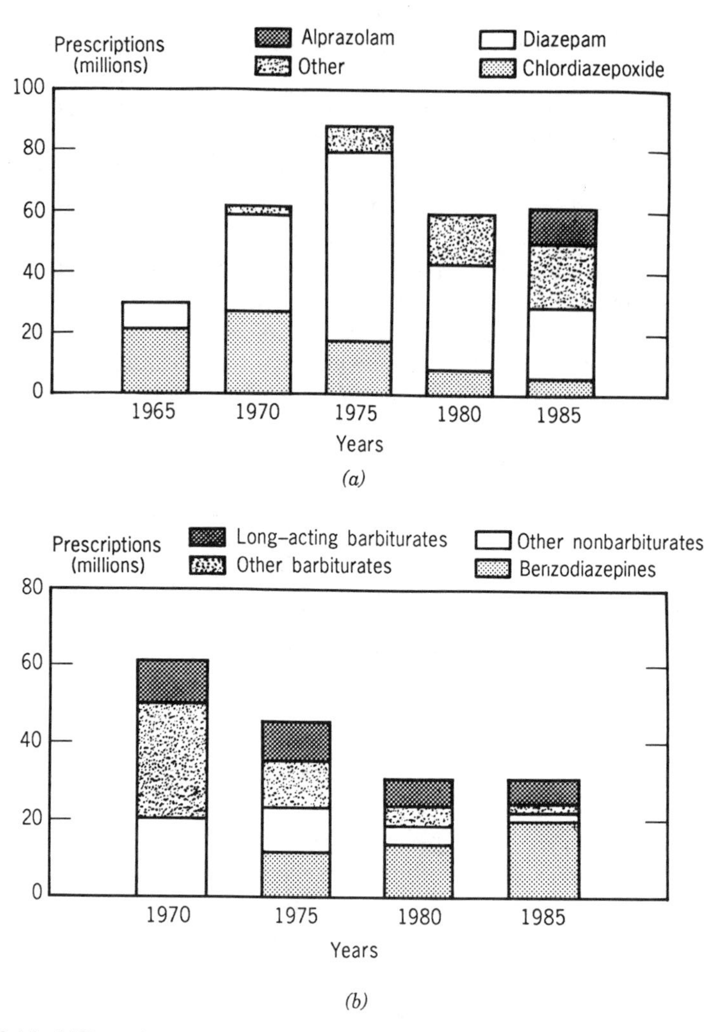

Figure 8.11. (a)Prescriptions for benzodiazepine tranquilizers. (b) Prescriptions for sedatives. Source: U.S. Food and Drugs Administration. Reference 8 / 12.

another quarter. The peak in demand was from 1973 to 1975 with about 90 million prescriptions/year. Diazepam was #1 in the Pharmacy Times Top-100 for many years.

In a market of this size and value, the competition between the various entities is of interest. Figure 8.11a shows the benzodiazepine anxiolytic market and Figure 8.11b shows the sedative market. Chlordiazepoxide was the original benzodiazepine and dominated the market in the 1960s. Diazepam dominated the 1970s with a few specialized compounds taking an increasing share. By 1985, lorazepam, clorazepate, and other benzodiazepines were collectively as important as diazepam. The most remarkable trend has been the rise of

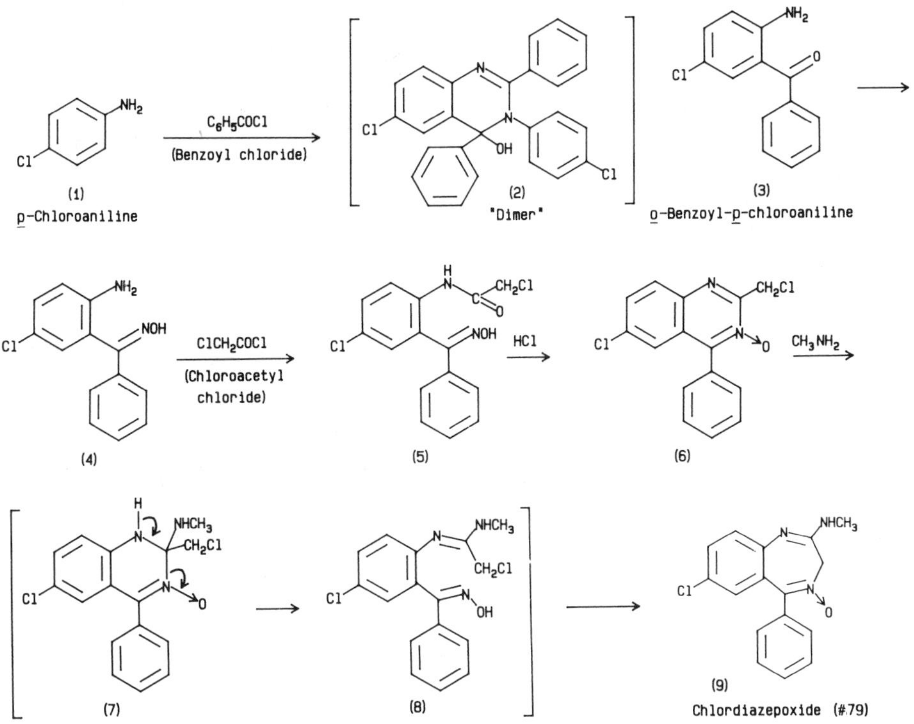

(1) p-Chloroaniline

C_6H_5COCl (Benzoyl chloride)

(2) "Dimer"

(3) o-Benzoyl-p-chloroaniline

(4)

$ClCH_2COCl$ (Chloroacetyl chloride)

(5)

HCl

(6)

CH_3NH_2

(7)

(8)

(9) Chlordiazepoxide (#79)

Figure 8.12. *Synthesis of chlordiazepoxide.*

alprazolam, which first appeared in the Pharmacy Times Top-200 in 1983 and had risen to #19 (15 million prescriptions) by 1986, only four places below diazepam.

8.3.2 Benzodiazepine Syntheses

The syntheses of benzodiazepines are complex and details of industrial processes are not disclosed. One synthesis of chlordiazepoxide, the earliest benzodiazepine, is shown in Figure 8.12. *p*-Chloroaniline (1) is treated with benzoyl chloride. Normally the nitrogen atom would be acylated, but, under the conditions used, a dimer (2) is formed which on hydrolysis gives *o*-benzoyl-*p*-chloroaniline (3). Yields are poor and the commercial process involves a different route. The ketone (3) is converted to an oxime (4) and the amino group is then reacted with chloroacetyl chloride to give (5), which may be cyclized and dehydrated with hydrogen chloride to give the quinoxaline *N*-oxide (6). Reaction of this compound with methylamine does not lead to simple replacement of the chlorine but to ring expansion. The methylamine is thought to add to a ring carbon on the quinoxaline ring to give an intermediate (7). The ring then opens to give the intermediate (8). The chlorine atom then combines with the oxime hydrogen to give chlordiazepoxide (9).

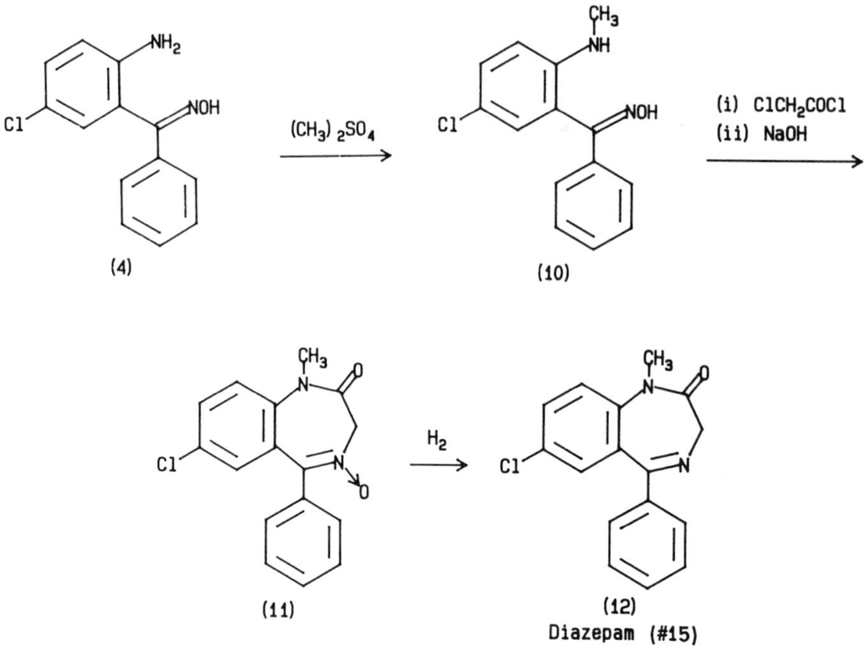

NH_2 NOH Cl (4) $\xrightarrow{(CH_3)_2SO_4}$ CH_3 NH NOH Cl (10) (i) $ClCH_2COCl$ (ii) NaOH →

CH_3 N O Cl N O (11) $\xrightarrow{H_2}$ CH_3 N O Cl N (12) Diazepam (#15)

Figure 8.13. *Synthesis of diazepam.*

Figure 8.13 shows a synthesis of diazepam. It is made from the same oxime (4) as chlordiazepoxide. Methylation gives the *N*-methyl analogue (10). The reaction is carried out under conditions such that only one hydrogen of the primary amine group is replaced. Chloroacetyl chloride reacts with the secondary amine to give an intermediate, which on treatment with sodium hydroxide undergoes ring closure to (11). Reduction of the oxide function gives diazepam (12).

Metabolic studies of chlordiazepoxide and diazepam showed that both were quickly converted in the body to oxazepam, whose structure and synthesis are given in Figure 8.14. Oxazepam turned out to be an anxiolytic in its own right, more potent but shorter-acting than the other two. It is made from the chloromethylquinazoline (1). Treatment with sodium hydroxide expands the ring to give nordiazepam *N*-oxide (2). This is similar to compound 11 in Figure 8.13 but lacks the *N*-methyl group. Treatment with acetic anhydride brings about the Polonovski rearrangement to give an ester (3). Saponification of the ester with sodium hydroxide in ethanol gives oxazepam (4).

Lorazepam is the *o*-chloro derivative of oxazepam but is prepared by a different route, shown in Figure 8.15. The substituted benzophenone (1) is condensed with diethyl aminomalonate (2) and gives a benzodiazepine (3) directly. The 3-carbon is activated by two carbonyl groups and brominates easily to give (4). Treatment with methanol gives the methyl ether (5). Alkaline

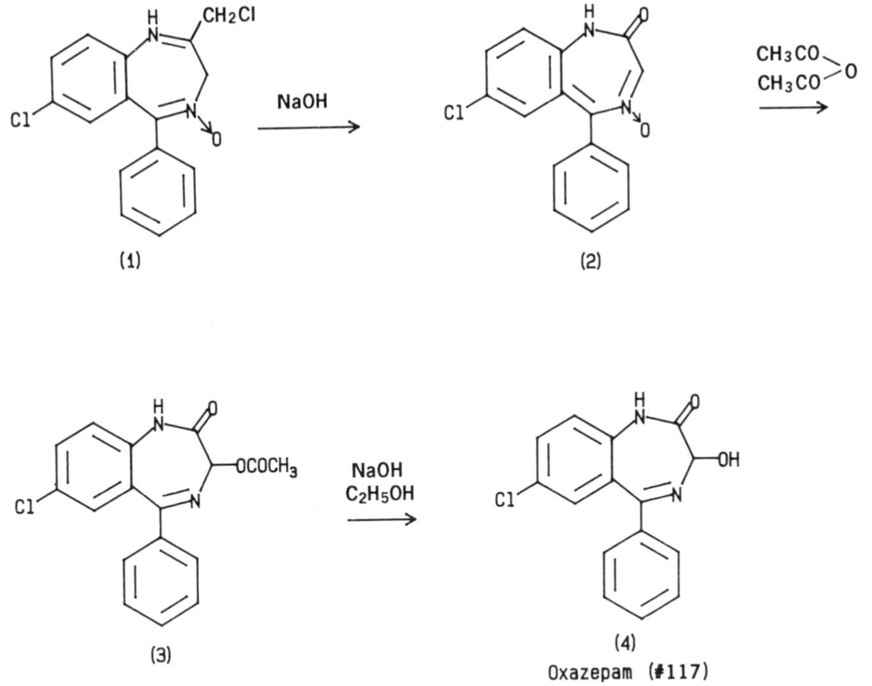

(1) NaOH → (2) CH_3CO/CH_3CO >O →

(3) NaOH, C_2H_5OH → (4) Oxazepam (#117)

Figure 8.14. Synthesis of oxazepam.

hydrolysis of the ester group gives a ketoacid, which decarboxylates to give the methyl ether of lorazepam (6). Cleavage of the methyl ether gives lorazepam (7).

There are a number of routes to alprazolam, all going via benzophenones but with the side chain built up in different ways. One route is shown in Figure 8.16. 2,6-Dichloro-4-phenylquinoline (1) is heated with hydrazine hydrate to give the benzodiazepine synthon (2). This reacts with triethyl orthoacetate in xylene to give the triazole (3). Sodium periodate and ruthenium oxide open the quinoline ring to give the substituted benzophenone (4). Formaldehyde followed by phosphorus tribromide in chloroform adds a bromomethyl substituent to the triazole ring to give (5). Cyclization with ammonia creates the benzodiazepine ring giving alprazolam (6).

A route to dipotassium clorazepate is shown in Figure 8.17. 2-Amino-5-chlorobenzonitrile (1) reacts with phenylmagnesium bromide to give the imine (2). This reacts with diethyl aminomalonate (as its hydrochloride) in a reaction analogous to the first reaction in Figure 8.15. The product is a benzodiazepine ester (3), which is converted by potassium hydroxide in ethanol to dipotassium clorazepate (4).

Nitrazepam, flurazepam, temazepam, and triazolam are hypnotics and sedatives but they are also benzodiazepines and will be considered here.

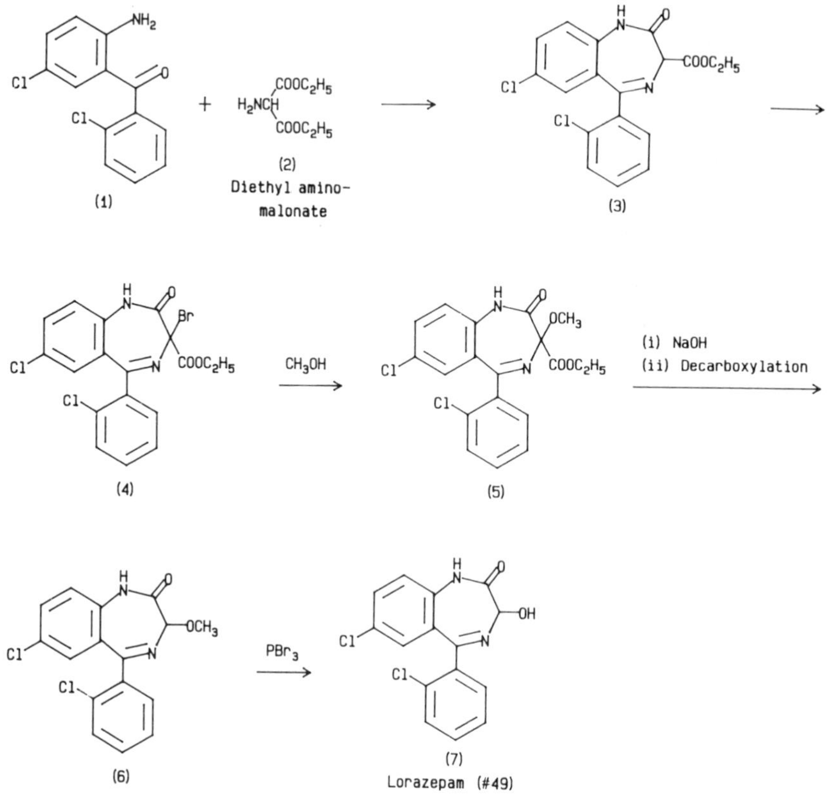

(1) + (2) Diethyl aminomalonate → (3) →

(4) CH_3OH → (5) (i) NaOH (ii) Decarboxylation →

(6) PBr_3 → (7) Lorazepam (#49)

Figure 8.15. *Synthesis of lorazepam.*

Figure 8.18 shows the synthesis of nitrazepam. It starts with 2-amino-5-nitrobenzophenone (1), which reacts with bromoacetyl bromide to give an amide (2). Ring closure in liquid ammonia gives nitrazepam (3). An alternative synthesis starts with 2-aminobenzophenone, that is, compound (1) without the nitro group. It is subjected to the same series of reactions to give the benzodiazepine (3) without the nitro group and this can then be nitrated.

Figure 8.19 shows the synthesis of flurazepam. It resembles the nitrazepam synthesis in that a substituted aminoketone is cyclized in a 7-membered ring. The starting material, however, is the fluorinated benzophenone (1). The reaction series used for diazepam eventually gives a fluorinated desmethyl diazepam (2). Reaction with 2-chloroethyldiethylamine gives flurazepam (3).

Of the two medium-acting soporifics, temazepam is the *N*-methyl analogue of oxazepam and is prepared by the route shown in Figure 8.14 except that the starting material is the *N*-methyl analogue of compound (1).

Triazolam is prepared from the chlorinated derivative of diazepam (1) as indicated in Figure 8.20. Treatment with phosphorus pentasulfide gives the thioamide (2). Condensation of (2) with acetylhydrazide yields triazolam (3).

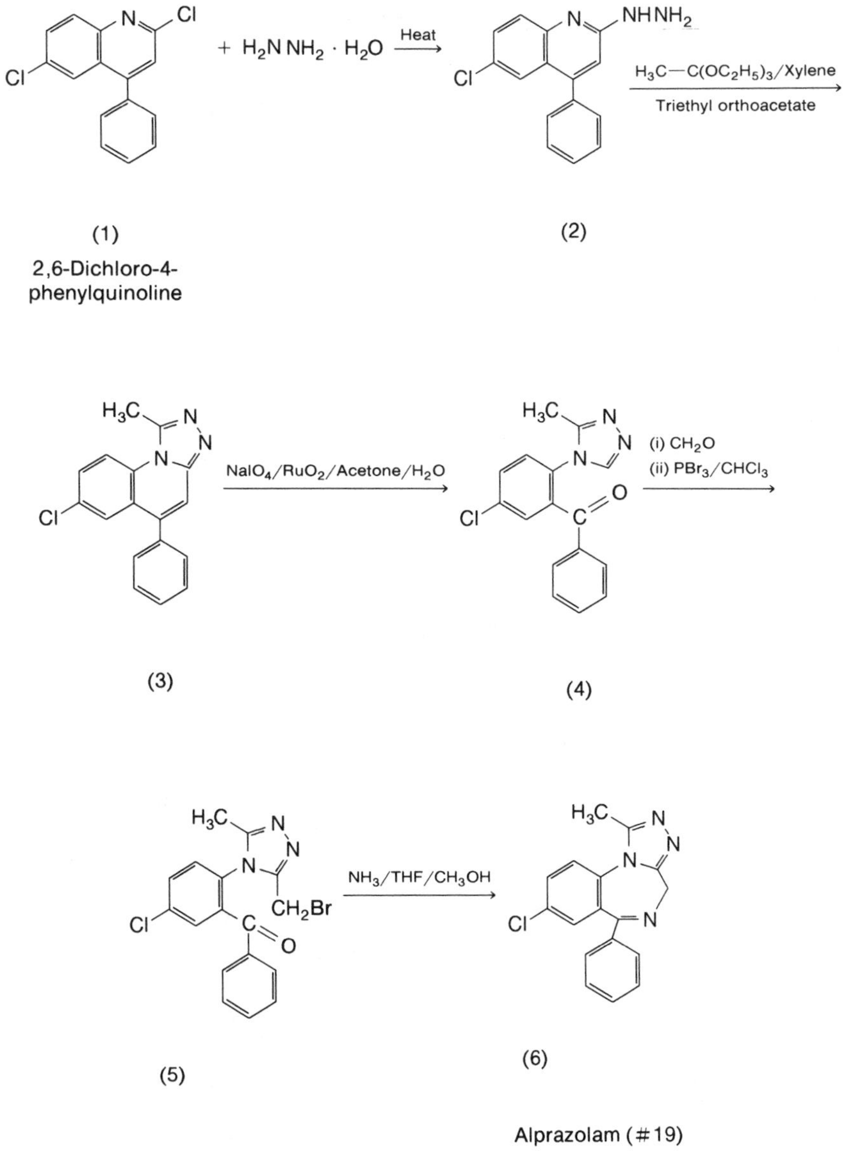

N Cl

Cl

$+ H_2NNH_2 \cdot H_2O \xrightarrow{\text{Heat}}$

N $NHNH_2$

Cl

$H_3C—C(OC_2H_5)_3$/Xylene

Triethyl orthoacetate

(1)

2,6-Dichloro-4-phenylquinoline

(2)

H_3C N N N

Cl

$NaIO_4/RuO_2$/Acetone/H_2O

H_3C N N N

Cl C O

(i) CH_2O

(ii) $PBr_3/CHCl_3$

(3)

(4)

H_3C N N N

CH_2Br

Cl C O

NH_3/THF/CH_3OH

H_3C N N N

Cl N

(5)

(6)

Alprazolam (#19)

Figure 8.16. *Synthesis of alprazolam.*

8.3.3 Meprobamate, Hydroxyzine, and Buspirone

Meprobamate, hydroxyzine, and buspirone are nonbenzodiazepine anxiolytics:

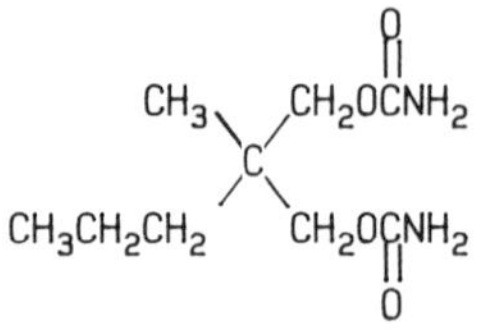

Meprobamate

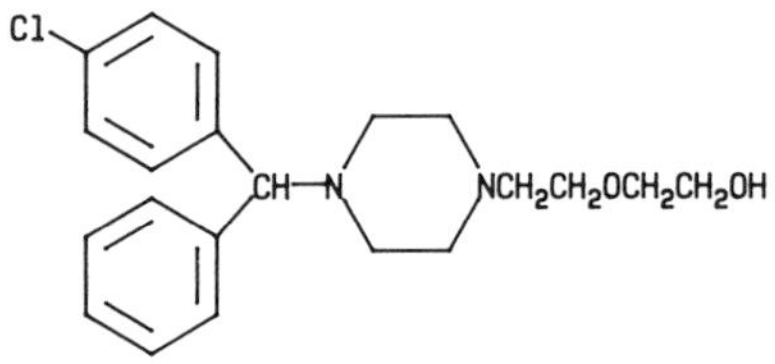

Hydroxyzine (#108)

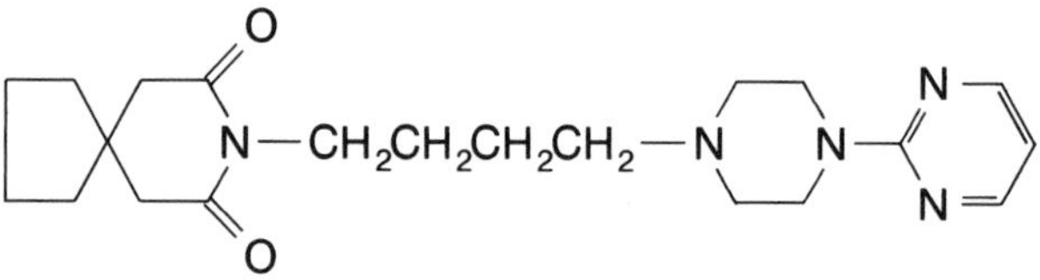

Buspirone

Meprobamate has been on the market since 1954, longer than any of the benzodiazepines. It was famous a generation ago as Miltown. It causes greater drowsiness than the benzodiazepines but less than the barbiturates. It is less effective than the benzodiazepines, more hazardous in overdosage, and more likely to induce dependence. It has dropped out of the Top-100 in recent years.

Hydroxyzine is an antihistamine. These are known not only to be antiallergic but also to cause drowsiness and to allay anxiety. Hydroxyzine is the antihistamine of choice for the last application. Its synthesis is given in Section 9.4 along with the other antihistamines.

Buspirone is a new nonbenzodiazepine anxiolytic launched in West Germany in 1985. It does not have muscle relaxant or anticonvulsant properties, but it does not cause drowsiness and it is said not to interact with alcohol.

Buspirone does not interact with benzodiazepine receptors but has a high affinity for serotonin receptors (Sections 8.4.1 and 8.4.4). Serotonin is a neurotransmitter that seems to be involved with the depressive state. It remains to be seen whether this new drug has sufficient advantages to displace the tried and tested benzodiazepines.

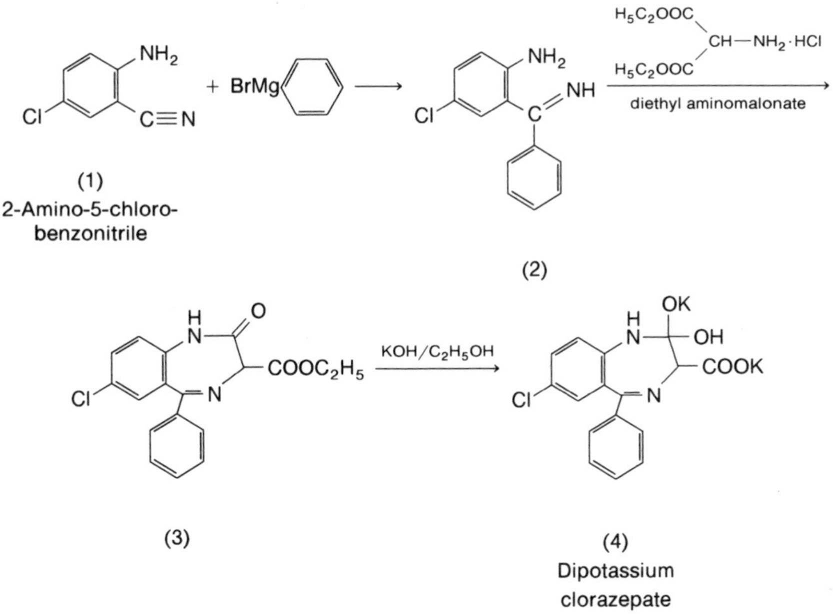

diethyl aminomalonate

(1)
2-Amino-5-chloro-
benzonitrile

(2)

KOH/C_2H_5OH

(3)

(4)
Dipotassium
clorazepate

Figure 8.17. *Synthesis of dipotassium clorazepate.*

8.3.4 Benzodiazepine Antagonists and Mode of Action

It is difficult to find the biochemical inbalances involved in so diffuse an illness as anxiety. Nonetheless, a tentative mechanism for benzodiazepine action has been proposed and the discovery of benzodiazepine antagonists has lent it credence.

γ-Aminobutyric acid (GABA), $H_2NCH_2CH_2CH_2COOH$, is an inhibitory neurotransmitter in the central nervous system. A subgroup of GABA receptors controls flows of chloride ions in and out of cells just as other receptors control the flow of calcium ions (Section 7.5.2). Benzodiazepines appear to act on both pre- and postsynaptic receptor sites to facilitate the effect of GABA. These benzoreceptors are part of a complex GABA receptor system shown diagrammatically in Figure 8.21. As indicated in the figure, three classes of compound are known to interact with the benzodiazepine receptor:

1. *Benzodiazepine agonists, such as diazepam.* These produce the expected anxiolytic effects, anticonvulsant action, muscle relaxation, and sedation.

2. *Benzodiazepine antagonists, such as flumazenil.* These have a high affinity to the binding site on the benzodiazepine receptor but have no action on their own. Flumazenil is similar in structure to triazolam, but in low doses blocks all the typical benzodiazepine effects. The benzodiazepine is displaced from the receptor and is destroyed in the body by normal processes. Flumazenil may be

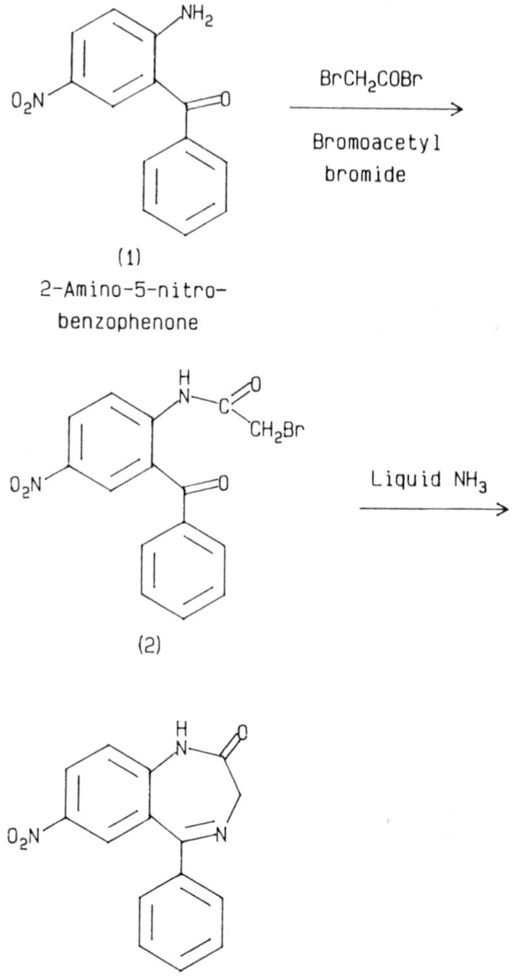

NH_2

O_2N

$BrCH_2COBr$

Bromoacetyl bromide

(1)
2-Amino-5-nitro-benzophenone

CH_2Br

Liquid NH_3

(2)

(3)
Nitrazepam

Figure 8.18. *Synthesis of nitrazepam.*

used for treatment of benzodiazepine overdosage and also for the rapid reversal of the anesthetic effects of benzodiazepines when the need for anesthesia is over. It can also be used rapidly to diagnose benzodiazepine intoxication, for example, after a suicide attempt.

3. *Inverse benzodiazepine agonists, such as the β-carboline esters.* These bind to benzodiazepine receptors but produce the opposite effects. That is, they cause anxiety, convulsions, increased muscle tone, and stimulation. The compound shown in the figure is dimethoxy-β-carboline methyl ester. The simpler

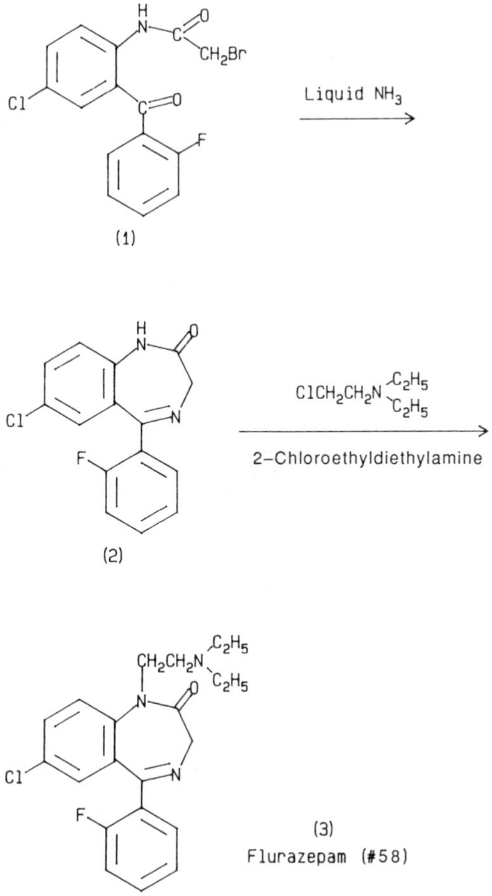

H N C O CH_2Br Cl C O F

(1)

Liquid NH_3

H N O Cl N F

(2)

$ClCH_2CH_2N(C_2H_5)_2$

2-Chloroethyldiethylamine

$CH_2CH_2N(C_2H_5)_2$ N O Cl N F

(3)
Flurazepam (#58)

Figure 8.19. *Synthesis of flurazepam.*

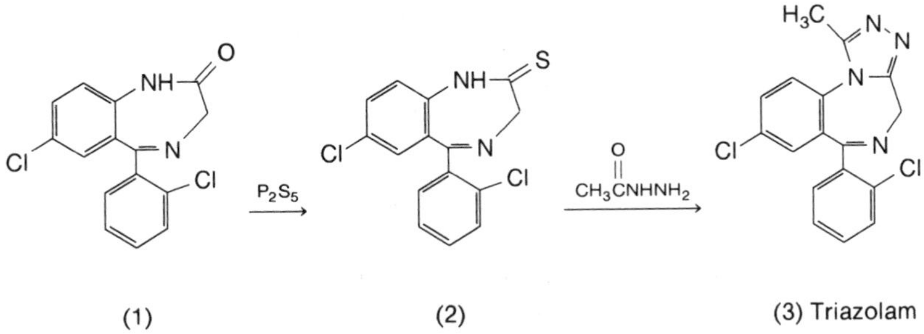

NH O Cl N Cl

P_2S_5

NH S Cl N Cl

$CH_3\overset{O}{\overset{\|}{C}}NHNH_2$

H_3C N N N Cl N Cl

(1) (2) (3) Triazolam

Figure 8.20. *Synthesis of triazolam.*

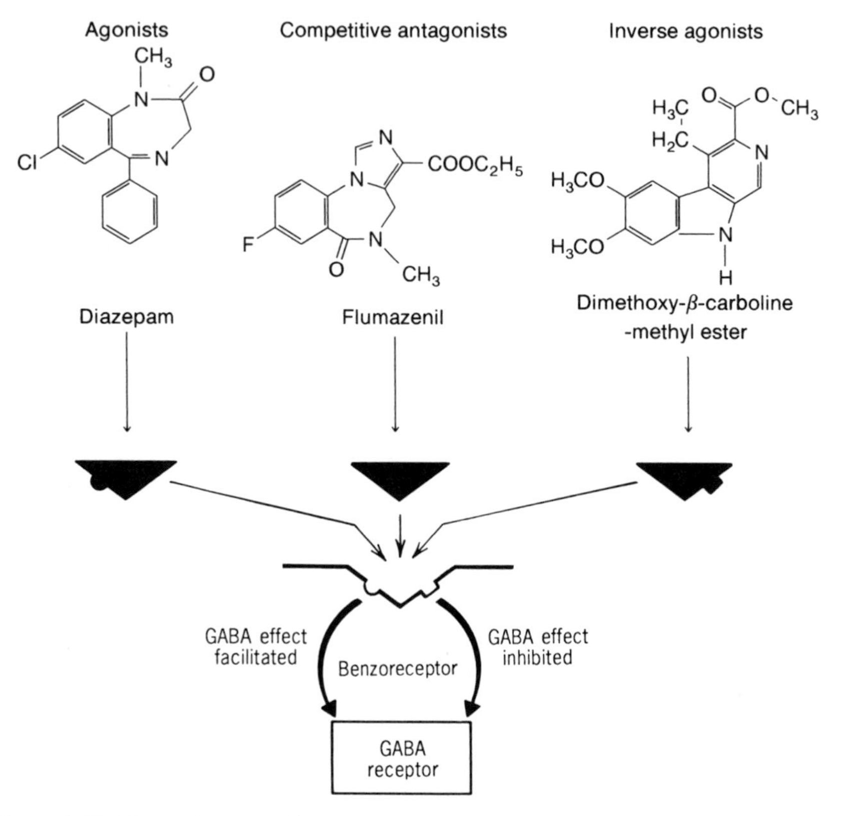

Figure 8.21. *Mode of action of benzodiazepines. The shapes in the middle of the diagram indicate how molecules might bind to the benzoreceptor to facilitate or inhibit the effects of GABA and how an antagonist might block the site.*

compounds lacking the methoxy and ethyl groups are of unusual interest because of the difference in action of the ethyl and propyl esters.

β-Carboline esters

The ethyl ester is anxiogenic; that is, it produces anxiety, at least in tests on monkeys. The propyl ester, on the other hand, is an antagonist and blocks the action of the ethyl ester although it has no action on its own. There is some hope that the β-carboline esters will be of value in the treatment of anxiety states and also of cerebral disorders in old people such as Alzheimer's disease.

8.4 ANTIDEPRESSANTS

The minor tranquilizers discussed in the previous section are anxiolytic; their function is to relieve anxiety and not to reduce depression. There is, of course, a question as to what constitutes clinical depression. Everyone experiences brief episodes of depression at one time or another, but some people suffer from long-term depression, perhaps accompanied by suicidal tendencies and changes in behavior. These patients may withdraw socially and stay at home; sleep rhythms may be disturbed; they may develop hypochondria and suffer from all sorts of symptoms apparently without any physical basis.

To give tranquilizers to patients in such a state will only exacerbate the condition. Tranquilizers are in fact depressants and are useless for the depressed patient. It was noted in the early 1950s, however, that iproniazid produced a euphoric reaction among patients taking it.

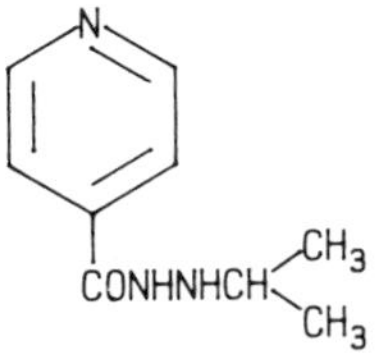

N

$CONHNHCH(CH_3)_2$

Iproniazid

Iproniazid is similar to isoniazid (Section 6.7.1), but has an isopropyl group in the side chain. At that time it was being used to treat tuberculosis.

8.4.1 Monoamine Oxidase Inhibitors

The reason iproniazid produces a euphoric reaction is well understood. In the discussion of α-blocking drugs for heart disease (Section 7.5.4), we mentioned that these materials displace norepinephrine from receptors and the norepinephrine is then either reabsorbed into the vesicles or destroyed by enzymes such as monoamine oxidase. The reduction in norepinephrine calms the patient and reduces the strain on the heart.

Iproniazid, on the other hand, was found to inhibit the action of monoamine oxidase; that is, it is a monoamine oxidase or MAO inhibitor. Monoamine oxidase destroys other neurotransmitters as well as norepinephrine. These include dopamine, and serotonin or 5-hydroxytryptamine:

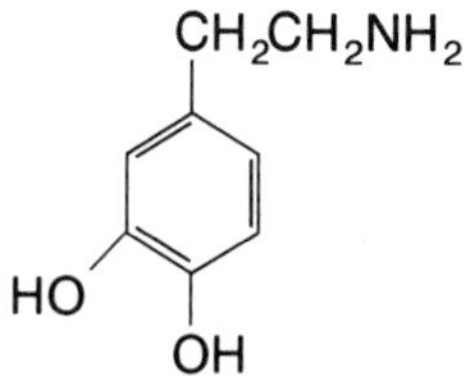

$CH_2CH_2NH_2$

HO

OH

Dopamine

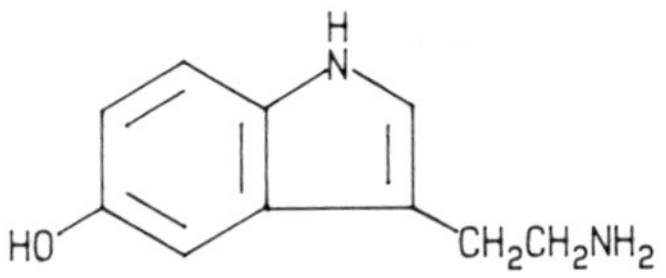

H
N

HO

$CH_2CH_2NH_2$

Serotonin
(5-Hydroxytryptamine)

If monoamine oxidase is inhibited, levels of these amines in the brain rise. There appears to be a relation between the depressive state and the lack of these amines in certain parts of the brain. The mechanism is uncertain, but it seems reasonable that high concentrations of materials known to be stimulants can counter depression.

The MAO inhibitors were the first drugs to be used in depressive illness. Iproniazid is rarely used today because it was found occasionally to cause fatal liver damage. Two safer MAO inhibitors are isocarboxazid and phenelzine:

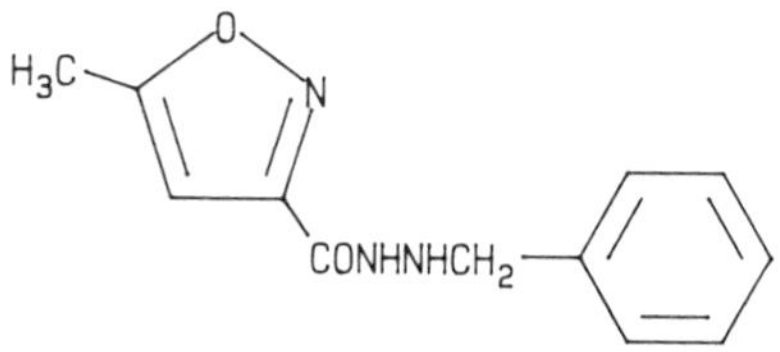

H_3C O N $CONHNHCH_2$

Isocarboxazid

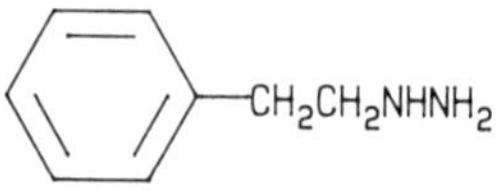

$CH_2CH_2NHNH_2$

Phenelzine

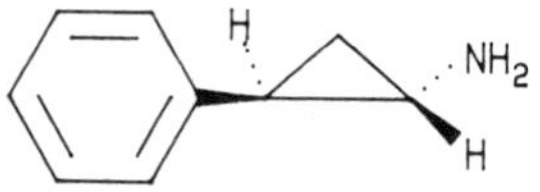

H NH_2 H

Tranylcypromine

Even these can lead to unpleasant side effects, including agitation, convulsions, hallucinations, and jerky muscular movements. MAO inhibitors are now usually reserved for patients who do not respond to other antidepressants, and accordingly they do not appear in our Top-100. On the other hand, with the recognition of two kinds of monoamine oxidases, interest in this class of compounds is increasing again. Monoamine oxidase A preferentially deaminates norepinephrine and serotonin, while monoamine oxidase B selectively degrades benzylamine and phenethylamine:

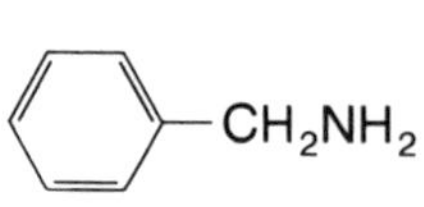

CH_2NH_2 $CH_2CH_2NH_2$ HO $CH_2CH_2NH_2$

Benzylamine 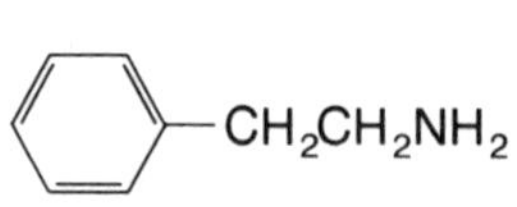 Phenethylamine 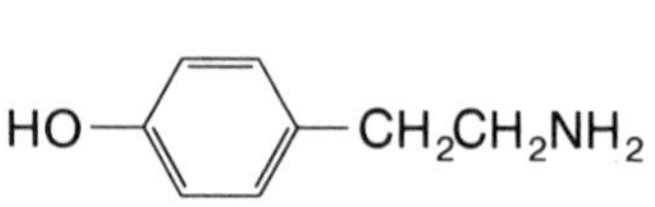 Tyramine

The MAO inhibitors mentioned so far all contain a hydrazine grouping, which is thought to be the effective part of the molecule. Hydrazine itself, however, is poisonous. Other MAO inhibitors include the cyclopropane derivative tranylcypromine, shown above. Tranylcypromine may be dangerously overstimulant and its use is permitted only under close supervision.

A curious aspect of MAO inhibitors is that they can lead to acute high blood pressure if taken with tyramine (see above), a chemical found in Cheddar, Camembert, and Stilton cheeses, beer, wine, pickled herring, chicken liver, yeast, broad beans, canned figs, and coffee. The effect is presumably related to the structural resemblances between tyramine, benzylamine, and phenethylamine.

8.4.2 Tricyclic Antidepressants

The role of MAO inhibitors has now been taken over by the tricyclic antidepressants and materials related to them (Figure 8.22). Imipramine, discovered in 1958, was the first. Amitriptyline (#34) and doxepin (#63) occur in the Top-100. Nortriptyline and desipramine are metabolites of amitriptyline and imipramine, respectively, and are drugs in their own right. Clomipramine is the second most frequently prescribed antidepressant in Europe but is not yet licensed in the United States. All six of these drugs have

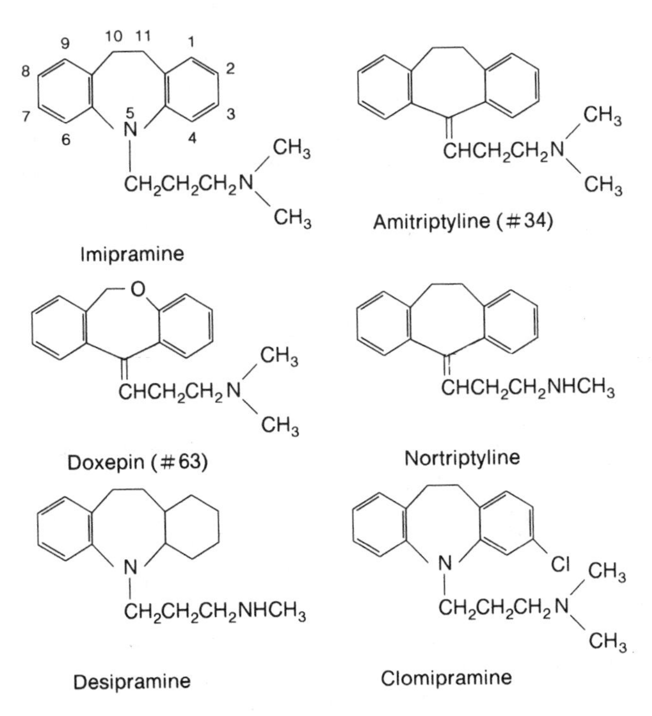

Figure 8.22. *Tricyclic antidepressants.*

three-ring structures in which the middle ring is alicyclic and contains seven atoms. In doxepin and imipramine the middle ring is heterocyclic, but in amitriptyline it is not. The side chain is derived from *N*-propyldimethylamine in all cases. Compare these structures with the phenothiazines in Figure 8.4. There are remarkable similarities. The *N*-propyldimethylamino side chain occurs in chlorpromazine, and the only significant difference is that the middle ring is six-membered and contains sulfur. There must be some reason for the opposite actions of the two groups of drugs but it is not clear what it is. There are also presumably some biochemical relationships between mania, depression, and anxiety.

Neither is the mode of action of the tricyclics well understood. They may produce an increased quantity of norepinephrine and serotonin at nerve endings in the brain by blocking the reuptake of these materials into the vesicles. There is, however, a crucial dependence on structure. Antidepressant activity in the tricyclics is limited to compounds having two or three carbon atoms in the side chain. Compounds lacking the side chain or with a branched side chain or with a chain of more than four carbon atoms are inactive. The nitrogen atom must have either methyl substituents or none at all. If the nitrogen substituent contains more than two carbon atoms, the compound is toxic and this increases with chain length. Tertiary amines are indeed active, as in imipramine, but this is attributed to the fact that the tertiary amine is rapidly converted to a secondary analogue in the body. Also the tertiary amines show sedative properties and, thus, imipramine is a weak tranquilizer and amitriptyline a stronger one.

The secondary amines are all much stronger inhibitors of norepinephrine take-up than primary or tertiary amines. Indeed, the secondary amine may be 70 times as effective. Nortriptyline, the amitriptyline metabolite, is a secondary amine that is two to five times more effective than its parent. It, too, is used clinically but does not appear in the Top-100.

Why are the preferred drugs tertiary amines when the secondary amines are more active? The reason is that the secondary amine is metabolized and excreted even more slowly than the tertiary amine even though the former is a metabolite of the latter. This is difficult to understand, but has been demonstrated conclusively in clinical tests.

The tricyclic antidepressants can take several weeks to have an effect and this is attributed to the necessity for the secondary amine metabolites to accumulate in the body. It is believed, incidentally, that the 10–11 bridge in the structures in Figure 8.22 is not necessary for antidepressant activity but seems to enhance the effect, perhaps by making the molecule coplanar.

The straightforward synthesis of imipramine is shown in Figure 8.23. Two moles of *o*-nitrobenzyl chloride (1) condense under alkaline conditions to give a dinitrostilbene (2). The nitro groups are reduced conventionally to amino groups (3). The styrene double bond is resonance-stablized by the aromatic rings and requires sodium and amyl alcohol to reduce it to (4). Strong heating of (4) leads to elimination of ammonia and cyclization to (5). Alkylation with

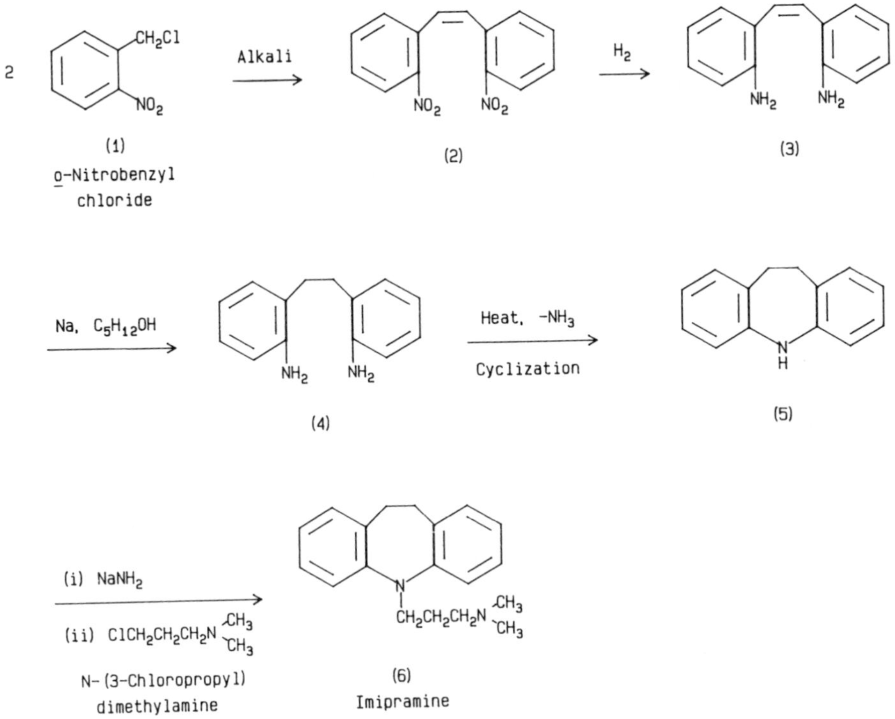

2 (1) o-Nitrobenzyl chloride $\xrightarrow{\text{Alkali}}$ (2) $\xrightarrow{H_2}$ (3)

$\xrightarrow{\text{Na, } C_5H_{12}OH}$ (4) $\xrightarrow[\text{Cyclization}]{\text{Heat, } -NH_3}$ (5)

$\xrightarrow[\text{(ii) } ClCH_2CH_2CH_2N(CH_3)_2]{\text{(i) } NaNH_2}$ (6) Imipramine

N-(3-Chloropropyl) dimethylamine

Figure 8.23. *Synthesis of imipramine.*

N-(3-chloropropyl)dimethylamine in the presence of sodium amide gives imipramine (6).

The synthesis of amitriptyline is shown in Figure 8.24. In the presence of sodium acetate, phthalic anhydride (1) and phenylacetic acid (2) undergo a reaction resembling an aldol condensation with elimination of both carbon dioxide and water. The methylenephthalide (3) that results is a useful intermediate for various drug syntheses. Treatment with hydriodic acid opens the lactone ring and removes the hydroxyl to give the acid (4). Cyclization gives another useful intermediate, a tricyclic ketone (5). This is condensed with cyclopropylmagnesium bromide to give a tertiary alcohol (6), which on treatment with hydrogen bromide goes via the cyclopropylcarbinyl cation with the elimination of water to the allylic halide (7). Reaction with dimethylamine gives amitriptyline (8).

The third important tricyclic, doxepin, is prepared as shown in Figure 8.25. 2-Benzyloxybenzoic acid (1) is cyclized with polyphosphoric acid to give (2). When (2) is treated with a Grignard reagent derived from *N*-(3-chloropropyl)dimethylamine, an alcohol (3) results, which on dehydration gives doxepin (4).

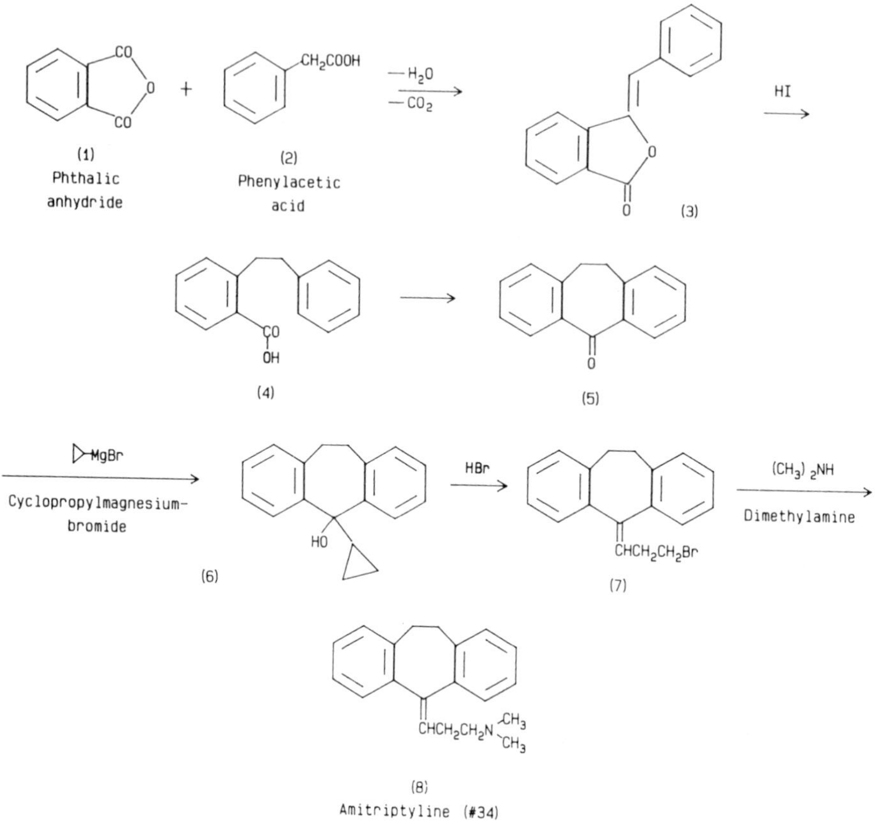

***Figure 8.24**. Synthesis of amitriptyline.*

The MAO inhibitors and the tricyclics are the major groups of antidepressants. In addition, there are a few compounds related to the amphetamine stimulants as well as lithium carbonate, which was mentioned in the introduction to this chapter, and which is used in treatment of manic illness and prevention of manic-depressive attacks. Lithium ions replace potassium ions inside nerve cells thus disrupting their functions, and lithium also alters the characteristics of membranes around cells. This is known to have an effect on excitability. It is possible that it also has an effect on monoamine oxidase activity.

8.4.3 Tetracyclic Antidepressants

The tricyclic antidepressants, as a class, were known in the 1960s. Recently there have been important discoveries relating to the mode of action of antidepressant treatment. The three conventional modes of treatment are

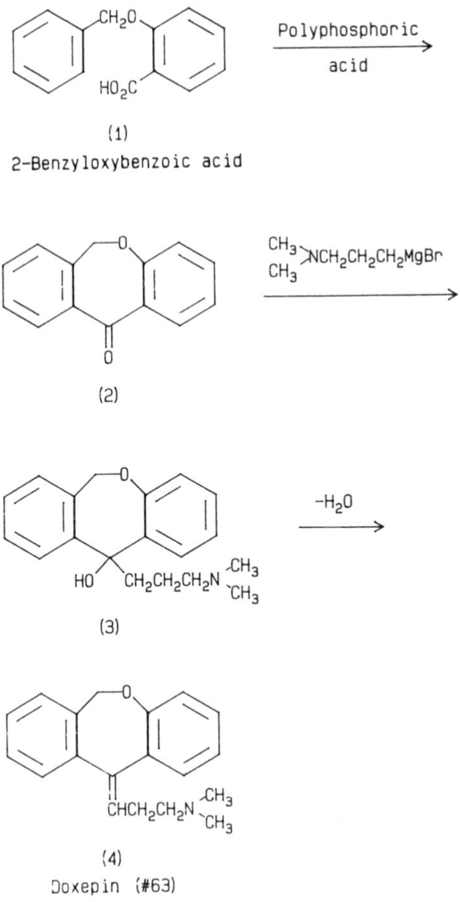

Polyphosphoric acid

(1)
2-Benzyloxybenzoic acid

$(CH_3)_2NCH_2CH_2CH_2MgBr$

(2)

$-H_2O$

(3)

(4)
Doxepin (#63)

Figure 8.25. *Synthesis of doxepin.*

MAO inhibitors, tricyclic antidepressants, and electroconvulsant shock treatment, a form of therapy involving the administration of electric shocks. All of them seem to cause a down-regulation of β receptors, that is, the number of β receptors is reduced in a way which correlates with the treatment starting to have an effect.

Whether this correlation is relevant should be decided soon in that a number of companies are searching for α_2-presynaptic antagonists, which are compounds that might reduce the number of β receptors. The presynaptic α receptor sites on the presynaptic nerve cell were shown in Figure 7.13. Clonidine (Figure 7.15) was found to have an agonist action on these receptors and, thus, reduces norepinephrine output. An α_2-presynaptic antagonist would have the opposite effect and would increase norepinephrine output. This, in turn, might down-regulate the β receptors.

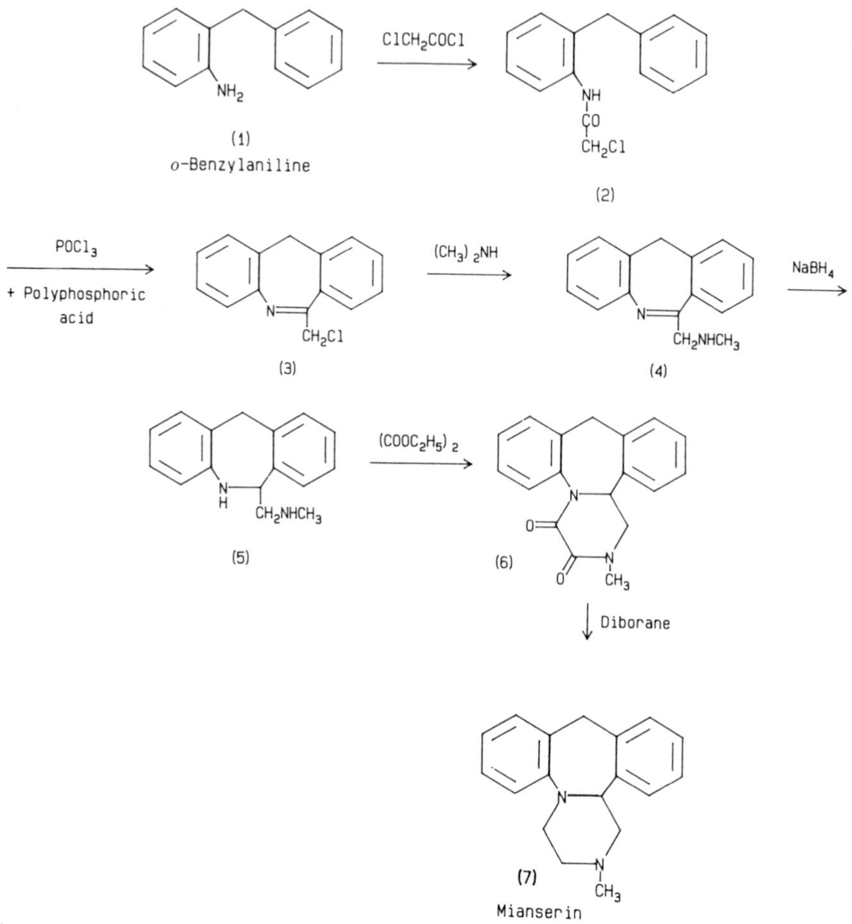

Figure 8.26. *Synthesis of mianserin.*

Mianserin (Figure 8.26, compound 7) is a tetracyclic antidepressant already in use which is believed to act exclusively in this way and not to interfere with reabsorption of neurotransmitters. If this is true, one might expect its action to be antagonized by clonidine, and such an effect is indeed observed.

The synthesis of mianserin (Figure 8.26) illustrates the introduction of a fourth ring. It starts with the benzylaniline (1). Reaction with chloroacetyl chloride gives the chloroamide (2). Compound (2) is simultaneously cyclized and dehydrated by phosphorus oxychloride and polyphosphoric acid to give (3). The chlorine atom is now allylic and reacts readily with dimethylamine to give a secondary amine (4). Reduction of the double bond with sodium borohydride gives (5). Ester interchange and condensation with diethyl oxalate forms the fourth ring as a cyclic diamide (6) and reduction of the keto groups with diborane gives mianserin.

Maprotiline was the first tetracyclic antidepressant to be marketed in the United States. It has a bridged ring structure and a methylaminopropyl side chain:

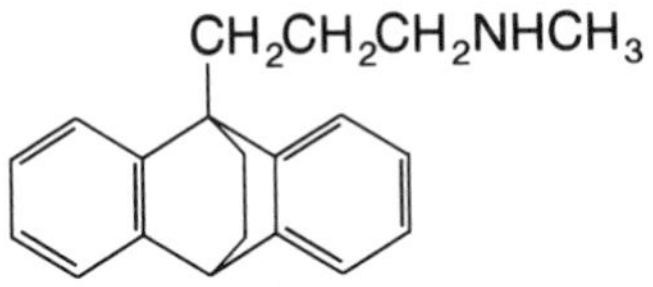

$CH_2CH_2CH_2NHCH_3$

Maprotiline

Clinically it resembles imipramine and it inhibits the reuptake of norepinephrine in the synapses.

8.4.4 Serotonin Reuptake Inhibitors

Antidepressant research has also focused on selectivity. The tricyclic antidepressants block the reuptake of both norepinephrine and serotonin. Some affect norepinephrine more than serotonin and vice versa. These drugs also affect other receptors and, therefore, have side effects. Desipramine is the drug that blocks norepinephrine reuptake the most in relation to serotonin and it appears an effective compound. The tetracyclic compound maprotiline also favors norepinephrine.

Nonetheless, the emphasis is currently on a range of selective serotonin reuptake inhibitors. Some are shown in Figure 8.27. Trazodone was the only one already on the U.S. market in mid 1986 and was (#83) in the Top-100. It specifically blocks serotonin reuptake and has little effect on norepinephrine. Its structure is unrelated to the tri- or tetracyclics.

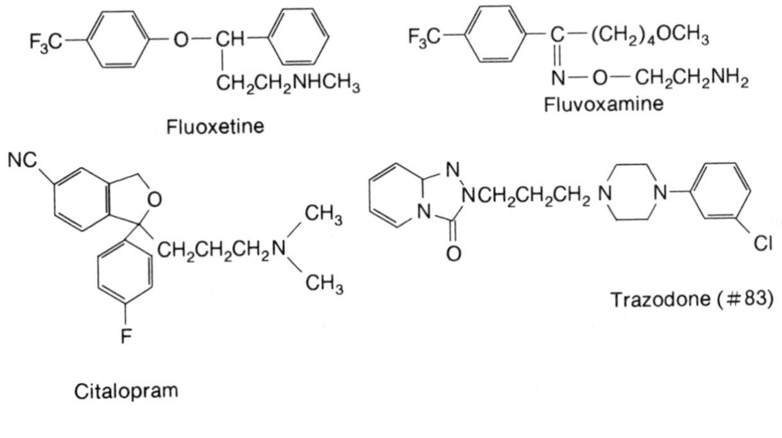

F_3C—C₆H₄—O—CH(C₆H₅)—$CH_2CH_2NHCH_3$

Fluoxetine

F_3C—C₆H₄—C(=N—O—$CH_2CH_2NH_2$)—$(CH_2)_4OCH_3$

Fluvoxamine

NC, O, F, $CH_2CH_2CH_2N(CH_3)_2$

Citalopram

N, $NCH_2CH_2CH_2$ N N, O, Cl

Trazodone (#83)

Figure 8.27. Serotonin reuptake inhibitors.

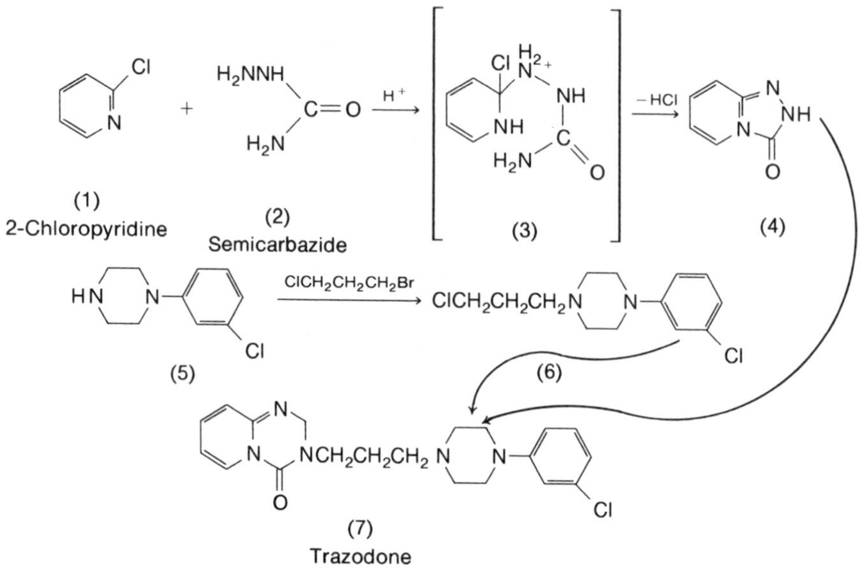

(1) 2-Chloropyridine

(2) Semicarbazide

(3)

(4)

(5)

(6)

(7) Trazodone

Figure 8.28. *Synthesis of trazodone.*

The synthesis of trazodone is shown in Figure 8.28. 2-Chloropyridine (1) is protonated, and the carbenium ion reacts with semicarbazide (2) to give the intermediate (3). This eliminates hydrogen chloride, aromaticity is restored, and the pyridine nitrogen attacks the semicarbazide carbonyl to give the fused triazole (4). Meanwhile, alkylation of the substituted piperazine (5) with 1-bromo-3-chloropropane leads to (6), which has a substituent on the second nitrogen atom. This reacts with the triazole (4) to give trazodone (7).

The newer drugs are said to be even more selective. Fluoxetine, for example, is a potent serotonin reuptake inhibitor but does not affect norepinephrine reuptake. Nor does it have affinity for α-noradrenergic, acetylcholine or histamine (H_2) receptors (Sections 7.6.1, 14 and 9.6). It is expected to provide antidepressant therapy with fewer side effects than currently available materials.

Although fluoxetine is the front-runner for FDA approval, at least five other serotonin-specific antidepressants are at an advanced stage of testing. Most of them are fluorine derivatives, and fluoxetine, fluvoxamine, and citalopram are shown in Figure 8.27. The most curious aspect is that while these serotonin-specific materials offer great promise for more effective treatment of depression, the drugs blocking norepinephrine reuptake seem to work very well and it is apparent that much remains to be learned about the biochemistry of depressive illness.

In addition to its antidepressant action, fluoxetine may well help with other illnesses such as obsessive–compulsive disorders (a use it shares with clomipramine, Figure 8.22) and obesity. Most antidepressants lead to the patient's gaining weight. Fluoxetine, on the other hand, is related to the appetite suppressant drug fenfluramine (Section 8.5.1) and appears to encourage weight loss. Thus, while the conventional antidepressants turn thin miserable people into fat happy people, fluoxetine might well turn fat miserable people into thin happy ones.

8.5 STIMULANTS AND APPETITE SUPPRESSANTS

The next group of central nervous system drugs in Table 8.1 is the stimulants and appetite suppressants, otherwise known as anorectics. Changing views on the role of drugs in the treatment of obesity have led to the disappearance of all these drugs except caffeine and phenylpropanolamine from the Top-100, but Figure 8.29 shows diethylpropion, phentermine, and fenfluramine, which are still of some importance, together with amphetamine, which was the first of these materials to be discovered. Caffeine is a mild central nervous system stimulant and is included in several Top-100 formulations to counter drowsiness. Cocaine is a local anesthetic, a stimulant, and a currently fashionable drug of abuse. These two will be discussed later.

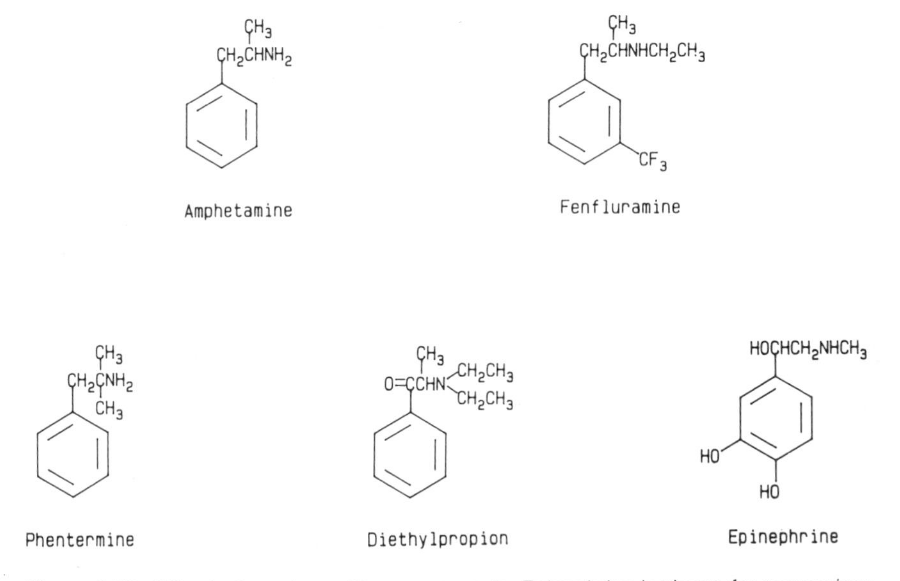

Figure 8.29. *Stimulants and appetite suppressants. Epinephrine is shown for comparison.*

8.5.1 Amphetamine-like Anorectics

The structures of the four anorectics in the figure closely resemble that of the hormone epinephrine, which is also shown. The amphetamines are sympathomimetic amines and stimulate the adrenergic nervous system, acting in the same way as epinephrine, the body's own stimulant. They are vasoconstrictors and, as a consequence, they raise blood pressure and stimulate respiratory action. Although their effect varies from person to person, they generally produce increased activity and alertness. They were used by fighter pilots during World War II to prevent their falling asleep at the controls. In the 1950s they became the so-called "uppers" of drug culture. They were not too difficult to obtain at the time because amphetamine in the form of Benzedrine inhalers was widely used to counter nasal congestion, and amphetamines were also used to promote weight loss. A mixture of a barbiturate, sodium amytal, with amphetamine was also marketed as an early antidepressant, the amphetamine neutralizing the soporific effects of the barbiturate and leading to mild euphoria. The tablets became known as purple hearts because of their color and shape. Dextroamphetamine, known as Dexedrine, and methylamphetamine, known as Methedrine, were used and abused similarly. At present, amphetamines are recommended only for narcolepsy, a condition characterized by frequent and overwhelming desire for sleep.

The increased activity brought about by amphetamines makes demands upon the stored energy reserves of the body, leading to a decrease in body weight and reduction in appetite. This use as appetite suppressants is now the major application of drugs related to amphetamine. Obesity is seen in Western society not only as unattractive but also as unhealthy. Overweight people often have to pay higher insurance premiums and may be regarded as ugly. Not surprisingly, dieting is a cult in our society and painless ways of losing weight are eagerly sought. Of course, the only way to lose weight is to eat less, but people's eating habits are established early in life and many find it difficult to summon up the willpower—or won't power—to change. Hence, the demand for drugs to aid weight loss.

There are other types of drugs used to aid loss of weight. Some nonprescription compounds are laxatives and cause loss of weight through loss of body fluid. Other weight loss preparations are diuretics and again cause loss of fluid. The weight loss in both cases is temporary. Another group of patent weight loss medicines contains materials like methylcellulose, which, when taken before meals, are supposed to reduce appetite. They swell up in the stomach and induce a "full" feeling and also absorb water and carry it out of the system. Again, there is little evidence that they bring any long-term benefit.

The use of amphetamine-like drugs as appetite suppressants is more effective than the above remedies. Nonetheless, they lose their effect after a few weeks, and increased doses may lead to dependence. One such compound is phenylpropanolamine, an over-the-counter drug and a constituent of many "cold cures" (Section 9.5).

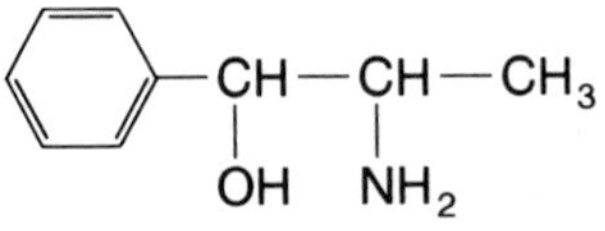

Phenylpropanolamine

It is also a component of various proprietary weight-reduction products. An FDA advisory panel has concluded that it is safe and effective as an aid to weight reduction, but there is also evidence that it is being ingested in excessive doses both as an anorectic and as a stimulant. The value of drugs in control of weight is trivial compared with change in eating habits. One cannot take appetite-suppressant drugs all one's life; when one stops, unless there has been a change in diet, the weight lost will be regained.

Amphetamine is obtained by reductive amination of phenylacetone with ammonia and hydrogen.

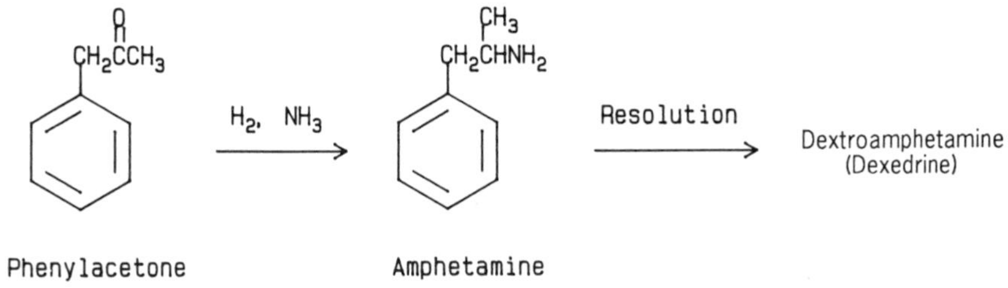

Phenylacetone —(H_2, NH_3)→ Amphetamine —(Resolution)→ Dextroamphetamine (Dexedrine)

Resolution of the racemic mixture gives dextroamphetamine, which is somewhat more potent.

Various central nervous system drugs can be usefully modified by incorporation of a trifluoromethyl group. A trifluoromethyl compound related to amphetamine is called fenfluramine (Figure 8.29). Fenfluramine retains the appetite suppressant properties of amphetamine but has a slight depressant rather than a stimulant effect and so is not a drug of abuse. A recent development in the field of antidepressants is the development of the serotonin reuptake inhibitors (Section 8.4.4). Fluoxetine, in particular, has a structural resemblance to fenfluramine and, unlike most antidepressants, promotes weight loss rather than discouraging it.

Amphetamine-like substances are believed to be eliminated from the body by oxidation and deamination, the first stage being the oxidation of the amino group to an imine:

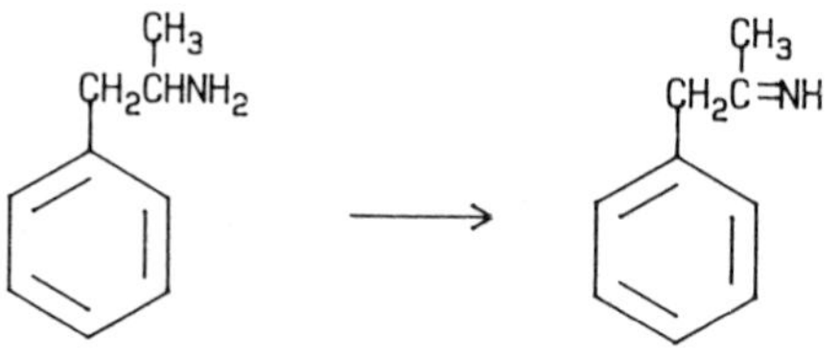

CH_3
CH_2CHNH_2

CH_3
$CH_2C{=}NH$

Amphetamine

Schiff's base

Without hydrogen atoms alpha to the amino group, this conversion is impossible, consequently a compound with an amine group on a tertiary carbon should be longer acting than amphetamine. Phentermine is such a compound. Another appetite suppressant is diethylpropion and, again, it is related to amphetamine.

These drugs all act on the central nervous system and are liable to abuse for this reason. Under development are compounds related to 2,5-anhydro-D-mannitol, a fructose analogue, which appears to act on the liver. These may prove to be genuinely nonaddictive nonstimulant appetite controllers.

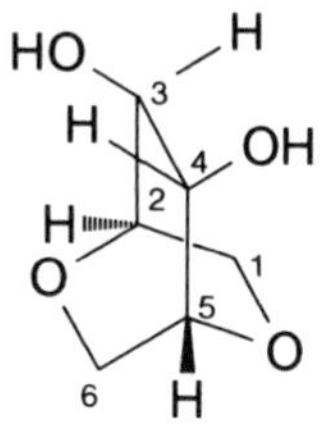

HO H 3 H 4 OH 2 H O 1 5 O 6 H

2,5-Anhydro-**D**-mannitol

8.5.2 Caffeine and Cocaine

Caffeine and cocaine, the two remaining stimulants, are unrelated to amphetamine:

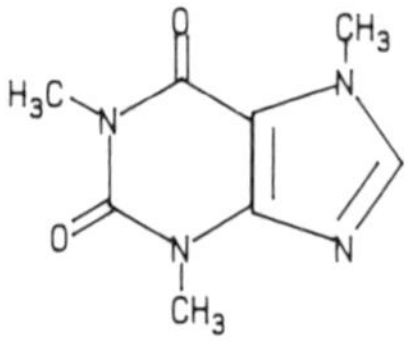

O CH_3 H_3C N N O N N CH_3

Caffeine

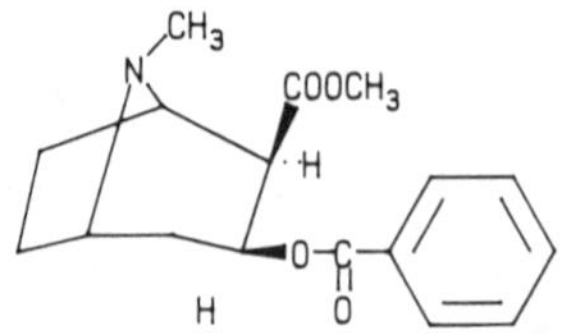

CH_3 N $COOCH_3$ H O C O H

Cocaine

TABLE 8.2. Typical Caffeine Contents of Drinks and Over-the-Counter Pharmaceuticals[a]

Substance	Caffeine Content (mg)
Coffee (per 5 fl. oz cup)	
Percolated	110
Instant	53
Decaffeinated	2
Cocoa (per 8 fl. oz mug)	13
Milk chocolate (per 4 oz. bar)	24
Tea (per 5 fl. oz)	
3-minute brew	20–46
Soft drinks (per 12 fl. oz can)	
Coca-Cola, Diet Rite, Diet Pepsi	34
Tab	44
Pepsi Cola	37
Drugs (per tablet (check label))	
Pain relievers	30–130
Cold remedies	30–130
Premenstrual diuretics	100–200
Caffeine tablets	Up to 200

[a]A rough guide based on U.S. Consumer Reports, 1981. The brain and central nervous system are stimulated by 100–300 mg. Adverse effects arise with doses larger than 1000 mg.

Caffeine is a constituent of Fiorinal (butalbital/aspirin/caffeine), Synalgos (dihydrocodeine/aspirin/caffeine), and Darvon 65 (propoxyphene/aspirin/caffeine). In addition, most people consume large quantities of caffeine every day as a constituent of tea, coffee, cocoa, or cola. The quantity in an average drink or over-the-counter pharmaceutical in 1981 is shown in Table 8.2. As a result of adverse publicity, quantities may have been reduced recently. The products named above all come at the low end of the range.

Approximately half of U.S. caffeine is synthesized. The method is similar to that for theophylline and is shown in Figure 15.5. The caffeine in soft drinks is a recycle of the caffeine removed from decaffeinated coffee.

The FDA has taken an interest in caffeine since 1978 and there has been lobbying to remove it from the "generally regarded as safe" category of food additives. In particular, there have been worries about the consumption of relatively large quantities of caffeine in soft drinks by growing children, and several manufacturers are marketing caffeine-free products. In the United Kingdom, a limit of 125 mg/L has been set for the caffeine content of soft drinks. Nonetheless, here is a widespread drug problem with a material that is, at worst, only mildly addictive and slightly dangerous.

Far fewer people take cocaine, but it is far more dangerous and addictive. The Indians of the high Andes in Ecuador and Peru have traditionally chewed the leaves of the coca bush to combat fatigue and deaden the pangs of hunger, and the Bolivian tin miners do so even today to enable them to work under miserable conditions at high altitudes. The active ingredient is cocaine.

Cocaine is of some historical interest in that it was pioneered by the young Sigmund Freud as a treatment, in hindsight a very dangerous one, for postnatal depression. It is also of literary interest because it was supposedly taken by Sherlock Holmes.

It readily penetrates the mucous membranes, which is a help in anesthesia and also enables addicts to inhale it. Apart from the dangers of addiction, inhalation leads to necrosis of the nose. In view of this mutilating side effect, the emergence of cocaine as a currently popular drug of abuse is difficult to understand, but apparently it is now marketed in a form that can be smoked, and is becoming a major problem in the United States.

The only clinical use of cocaine at present is as a topical local anesthetic, although even there it is being replaced by less toxic materials. It is, however, of tremendous clinical significance as the progenitor of the whole range of local anesthetics in current use. We have ignored these in our discussion because they are not widely prescribed, but they are nevertheless important and there is a fascinating case study of their development by Lednicer and Mitscher, and Gilbert and Sharp (see Bibliography).

8.6 PSYCHOTOMIMETICS

The final category of central nervous system drugs in Table 8.1 is the psychotomimetics or psychodysleptics. They are frequently called hallucinogens. A number are shown in Figure 8.30. They cause mental and physical disturbance, including vivid visual hallucinations, anxiety, and delusions. The responses mimic some of the symptoms of psychotic states. Some psychotomimetics such as lysergic acid diethylamide (LSD) have been used as adjuncts to psychotherapy although there are risks attached to such use and the therapeutic value is unproved. This class of drugs has few legitimate medical uses and none of these drugs appears in the Top-100. We list them, however, because of the part they have played in drug culture.

Psychotomimetics are not a clear-cut group of drugs either in terms of chemical structure or their pharmacological effects, but all of them appear to act in the brain. Three types of chemical compound appear most likely to lead to hallucinations: cannabis, amphetamine-like drugs such as mescaline, and indoleamines such as LSD.

Mescaline is the active ingredient of the peyote cactus and has been used in religious ceremonies for centuries by American and Mexican Indians. It is not particularly potent nor addictive and does not appear to cause withdrawal symptoms. Its structure is related to the sympathomimetic amines such as epinephrine and amphetamine, and various other hallucinogenic drugs can be synthesized by addition of methoxy groups to amphetamine. Indeed, there is

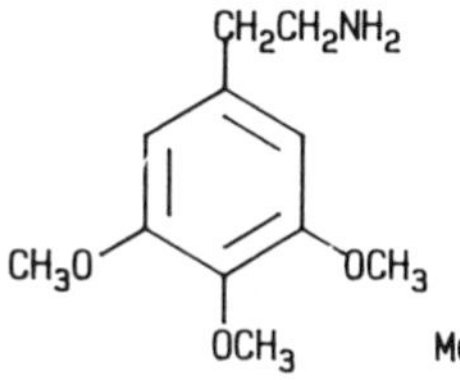

Mescaline

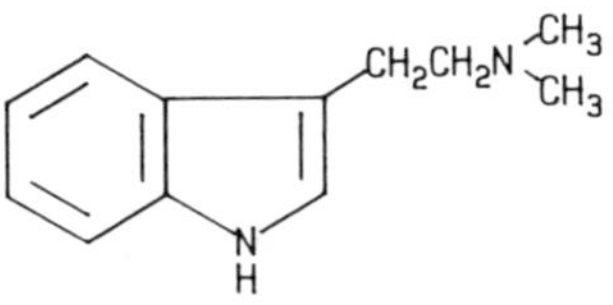

Dimethyltryptamine

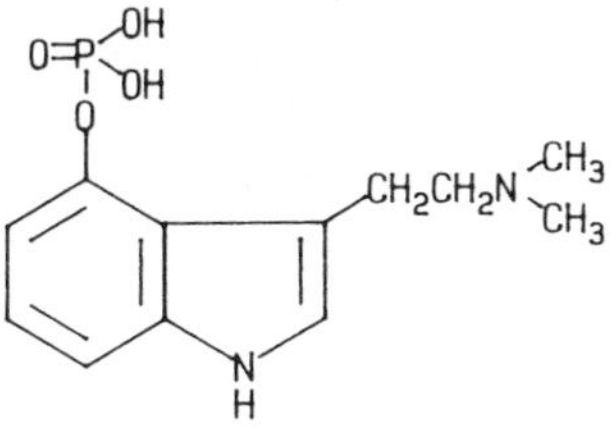

Psilocybin

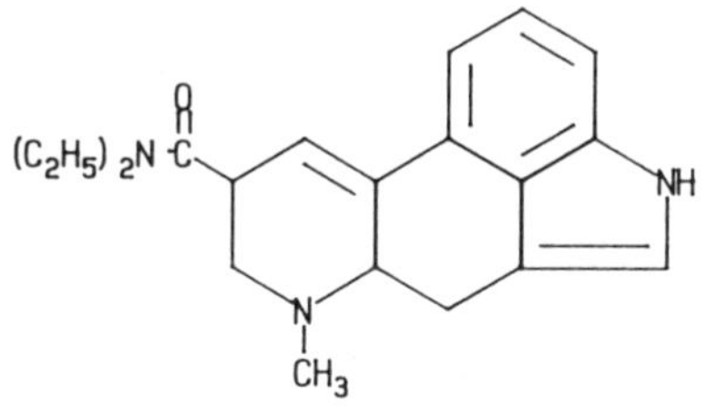

Lysergic acid diethylamide
(LSD)

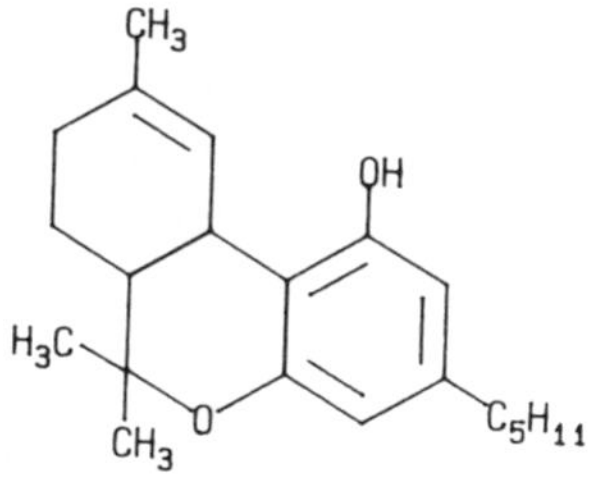

Tetrahydrocannabinol
(THC)

Figure 8.30. *Psychotomimetics or psychodysleptics (hallucinogens).*

some evidence that the psychotic effects that may result from amphetamine are due to its conversion in the liver to mescaline or a mescaline-like substance.

In the same way that mescaline is a methoxylated variant of norepinephrine-like molecules, many hallucinogens are methylated versions of the indoleamines, which are also important biological substances. The methyl groups make the molecule more lipophilic and, hence, better able to penetrate the central nervous system. An example is shown in Figure 8.30. Dimethyltryptamine is a methylated dehydroxylated analogue of the neurotransmitter, serotonin. These drugs, which resemble neurotransmitters, appear to delay metabolic oxidative deamination reactions and thus prolong the existence of active amines in the body. Dimethyltryptamine is not a drug of abuse because it is not very potent and not readily available.

Another more potent drug of this type shown in Figure 8.30 is psilocybin, found in the mushroom *Psilocybe mexicana*. This mushroom is used in religious ceremonies by Mexican Indians. In the body, the phosphoric acid group is hydrolyzed leaving the 4-hydroxy compound, which is equally active.

Lysergic acid diethylamide is more widely used. It is also an indoleamine and is believed to interfere with the natural function of serotonin. LSD has been associated with the development of acute leukemia and has also been said to cause chromosomal damage, although the evidence on this latter point is thin. Suicides and murders have undoubtedly been committed on LSD trips but it is possible that these were a function of the undrugged personality, exacerbated by the drug's influence on decision-making. Lysergic acid results from hydrolysis of the alkaloids found in ergot.

Cannabis is one of the oldest hallucinogens known and comes from the flowering tops and leaves of hemp, *Cannabis sativa*. The word marijuana applies to the whole plant or the resin from it and there is a language differing from culture to culture to describe the different kinds and potencies of cannabis. The active ingredient is tetrahydrocannabinol or THC, shown in Figure 8.30. It has been used clinically as an antiemetic, that is, a drug to inhibit vomiting. The mode of action of cannabis is not known and repeated experiments have failed to show any major short-term dangers, although equally it has certainly not been shown to be safe in the pharmacological sense. It is not addictive and there are no withdrawal symptoms. The long-term medical or social effects are still unknown.

Morphine and heroin are both alkaloids derived from opium from the poppy, *Papaver somniferum*. They are much more dangerous drugs, rapidly producing addiction, with the concomitant withdrawal symptoms—the so-called "cold turkey"—when the drug is withheld. They will be discussed in Chapter 10.

The availability of psychotomimetic drugs has influenced life-styles markedly. The drug culture had its heyday in the sixties when it affected language, dress, music, and morals.

8.7 USE AND ABUSE

Like all other discoveries, psychotropic drugs can be used or abused and can lead to benefits or disadvantages. In favor of the major tranquilizers, it must be pointed out that they have revolutionized the treatment of serious mental illness. Figure 8.31 shows their effect on the populations of mental hospitals in

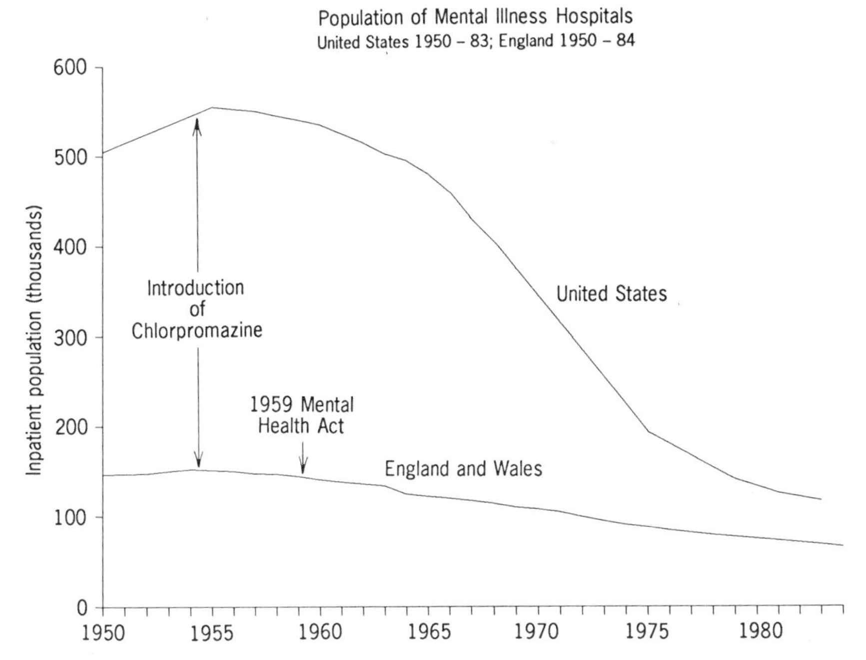

Figure 8.31. *Population of mental illness hospitals (United States 1950– 1983; England 1960– 1984). Source: U.S. National Institute of Mental Health Reports, Statistical Abstract of the Compendium of Health Statistics, Office of Health Economics. Data after 1968 refer to England only, not England and Wales.*

the United Kingdom and the United States. In the 18 years after the introduction of chlorpromazine, there was a drop of about a third. These drugs have allowed many people to function in the community instead of being confined. The picture is not as rosy as it might appear because sometimes patients discharged from hospital lack a support system at home (which reminds them, among other things, to take the medicines) and they finish up as "street people". Estimates from both London and New York state that between a quarter and a third of vagrants are long-term patients discharged from mental hospitals and showing some symptoms of schizophrenia.

Nonetheless, the drugs have helped most of those discharged from hospitals and have also diminished the demeaning effects of confinement. Instead of mental asylums being places where incurably insane patients were kept for the rest of their lives, they are now places where the mentally sick can be nursed back to normality.

People use the word "bedlam" today without realizing that it is the corruption of the name of the Bethlehem Asylum for the Insane in London where, in Tudor times, lunatics were confined in cages and exhibited for the amusement of the public. Thanks to psychotropic drugs, there is no place for bedlams in our society and indeed the Tudor building in South London that once housed it is now the Imperial War Museum, perhaps a monument to another form of human insanity.

Having said this, we must mention the situations where major tranquilizers are abused. Their use on political prisoners in the Soviet Union is well documented and there is no doubt that every major power has investigated brainwashing techniques based on these materials. There is room for debate on marginal cases, for example, the use of drugs to quiet violent prisoners or the extent to which patients confined to mental institutions are kept highly medicated because that makes life easier for the staff.

Similarly, the anxiolytics can be used sensibly or foolishly. At best, they can enable people to cope with crises in their lives—bereavement perhaps—that would otherwise make them unable to perform their jobs and support their

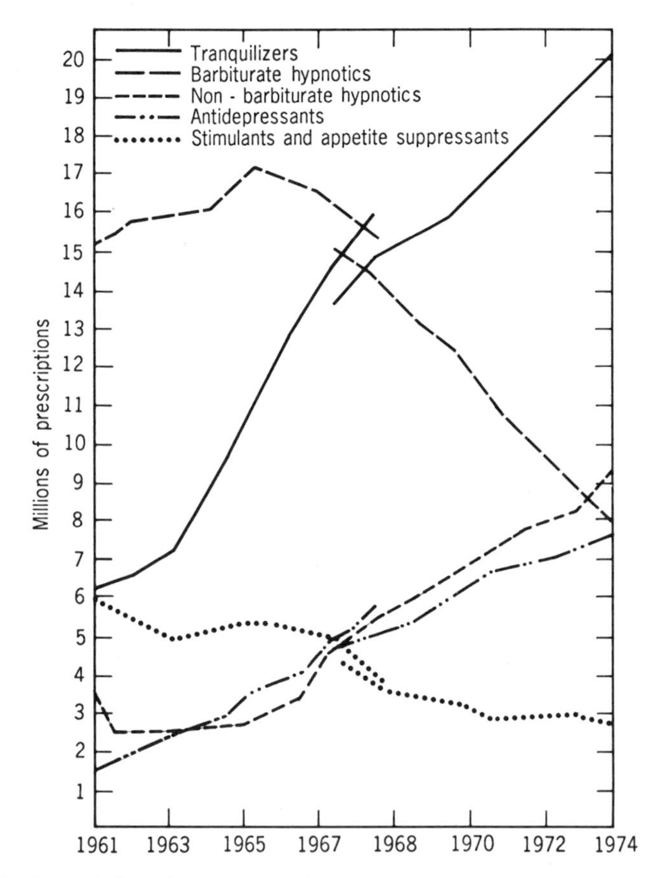

__Figure 8.32.__ Prescriptions for psychotropic medicines (England and Wales: 1961 – 1968, England: 1967 – 1974). Note: Certain of these classifications are not fully comparable over time. In particular, the group "stimulants and appetite suppressants" changed markedly in composition over the period. Source: U.K. Department of Health and Social Security, Office of Health Economics, London, 1975.

families, which, in turn, may lead to irremediable breakdown. On the other hand, many people take minor tranquilizers continually to enable them to cope with an impossible life-style and here the conscientious physician should recommend change of life-style rather than a regimen of drugs. Equally, there are many people taking tranquilizers whose problems are not anxiety or depression but boredom and unhappiness, and again the drugs are valueless without an attempt to identify and remove the causes.

Not all psychotropic drugs have the same effects and Figure 8.32 shows the changes in prescribing patterns over the crucial period 1961–1974 as benzodiazepine tranquilizers, nonbarbiturate hypnotics, and tricyclic antidepressants have entered the market and barbiturates and amphetamines have lost favor in medical circles.

It is possible to discuss endlessly and emotionally the differences between the modern drugs that are used and abused to combat unhappiness and depression, and the traditional psychtropics such as nicotine, alcohol, and caffeine. In Western society, alcohol is accepted while marijuana is not. In Islamic societies, both alchohol and psychotomimetic drugs are forbidden but hashish and opium have a long history of use. We leave the assessment of the moral problem to the reader, but it is worth commenting that of the order of one in ten male deaths in Western society is due to a disease related to smoking. About one person in a hundred has a serious drinking problem, and a recent Swedish survey showed that, among males who died before the age of 50, alcohol consumption was the most important single factor distinguishing them from the rest of the population.

Statistics on the proportion of the population taking other drugs are less easy to come by. It would be fascinating to know whether or not the individual unable to cope does better to soothe himself with whiskey or Valium, but worrying about the relatively recent problems presented by the latter should not cause us to overlook the long-standing abuse of the former.

9

Antihistamines

Antihistamines are divided into two groups, the H_1 and H_2 blockers. H_2 blockers are used against stomach ulcers and will be discussed at the end of this chapter. Conventional antihistamines are H_1 blockers and are principally used to counter allergic conditions such as hay fever and contact dermatitis. They are also constituents of formulations that alleviate the symptoms of coughs and colds, and such formulations involve another three compounds in our Top-100. Although they are not antihistamines, they have been included in this chapter in Section 9.5. The antiallergic application is, however, the most significant and we therefore begin by summarizing what is meant by allergy.

9.1 ALLERGY

Allergy takes various forms, the commonest being hay fever, hives, asthma, childhood eczema, and food allergies. About 10% of the population suffers from them. The tendency toward allergy is inherited. The severity of allergic attacks varies widely, some people being completely disabled and others only mildly inconvenienced.

Allergies arise from contact with foreign proteins called allergens, which occur, for example, in pollen, milk, dust, eggs, and strawberries. The body is organized to resist such foreign proteins as part of its defense mechanism. It produces antibodies which combine with the foreign proteins to neutralize their effect. When the body is exposed to the proteins for a second time, the antibodies, which are themselves proteins, may give permanent protection. Allergens are not harmful to most people but allergy sufferers overreact and develop the characteristic allergic symptoms.

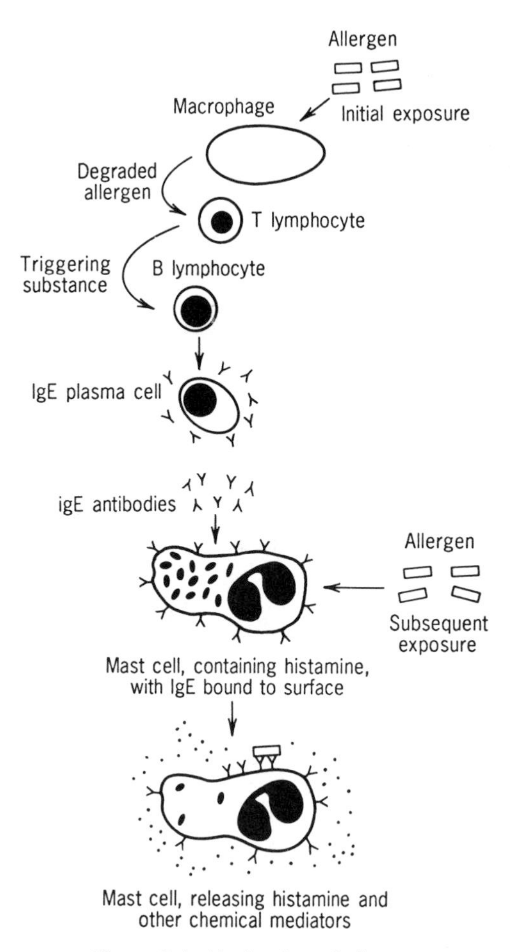

Figure 9.1. *Mechanism of allergy.*

The process is illustrated in Figure 9.1. On the first exposure, the allergen is picked up by a scavenger cell—a macrophage—which degrades it and presents it to a T lymphocyte. T lymphocytes are white blood cells that can destroy other cells (cancer cells, transplanted tissues) and also affect the functioning of other white blood cells called B lymphocytes, either suppressing them or aiding them in the production of antibodies.

In this case, the T lymphocytes secrete a substance that triggers the B lymphocytes to produce Y-shaped allergic antibodies called IgEs. These are different from other antibodies and their special characteristic is that they attach themselves, prong-like, to receptors on the surface of mast cells lining the skin and mucous membranes. Mast cells are a distinctive type of cell found primarily in connective tissue. They are also the cells that synthesize heparin, an anticoagulant mentioned in Section 7.7.

The body has now been sensitized. When the foreign protein enters the body for a second time, the allergen cross-links two or more IgE molecules, which in turn pulls the receptors together. A channel is formed and calcium ions rush through it into the cell. That causes the mast cell to degranulate and to release a number of substances into the bloodstream of which the first to be recognized was histamine:

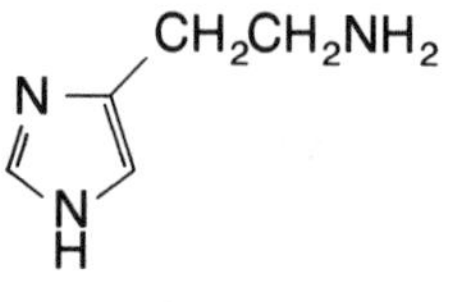

$CH_2CH_2NH_2$

Histamine

Histamine is released when cells are damaged. It dilates the small blood vessels near the surface of the skin and makes them more permeable so that plasma flows into the surrounding tissues, making them red and swollen. This is called edema, like the fluid retention in heart disease. Dilation of blood vessels in the brain can lead to headache, and in other regions can lead to the runny eyes and nose, and sore throat of hay fever, or to allergic rashes. In extreme cases, the patient's circulation collapses; blood pressure falls dramatically and acute asthma ensues. This is called anaphylactic shock and can be fatal. Such symptoms can, of course, have other causes.

9.2 ANTIALLERGY AND ANTINAUSEA COMPOUNDS

Antihistamines are histamine blockers, that is, they occupy the sites at which histamine would normally act to produce allergic reactions. There are now known to be two kinds of histamine receptors, H_1 and H_2. Conventional antihistamines that counter allergy are H_1 blockers. They share the grouping

$$R—X—C—C—N<$$

where X is nitrogen, oxygen, or carbon. It is presumably this grouping that bonds to the receptor.

There are three H_1-blocking antihistamines in the Top-100. The number has dropped sharply in recent years and structures of some of the older important antihistamines are given in Figure 9.2. The structure of the anxiolytic hydroxyzine is shown because it resembles an antihistamine more than an anxiolytic. Apart from the grouping mentioned above, the antihistamines are structurally unlike histamine, but have something in common with many of the drugs discussed so far. Note the resemblances between many of the substances shown and the amphetamines and epinephrine (Figure 8.29).

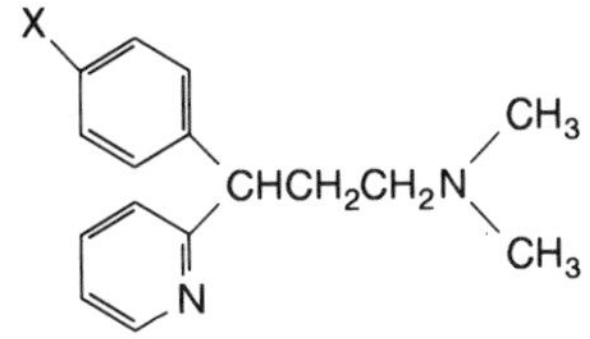

X = Cl Chlorpheniramine (#67)
X = Br Brompheniramine

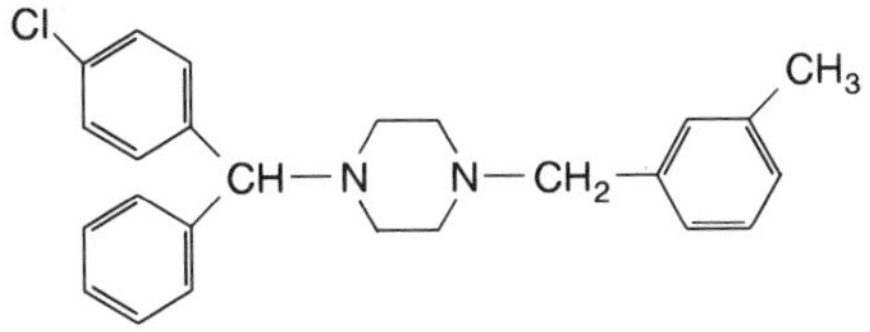

Meclizine (#88)

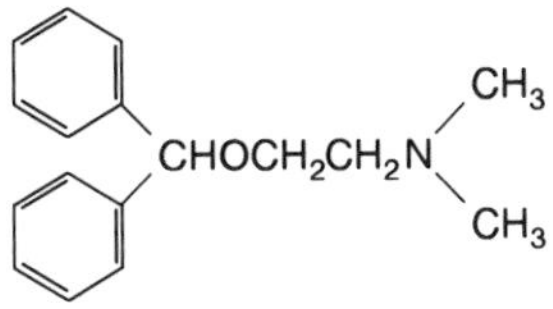

Diphenhydramine

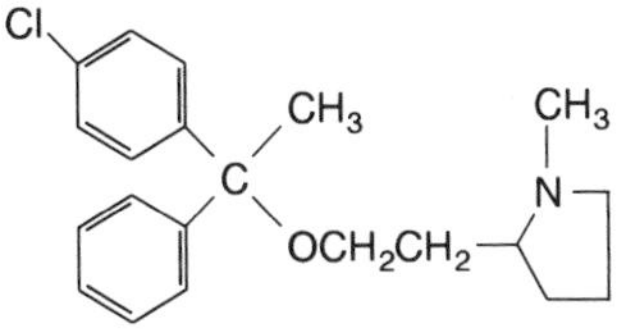

Clemastine (#102)

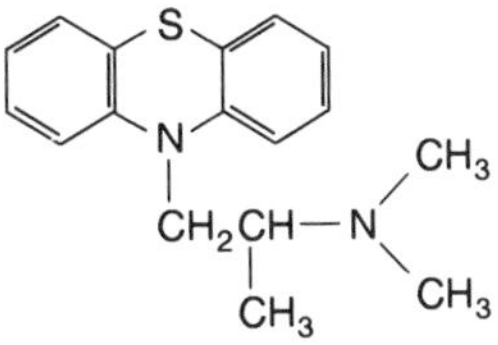

Promethazine (#109)

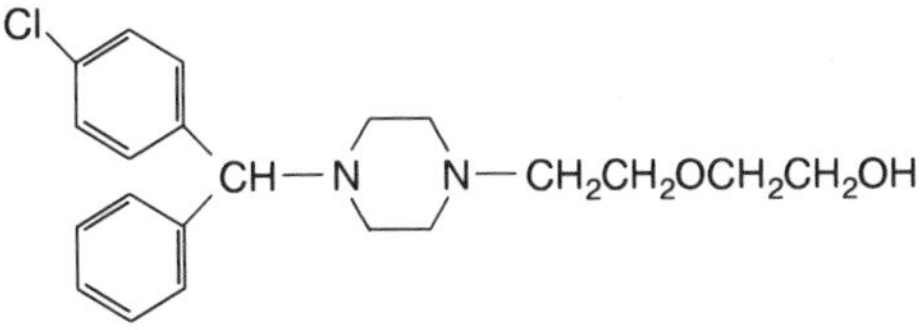

Hydroxyzine (#108)

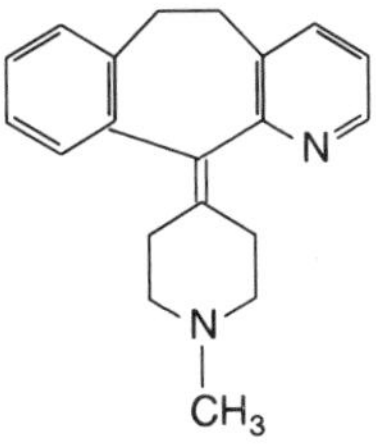

Azatadine (#104)

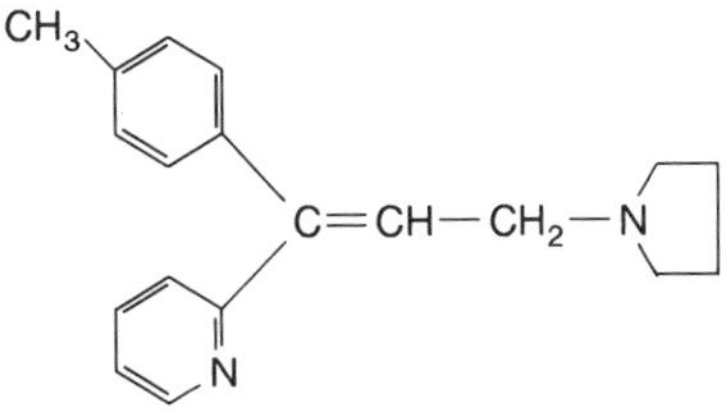

Triprolidine

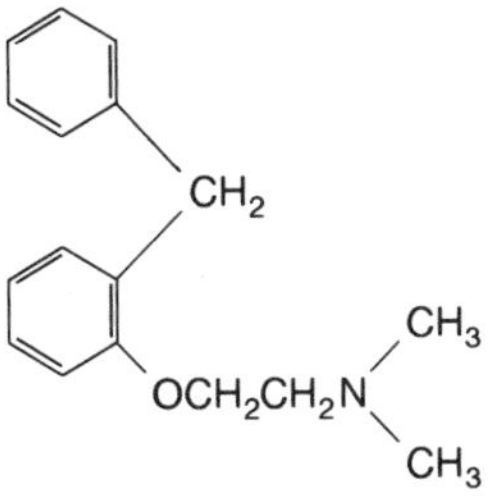

Phenyltoloxamine

Figure 9.2. *H_1 blocking antihistamines.*

Meclizine resembles the minor tranquilizer hydroxyzine; both are shown in Figure 9.2. Note also the similarity between promethazine in Figure 9.2 and the phenothiazine major tranquilizers shown in Figure 8.5. There is little obvious resemblance between diphenhydramine and the phenothiazines, but it was structural modifications of diphenhydramine and related compounds that led to the discovery of chlorpromazine, a historically important major tranquilizer (Section 8.2). Azatadine is clearly related to the tricyclic antidepressants (Section 8.4.2).

Noteworthy, too, is the large number of drugs containing two phenyl groups attached to one carbon atom. Sometimes one ring is substituted and sometimes there is a pyridine ring instead of phenyl. There are six examples in Figure 9.2.

Brompheniramine (Dimetapp), triprolidine, diphenhydramine (Benadryl), promethazine (Phenergan), and chlorpheniramine were the main antiallergic antihistamines on the market for many years. In addition to countering allergy, they produce drowsiness and this is made use of in a number of over-the-counter sleeping products. It is a disadvantage, however, to an allergy sufferer who wishes to stay awake. Everyone who takes these antihistamines is warned to be careful if driving or operating machinery.

Antihistamines also counter nausea, vomiting, and motion sickness. The preferred antihistamines for this application are diphenhydramine, meclizine, promethazine, and dimenhydrinate (see below). Meclizine is used to counter the vertigo due to inner ear defects such as Meniere's disease. Promethazine is an in-between compound and is quite close in its action to the related phenothiazine tranquilizer, perchlorperazine, which is also used for nausea and vertigo. Sometimes drugs do not fit neatly into a therapeutic class and here is an excellent example.

The anti-motion-sickness drug dimenhydrinate is sold over the counter under the name of Dramamine. It is the 8-chlorotheophyllinate salt of diphenhydrammonium, the cation from diphenhydramine:

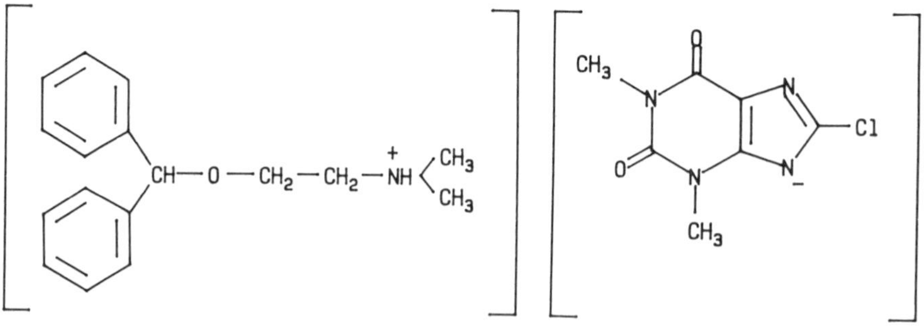

Dimenhydrinate
(Dramamine)

The original idea was to limit the drowsiness caused by diphenhydramine by converting it into a salt with a purine anion which would have a stimulant effect. Only 8-chlorotheophyllinic acid was sufficiently acidic to give an isolable salt and it turned out to counter motion sickness.

9.3 NONSOPORIFIC ANTIHISTAMINES

The disappearance of so many antihistamines from the Top-100 after they had occupied their positions for decades is partly because of the development of nonsoporific antihistamines, which provide relief from allergic disorders but do not cross the blood–brain barrier (Section 16) in sufficient quantity to cause drowsiness. Terfenadine, the first such drug on the market, was an immediate success. It received FDA approval on 8 May 1985 and by the year's end had become the top-selling U.S. prescription antihistamine with sales of $25 million. Fifty-eight thousand prescriptions were written in a single week in August 1985.

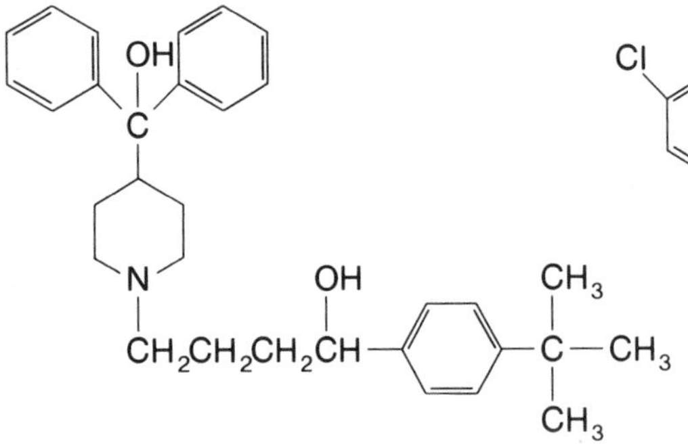

Terfenadine (#51)

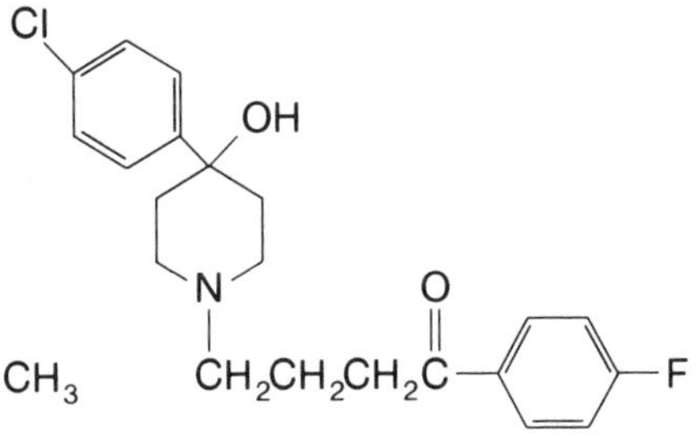

Haloperidol

Terfenadine was the outcome of a search for butyrophenone neuroleptics related to haloperidol (Section 8.2), the structure of which is shown above. This is discussed further in Section 10.1.1. Compared with terfenadine, the original compound synthesized had a fluorine instead of a *tert*-butyl and a keto group rather than a hydroxyl on the side chain. It also had a chlorine atom, as in haloperidol. This compound turned out to have weak CNS activity —about 1% that of haloperidol—but it blocked H_1 receptors, which haloperidol does not. The later structural modifications reduced CNS activity even further.

A second nonsoporific antihistamine, not yet on the U.S. market but already launched in the United Kingdom, is astemizole. A further advantage of these new compounds is that they do not appear to be potentiated by alcohol.

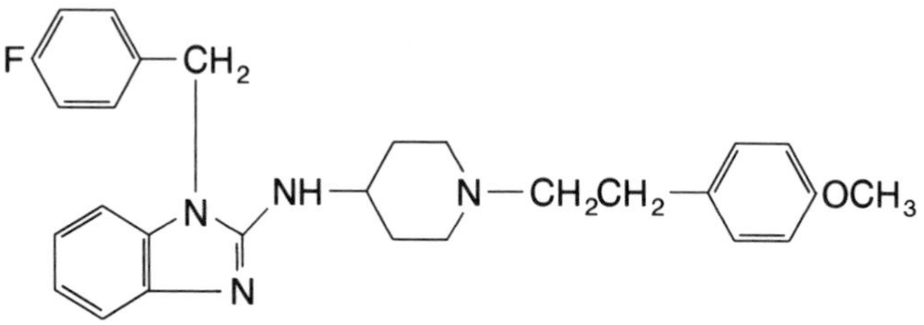

Astemizole

Terfenadine and astemizole act in the same way as the older antihistamines by blocking the action of histamine on tissues. Current research also aims at blocking the allergy chain at an earlier stage—at degranulation or even at the initial T and B cell interaction. An immunization product based on allergens attached to polymers is nearing launch. In another project, the suppressive factor of allergy (SFA) produced by lymphocytes, which helps regulate production of IgE, has been identified. A gene-cloning process might produce SFA, which should then reduce cell IgE production and counter allergies.

9.4 SYNTHESES OF H_1 BLOCKERS

The synthesis of terfenadine is shown in Figure 9.3. Azacyclonol (1) is alkylated with 1-[4-(1,1-dimethylethyl)-phenyl]-4-chloro-1-butanone in refluxing toluene with potassium bicarbonate and potassium iodide catalyst. Acid treatment then gives the hydrochloride (2), in which form the compound can be purified. Potassium hydroxide in methanol regenerates the free base, which is reduced with potassium borohydride to give terfenadine (3).

Two of the most widely prescribed antihistamines were chlorpheniramine and its bromine analogue brompheniramine. The former still ranks #67 and both syntheses are shown in Figure 9.4. The route to chlorpheniramine starts with the appropriate chlorophenylpyridylcarbinol (1). Reaction with thionyl chloride replaces the hydroxyl group by a halogen atom to give compound (2). The halogen atom is readily removed by reduction to give (3). The acidic hydrogen reacts with sodium amide, which makes alkylation with dimethylaminoethyl chloride facile, and chlorpheniramine (4) results. Brompheniramine is synthesized similarly.

The synthesis of meclizine is shown in Figure 9.5. It starts with the *p*-chlorobenzhydryl chloride (1). Treatment with excess piperazine (2) gives (3). Further reaction with *m*-methylbenzyl chloride (4) gives meclizine (5).

Figure 9.8 also shows the synthesis of the minor tranquilizer hydroxyzine, which we omitted from our section on central nervous system drugs. It starts

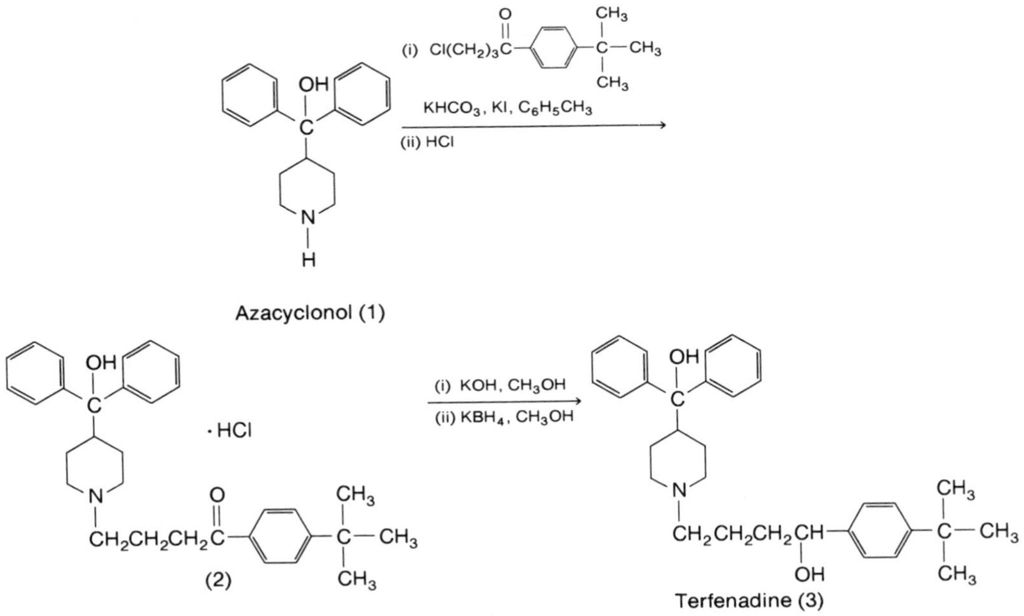

(i) $Cl(CH_2)_3C(=O)$–C_6H_4–$C(CH_3)_3$

$KHCO_3$, KI, $C_6H_5CH_3$

(ii) HCl

Azacyclonol (1)

· HCl

(2)

(i) KOH, CH_3OH

(ii) KBH_4, CH_3OH

Terfenadine (3)

Figure 9.3. *Synthesis of terfenadine.*

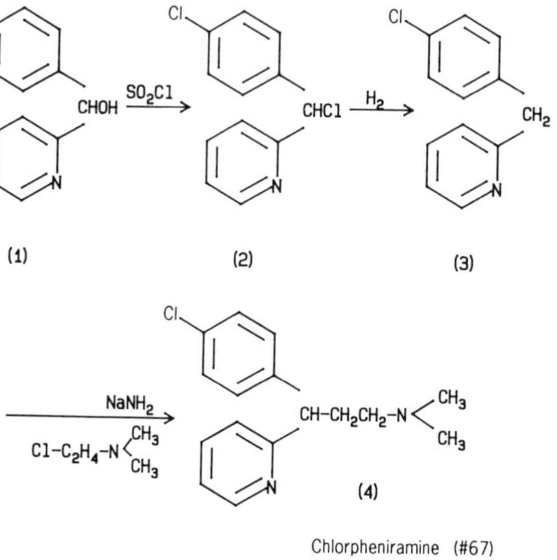

(1) (2) (3)

(4)

Chlorpheniramine (#67)

Figure 9.4. Synthesis of chlorpheniramine.

with the intermediate (3) from the meclizine synthesis and a single reaction with β-chloroethoxyethyl alcohol gives hydroxyzine.

9.5 "COLD CURES"

Many antihistamines are available without prescription and, therefore, their sales are not represented in the Top-100. In addition, various of them are widely prescribed and their limitations are not always recognized. Certainly they are of value in mild allergic conditions and to counter nausea, vomiting, and motion sickness, but not in cases of anaphylaxis as described earlier. They are of some help in allergic rashes and for insect bites, but not in long-term dermatitis, and there is the danger of the patient becoming sensitized and, hence, allergic to the antihistamines themselves. They are useless for asthma and for allergic reactions of the stomach or gut.

Antihistamines are a constituent of the most widely sold cough and cold medicines, yet, apart from putting the patient to sleep, there is little evidence that they have any beneficial effect. Indeed, by drying the lining of the nose and throat they may possibly impair the body's natural defenses. And, finally, since they have no effect on viruses, they are useless against colds apart from providing some relief from a running nose and a degree of sedation at bedtime. This convenience factor explains the high level of consumption.

The constituents of prescription "cold cures" are represented in the Top-100. The cold cures consist of an antihistamine mixed with a nasal decongestant.

1. Synthesis of Meclizine:

(1) p-Chlorobenzhydryl chloride + (2) Piperazine → (3)

(3) + (4) → (5) Meclizine (#88)

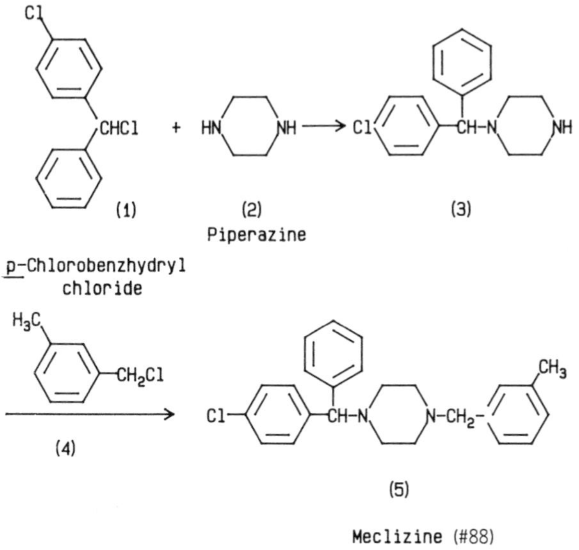

2. Synthesis of Hydroxyzine:

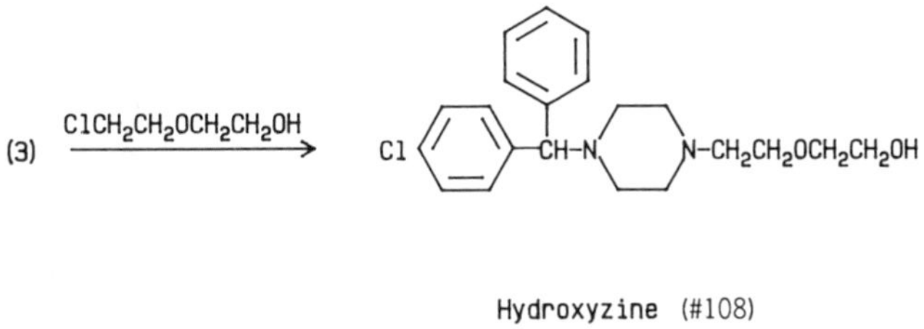

(3) $\xrightarrow{ClCH_2CH_2OCH_2CH_2OH}$ Hydroxyzine (#108)

Figure 9.5. Synthesis of meclizine and hydroxyzine.

The antihistamines dry nasal secretions as do the anticholinergic drugs (Chapter 14) and cause sedation. The decongestants cause vasoconstriction in the blood vessels of the nose, hence enlarging the air passages.

Contac is a typical and successful over-the-counter product. It consists of a slow-release mixture of chlorpheniramine (Figure 9.2) and phenylpropanolamine. Phenylpropanolamine is an α- and β-adrenergic stimulant, which causes the blood vessels in the nose to contract. Ornade is a prescription drug with a similar formulation to Contac. Phenylpropanolamine is a component of so many formulations of this type that it appears at #38 in the Top-100. It is also a constituent of many over-the-counter anorectics (Section 8.5). These applications mean that it is manufactured on a scale of about 400,000 pounds per year, which is large by pharmaceutical standards. Its structure and synthesis are shown in Figure 9.6. Propiophenone (1) reacts with

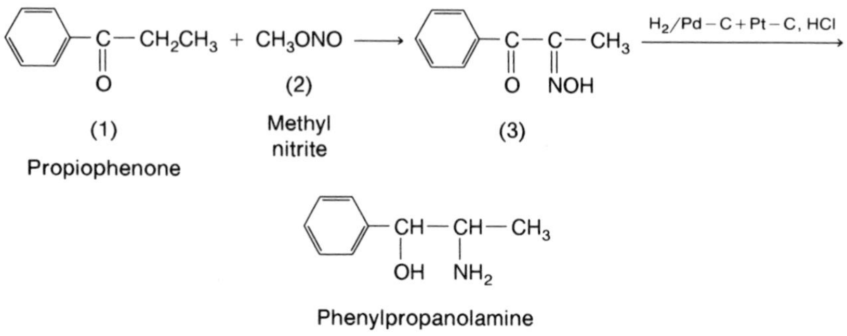

$$C_6H_5-C(=O)-CH_2CH_3 + CH_3ONO \longrightarrow C_6H_5-C(=O)-C(=NOH)-CH_3 \xrightarrow{H_2/Pd-C+Pt-C,\ HCl} C_6H_5-CH(OH)-CH(NH_2)-CH_3$$

(1) Propiophenone; (2) Methyl nitrite; (3); Phenylpropanolamine

Figure 9.6. Synthesis of phenylpropanolamine.

methyl nitrite (2) to give an intermediate (3), which is reduced by hydrogen over palladium and platinum on charcoal to give phenylpropanolamine (4).

Naldecon, a prescription product, also contains chlorpheniramine and phenylpropanolamine but also phenylephrine and phenyltoloxamine:

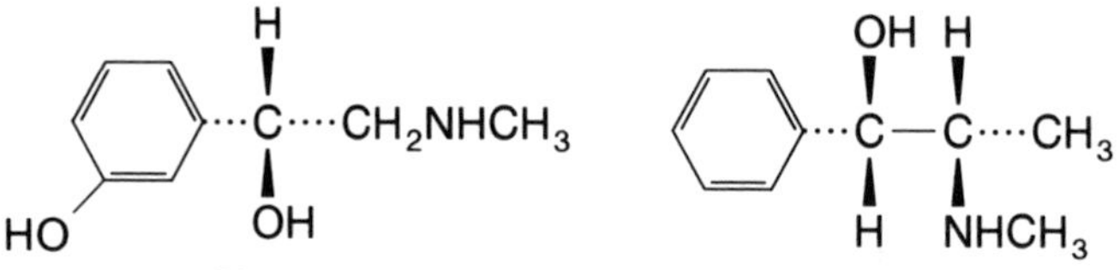

Phenylephrine: $HO-C_6H_4-C(H)(OH)-CH_2NHCH_3$

Pseudoephedrine: $C_6H_5-C(OH)(H)-C(H)(NHCH_3)-CH_3$

Phenylephrine is a decongestant, like phenylpropanolamine, but has no effect on the β receptors of the heart. Phenyltoloxamine (Figure 9.2) is another antihistamine.

Trinalin consists of a mixture of the antihistamine azatadine (#104, Figure 9.2) with the decongestant, pseudoephedrine (see above).

Tavist-D contains clemastine (#102, Figure 9.2) and phenylpropanolamine. Phenergan/codeine is promethazine (#109, Figure 9.2) plus codeine (Section 10.1), an analgesic and cough inhibitor. Entex contains phenylpropanolamine and phenylephrine, both decongestants, plus guaifenesin, an expectorant (Section 10.1.7).

9.6 H_2-BLOCKING ANTIHISTAMINES AND ANTIULCER DRUGS

The new H_1 blockers, terfenadine and astemizole, have been a remarkable pharmacological and commercial success. They have still a long way to go before matching the H_2 blockers, which came on the market in 1977. The first of them was cimetidine:

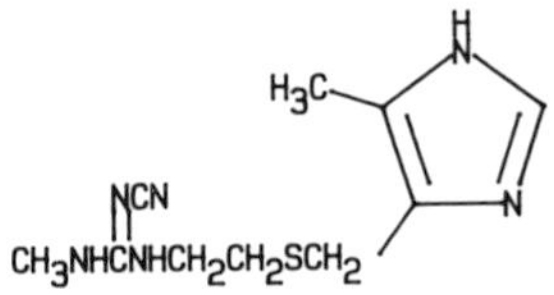

H_3C

NCN

$CH_3NHCNHCH_2CH_2SCH_2$

Cimetidine (#14)

Cimetidine is quite different from the other antihistamines and is used for treatment of gastric and duodenal ulcers. Conventional antihistamines are useless for this purpose. The question is why this should be so. It was known that excess hydrochloric acid in the stomach was an exacerbating factor in ulcers, and it was also known that histamine stimulated the secretion of acid. Why do antihistamines not help?

It was eventually realized that there are two types of histamine receptor sites in the body. The sites involved in hay fever and other allergies were designated H_1. They are the ones that are blocked by conventional antihistamines. The sites in the gastric mucosa responsible for hydrochloric acid secretion were designated H_2.

The head of the SmithKline research team seeking a blocking agent for the H_2 receptors was J. W. Black who, when he was with ICI, had discovered the β-adrenergic blocking agent, propranolol, that was discussed in Section 7.5.6. In the case of the β blockers, there were two kinds of receptor sites for epinephrine and norepinephrine and they could be blocked by the α and β blockers. The α blockers bore no structural resemblance to epinephrine, but the β blockers did. This can be seen from a comparison of epinephrine with prazosin and propanolol in Figure 7.14.

Conventional antihistamines do not resemble histamine. Black argued therefore that the H_2 blocker he was seeking would indeed resemble histamine. By analogy with his work on the heart drugs, he initially felt he should modify the ring structure of histamine while leaving the side chain unchanged, but this approach failed and the opposite was tried. Several hundred compounds later a material with mild H_2-blocking activity was found and this was refined until cimetidine was discovered.

After the usual long-drawn-out clinical trials, cimetidine was introduced in the United States and the United Kingdom in 1977. It has a dramatic effect on ulcers and the few side effects do not seem serious, although it can cause impotence in men. The other drawback is that, once the medication is withdrawn, the H_2 receptors are no longer blocked and the ulcer may recur.

Cimetidine was developed by the British end of the American SmithKline Corporation and provides an interesting success story. The market is huge. In 1970 there were about 3.5 million peptic ulcer sufferers in the United States, of whom 8600 died. In 1976, SmithKline's total pharmaceutical sales amounted to $386 million. Sales of Cimetidine in 1980 were estimated at $580 million and have since risen above the billion dollar mark. Even allowing for inflation, this means that a single drug contributed more business to SmithKline than did all their drugs four years previously.

This story illustrates the great potential for profitability in the drug industry. Of course, there are not many products to match this one, but it is pleasant when profits are associated with alleviation of suffering. A measure of cimetidine's medical success is that, in the United Kingdom, operations for duodenal ulcer dropped by 39% between 1976, when the drug was introduced, and 1981. Cimetidine was a unique and novel drug and was put on the market at a high price. The manufacturers claimed that, even at that price, its benefits were far greater than its costs. They commissioned a number of cost–benefit studies and, partly as a result of these, cimetidine has been subject to more economic evaluation than any other drug. The studies examined not only health expenditures, but also the general macroeconomic picture of the costs of the disease.

Medicaid expenditures in Michigan are shown in Figure 9.7, which indicates that the major saving for cimetidine-treated patients is in hospitalization costs. It was also shown that, among nonhospitalized patients who were on

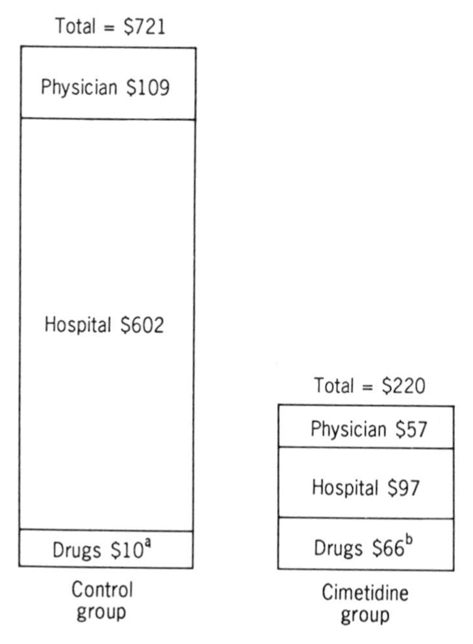

Figure 9.7. *Costs and benefits of cimetidine. Average annual Michigan Medicaid expenditures per patient with duodenal ulcer. [a]Does not include antacids that are excluded from Michigan Medicaid; [b]includes cost of cimetidine therapy. Source: M. L. Patterson, Managerial and Decision Economics 4, 50 (1983).*

older drugs, two days of work per week were missed on average, while among the patients taking cimetidine only a single day was missed. More details of these economic studies are given in the bibliography.

The success of cimetidine induced many other drug firms to search for their own H_2 blockers. Indeed, there was a race to make cimetidine obsolete and to achieve a share of the wealth that it was generating. The most successful attempt so far is ranitidine, developed by the British firm Glaxo:

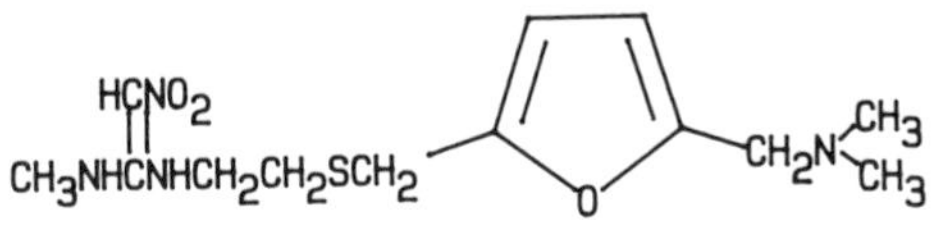

$HCNO_2$
$CH_3NHCNHCH_2CH_2SCH_2$ O $CH_2N(CH_3)_2$

Ranitidine

Its side chain is similar to cimetidine, except that a nitro group replaces the nitrile, but the imidazole ring has been replaced by furan. Ranitidine is claimed to have fewer side effects, to be more specific, and to be effective in smaller doses than cimetidine. It does not bind to androgen receptors and, consequently, does not cause impotence. It has risen to #30 in the Top-100, and 1986 U.S. sales topped the billion dollar mark. Thus, cimetidine and ranitidine are vying for the position of the most successful commercial drug of all time.

An important aspect of the cimetidine story is that to some extent the drug was designed rather than discovered. Black had some idea of what he was looking for and the biochemical effect he hoped it would have. This illustrates an important shift from pure empiricism in the drug industry toward drug design based on a degree of understanding of biochemical processes. Black was rewarded for his discovery of two blocking agents—one for hearts and one for stomachs—with a knighthood and a share in the 1988 Nobel Prize. It would be nice if our understanding of receptors in the heart and the stomach could be extended to the mind, so that drugs could be designed for that.

Many other companies have H_2 blockers either recently launched or under development. It is difficult to predict which will ultimately be successful, but an example of the new drugs is famotidine which was approved in the United States at the end of 1986:

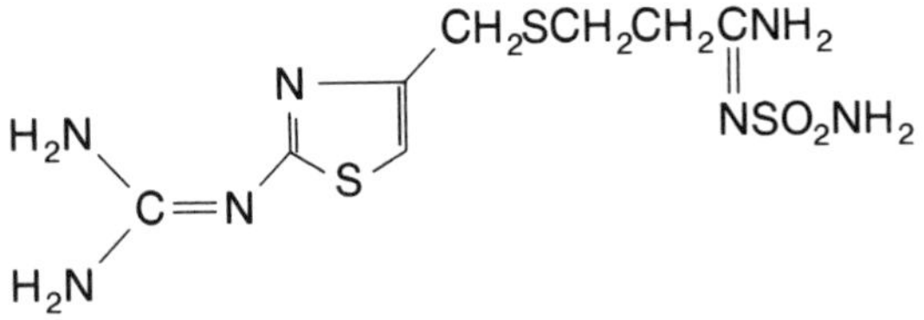

$CH_2SCH_2CH_2CNH_2$
N
H_2N NSO_2NH_2
S
C=N
H_2N

Famotidine

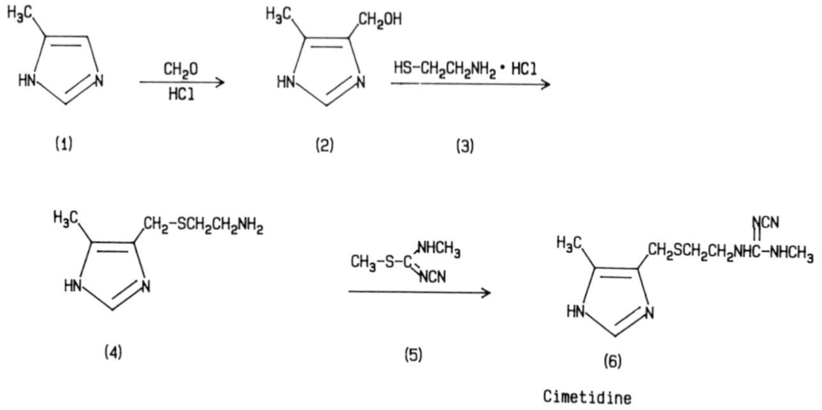

H_3C HN N (1) $\xrightarrow[HCl]{CH_2O}$ H_3C CH_2OH HN N (2) $\xrightarrow{HS\text{-}CH_2CH_2NH_2 \cdot HCl}$ (3)

H_3C $CH_2\text{-}SCH_2CH_2NH_2$ HN N (4) $\xrightarrow{CH_3\text{-}S\text{-}C(=NCN)NHCH_3}$ (5) H_3C $CH_2SCH_2CH_2NHC(=NCN)\text{-}NHCH_3$ HN N (6)

Cimetidine

Figure 9.8. *Synthesis of cimetidine.*

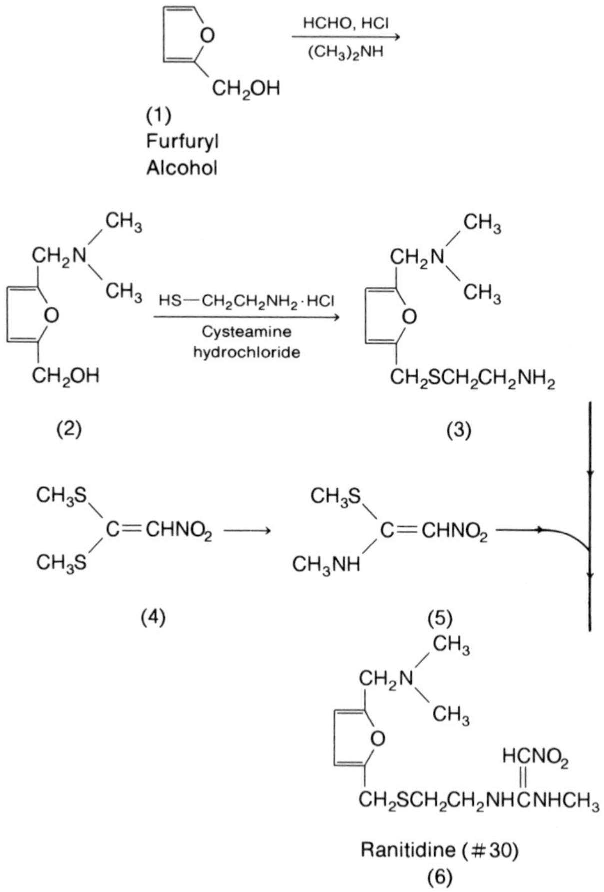

(1) Furfuryl Alcohol, CH_2OH $\xrightarrow[(CH_3)_2NH]{HCHO,\ HCl}$

(2) $CH_2N(CH_3)_2$, O, CH_2OH $\xrightarrow[\text{Cysteamine hydrochloride}]{HS\text{—}CH_2CH_2NH_2 \cdot HCl}$ (3) $CH_2N(CH_3)_2$, O, $CH_2SCH_2CH_2NH_2$

(4) $(CH_3S)_2C{=}CHNO_2$ $\longrightarrow$ (5) $CH_3S(CH_3NH)C{=}CHNO_2$

$CH_2N(CH_3)_2$, O, $CH_2SCH_2CH_2NHC(=HCNO_2)NHCH_3$

Ranitidine (#30)
(6)

Figure 9.9. *Synthesis of ranitidine.*

It is effective at far lower doses than the other drugs. The standard dose of cimetidine is 800 mg/day, of ranitidine 300 mg/day, and of famotidine 40 mg/day. Famotidine is said to have a longer duration of action and to dissociate more slowly from the H_2 receptor.

9.6.1 Syntheses of H_2 Blockers

The synthesis of cimetidine is shown in Figure 9.8. It starts with the imidazole (1). Treatment with formaldehyde and hydrochloric acid leads to an alcohol (2), which is converted to an aminoethylthiomethyl derivative (4) by reaction with 2-mercaptoethylamine hydrochloride (3). This, in turn, is reacted with the thionitrile (5) to give cimetidine (6).

The synthesis of ranitidine is shown in Figure 9.9. Furfuryl alcohol (1) is treated with formaldehyde, hydrochloric acid, and dimethylamine to give the furan (2). This is reacted with cysteamine hydrochloride in the presence of an acid catalyst. The primary alcohol group is displaced and (3) results. A synthon (5) is prepared from (4) by an addition–elimination reaction and this reacts with (3) to give ranitidine (6).

The synthesis of famotidine is shown in Figure 9.10. The aminothiazole derivative (1) is treated with β-chloropropionitrile (2) to give the nitrile (3). Benzoyl isothiocyanate (4) in the presence of a base gives (5) plus benzoic acid. Methyl iodide methylates the sulfur to give (6), and ammonia removes the sulfur to leave a compound with a guanidine side chain (7). Methanol in the presence of hydrogen chloride then adds to the nitrile group on the other side chain to give (8). Treatment with sulfamide results in famotidine (9).

9.6.2 Other Antiulcer Drugs

Although cimetidine and ranitidine have overshadowed them, there are two more drugs for treatment of ulcers in the Top-100 and another at #119. The older ones are atropine (plus other solanaceous alkaloids) at #62 and dicyclomine at #119. They are anticholinergic drugs and will be discussed in Chapter 14. The newer one is sucralfate (#99). It is a complex of sulfated sucrose and aluminum hydroxide and is said to be cytoprotective; that is, it forms a complex with proteins—primarily albumin and fibrinogen—that adhere to acids and forms a barrier against stomach and bile acids, and pepsin.

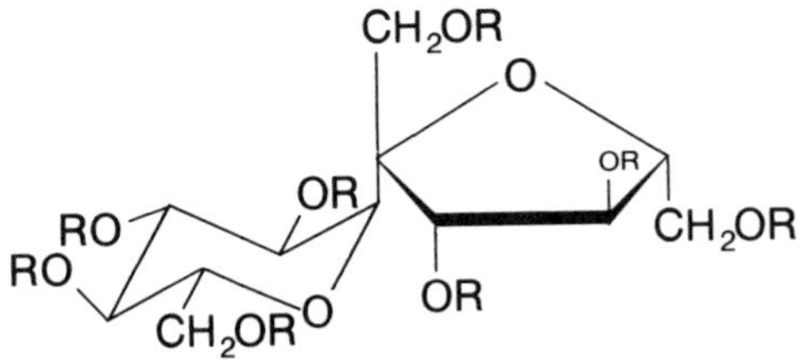

CH₂OR, O, OR, OR, RO, RO, CH₂OR, O, OR, CH₂OR

$R = SO_3[Al_2(OH)_2(H_2O)_y]$

Sucralfate (#99)

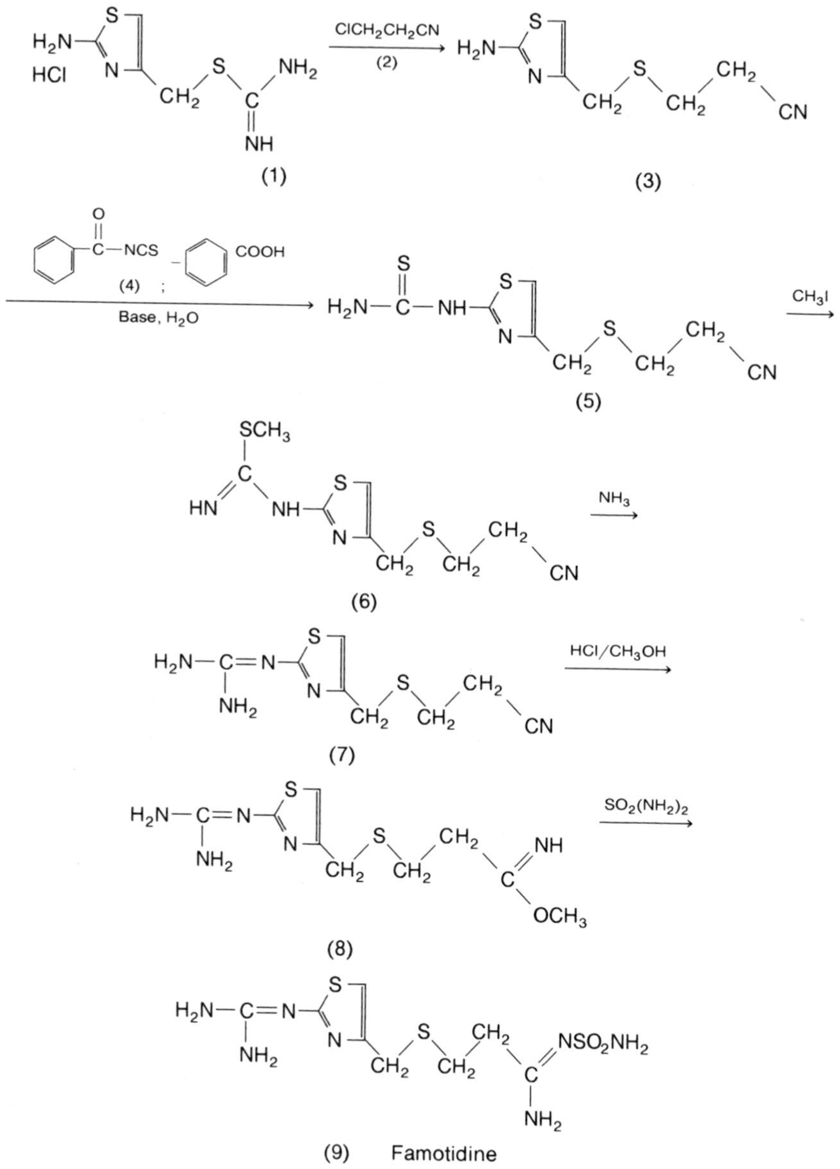

$ClCH_2CH_2CN$ (2)

(1) (3)

(4) ; Base, H_2O

(5)

CH_3I

(6)

NH_3

(7)

HCl/CH_3OH

(8)

$SO_2(NH_2)_2$

(9) Famotidine

Figure 9.10. *Synthesis of famotidine.*

Yet another type of antiulcer drug, omeprazole, was launched on the French market in 1987. It is said to be a proton pump inhibitor. It inhibits secretion of stomach acid by inhibiting the acid-producing mechanism in the parietal cells at the top (the fundus) and in the body of the stomach.

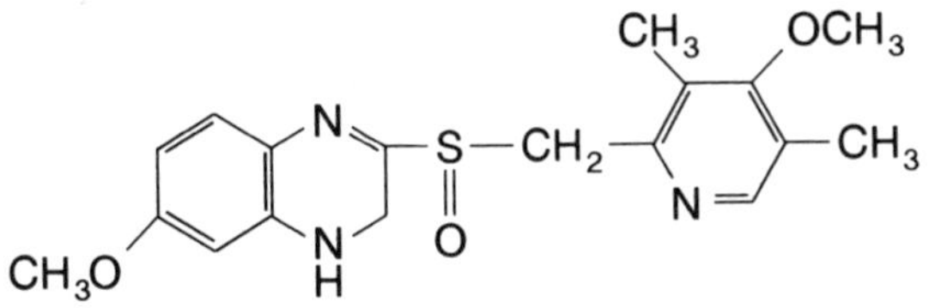

Omeprazole

In addition, a new group of antiulcer drugs has been discovered based on prostaglandins. They will be discussed with the other prostaglandins in Chapter 19.

Because of the H_2 blockers, the cytoprotectives, the proton pump inhibitors, and the prostaglandins, research into drugs for the treatment of ulcers has been one of the most exciting and successful areas of pharmaceutical research in the past two decades.

10

Analgesics

Pain accompanies most disease and the relief of pain is an important, if secondary, role of medical science. Pain relievers may be divided into two groups, narcotic and nonnarcotic analgesics. Narcotic analgesics prevent the appreciation of pain by the nerve centers in the brain. Morphine is an example. Narcotic analgesics may produce addiction. They overlap to some extent with anesthetics. Nonnarcotic analgesics block the pain at its site and prevent pain impulses from being sent to the brain. Aspirin falls into this class. Nonnarcotic analgesics are not addictive and overlap with nonsteroid anti-inflammatory drugs.

Many mixtures of narcotic and nonnarcotic pain relievers are on the market, but the British National Formulary declares their use to be undesirable. In spite of that, most of the widely prescribed or over-the-counter analgesics contain two or three active ingredients. We have attempted to deconvolute the data on these in our Top-100 compilation. Aspirin and acetaminophen are the two best-known over-the-counter, nonnarcotic analgesics. They can be bought as individual drugs and also occur as a component of almost all formulations. By quantity consumed, they are the most important drugs of all. In formulations, they are mixed with a narcotic analgesic of which there are four in the Top-100 plus another at #131.

10.1 NARCOTIC ANALGESICS

The narcotic analgesics in the Top-100 are codeine (#3), propoxyphene (#23), oxycodone (#46), and hydrocodone (#81). Related compounds are dihydrocodeine (#131), the antidiarrhea compounds, diphenoxylate (#110) and loperamide (#130), and the cough inhibitor, dextromethorphan (#127).

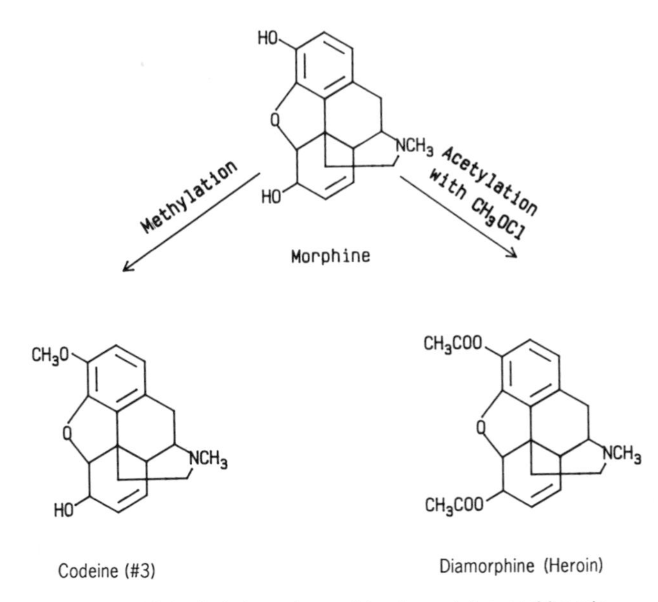

Figure 10.1. *Relation of morphine to codeine and heroin.*

The first narcotic analgesic to be discovered was morphine (Figure 10.1) and it is still used today for shock and, combined with the alkaloids hyoscine or atropine, for preoperative sedation. It is also used to relieve pain in terminal cancer. It is more effective when injected than when taken by mouth. Recent development of a slow release formulation permits pain to be "titrated" with an exact dose, and this has increased morphine usage dramatically. Physical dependence can nonetheless be developed within 24 hours if morphine is given every four hours, and tolerance also develops. It is a truly dangerous drug.

It is obtained with many other so-called opium alkaloids from the juice of the poppy, *Papaver somniferum*. Codeine, an O-methyl derivative of morphine, may be obtained either by methylation of morphine, as shown in Figure 10.1, or by extraction from the opium alkaloids. It is a weaker analgesic than morphine, and is more effective by mouth than by injection. It is particularly useful as an antitussive, a medication that depresses a dry cough. It is widely used in prescription pain killers and appears at #3 in the Top-100. Also shown in Figure 10.1 is the acetylation of morphine to yield the diacetyl compound, heroin, which produces euphoria when injected and has been illegal in the United States for many years.

An important objective of early twentieth century chemists was to modify morphine in such a way that its analgesic properties would be retained but the risk of addiction decreased. They modified the morphine molecule in many ways but little of medical value resulted. After the synthesis of several hundred derivatives, they concluded that little could be done to make morphine modifications both pain killing and nonaddictive.

These researchers were inhibited by their assumption that morphine was a phenanthrene derivative. This was an oversimplification. It is indeed a phenanthrene derivative as can be seen:

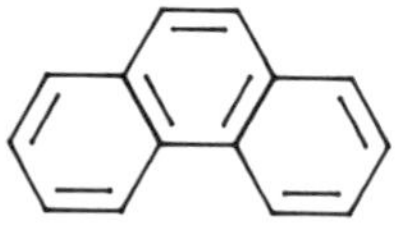

Phenanthrene 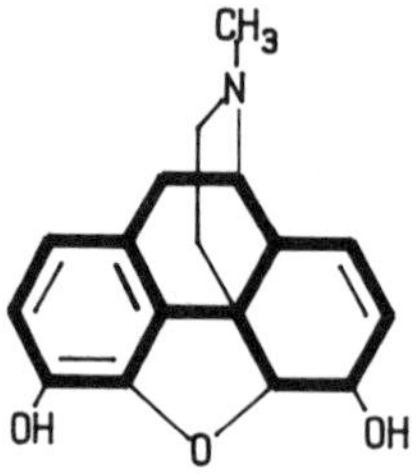 Morphine

But this does not mean that it cannot also be a derivative of something else. In 1939, the molecule now called meperidine was synthesized. Trade names included Pethidine and Demerol. The synthesis was motivated by a need for antispasmodics. Meperidine turned out not only to be a good antispasmodic but also to relieve pain. In Figure 10.2 the structure of meperidine is written in two ways to emphasize the fact that it is a 4-phenylpiperidine derivative. The structure of morphine has also been redrawn to show that it, too, is a 4-phenylpiperidine derivative. Meperidine and morphine thus have a structural relationship.

The hopes raised by the realization of the importance of the 4-phenylpiperidine moiety were only slightly diminished by the fact that meperidine was also addictive. Thousands of derivatives and related molecules were created in a remarkably short time and a number of these were more effective than meperidine itself.

10.1.1 Antidiarrhea Drugs

One new series of compounds involved replacement of the *N*-methyl group in meperidine. The most interesting compound to result was diphenoxylate or Lomotil, shown at the bottom of Figure 10.3. Diphenoxylate is meperidine whose methyl side chain has been replaced by one containing a cyano group and two phenyl groups. Although diphenoxylate relieves pain, it is better known for the treatment of diarrhea. Other narcotic analgesics also have a constipating effect as drug addicts know. Indeed, opium is still used for diarrhea under the name paregoric, which is a mixture of opium and camphor.

Diphenoxylate reduces the activity of the bowel by acting on the wall of the gut. Lomotil, the brand name of the formulation containing diphenoxylate, also contains a small amount of atropine. The anticholinergic properties of atropine will be discussed in Chapter 14 but in Lomotil the quantity of

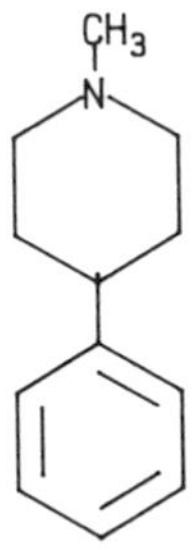

1-Methyl-4-phenylpiperidine

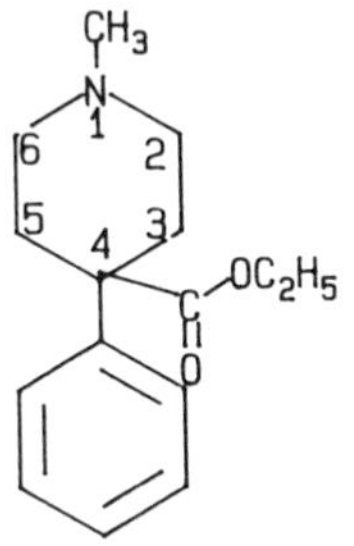

Meperidine
(Pethedine, Demerol)

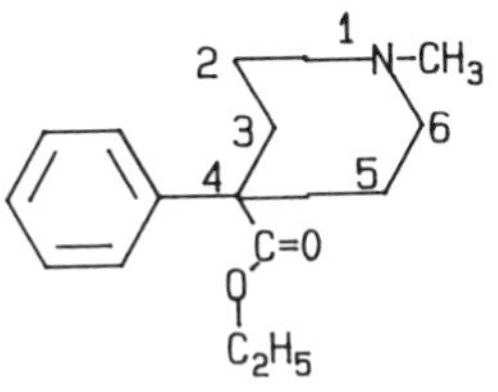

Meperidine

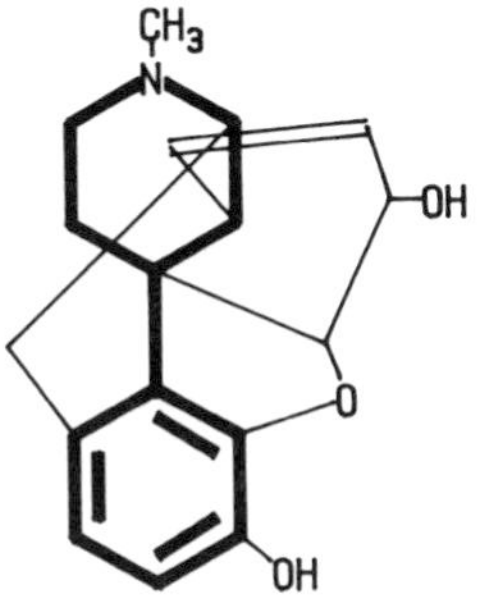

Morphine

Figure 10.2. *The relation of morphine and meperidine to 1-methyl-4-phenylpiperidine.*

atropine is too small for it to have a pharmacological effect; it is added to discourage excessive use of the formulation.

The synthesis of diphenoxylate is shown in Figure 10.3. It starts with phenylacetonitrile (1), which on reaction with the toluenesulfonamide of bis-β-chloroethylamine (2) gives (3). Treatment of (3) with alkali converts the nitrile to a carboxyl (4). Further treatment with sulfuric acid and ethanol cleaves the toluenesulfonamide to yield the key intermediate (5). This, in turn, is reacted with β-bromoethylacetonitrile (7) to give diphenoxylate (8). The nitrile (7) results as shown from the alkylation of diphenylacetonitrile (6) with dibromoethane.

Another antidiarrhea compound related to meperidine is loperamide (#130). This is also related to the neuroleptic, haloperidol (Section 8.2), which, in turn, is related to the antihistamine, terfenadine (Section 9.3). The resemblances are clear if the structures are placed side by side, as in Figure 10.4. There is, thus, a link between an opiate, an antidiarrheal, a neuroleptic, and an antihistamine. The figure also shows that haloperidol, usually described

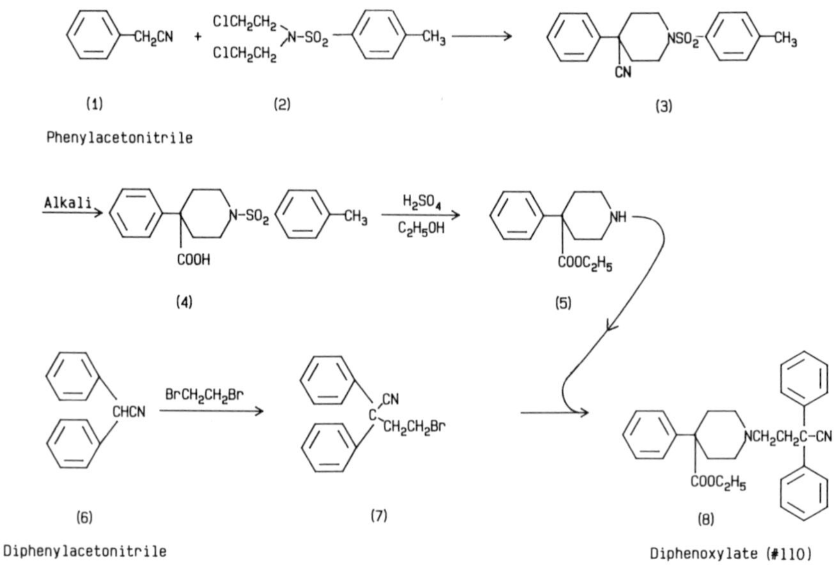

CH_2CN + $ClCH_2CH_2$ / $ClCH_2CH_2$ $N-SO_2$ CH_3 → NSO_2 CH_3 CN

(1) (2) (3)

Phenylacetonitrile

Alkali → $N-SO_2$ CH_3 COOH $\xrightarrow[C_2H_5OH]{H_2SO_4}$ NH $COOC_2H_5$

(4) (5)

CHCN $\xrightarrow{BrCH_2CH_2Br}$ C CN CH_2CH_2Br → NCH_2CH_2C-CN $COOC_2H_5$

(6) (7) (8)

Diphenylacetonitrile

Diphenoxylate (#110)

Figure 10.3. *Synthesis of diphenoxylate.*

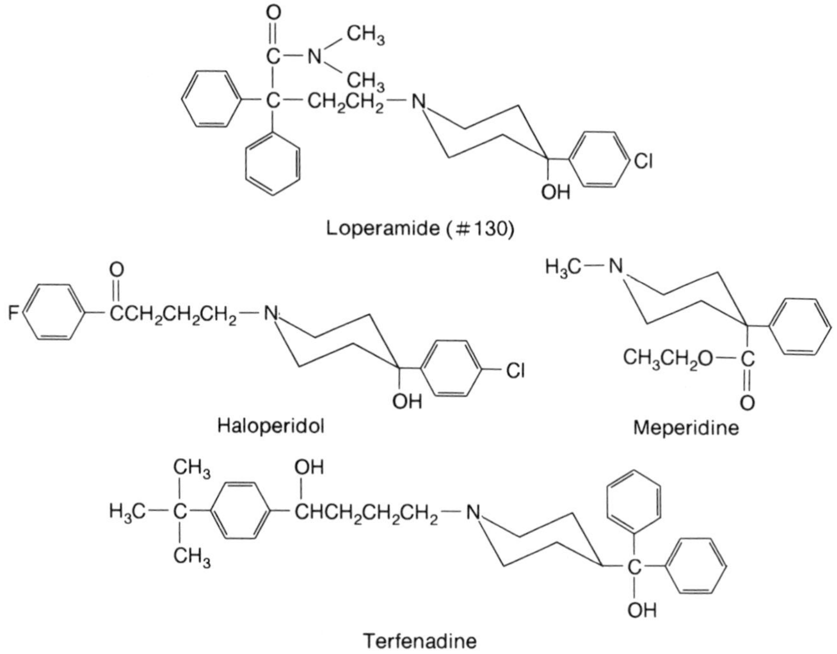

O C N CH_3 CH_3 C CH_2CH_2 N OH Cl

Loperamide (#130)

F O $CCH_2CH_2CH_2$ N OH Cl

Haloperidol

H_3C N CH_3CH_2O C O

Meperidine

CH_3 H_3C C CH_3 OH $CHCH_2CH_2CH_2$ N C OH

Terfenadine

Figure 10.4. *Loperamide and related drugs.*

as a butyrophenone, is also a phenylpiperidine related to the narcotic analgesics.

10.1.2 Methadone and Propoxyphene

A further result of the synthetic work on 4-phenylpiperidine derivatives was the discovery that the "reversed" esters were analgesics just as effective as meperidine. In a reversed ester, the sources of the hydroxyl and the carboxyl needed to form the ester are interchanged. Alphaprodine is the reversed ester of meperidine:

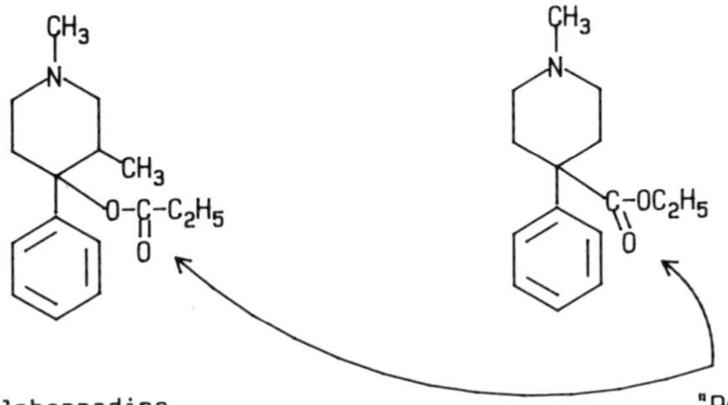

Alphaprodine Meperidine "Reversed Esters"

In meperidine, the carboxyl comes from the piperidine ring, whereas in alphaprodine it is on the side chain, which is derived from propionic acid. The ring structure, however, remains unaltered.

Further synthetic work led to the discovery of methadone:

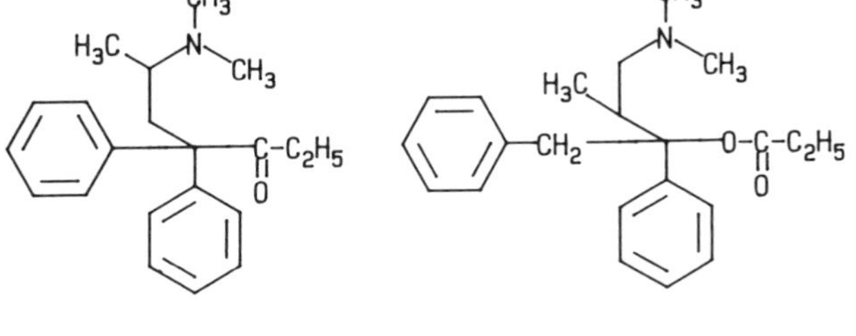

Methadone Propoxyphene

This is a more complex molecule related to meperidine. The piperidine ring has been broken, another phenyl group has been added, and the ester linkage has been replaced by a ketone.

Like morphine, methadone alleviates pain and is addictive. On the other hand, it is effective orally, it is long-lasting, and its withdrawal symptoms are less severe than those of morphine. Accordingly, one treatment for morphine addicts is substitution of methadone hydrochloride for morphine, followed by gradual reduction of dosage, so that the patient is cured without painful withdrawal symptoms. There is controversy as to how effective this has been, but the belief that it is a way to cure morphine addiction has led to the

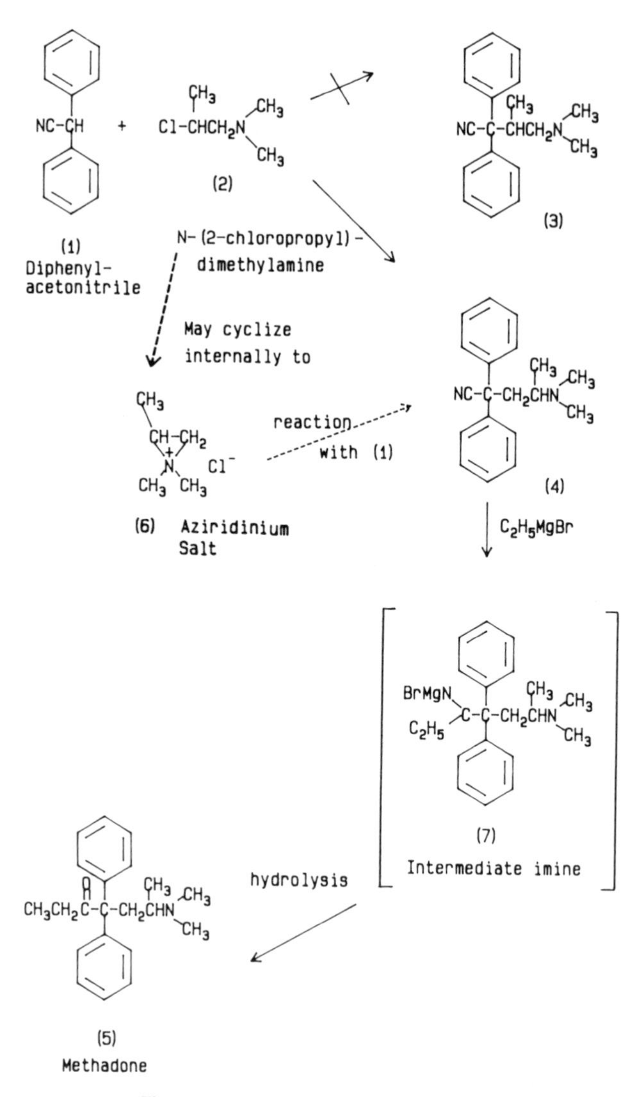

Figure 10.5. Synthesis of methadone.

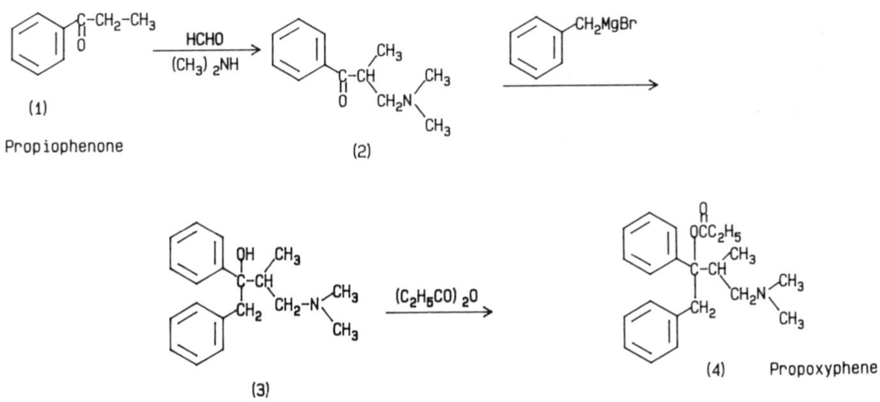

(1) Propiophenone

HCHO, $(CH_3)_2NH$

(2)

$C_6H_5CH_2MgBr$

(3)

$(C_2H_5CO)_2O$

(4) Propoxyphene

Figure 10.6. *Synthesis of propoxyphene.*

establishment of hundreds of methadone clinics throughout the Western World.

Methadone is synthesized as shown in Figure 10.5 by reaction of diphenylacetonitrile (1) with *N*-(2-chloropropyl) dimethylamine (2). Two possible isomeric amines (3 and 4) may result. The less likely appears to be (4) but is, in fact, the one that is produced. Possibly the haloamine (2) cyclizes internally to an aziridinium salt (6). Reaction of this salt with (1) can be visualized to give (4). When (4) is treated with ethylmagnesium bromide, an intermediate imine (7) results, which on hydrolysis yields methadone (5).

Propoxyphene, shown on p. 297 next to methadone, is another related compound and the structural similarity is evident. It may be given orally or parenterally. It is addictive but to a much lesser extent than methadone. It is widely prescribed (#23 in the Top-100) and is often regarded as little more than a potent form of aspirin. This is not so. It is a powerful analgesic affecting the brain. Overdosage is often fatal as a result of respiratory depression. The drug should never be taken with central nervous system depressants, for the combination can be fatal.

The synthesis of propoxyphene is shown in Figure 10.6. It starts with the Mannich reaction of propiophenone (1) with formaldehyde and dimethylamine. The ketoamine (2) is then treated with benzylmagnesium bromide to yield the aminoalcohol (3). This has no analgesic activity. Esterification of its alcohol group with propionic anhydride yields propoxyphene (4).

Propoxyphene contains two asymmetric carbon atoms and, hence, two sets of diastereoisomers can exist. The less soluble pair is designated α and the more soluble β. The β stereoisomers are inactive for practical purposes and the α-D-diastereoisomer is the analgesic. The drug is properly called dextropropoxyphene for this reason.

10.1.3 Morphine Antagonists

In the course of modification of the morphine molecule, its *N*-allyl derivative was synthesized; that is, the methyl group was replaced by an allyl group to give *N*-allylnormorphine. The resulting compound, nalorphine, was found to be a morphine antagonist:

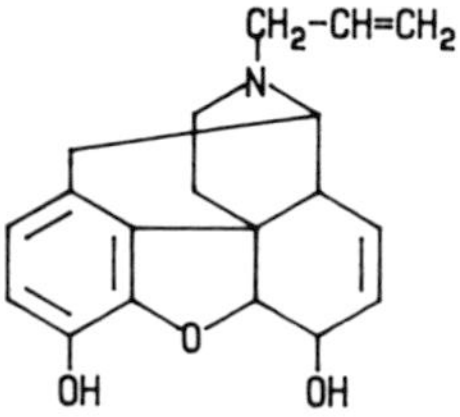

Nalorphine

(N-allyl-normorphine)

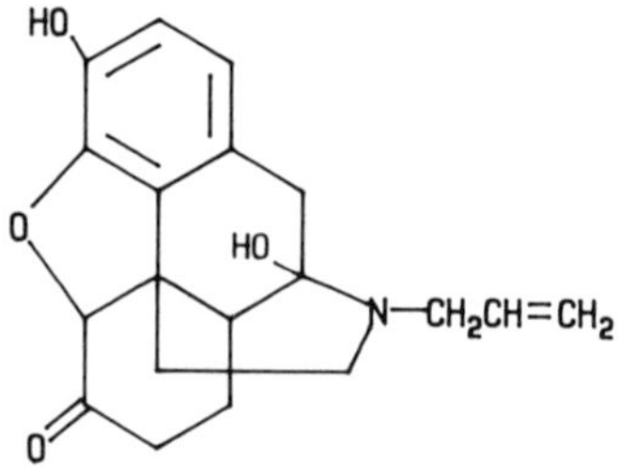

Naloxone

Thus nalorphine will counteract the adverse effects of an overdose of morphine, meperidine, or related molecules that attach to the morphine receptor sites. If nalorphine is given to a patient in a state of narcosis from an overdose of morphine, the morphine is displaced from the receptor sites, and normal breathing and blood pressure are restored within seconds.

Nalorphine, unfortunately, often gave rise to severe psychotomimetic side effects. A more effective drug with fewer side effects is naloxone (see above) and this is the drug of choice today.

Nalorphine is also useful as a test for drug addiction because a drug addict who receives it will exhibit the withdrawal symptoms associated with lack of morphine. Because the morphine is displaced from its receptor site, it can no longer create the euphoria the addict craves.

The unexpected property of nalorphine is that it has analgesic properties in spite of being a morphine antagonist. It displaces morphine from the receptor but, in binding to it, exerts an analgesic effect of its own. Although the binding is stronger, the analgesic effect is less and nalorphine is not addictive. It is unsuitable for clinical use as an analgesic because of its side effects. The remarkable nonaddictive property, however, led to the search for other narcotic antagonists that might share morphine's good (analgesic) properties but not its bad (addictive) ones. Numerous related compounds were discovered, of which we shall describe pentazocine, etorphine, and buprenorphine.

10.1.4 Opioid Receptors

Pentazocine is important because it was involved in the identification of different kinds of opioid receptors. Its structure is

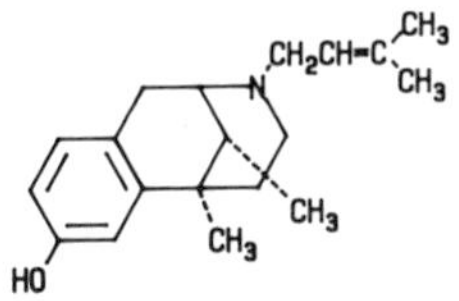

(10)
Pentazocine

The existence of three types of opioid receptor has now been established and others have been proposed. The μ receptor mediates morphine-like analgesia and euphoria; the κ receptor mediates pentazocine-like analgesia, sedation, and constriction of the pupil of the eye; the σ receptor mediates dysphoria and psychotomimetic effects produced by pentazocine. δ And ε receptors also seem likely. The δ receptor is selective for enkephalins (Section 10.3). The ε receptor is selective for β-endorphin (Section 10.3) but not for enkephalins. There seems some doubt as to whether the μ and κ or the δ receptors are responsible for respiratory depression. The upshot of this is that different narcotic analgesics have different affinities for the different receptor sites and, once adsorbed, they have different potencies.

Pentazocine is an *N*-methylpiperidine derivative. It was developed on the assumption that a smaller molecule, properly structured, would adsorb only onto the pain-relieving receptor site and not onto the site that facilitates addiction. It turned out to be agonist at the κ and σ receptors and weakly antagonist at the μ receptors. It was the first mixed agonist–antagonist analgesic to be marketed. Because it does not cause euphoria, it is less addictive than morphine, and it has fewer psychotomimetic side effects than nalorphine. It is a moderately potent analgesic and is particularly suitable for patients sensitive to aspirin. It occasionally causes hallucinations.

10.1.5 Etorphine

Much of the synthetic work on opioids was based on the observation that thebaine, a minor constituent of poppy juice, is an ideal dienophile for Diels–Alder reactions. In particular, thebaine (Figure 10.7 (1)) reacts with methyl vinyl ketone (2) to yield an adduct (3), which is just as potent as morphine. When the ketone was converted by a Grignard reagent to a tertiary alcohol, a stronger analgesic (4) resulted. For example, in compound (4), when $R = -CH_2CH_2C_6H_5$, it is 300 times as potent as morphine.

All these materials have a methoxy group, as does codeine. Since morphine, with its 3-hydroxy group, is far more potent than codeine, it was reasoned that

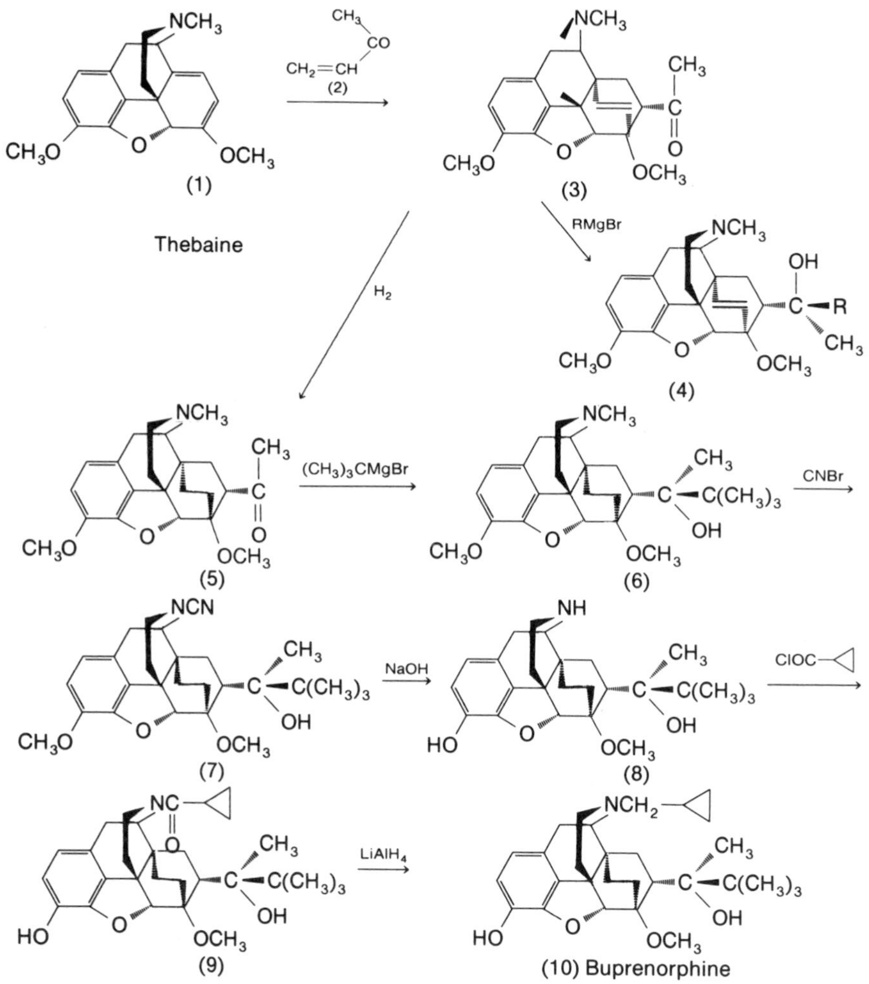

(1) Thebaine

$CH_2{=}CH{-}CO{-}CH_3$ (2)

(3)

RMgBr

(4)

H_2

(5)

$(CH_3)_3CMgBr$

(6)

CNBr

(7)

NaOH

(8)

ClOC–cyclopropyl

(9)

$LiAlH_4$

(10) Buprenorphine

Figure 10.7. Thebaine derivatives.

the conversion of the methoxy group in compound (4) to a hydroxyl would markedly increase potency. And indeed it did. Etorphine is an example:

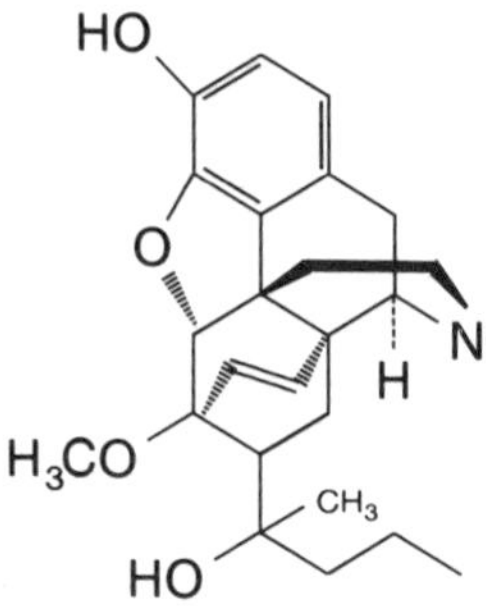

Etorphine

It is 2000–10,000 times as potent as morphine and is used by game conservationists to immobilize large animals. A tiny dose dissolved in less than 1 cm^3 of water is shot into the animal with a crossbow dart, which is able to penetrate its hide. The animal is immobilized for tagging or other studies.

It has been shown that, in the drug–receptor interaction, the key functional group is not the 3-hydroxyl but the R group in structure (4). Activity is apparently associated with the lipophilicity of this group. Activity is also associated with stereochemical configuration and is shown by the enantiomer related to natural morphine. The mirror image of morphine has no analgesic properties.

The difficulties in handling these materials should be stressed. Lewis (see bibliography) notes that on the first occasion that one of the superpotent analgesics was prepared, a laboratory-full of chemists was rendered insensible by drinking tea brewed in a beaker that had been used for it, in spite of the beaker having been rinsed.

10.1.6 Buprenorphine

Buprenorphine (Figure 10.7, (10)) is a modern agonist–antagonist analgesic which acts in a similar way to pentazocine. It has a high affinity for μ receptors, but its activity when it arrives there is much less than that of morphine. Thus, a very low dose of buprenorphine is required to match the pain killing effects of a moderate dose of morphine, but morphine will alleviate high levels of pain that buprenorphine will not. Buprenorphine's low potency at the μ receptor means that it is not addictive. For the same reason it is safe and overdoses are not dangerous. Buprenorphine also has a high affinity for κ and δ receptors but appears to motivate a very low physiological response there.

Buprenorphine is close to the ideal analgesic and the fascinating story of its development has been told by Lewis (see bibliography). Its usefulness is limited only by certain morphine-like side effects such as nausea and vomiting. It is not supplied to the general public even on prescription but is probably the most important analgesic used in hospitals for the pain of carcinoma and postsurgical discomfort. It is also used as an adjunct to anesthesia. It is usually given by injection but is also well absorbed from under the tongue.

The commercial synthesis of buprenorphine is shown in Figure 10.7. It starts with the methyl vinyl ketone adduct of thebaine, but this is hydrogenated to (5). Reaction with a Grignard reagent converts the ketone group to a tertiary hydroxyl (6). Reaction with cyanogen bromide converts the *N*-methyl to NCN (7), which, in turn, becomes a secondary amine (8) on reaction with sodium hydroxide. The nitrogen atom is then acylated with cyclopropyl chloroformate to yield an amide (9), which, on hydrogenation with lithium aluminum hydride, gives buprenorphine (10).

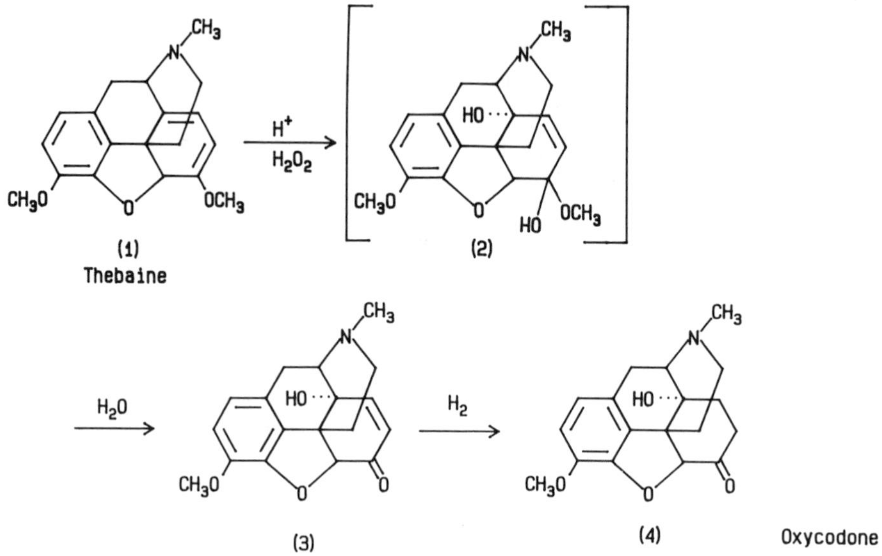

CH_3 N CH_3O O OCH_3

(1) Thebaine

$\xrightarrow[H_2O_2]{H^+}$

CH_3 N HO··· CH_3O O HO OCH_3

(2)

$\xrightarrow{H_2O}$

CH_3 N HO··· CH_3O O O

(3)

$\xrightarrow{H_2}$

CH_3 N HO··· CH_3O O O

(4) Oxycodone

Figure 10.8. Synthesis of oxycodone.

10.1.7 Oxycodone, Hydrocodone, Dihydrocodeine, and Dextromethorphan

The remaining narcotic analgesics in the Top-100 are oxycodone and hydrocodone. Both are phenanthrene-related drugs; that is, they resemble morphine. Oxycodone ranks #46. Its synthesis from thebaine (1) is shown in Figure 10.8. Thebaine's internal diene system undergoes the Diels–Alder reaction as well as other addition reactions. Just as 1,3-butadiene adds

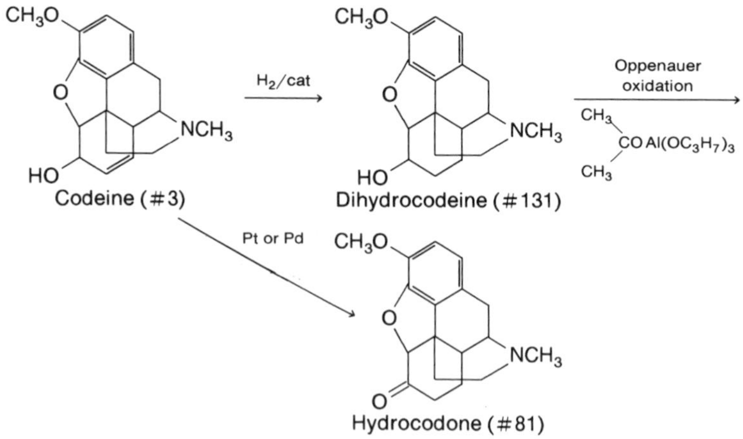

CH_3O O NCH_3 HO

Codeine (#3)

$\xrightarrow{H_2/cat}$

CH_3O O NCH_3 HO

Dihydrocodeine (#131)

Oppenauer oxidation

$(CH_3)_2CO$ $Al(OC_3H_7)_3$

Pt or Pd

CH_3O O NCH_3 O

Hydrocodone (#81)

Figure 10.9. Syntheses of hydrocodone and dihydrocodeine.

chlorine in the 1,4-positions with the double bond shifting, thebaine adds two hydroxyls from acidified hydrogen peroxide to give first the 14-hydroxy-6-hemiketal (2). On hydrolysis (2) yields the unsaturated ketone (3), which may be catalytically reduced to oxycodone (4).

Hydrocodone (#81) is prepared from codeine via another analgesic, dihydrocodeine (#131), as shown in Figure 10.9. Catalytic reduction of codeine gives dihydrocodeine and the Oppenauer oxidation (a ketone plus aluminum alkoxide, the ketone being reduced to an alcohol) gives hydrocodone. A direct method to hydrocodone involves the heating of codeine with a noble metal (Pt or Pd), which isomerizes the allylic alcohol section of the molecule to a ketone.

Dextromethorphan (#127) is a safe over-the-counter antitussive. It is nonnarcotic but its structure resembles the opiates:

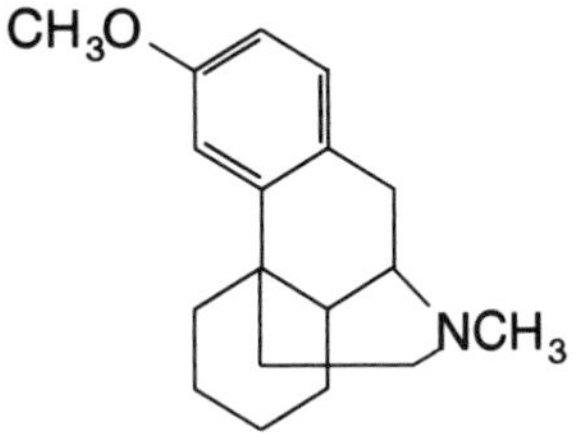

CH_3O NCH_3

Dextromethorphan (+ epimer)

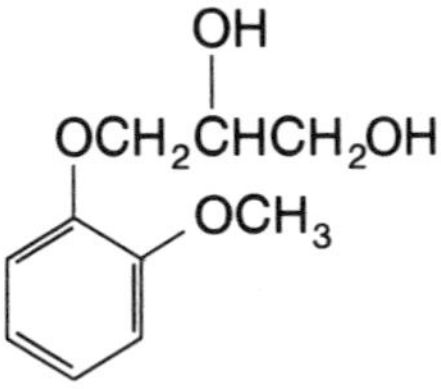

OH
OCH_2CHCH_2OH
OCH_3

Guaifenesin

Other components of widely prescribed cough mixtures are Organidin (#78), an iodinated glycerol, and guaifenesin (#84 see above), which is an expectorant, that is, it helps the expulsion of mucus. It should be obvious that a compound designed to help a sufferer cough up mucus should not be combined with another designed to inhibit coughing altogether, but many over-the-counter cough preparations are just such mixtures. The cautious layman should decide what sort of a cough he has and then buy a single-entity preparation.

10.2 NONNARCOTIC PAIN RELIEVERS

The two main nonnarcotic pain relievers are aspirin and acetaminophen (known as paracetamol in the United Kingdom):

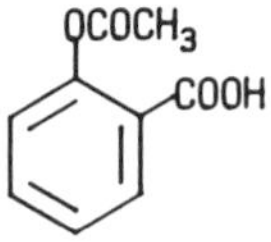

$OCOCH_3$
$COOH$

Aspirin

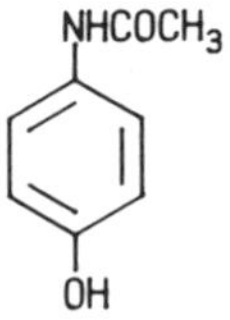

$NHCOCH_3$
OH

Acetaminophen

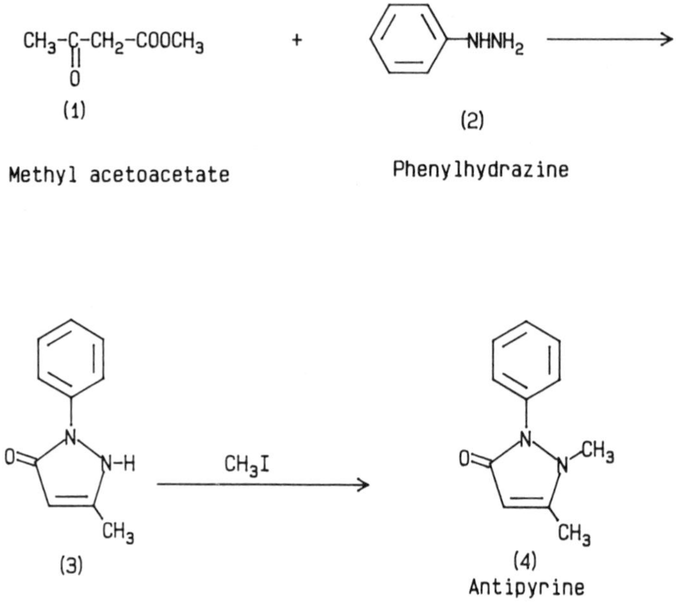

$CH_3-C(=O)-CH_2-COOCH_3$ + $C_6H_5-NHNH_2$ →

(1) Methyl acetoacetate

(2) Phenylhydrazine

(3) —CH_3I→ (4) Antipyrine

Figure 10.10. Synthesis of antipyrine (1887, Knorr).

A milestone in the development of the nonnarcotic pain relievers was the synthesis of salicylic acid by Kolbe in 1839. This was later to be the starting material for aspirin. In 1887 Knorr synthesized antipyrine as shown in Figure 10.10. Acetoacetic methyl ester (1) is condensed with phenylhydrazine (2) to give an intermediate pyrazolone (3), which is methylated to give antipyrine (4). As its name implies, it reduced fever. It is no longer used except in over-the-counter preparations for the alleviation of earache but is important because it helped lay the foundation for chemotherapy. It is one of the group of aniline-related analgesics which includes acetaminophen. Related to antipyrine is dipyrone, an analgesic used widely in Germany:

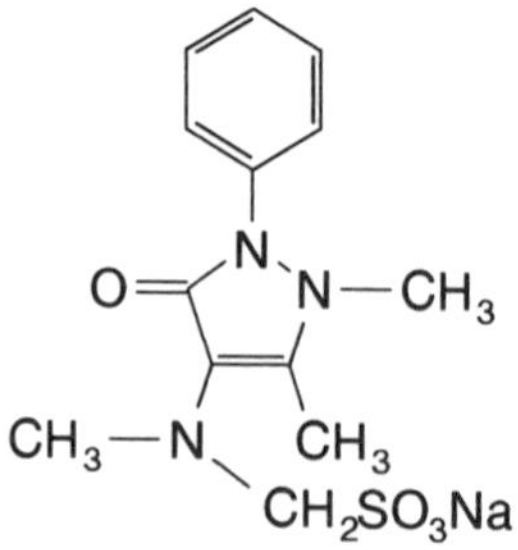

$O=$, $N-CH_3$, CH_3-N, CH_3, CH_2SO_3Na

Dipyrone

10.2.1 Aspirin

In 1899 Bayer discovered aspirin and laid the foundation for the giant chemical company that still bears his name. Aspirin, along with acetaminophen and vitamin C, are the only pharmaceuticals manufactured on the scale of an industrial chemical. It is a proprietary drug, available freely without prescription, and is without question the most widely used drug in the world. It has alleviated aches and pains that might otherwise have kept individuals from work and thus has increased their productivity. In the developed world the average man, woman, and child buys, even if he or she does not consume, about 200 aspirin or aspirin-containing tablets per year. Although available freely as a pure substance in the United States, aspirin is sold mixed with codeine and related narcotic analgesics only on prescription.

Compared with other pharmaceuticals, aspirin is safe and well tolerated. Even so it has side effects; in particular, it irritates the stomach wall and increases the secretion of acid by the stomach. This has one advantage in that many suicide attempts fail because the victim vomits before the dose is absorbed. For most patients, however, it is a problem, and many cannot take aspirin at all. There is also evidence of links between Reye's syndrome and aspirin consumption by feverish children. Reye's syndrome is a brain disease that also causes fatty degeneration of organs such as the liver. It can occur during recovery from chickenpox or flu. Coma and death sometimes result and the survivors may suffer brain damage. There are 250–650 cases/year in the United States, of which about a quarter are fatal. Acetaminophen is now the preferred analgesic for children under twelve years old.

Until recently, a simple explanation of sensitivity to aspirin was accepted. Aspirin is a sparingly soluble acid and, when swallowed, lies on the stomach lining, causing bleeding. The problem with this explanation is that hydrochloric acid, which is found in the healthy stomach, is a stronger acid than aspirin and the stomach wall seems unaffected by it. In 1971 some insight into the mode of action of aspirin was obtained. Aspirin is a receptor antagonist, blocking the synthesis of the inflammatory prostaglandins E_2 and I_2. The inhibition of these prostaglandins relieves pain, but in the longer term it leads to inhibition of platelet aggregation (hence prolonging bleeding times) and inhibition of the formation of gastric mucosa, so that the stomach wall is exposed to attack by acids and pepsin. The inbition of platelet aggregation is of value in the prevention of heart disease, and aspirin has a role in this application. This will be discussed further in Chapter 19.

One solution to the problem of stomach irritation is to formulate aspirin as the soluble calcium salt of acetylsalicylic acid. When it reaches the stomach, however, the stomach acid precipitates the insoluble acetylsalicylic acid. Thus, a widespread mild irritation replaces the localized bleeding. In spite of this and the development of various buffered forms of aspirin, there is a trend toward related compounds, especially acetaminophen, which is a weaker acid and is less irritating. In some countries, acetaminophen has overtaken aspirin.

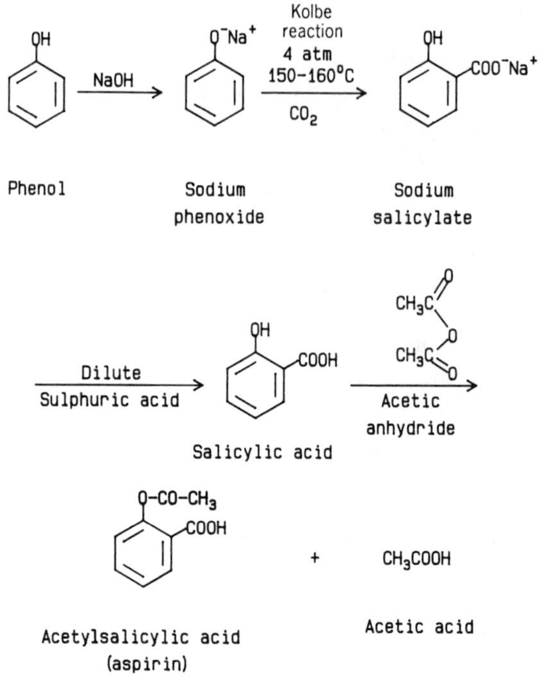

Kolbe reaction
4 atm
150–160°C
CO_2

NaOH

Phenol

Sodium phenoxide

Sodium salicylate

Dilute Sulphuric acid

Salicylic acid

Acetic anhydride

+ CH_3COOH

Acetylsalicylic acid (aspirin)

Acetic acid

Figure 10.11. Synthesis of aspirin.

The industrial synthesis of aspirin is shown in Figure 10.11. It is often an experiment in beginning organic chemistry and will be familiar to many readers.

10.2.2 Acetaminophen (Paracetamol)

Acetaminophen, when mixed with codeine in the branded product Tylenol/codeine, was #4 in the 1986 list of branded drugs, with about 19 million prescriptions. It can also be bought freely without prescription. Like antipyrine, it is an aniline derivative and together with aspirin it is antipyretic, that is, it reduces fever as well as relieving pain. Unlike aspirin, it is not anti-inflammatory. It is less irritating than aspirin but carries with it the danger of liver failure in cases of overdose. Suicide deaths from aspirin in Britain are still higher than from acetaminophen, but there have been many cases where attempted suicides have awakened from an overdose of acetaminophen and changed their minds, yet still died a few days later from liver damage.

The process by which acetaminophen is eliminated from the body and, hence, the mechanism of poisoning are reasonably well understood. Acetaminophen is probably metabolized to a quinone imine:

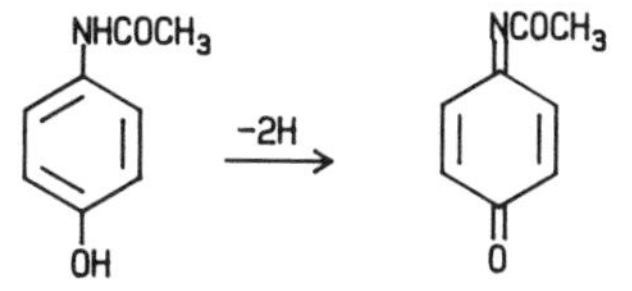

Acetaminophen —(−2H)→ Quinone imine

The quinone imine is an electrophile and, like other potentially harmful electrophiles, is eliminated in the liver by reaction with a tripeptide, glutathione. The first stage of the process is

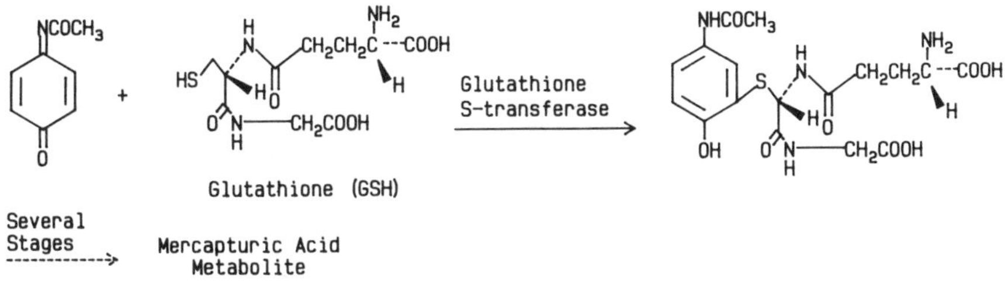

Quinone imine + Glutathione (GSH) —(Glutathione S-transferase)→

Several Stages ⇢ Mercapturic Acid Metabolite

The remaining stages will be found in Burger (see bibliography). The acetaminophen is eventually excreted as a mercapturic acid metabolite. If insufficient glutathione is available, the toxic electrophile will not be eliminated and will react with toxic nucleophilic groups in vital cellular proteins and nucleic acids. Methionine and *N*-acetylcysteine boost levels of glutathione in the liver and, consequently, they are antidotes for acetaminophen poisoning if the overdose is discovered in time.

$CH_3SCH_2CH_2CH(NH_2)COOH$ — Methionine

$HSCH_2CH(NHCOCH_3)COOH$ — N-Acetylcysteine

A new formulation is being marketed in the United Kingdom which incorporates methionine into acetaminophen tablets so the drug carries its own antidote with it. *N*-Acetylcysteine is marketed both as an acetaminophen antidote and also as a mucolytic (to thin the mucus) in bronchitis.

Acetaminophen is manufactured by the simple acetylation of *p*-aminophenol:

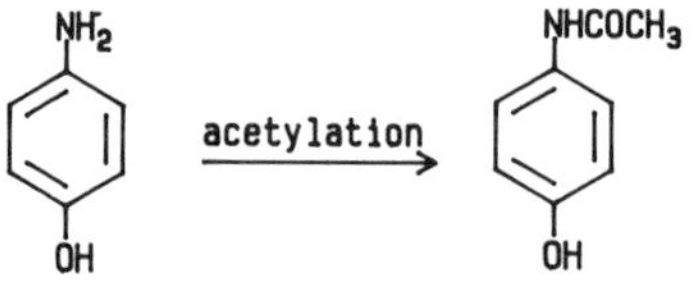

NH2 / OH → acetylation → NHCOCH3 / OH

p-Aminophenol   Acetaminophen

The *p*-aminophenol is made by nitration of phenol to give the ortho-para mixture. The ortho isomer is removed by steam distillation and the *p*-nitro group reduced.

Phenacetin is another pain reliever, related to aniline, that used to feature in the Top-100.

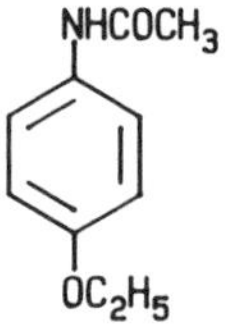

NHCOCH3 / OC2H5

Phenacetin

It is metabolized in the body to acetaminophen and there is some evidence that various of its side effects are due to the conversion of a small part of the dose to aniline. It is possible that the drug causes kidney damage and it was deleted from the *British Pharmacopeia* in the early 1980s and has also been outlawed in the United States.

There has been much research into nonnarcotic analgesics in recent years but the drugs that have emerged fit more conveniently into the categories of nonsteroid anti-inflammatory drugs (Chapter 12) or materials that are involved in prostaglandin chemistry (Chapter 19).

10.3 ENDORPHINS AND ENKEPHALINS

The structure of nalorphine was shown in Section 10.1.3. This drug has the ability to replace morphine on a morphine receptor site. Once displaced, the morphine is destroyed by normal bodily processes. Thus, nalorphine is a morphine antagonist. This seems strange, but is no stranger than the fact that the body contains receptors for morphine, a molecule produced by a poppy.

It was puzzles such as these that initiated the quest for morphine-like substances in the body. The quest culminated in 1975 when three laboratories simultaneously announced the discovery of the endorphins (*endo*genously produced m*orphin*e) and enkephalins (derived from the Greek and meaning "in the head"). The discovery generated great excitement in the scientific and the popular press. These compounds, which exist in minute quantities, are the body's own morphine, its endogenous pain killers. It was believed that they were not addictive, on the grounds that the body cannot become addicted to its own chemicals, although this belief later turned out to be incorrect.

The endorphins attach to the same receptor sites in the brain as morphine and they kill pain. This explains the results of acupuncture. When a needle is inserted into an acupuncture point, and a stimulus applied either electrically or by twisting, endorphins are produced and the body ceases to feel pain. This theory is supported by the fact that acupuncture-induced analgesia can be reversed by naloxone, a morphine antagonist.

Surprisingly, the endorphins and enkephalins do not have a morphine-like structure. Enkephalins turned out to be pentapeptides:

tyrosine-glycine-glycine-phenylalanine-leucine

Leu-enkephalin

tyrosine-glycine-glycine-phenylalanine-methionine

Met-enkephalin

These peptides differed only because the carboxyl-terminated amino acid in the end position was leucine in one molecule and methionine in the other. The molecule terminated with leucine was consequently called leu-enkephalin and the one with methionine met-enkephalin.

Simultaneously, opiates of higher molecular weight were found in the pituitary gland. These came to be known as endorphins which is now a generic term for peptides with opiate properties. The term enkephalin is reserved for opiate peptides with five amino acids. Two of the endorphins containing 16 and 17 amino acids* were named α- and γ-endorphins:

H-Tyr-Gly-Gly-Phe-Met-Thr-Ser-Glu-Lys-Ser-Gln-Thr-Pro-Leu-Val-Thr-OH

α-Endorphin

H-Tyr-Gly-Gly-Phe-Met-Thr-Ser-Glu-Lys-Ser-Gln-
Thr-Pro-Leu-Val-Thr-Leu-OH

γ-Endorphin

It was very early observed that the sequence of amino acids in met-enkephalin is identical with a residue containing five amino acids in the hormone β-lipotropin, first isolated in 1964. This so-called C fragment, which contains the five amino acids found in enkephalin, has potent opiate activity and is

*The names of the amino acid residues are abbreviated; an H is used at one end to indicate a terminal amino group and an OH at the other to signify a terminal carboxyl group.

called β-endorphin:

H-Tyr-Gly-Gly-Phe-Met-Thr-Ser-Glu-Lys-Ser-Gln-Thr-Pro-Leu-
Val-Thr-Leu-Phe-Lys-Asn-Ala-Ile-Ile-Lys-Asn-Ala-
Tyr-Lys-Lys-Gly-Glu-OH

β-Endorphin

The 31 amino acids that it contains start with the 16 in the endorphins and the five in met-enkephalin. These five are also the 61–65 amino acid sequence of β-lipotropin, a pituitary hormone.

Although enkephalins bear little chemical resemblance to morphine, it is possible that they attach to the same receptors because of a topographical similarity. Various ways have been devised to fold the amino acid chains in the above compounds to correspond to morphine and related substances. Figure 10.12 shows the topographical relation between unsubstituted oripavine (an opioid) and met-enkephalin. Note also that the phenyl group of phenylalanine in enkephalins is oriented relative to the tyrosine residue in almost exactly the same way that the phenyl group in the etorphine nucleus (Section 10.1.5) is oriented to the phenolic ring. The discovery of tiny quantities of morphine in the brains of mammals threw some doubts on the theory and it has not yet been proved.

Relief from pain is not the only function of the endorphins. It has been shown that they are also neurotransmitters. As such, they interact with a large number of systems and mediate not only pain relief but also sexual behavior, hormonal balance, and schizophrenia. The injection of endorphins into rats produces profound behavioral changes and the main interest in endorphins today is in this effect on behavior. On the other hand, the enkephalins, which are simpler molecules, are being studied as potential analgesics with the aim of achieving high potency, long duration of action, and lack of addiction.

For several years after their discovery, the endorphins and enkephalins were the subject of the largest number of scientific papers in any field. A massive amount of research was aimed at devising opiate-like peptides but so far without success. Enkephalins are unstable in all mammals and are destroyed before they can reach the relevant receptors, so their effects are difficult to observe. β-Endorphin is hard to prepare, costly, and available only in small quantities, so that only pilot studies have been possible. Synthetic peptides are either destroyed quickly by the body or produce addiction and depressed respiration like the opiates. Metkephamid, a derivative of enkephalin, was said in 1982 to be better in these respects:

H-Tyr-D-Ala-Gly-Phe-MeMet-NH_2

Metkephamid

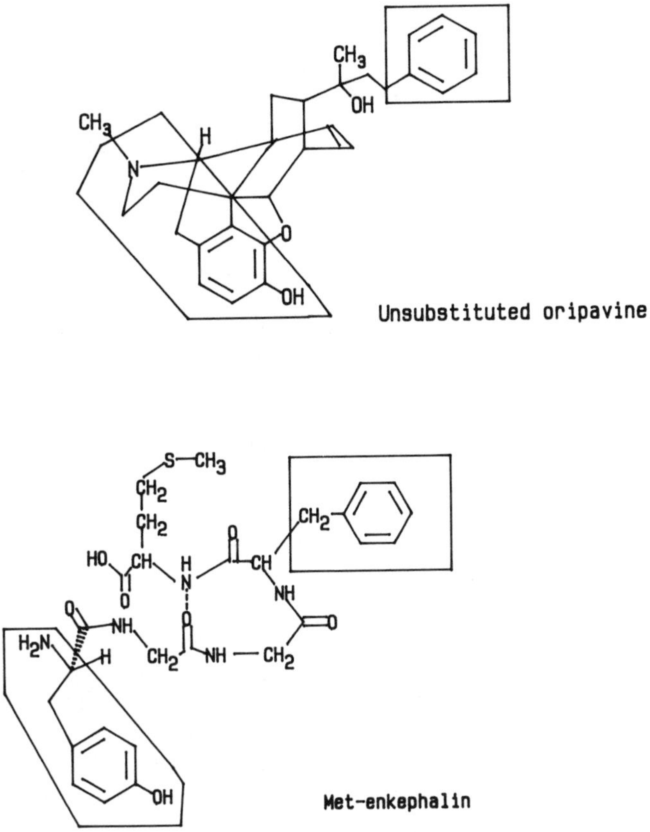

Unsubstituted oripavine

Met-enkephalin

Figure 10.12. *Topographical relation between unsubstituted oripavine and Met-enkephalin.*

Meanwhile, automated methods for peptide synthesis are being developed, and reasonable supplies of peptides may speed evaluation. Schering-Plough is also developing an enkephalinase inhibitor which might act as an analgesic by slowing the body's destruction of its own enkephalins. Nonetheless, in the past few years, emphasis has switched to peptides, such as bradykinin (Section 19.3), which act outside the central nervous system.

The enkephalin/endorphin story is still a remarkable one. Because pharmacologists discovered receptor sites for morphine, they were able to predict the existence of endogenous morphine-like materials. The eventual discovery of these was a triumph for the scientific method which encompasses generation, testing, and ultimate proof of a hypothesis.

11

Steroid Drugs

Thus far this book has dealt with diseases and the drugs that have been developed to cure them; drugs have not been categorized by chemical class. This chapter, devoted to steroid drugs, departs from that principle. Only rarely do medicinal chemists find a single group of compounds that merits intensive study. The examples can be counted on the fingers of one hand—the sulfonamides, epinephrine and its related compounds, the steroids, the 4-phenylpiperidines, and the prostaglandins. Each of these classes of compounds cures several ailments and the fact that the classes exist justifies their discussion as such.

There are eight steroids in the Top-100 drugs and they are listed in Table 11.1 together with other important steroids. Conjugated estrogenic hormone is used for hormone therapy to ease the symptoms of menopause. Mixtures of norethindrone–mestranol, ethynylestradiol–norethindrone, and ethynylestradiol–norgestrel are oral contraceptives. Prednisone, betamethasone, and methylprednisolone together with the other compounds listed are used as anti-inflammatories, among other things for the treatment of severe arthritis which does not respond to other drugs, and in ointments for various skin conditions. Other steroids are used as sex hormones, adrenocortical hormones, and anabolic agents.

11.1 NATURAL STEROIDS

The steroid drugs are all agonists or modified agonists, that is, they mimic the effect of hormones normally produced by the body. These natural steroids will therefore be described first. They are all based on the 17-carbon perhydro-

TABLE 11.1. Important Steroid Drugs

Drug	Use
Conjugated estrogenic hormone (#27)	Alleviation of menopausal symptoms
Norethindrone (#12)	Constituents of oral contraceptives
Ethynylestradiol (#16)	
Mestranol (#24)	
Norgestrel (#28)	
Medroxyprogesterone (#80)	For secondary amenorrhea
Prednisone (#35)	Anti-inflammatory steroids
Betamethasone (#77)	
Fluocinonide (#111)	
Methylprednisolone (#112)	
Flunisolide (#133)	
Cortisone	
Hydrocortisone (component of #121)	
Beclomethasone dipropionate	
Budesonide	

cyclopentanophenanthrene ring system:

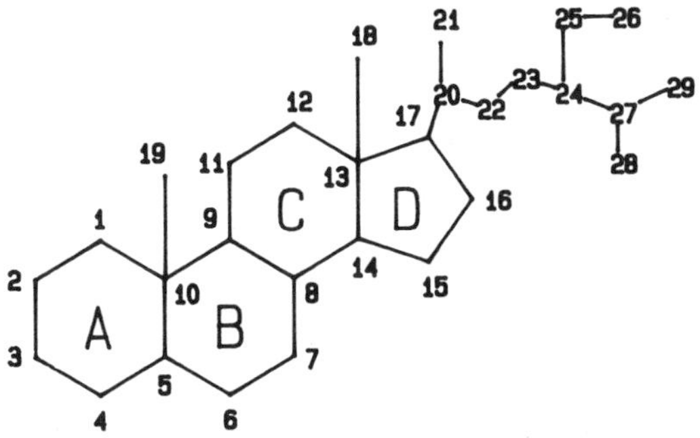

The ring carbons are numbered 1 to 17 and the four rings are labeled A, B, C, and D as indicated. The biological activity of the individual steroid hormones depends on the substituents attached to this basic structure.

Figure 11.1 shows the four main groups of steroid hormones. They are called the estrogens, the androgens, the progestogens, and the corticosteroids.

The estrogens are the female hormones. Compared with the basic steroid ring structure, they have a methyl group attached at C-13, making 18 carbons in all, and an aromatic A ring. Estrone and estradiol are the most important members of the group. They are produced mainly by the ovaries.

The androgens are male hormones. They have methyls attached to C-10 and C-13 of the basic structure, making 19 carbons in all, and a nonaromatic A ring. The most active androgen is testosterone and it is produced mainly by the testes.

1. Estrogens — Female hormones. Methyl is attached to C-13 (making 18 carbons) and the A ring is usually aromatic.

Estrone | Estradiol

2. Androgens — Male hormones. Methyls attached to C-10 and C-13 (making 19 carbons), and the A ring is aliphatic.

Testosterone

3. Progestogens — Methyls on C-10 and C-13, and $COCH_3$ on C-17 (making 21 carbons). The A ring is aliphatic.

Progesterone

4. Corticosteroids — Similar to progestogens but with $COCH_2OH$ side chain on C-17. A hydroxyl or ketone group is on C-11. This appears to be unique in steroid chemistry.

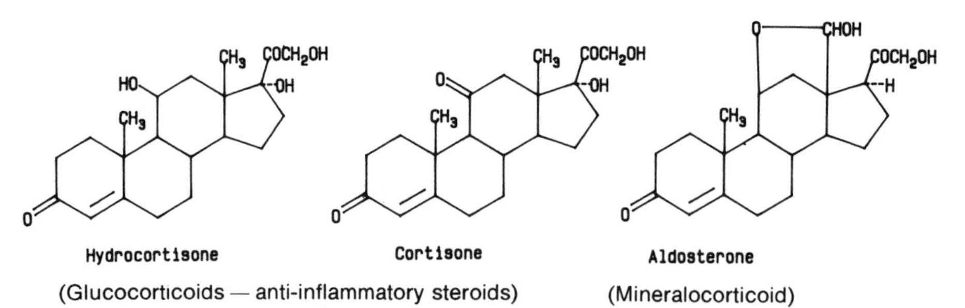

(Glucocorticoids — anti-inflammatory steroids) (Mineralocorticoid)

Figure 11.1. *Steroid hormones.*

The progestogens are similar to the androgens with methyls on C-10 and C-13 and a nonaromatic A ring, but they have an acetyl side chain on C-17 making 21 carbons in all. Progesterone is the most important member of the class. It is produced in the ovaries in the second half of the menstrual cycle and provides the conditions necessary for the developing ovum.

The above hormones are concerned with sexuality. Their release is stimulated primarily by two gonadotropic hormones, the follicle-stimulating hormone (FSH) and the luteinizing hormone (LH). In the female, FSH stimulates development of the follicles in the ovaries, causing the ova to mature and estrogens to be released. In the male, it stimulates development of the testes and manufacture of sperm. In the female, LH causes rupture and release of the ovum from the follicle; in the male, where it is called the interstitial cell-stimulating hormone, it stimulates the interstitial cells of the testes to secrete testosterone.

The corticosteroids are similar to the progestogens but have a hydroxyacetyl side chain and frequently also a hydroxyl or ketone group at C-11. In the body, corticosteroids are produced by the outer layer of the adrenal gland, known as the adrenal cortex. They fall into two groups. First, there are the glucocorticoids such as cortisone and hydrocortisone, which are anti-inflammatory steroids and affect protein, sugar, and calcium metabolism. Second, there are the mineralocorticoids which affect salt and water metabolism. The most important mineralocorticoid is aldosterone, which was mentioned in Section 7.4 on diuretics. Spironolactone is an aldosterone antagonist and its structure was given in Figure 7.4.

11.2 CONJUGATED ESTROGENIC HORMONE

One steroid drug that is a true agonist is conjugated estrogenic hormone, #27 in the Top-100. It is a mixture of female sex hormones that occur in the human body and may be extracted from the urine of pregnant mares. The main hormones in the mixture are estrone, equilin, and 17α-dihydroequilin:

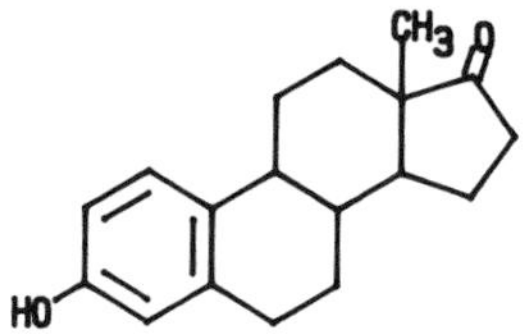

Estrone 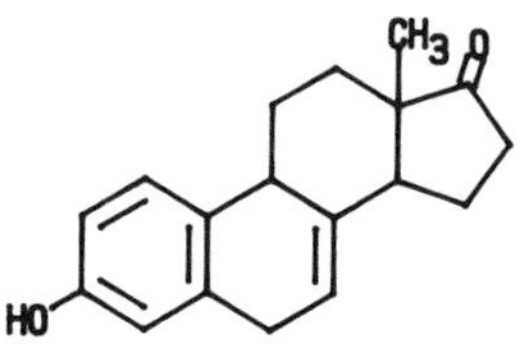 Equilin 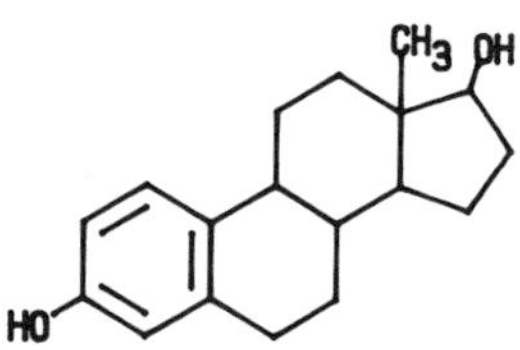 17α-Dihydroequilin

The estrogenic hormones and progesterone (Figure 11.1) control the balance in the female body between FSH and LH, the two gonadotropic hormones produced by the pituitary gland. These estrogens, released in response to FSH, thus have a feedback effect on the pituitary gland.

As the ovaries get older, the follicles respond less and less to FSH. There is a decreased production of estrogens in the body and a rise in FSH. This rise affects the other hormones produced by the pituitary and leads to the characteristic symptoms of menopause—increase in weight, hot flashes, and psychological problems. The lack of estrogens also leads to a decrease in feminine characteristics. These symptoms may be alleviated by consumption of estrogenic hormones, which take the place of those no longer produced by the body.

This therapy has been widely used since 1969, but recent studies have shown that there is an increased risk of cancer of the lining of the uterus in postmenopausal women who have taken estrogens for more than a year. These studies agree with morbidity data showing a sharp rise in cancer of the womb in the United States since 1969. The therapy has therefore been changed. The current recommendation is that the minimum dose of estrogen be taken that will cope with the symptoms and, possibly, that estrogens followed by progestogens be taken in a monthly cycle. This reduces side effects and the cancer risk. There is no evidence that natural estrogens are any less hazardous than synthetic ones.

11.3 ORAL CONTRACEPTIVES

One market for steroids is in oral contraceptives. Progesterone, shown below is one of the sex hormones mentioned earlier. Its most important role in the body is the maintenance of pregnancy. After 3 months, the placenta starts to produce it and goes on doing so until childbirth. A side effect is that it inhibits penetration of sperm through the neck of the womb and also inhibits the release of LH from the pituitary. This, in turn, prevents the rupture and release of the ovum from the follicle. Consequently, if a woman who is not pregnant has a raised progesterone level, ovulation will be blocked. Dosage with estrogens makes the blocking more effective.

The difficulty is that progesterone is metabolized in the liver to pregnanediol:

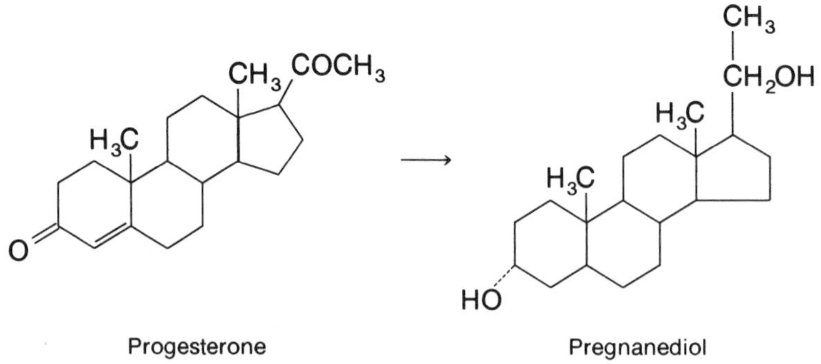

CH_3 $COCH_3$ H_3C O → CH_3 CH_2OH H_3C H_3C HO

Progesterone Pregnanediol

As a result, progesterone is inactive when taken by mouth. To make things worse, it is insoluble and cannot be injected into a vein. It must be given by intramuscular injection, by implantation of a pellet under the skin, or as a suppository.

These methods are inconvenient and a research effort was directed to modify the progesterone molecule so it could be taken orally. The first successful compound was 17α-ethynyl-19-nortestosterone known as norethindrone, #12 in the Top-100. The same compound with ethyl instead of methyl attached to C-13 is also active and is called norgestrel:

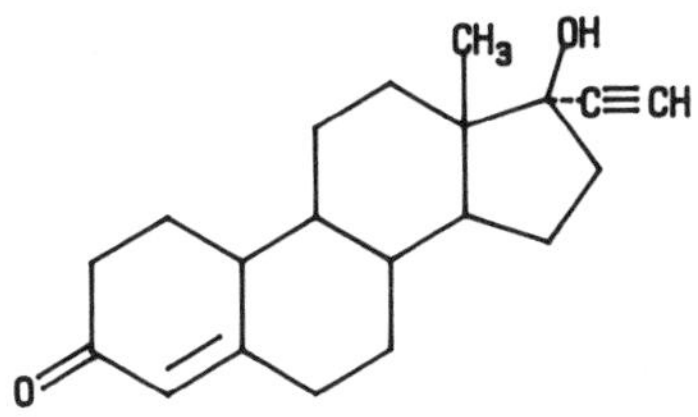

Norethindrone (#12)

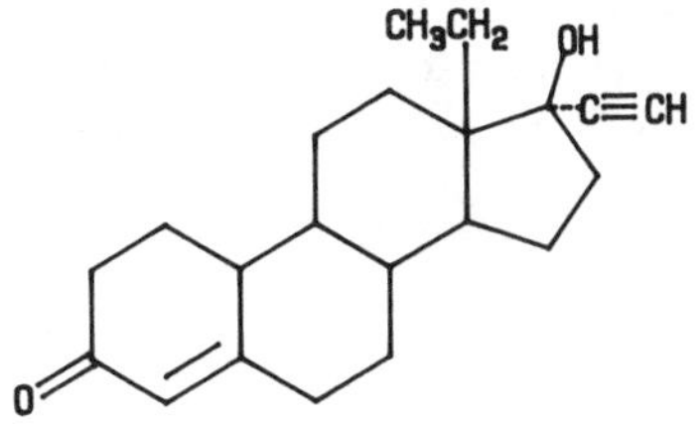

Norgestrel (#28)

Similarly, the first oral estrogen was 17α-ethynylestradiol. The 3-methyl ether of this compound is also effective and is called mestranol.

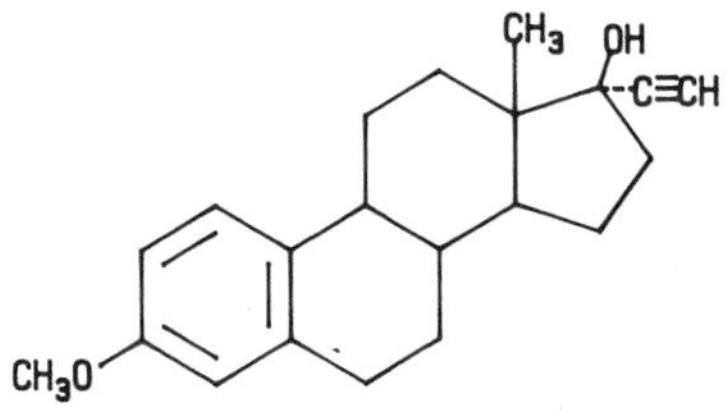

Mestranol (#24)

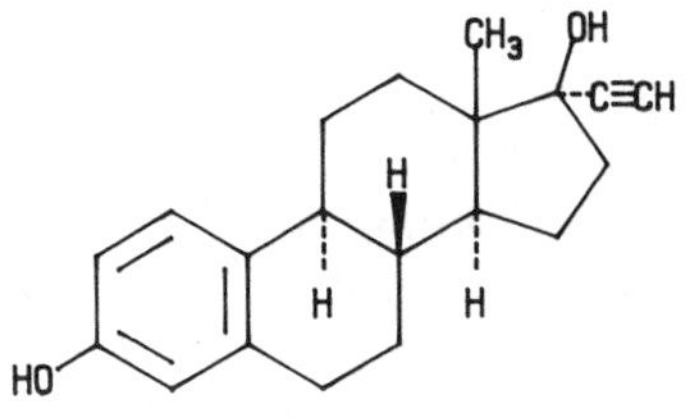

17α-Ethynylestradiol (#16)

Mestranol is a pro-drug and is efficiently converted to ethynylestradiol in the liver. Mixtures of norethindrone–mestranol, ethynylestradiol–norethindrone, and ethynylestradiol–norgestrel are the most widely used oral contraceptives. They are formulated in a number of ways. One reason is to aid compliance with an unusual dosage pattern. For example, seven inactive pills may be included in the dosage pack so that a woman has a pill to take every day instead of only the 5th through the 25th days of the menstrual cycle. Another reason may be to alter the estrogen and progestogen levels so that they follow the hormonal phases of the menstrual cycle. One new formulation is said not to affect metabolic parameters such as cholesterol levels. The variable dose products are said to be "biphasic" or "triphasic" depending on the number of dosage regimens in a cycle.

In addition, "mini-pills" have appeared, which contain a progestogen only and in a smaller dose. They are suitable for patients who are prepared to accept menstrual irregularities, do not require the most effective protection, and wish to avoid estrogens, perhaps because they are older or suffer from migraine, which is a side effect.

Medroxyprogesterone is a progesterone derivative, which is used to treat secondary amenorrhea (absence of menstrual discharge due to incorrect function of the adrenal cortex or thyroid, or to diabetes) and abnormal bleeding from the uterus. Like progesterone, it is a progestogen and could in principle fill that role in an oral contraceptive.

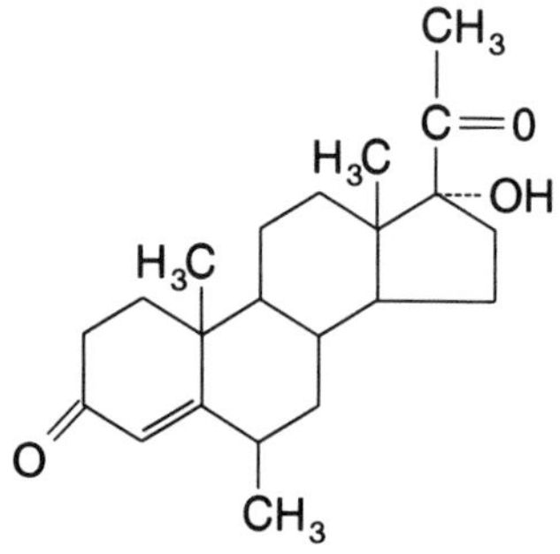

Medroxyprogesterone (#80)

New steroids for contraceptive pills have been introduced to the market but none of them has yet found a place in the Top-100. The reason is probably that the older materials have now been evaluated for 30 years. Experience with the newer materials is obviously less. A range of long-acting dosage forms has also been developed. Given the entirely understandable conservatism of the public, a new pill would have to offer substantial advantages over the older formulations before it could gain a large market.

11.3.1 Progesterone Antagonist

Progesterone is required to maintain pregnancy; hence a progesterone antagonist would be expected to be an abortifacient. Mifepristone, which blocks progesterone receptors in the lining of the womb, is the first such material to be clinically evaluated and was approved in China and France in 1988 as a nonsurgical alternative to abortion.

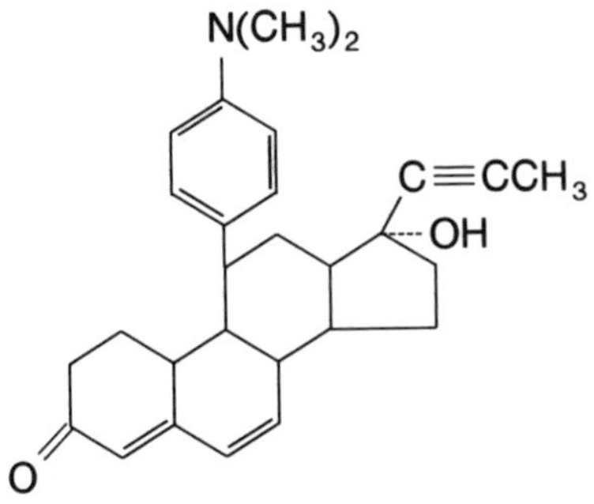

Mifepristone

If a dose of mifepristone is followed a day later by a prostaglandin pessary, a pregnancy of up to three months may be terminated.

Mifepristone holds out the possibility of cheaper, safer, and easier abortions. This is not universally agreed to be a good thing and the future of the drug depends as much on political as on medical factors.

11.3.2 Risks of Oral Contraceptives

Oral contraceptives find a huge market. In the United States approximately 11 million women have used them and, worldwide, about 50 million. They were tested comprehensively before they were put on the market and they appeared completely safe. Much longer-term trials, however, have shown that users are 4 to 11 times as likely as nonusers to develop thrombosis, myocardial infarction, hypertension, and liver and gall bladder problems. Data are given in Figure 11.2. The risk is a very small one for women between 15 and 34, the mortality rate being about 5 per 100,000 women at risk. It rises steadily with age and for the 40–44 age group reaches 25 per 100,000 women at risk. There is also a tenfold difference in the risk between smokers and nonsmokers in the older age group.

Set against these risks is the fact that women using less effective methods of contraception are more likely to get pregnant accidentally. Data for this are also shown in the figure. Both pregnancy and abortion carry with them a risk of death. When these risks are combined, the risk of death from them is of the same order as the risk of using contraceptive pills. Nonetheless, the situation differs from country to country and formulation to formulation. In 1980 one could say that the pill probably represented an acceptable risk for women under 35 especially if they were nonsmokers. Nonetheless, the risk was given widespread publicity and use of oral contraceptives dropped appreciably.

Studies since 1980 have shown the pill to be far less dangerous than previously portrayed, and indeed its use confers some benefits. The risk of cancer of the ovary or uterus is approximately halved and that of pelvic inflammatory disease is cut by a half to two-thirds. Risk of toxic shock

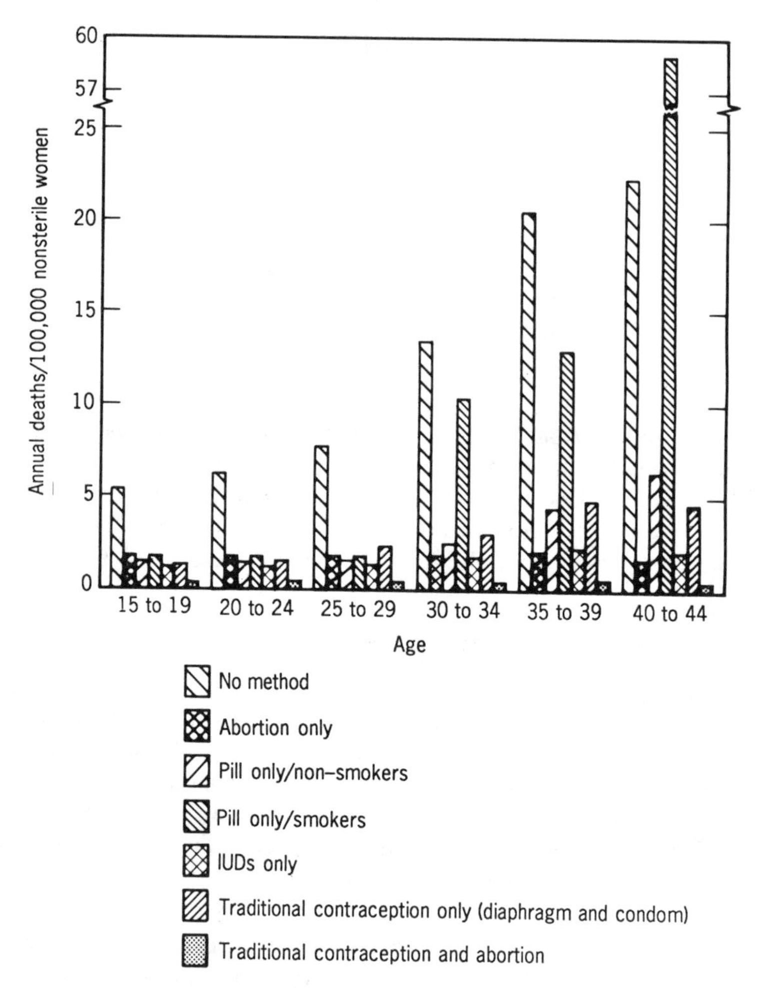

Figure 11.2. *Risks and benefits from oral contraceptives. Annual number of deaths associated with control of fertility and no control per 100,000 nonsterile women, by regimen of control and age of women. Source: Physicians' Desk Reference, 1985, Reproduced with permission (Reference 2)*

syndrome, a rare but acute illness suffered by young menstruating females who use certain types of tampons (no longer on the market), is cut by a factor of four and the risk of rheumatoid arthritis is appreciably reduced.

Like the mood-altering drugs discussed earlier, the oral contraceptives have helped to alter life-styles. They have removed a major risk associated with random and "spontaneous" sexual activity. Another risk, that of sexually transmitted disease, was reduced by the discovery of antibiotics but has now increased with the appearance of AIDS. It is not our task to moralize but we would be equally remiss if we were not to call attention to the role chemistry plays in influencing all aspects of the human process including morality. The fact that there is a risk associated with the use of the pill will do little to alter moral standards. Education and self-realization might.

11.4 ANTI-INFLAMMATORY STEROIDS

No discussion of steroids would be complete without mention of cortisone, which, together with earlier steroids, was associated with Nobel prize-winning chemistry. Its structure, together with those of other corticosteroids, is shown in Figure 11.3. It was the original steroid anti-inflammatory and was used to combat rheumatoid arthritis. Derivatives of it were soon found to be more effective, but all the compounds induced such severe side effects that they are used internally today only in severe cases.

Cortisone is produced in the body by the adrenal cortex. The other important natural glucocorticoid or anti-inflammatory hormone is hydrocortisone, also shown in Figure 11.3. In quantities far larger than those produced by the body, they have an anti-inflammatory, antiallergic, and antirheumatic effect. Cortisone was initially used against rheumatoid arthritis, as already mentioned, but the side effects limited its value. It is now used in replacement therapy when the adrenal cortex is malfunctioning and in a variety of serious situations such as polymyalgia rheumatica and inflammation of the heart sheath.

The aim of subsequent research was to modify the cortisone structure so that the beneficial effects were retained while the adverse effects on salts, sugar, and water metabolism were minimized. Prednisone, at #35 in the Top-100, is the most popular corticosteroid to be taken orally, and it is still high on the list in spite of the risks associated with its use. Methylprednisolone, at #112, is the other oral corticosteroid.

Betamethasone ranks #77 and fluocinonide #111 in the Top-100, but this underestimates the importance of steroids for topical application. Hydrocortisone ointments may be bought over the counter. They would certainly rank in the top ten pharmaceuticals if such sales were taken into account. As it is, they are represented at #135 by hydrocortisone cream and at #121 by Anusol-HC, an ointment for the relief of hemorrhoids. In addition to hydrocortisone acetate, the latter contains bismuth subgallate, bismuth resorcin compound, benzyl benzoate, Peruvian balsam, and zinc oxide. This is a mixture of protectants and emollients that relieves symptoms rather than curing the condition.

Flunisolide is used in nasal sprays to reduce the discomfort associated with the common cold when conventional remedies have proved ineffective.

Beclomethasone dipropionate and budesonide are steroids for treatment of asthma and are discussed in Section 15.4. Hydrocortisone is the preferred compound for injection in emergencies, such as severe asthma attacks, but these are not sufficiently frequent to bring it into the Top-100. Various of its uses are in hospitals rather than by prescription.

1. Natural glucocorticoids:

Cortisone

Hydrocortisone (#121)

2. Synthetic glucocorticoids:

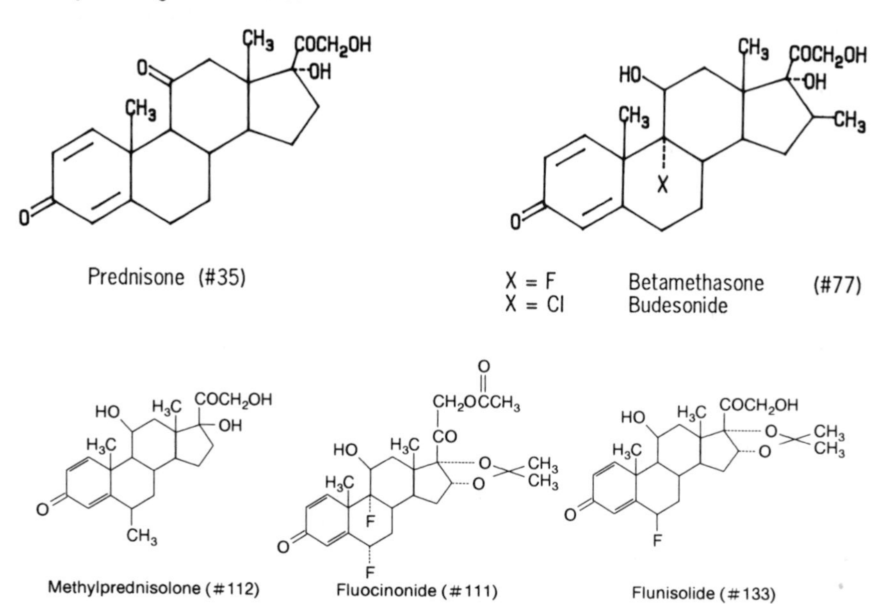

Figure 11.3. *Corticosteroids.*

As is implied above, many of the anti-inflammatory steroids used in ointments are first converted to their esters. Either the C-17 or C-21 hydroxyls or both may be esterified. The choice of esterifying agent is not arbitrary. Esterification increases the lipophilicity of the steroid and helps it to pass through the horny outer layers of the skin into the epidermis. It is not desirable, however, for it to pass further into the dermis and the bloodstream. Typically, C-21 acetates are ten times as active as the corresponding alcohols. 17-Monoesters are more active than 21-monoesters, probably because they are more resistant to chemical and enzymatic hydrolysis. The optimum betamethasone derivatives are the 17-valerate and the dipropionate. Flunisolide and fluocinonide are both acetonides. Hydrocortisone is marketed as the acetate, butyrate, or valerate, or as the free alcohol.

The anti-inflammatory action of corticosteroids is due to their "shutting down" the arachidonic acid cascade (Section 19.1) and blocking the synthesis of various prostaglandins and leukotrienes. The physiological effects are far-reaching and the use of corticosteroids has become less common as the non-steroid anti-inflammatories (Chapter 12), which shut down only part of the cascade, have been developed. The use of long-term corticosteroid therapy nowadays implies a difficult-to-cure and often intractable illness.

11.5 OTHER APPLICATIONS OF STEROIDS

In addition to being anti-inflammatory, the glucocorticoids are antiallergic. Their mode of action is complicated, but basically they interfere with macrophages and T lymphocytes so that B lymphocytes do not produce antibodies to foreign proteins (Section 9.1). Thus, the glucocorticoids diminish the body's immune reaction. They can be used not only to counter acute allergy but also to prevent rejection and prolong the life of transplanted organs. However, they also inhibit the body's defenses against infection. Thus, transplant patients have to be nursed in sterile units and there is a constant conflict between preventing rejection of the transplant and permitting infection of the patient.

In addition to the above applications, sex hormones have been used with great success to treat prostate cancer. Anabolic hormones have found a place in bodybuilding, perhaps after major surgery or a severe accident, or to counter debilitation due to disease or aging. Most of them are related to the male sex hormone testosterone (Figure 11.1). While anabolic hormones have been used cautiously by physicians, they have been adopted with great enthusiasm by athletes to improve their performances in such sports as weight lifting. Thus, we come up again against the problem of drug abuse and, this time, not among the dropouts from society but among those on whom society heaps some of its greatest honors.

11.6 MANUFACTURE OF STEROIDS

The manufacture of steroids provides an excellent case study of how the chemist and the microbiologist solve difficult problems to make health-giving drugs available. The total synthesis of steroid drugs is a challenging enterprise in itself. Cortisone synthesis won a Nobel prize for Hensch and Kendall in 1950. Most contraceptive drugs, and cortisone and its derivatives, are made by partial synthesis from the starting point of a naturally occurring steroid. Figure 11.4 shows the structure of cortisone and some naturally occurring steroids.

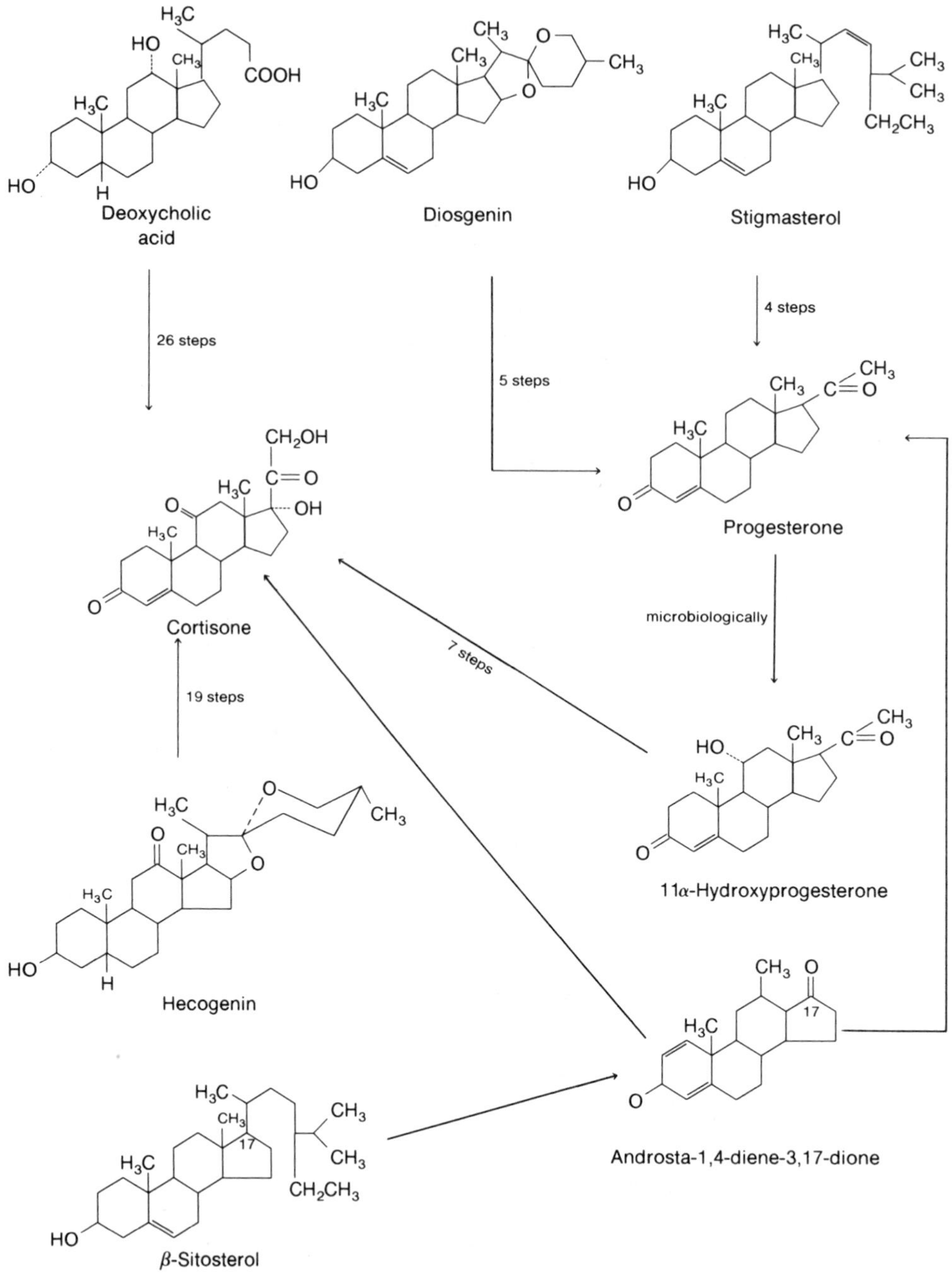

Deoxycholic acid

Diosgenin

Stigmasterol

26 steps

4 steps

5 steps

Cortisone

Progesterone

microbiologically

7 steps

19 steps

11α-Hydroxyprogesterone

Hecogenin

Androsta-1,4-diene-3,17-dione

β-Sitosterol

Figure 11.4. *Production of cortisone from natural sterols.*

The large-scale manufacture of steroid drugs from natural steroids involves numerous stages. The syntheses are too complicated to be described fully here and we will focus on a few of the problems. Two of them were crucial especially for the manufacture of cortisone. These crucial steps were the degradation of the C-17 side chain of naturally occurring sterols and the insertion of an oxygen atom at C-11.

The first route to cortisone started with deoxycholic acid from bile (Figure 11.4). It has two important structural features: a hydroxyl group at C-12 and a short side chain at C-17 with a carboxyl group on the end. It is tedious but possible to transfer the oxygen from C-12 to C-11. The side chain contains only five carbon atoms and the carboxyl group provides a handle for its degradation. Even so, the synthesis of cortisone from deoxycholic acid required 26 steps and gave a yield of 1%. In any case, deoxycholic acid comes from packing house wastes and is available in limited quantities, so the route is not attractive. Although it is said to be used in Germany, another source of raw material was clearly desirable.

11.6.1 Steroids from Diosgenin

A search for other naturally occurring steroids led to diosgenin. Diosgenin (Figure 11.4) is a component of a yam, the barbasco root, that grows wild in the jungles of Mexico. The side chain is readily degraded to give progesterone, an important component of oral contraceptives, as shown in Figure 11.5.

Diosgenin (1) is treated with acetic anhydride to give the acetole (2). Oxidation with chromium trioxide breaks the dihydrofuran ring and gives (3) with the desired acetyl group at position 17. This is an ester of a β-ketoalcohol and on treatment with acetic anhydride the ester grouping splits off to leave (4). Catalytic reduction of the conjugated double bond yields pregnenolone acetate (5) leaving the 5-double bond untouched. Saponification gives pregnenolone (6). Treatment with an aluminum alcoholate (the Oppenauer reaction) leads to the formation of a 3-one 5-ene, which under basic conditions rearranges to progesterone.

Both compound (4) and progesterone are useful intermediates for the manufacture of various steroids, but, if the aim of the synthesis is to make cortisone, there is still the problem of the 11-keto group. Microbiology sometimes accomplishes what is seemingly impossible in the laboratory. Out of the vast range of microorganisms, it is incredible that one was found that would oxygenate the 11-position of a steroid molecule. Called *Rhizopus nigricans*, it provided the vital link in the chain. The product, 11α-hydroxyprogesterone, can be converted to cortisone in seven steps by conventional chemistry.

Diosgenin thus became the major source of cortisone and the sex hormones in Europe. It was never found possible, however, to cultivate the barbasco root and its harvesting requires backbreaking work in steamy jungles. Because

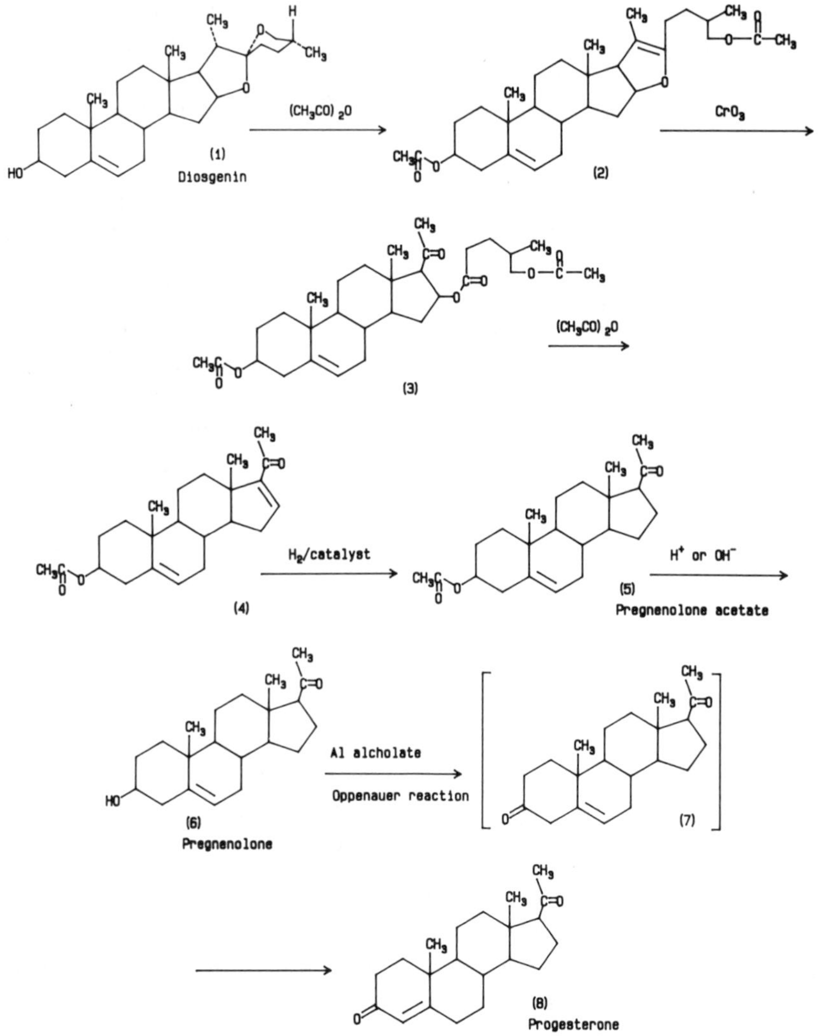

Figure 11.5. Synthesis of progesterone from diosgenin.

supplies were apparently being depleted there seemed to be a need to develop other cortisone precursors.

11.6.2 Steroids from Stigmasterol

Another raw material developed concurrently with diosgenin is stigmasterol, shown in Figure 11.4. It is the most important raw material in the United States. Stigmasterol is a component of the distillate from the steam deodoriza-

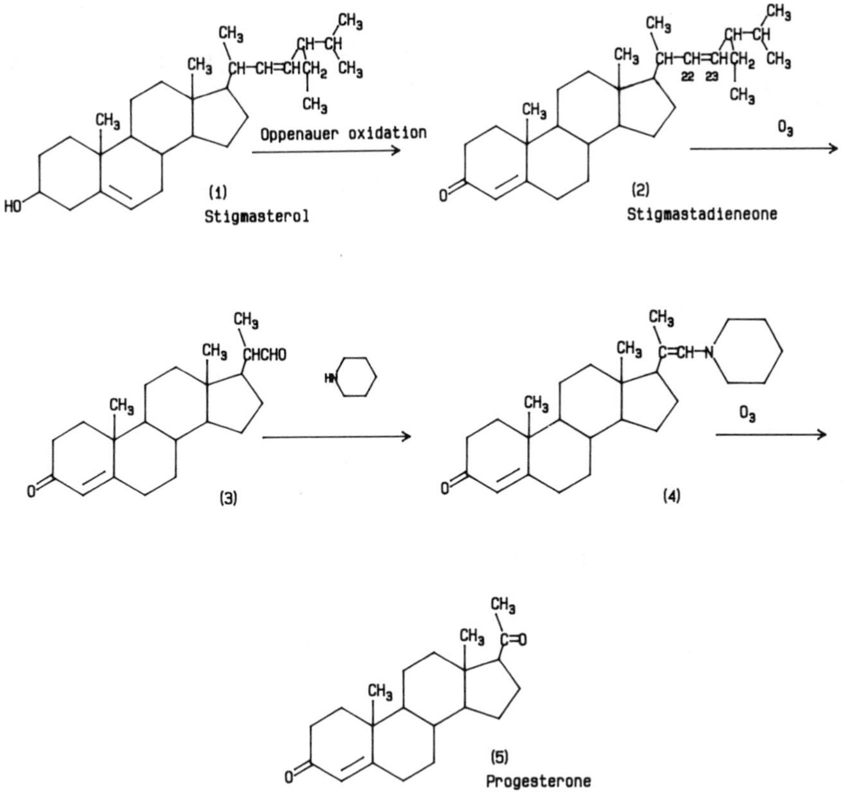

Oppenauer oxidation

(1) Stigmasterol

(2) Stigmastadieneone

(3)

(4)

(5) Progesterone

Figure 11.6. *Synthesis of progesterone from stigmasterol.*

tion of soybean oil intended for edible purposes. It occurs in a mixture with vitamin E precursors and with other steroids, primarily β-sitosterol, and must first be purified by tedious methods. It too provides a route to progesterone, shown in Figure 11.6. The Oppenauer oxidation of stigmasterol (1) gives stigmastadienone (2). Ozonolysis of the C-22–C-23 double bond removes most of the side chain and gives a ketoaldehyde (3) with only one carbon atom more than progesterone. This aldehyde is converted with piperidine to the 22-eneamine (4), which on ozonolysis gives progesterone (5).

This is still not a satisfactory solution to the problem because nature does not provide abundant stigmasterol. The extract from soybeans and other plant products indicated earlier contains other steroids, such as β-sitosterol shown in Figure 11.4, but none of them has an unsaturated side chain. It is the removal of this side chain that poses the problem and again microbiology comes to the rescue. Organisms of the genera *Mycobacterium*, *Anthrobacter*, *Brevibacterium*, and *Nocardia* degrade the side chain completely to the 17-carbon atom. This has made possible the use of sitosterol as a raw material for sex hormones such as the estrogens, norethindrone, and norgestrel.

11.6.3 Steroids from Sitosterol

The side-chain degradation of sitosterol can lead to several compounds including androsta-1,4-diene-3,17-dione (ADD):

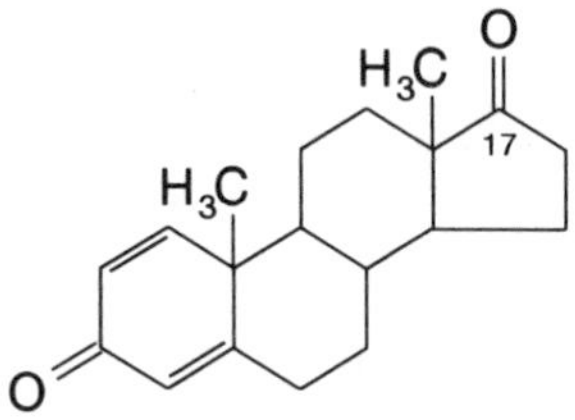

Androsta-1,4-diene-3,17-dione (ADD)

Unfortunately, ADD has no $COCH_3$ group at the 17-position that can be converted to the $COCH_2OH$ group necessary for cortisone. The building up of this side chain has occupied the attention of many chemists and a rich literature has accumulated. At least five routes have been proposed and these are described as the cyanohydrin method, methods using the Wittig, Grignard, and Reformatskii reactions, and an organic lithium method.

The cyanohydrin method is the oldest and there are many variations of it. Initially it comprised the reaction, shown in Figure 11.7, of the 17-one (1) with potassium cyanide and acetic acid to yield the cyanohydrin (2). Phosphorus oxychloride in pyridine is an effective dehydrating agent and gives the unsaturated nitrile (3), which, in a Grignard reaction with methylmagnesium bromide, provides the intermediate (4). This is a key intermediate for progesterone (Figure 11.5, (4)), so progesterone and, hence, cortisone are accessible from sitosterol by this route.

In the Wittig reaction route, shown in Figure 11.8, a 17-keto steroid is reacted with a phosphorus-containing ester, such as diethoxycarbethoxyphosphine oxide (2), to yield an ester (3), which, on hydrogenolysis with lithium aluminum hydride, provides the substituted allyl alcohol (4). This is esterified with acetic acid to (5) and converted to the cortisone side chain (6) by an oxidative hydroxylation with a mixture of osmium tetroxide and either chromic acid or a complex of the oxide and hydroperoxide of *N*-methylmorpholine.

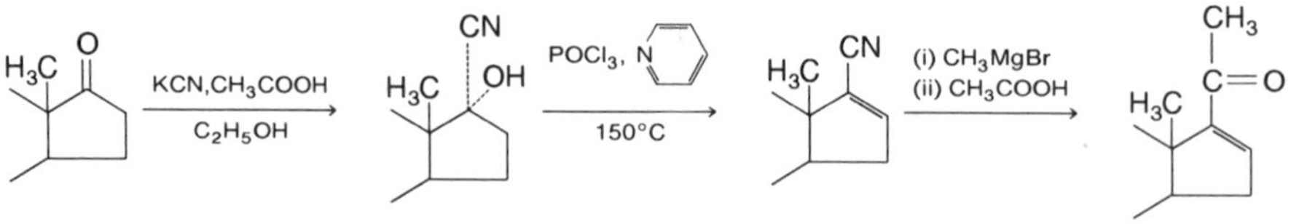

KCN, CH_3COOH / C_2H_5OH ; $POCl_3$, pyridine / 150°C ; (i) CH_3MgBr (ii) CH_3COOH

(1) (2) (3) (4)

Figure 11.7. *Building up the side chain on androsta-1,4-diene-3,17-dione — the cyanohydrin route.*

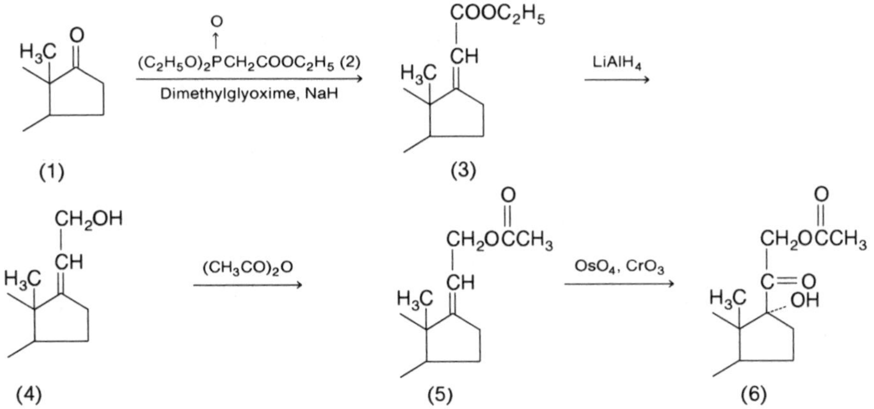

$(C_2H_5O)_2P(O)CH_2COOC_2H_5$ (2), Dimethylglyoxime, NaH; $LiAlH_4$; $(CH_3CO)_2O$; OsO_4, CrO_3

(1) (3) (4) (5) (6)

Figure 11.8. *Building up the side chain on androsta-1,4-diene-3,17-dione — the Wittig reaction.*

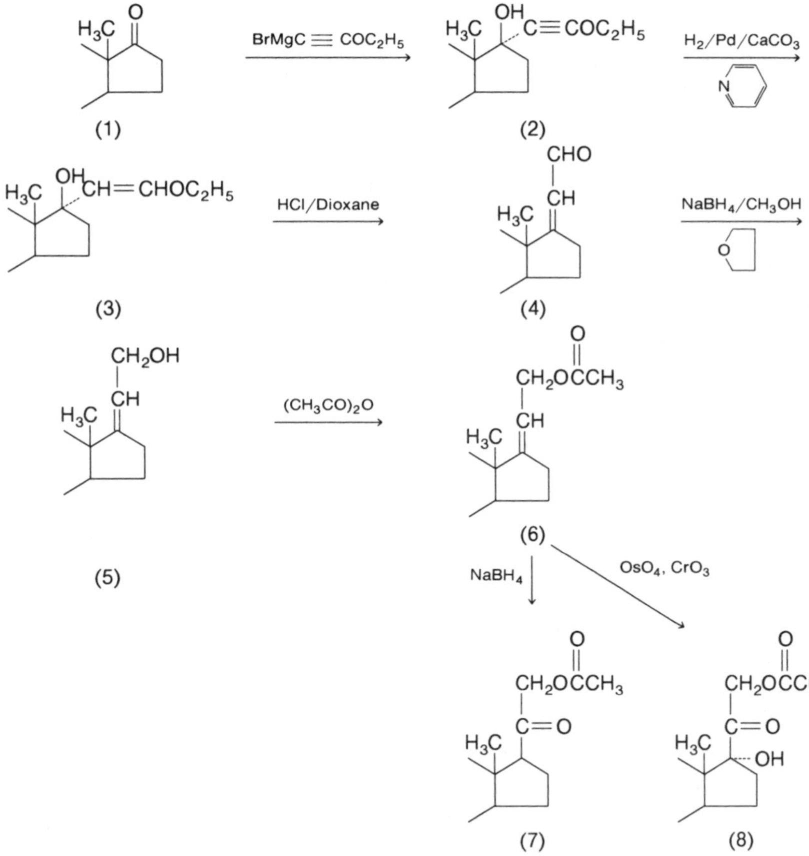

$BrMgC{\equiv}COC_2H_5$; $H_2/Pd/CaCO_3$, pyridine; HCl/Dioxane; $NaBH_4/CH_3OH$, THF; $(CH_3CO)_2O$; $NaBH_4$; OsO_4, CrO_3

(1) (2) (3) (4) (5) (6) (7) (8)

Figure 11.9. *Building up the side chain on androsta-1,4-diene-3,17-dione — the Grignard reaction.*

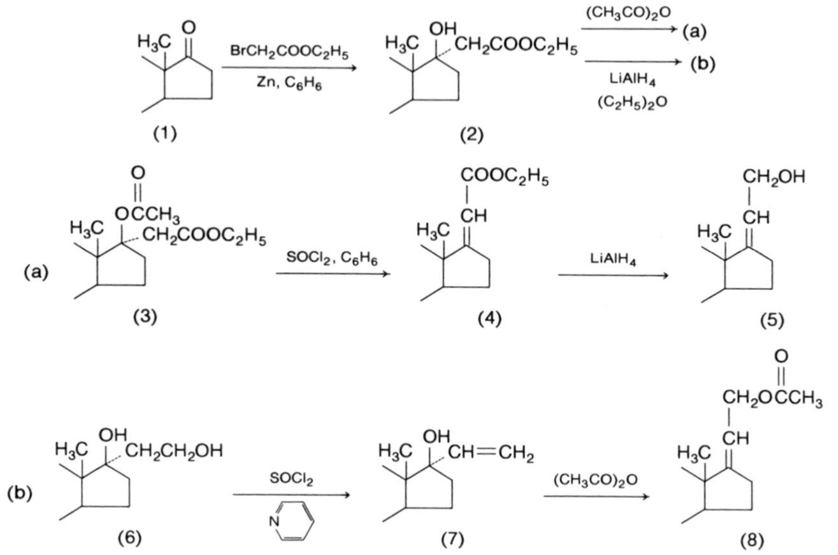

Figure 11.10. *Building up the side chain on androsta-1,4-diene-3,17-dione— the Reformatskii reaction.*

The Grignard method (note that a Grignard reagent is also used in the cyanohydrin route) offers another route from the 17-keto group to the cortisone side chain. Other functional groups in the molecule that can react with a Grignard reagent must be protected. As shown in Figure 11.9, a Grignard reagent from ethoxyacetylene adds directly to the 17-keto compound (1) to yield (2), which is transformed by partial hydrogenation to the ethoxyvinyl compound (3). Acid hydrolysis leads to the α, β-unsaturated aldehyde (4), which, on reduction with sodium borohydride, yields (5). Acetylation provides (6). Oxidative hydroxylation (see above) provides the cortisone side chain (8). Hydroboration of (6) yields (7) via an intermediate 20,21-diol.

In the Reformatskii reaction (Figure 11.10), the 17-ketosteroid (1) reacts with the ethyl ester of bromoacetic acid in the presence of zinc to yield the β-hydroxycarbonic acid ester (2), which is acetylated to (3). Removal of the acetic acid moiety with thionyl chloride yields the α, β-unsaturated ester (4) and this, on hydrogenolysis, provides the allyl alcohol (5). This chemistry was also part of the Wittig route (see above). As in both the Wittig and Grignard routes, oxidative hydroxylation converts (5) to the desired cortisone structure.

Alternatively, (2), on hydrogenolysis, yields (6), which can be dehydrated with thionyl chloride in pyridine to (7) and acetylated to (8).

The organic lithium procedure (Figure 11.11) is related to the Grignard method. Indeed, a lithium compound may be substituted for the Grignard reagent in the cyanohydrin method (see above). Another possibility is that the 17-keto compound (1) is reacted with the lithium salt of a Schiff's base,

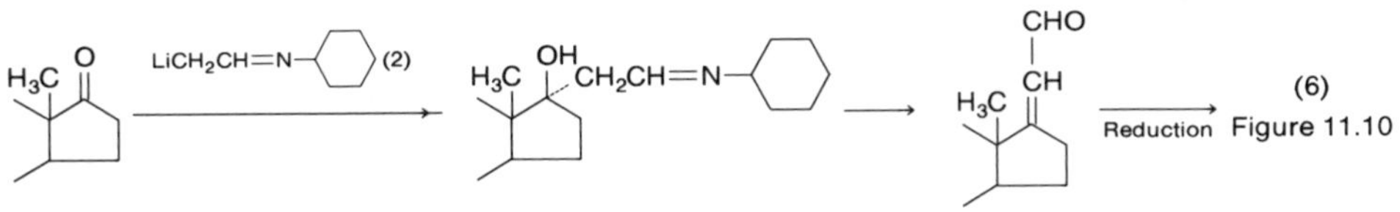

Figure 11.11. *Building up the side chain on androsta-1,4-diene-3,17-dione— the organic lithium route.*

prepared by reaction of acetaldehyde with cyclohexylamine (2). An intermediate (3) results, which is not isolated, but which reacts with acetic acid to yield the acrolein derivative (4). Reduction converts (4) to an intermediate (5) already encountered in the Reformatskii reaction as (6), Figure 11.10.

Cortisone can thus be made from sitosterol by a variety of methods. This discussion only skims the surface of this interesting chemistry and the reader is referred to the review articles listed in the Bibliography.

In 1984 at least two-thirds of U.S. steroids were made from stigmasterol via 11-hydroxylation. In particular, the only producer, Upjohn, uses this method. The remainder was made from sitosterol and diosgenin. More diosgenin has appeared on the market in recent years and the supply situation is not critical.

In Europe, soybeans are not grown and steroids come from a variety of sources. Upjohn sells sitosterol from stigmasterol separation to the German company Schering. Hoechst is said to use bile acids. Other European companies, notably Glaxo, use a route based on hecogenin, a steroid found as its glycoside in the juice expressed from the sisal plant in the course of the production of sisal fiber. Hecogenin already has a ketonic oxygen at C-12 and this can be transposed to C-11 in a 19-step preparation of cortisone. Its structure is given in Figure 11.4.

12

Nonsteroid Anti-inflammatory Agents

The terms rheumatism and arthritis are used to cover a multitude of aches and pains, from a minor pain and swelling in a single joint to a serious and disabling condition like rheumatoid arthritis. These illnesses are widespread; indeed, they are the main source of physical disability in adults. The sufferers rarely die and rarely get better, so there is a huge market for drugs that relieve pain and swelling and reduce inflammation. The search for drugs to do this has been described as the longest-run race in the history of the pharmaceutical industry. Some of the drugs used to reduce inflammation and the chemical groups into which they fall are shown in Figure 12.1. Seven of them—ibuprofen, naproxen, piroxicam, sulindac, indomethacin, diflunisal, and fenoprofen—feature in the Top-100. Aspirin should be included in this list but was discussed in Section 10.2 as a nonnarcotic analgesic. All of them are believed to function like aspirin by inhibiting the formation of inflammatory prostaglandins. They affect the stomach adversely because the prostaglandins they inhibit facilitate the formation of stomach mucus. This topic will be discussed in Chapter 19.

Aspirin was the only nonsteroid anti-inflammatory for over half a century. Arthritics take it in large doses, often for the rest of their lives. Under such conditions, the problems of stomach irritation and bleeding can become serious. Acetaminophen counters these problems to some extent, but only relieves pain and does not reduce inflammation. Cortisone was discovered in 1948 and was hailed as a miracle drug, but, as noted in Section 11.4, it has adverse effects on salt, sugar, and protein metabolism.

1. α-Arylpropionic acids:

Ibuprofen (#11)

Naproxen (#18)

Fenoprofen (#97)

Benoxaprofen

2. Substitute acetic acid derivatives:

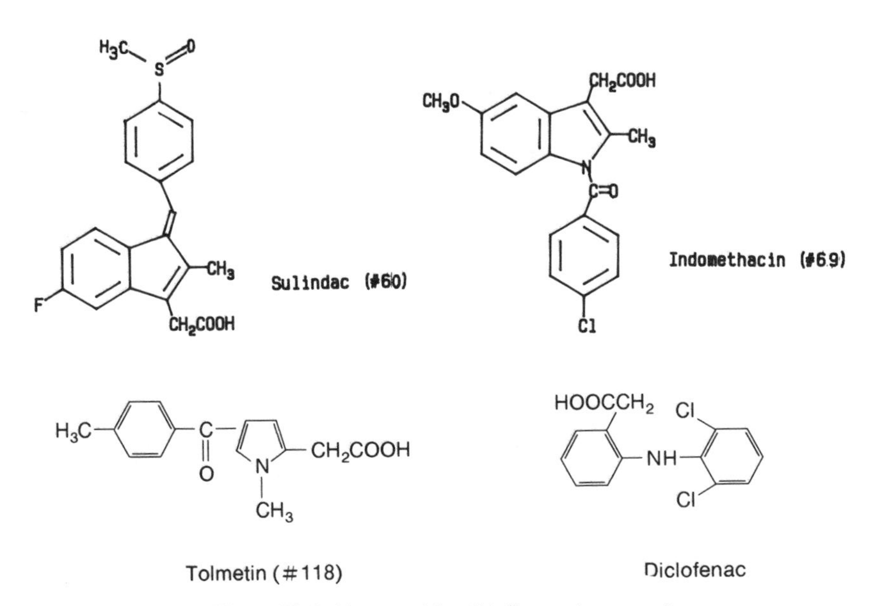

Figure 12.1. *Nonsteroid anti-inflammatory agents.*

Disillusion with the steroids led to a search for nonsteroid anti-inflammatories (NSAIs). Phenylbutazone was discovered in 1949 and indomethacin in 1963. Both were still in the Top-100 in 1983, but phenylbutazone was withdrawn from general use in 1984, primarily because better and safer drugs are now available. Many other nonsteroidal anti-inflammatories have been developed and fall mainly into the chemical classes shown in Figure 12.1. All the drugs are acidic and this seems to be an important structural requirement.

3. Enolic β-diketones:

Phenylbutazone

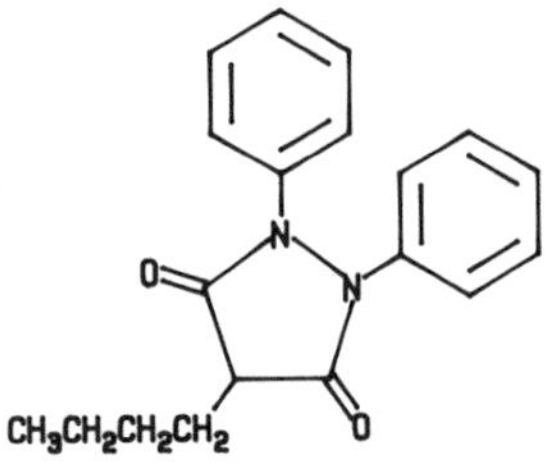

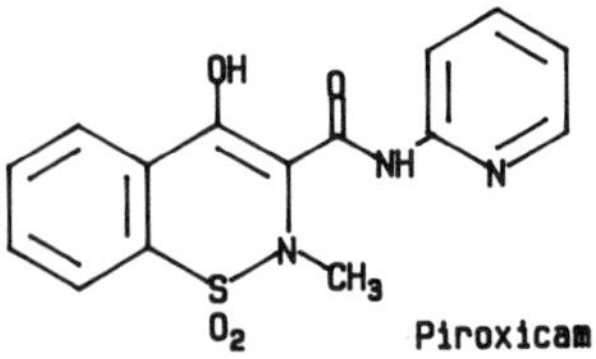

Piroxicam (#48)

4. Anthranilic / salicylic acid derivatives:

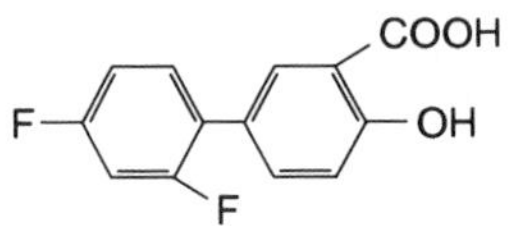

Diflunisal (#95)

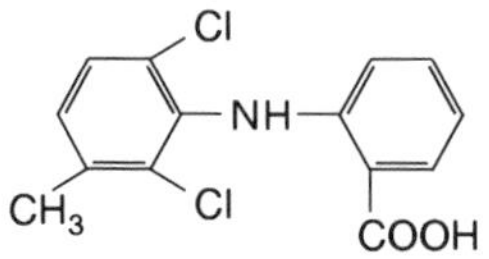

Meclofenamic acid (#120)

Figure 12.1. *Continued.*

12.1 α-Arylpropionic Acids

The largest group is the α-arylpropionic acids, and ibuprofen, the first compound shown, was #11 in the Top-100. It was authorized for over-the-counter sale in both the United States and United Kingdom a few years ago and was probably the first new chemical entity to be so authorized for about 20 years. Naproxen is a more powerful drug than ibuprofen and need be administered only twice a day. Fenoprofen is as effective as naproxen, although it has been suggested that it is more likely to upset the stomach.

Benoxaprofen, the fourth member of the α-arylpropionic acid group in Figure 12.1, promises to become a cause célèbre. It came onto the British market in 1980 and the U.S. market in 1982. An aggressive advertising campaign claimed it to be a novel anti-inflammatory, and only one tablet per day needed to be taken. Half a million patients took it in its first six weeks on the U.S. market. It was withdrawn later in 1982. Alleged side effects included internal bleeding, kidney and liver damage, skin rashes, sensitivity to sunlight, "ridging" of the finger nails, and bone marrow changes. Implication in 76 deaths in Britain and 26 in the United States was alleged. Among other problems, it was found that the half-life of benoxaprofen in elderly patients was five times that in younger patients. The slow rate at which the drug was metabolized presents a special problem in that most arthritic patients are indeed elderly, and it was tantamount to their taking excessive and harmful doses.

An award of $6 million damages was made in a single case in the U.S. courts, and compensation after out-of-court settlements was believed to have totalled $50 million. In the United Kingdom, class actions, in which the court disposes of a number of related claims by hearing a sample case, were not permitted. In addition, in the United Kingdom, victims of a disaster need to prove negligence, whereas in the United States they need only prove injury. Only about $5.4 million was offered in out-of-court settlements—far less than in the United States among more claimants—and few of the claimants have the resources to take the claim further. This has led to a campaign in the United Kingdom for the law to be changed.

12.1.1 Syntheses of Ibuprofen

Thus far we have given a single synthesis of most of the Top-100 drugs. Often it will not have been the route used by all manufacturers and possibly not the one used by any. This section details the many syntheses of ibuprofen and the next section deals with naproxen. They demonstrate the complex role organic chemistry plays in the pharmaceutical industry and illustrate the multiplicity of routes to a given compound and the difficulty even of knowing which technology a particular company uses.

One synthesis of ibuprofen is shown in Figure 12.2 on the left side. Isobutylbenzene (1) is acylated with acetyl chloride to give 4-isobutylacetophenone (2), which is converted with HCN to a cyanohydrin (3). Hydrogen and palladium reduce the benzylic hydroxyl. Hydrolysis of the nitrile group gives the substituted benzylic acid, ibuprofen (4).

The addition of HCN does not go well, however, and another possibility involves the use of diethylaluminum cyanide as shown in the center of Figure 12.2. It is believed that a considerably improved yield results.

Another approach is to treat the 4-isobutylacetophenone with isopropyl chloroacetate and sodium isopropoxide (Figure 12.2, right). This gives an aldehyde (6), which can be oxidized directly to ibuprofen.

A newer procedure, shown in Figure 12.3, involves condensation of 2 mol of isobutylbenzene (1) with acetaldehyde to give 1,1-(bisisobutylphenyl)ethane (2). This compound can be cleaved to isobutylbenzene and 3-(isobutylphenyl)propene (3). Treatment of the propene with carbon monoxide and hydrogen provides a propionaldehyde (4), which on mild oxidation with sodium hypochlorite yields ibuprofen.

A fifth synthesis, shown in Figure 12.4, involves the chloroformylation of isobutylbenzene to give the substituted benzyl chloride (2), which on reaction with sodium cyanide provides a nitrile (3). The carbon adjacent to the nitrile group is methylated with methyl iodide in the presence of a strong base, sodium amide. Hydrolysis of the nitrile group yields ibuprofen.

Still another synthesis (Figure 12.5) depends on the formation of a Grignard reagent (2) from 4-bromoisobutylbenzene (1). This reacts with sodium

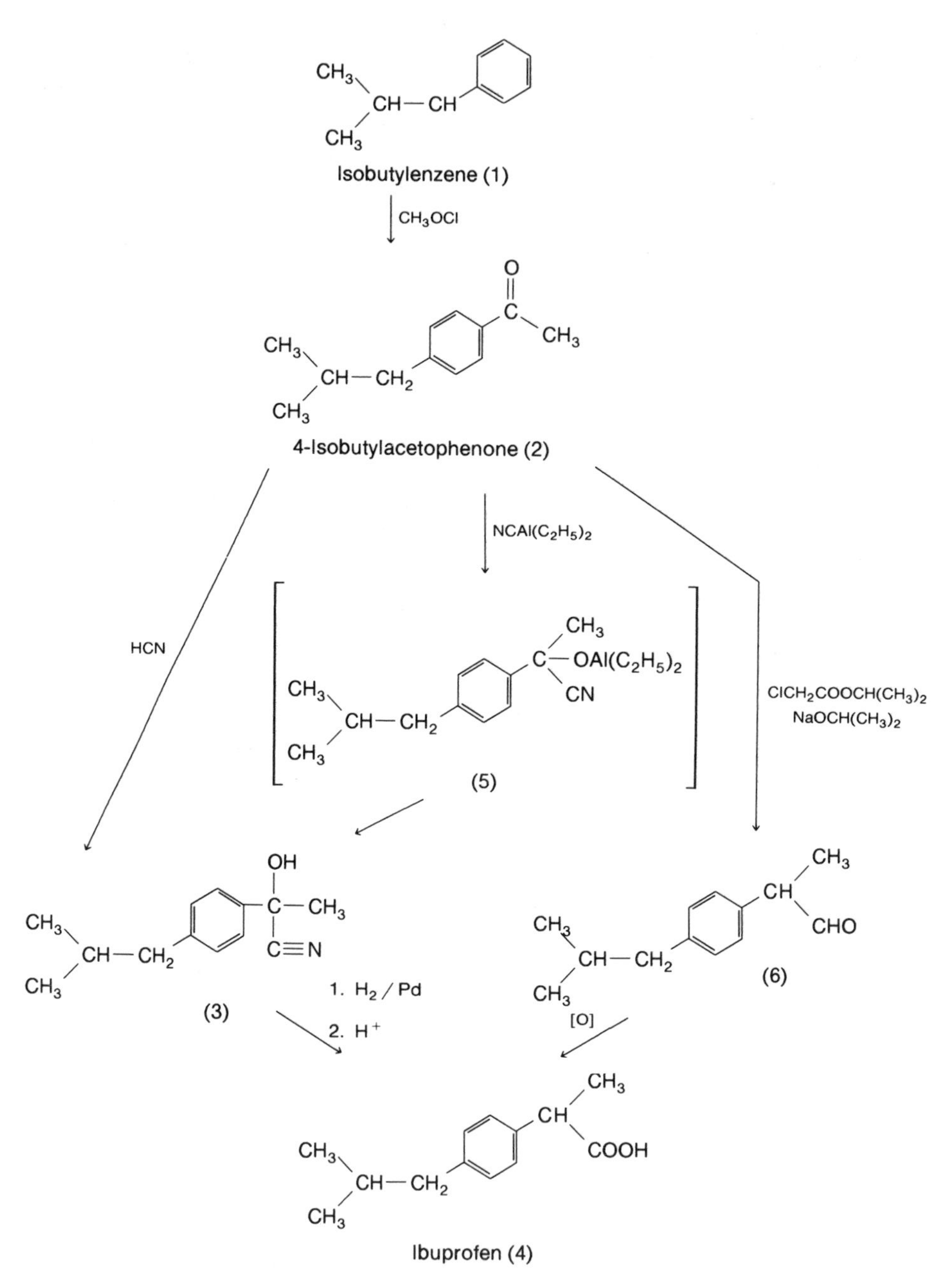

Figure 12.2. *Three syntheses of ibuprofen.*

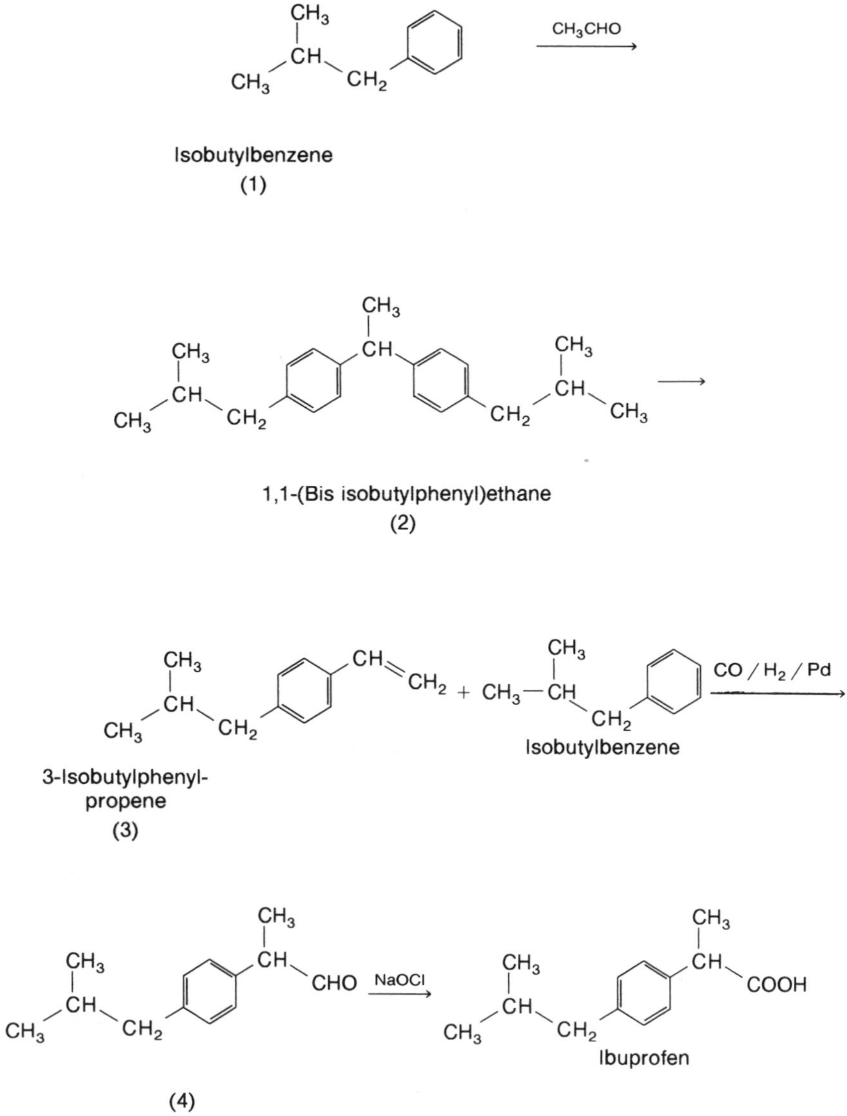

CH_3 | $CH_3-CH-CH_2-C_6H_5$ $\xrightarrow{CH_3CHO}$

Isobutylbenzene
(1)

CH_3 | $CH_3-CH-CH_2-C_6H_4-CH(CH_3)-C_6H_4-CH_2-CH(CH_3)-CH_3$ $\longrightarrow$

1,1-(Bis isobutylphenyl)ethane
(2)

$CH_3-CH(CH_3)-CH_2-C_6H_4-CH{=}CH_2$ + $CH_3-CH(CH_3)-CH_2-C_6H_5$ $\xrightarrow{CO / H_2 / Pd}$

3-Isobutylphenyl-
propene
(3)

Isobutylbenzene

$CH_3-CH(CH_3)-CH_2-C_6H_4-CH(CH_3)-CHO$ $\xrightarrow{NaOCl}$ $CH_3-CH(CH_3)-CH_2-C_6H_4-CH(CH_3)-COOH$

(4)

Ibuprofen

Figure 12.3. *A fourth route to ibuprofen.*

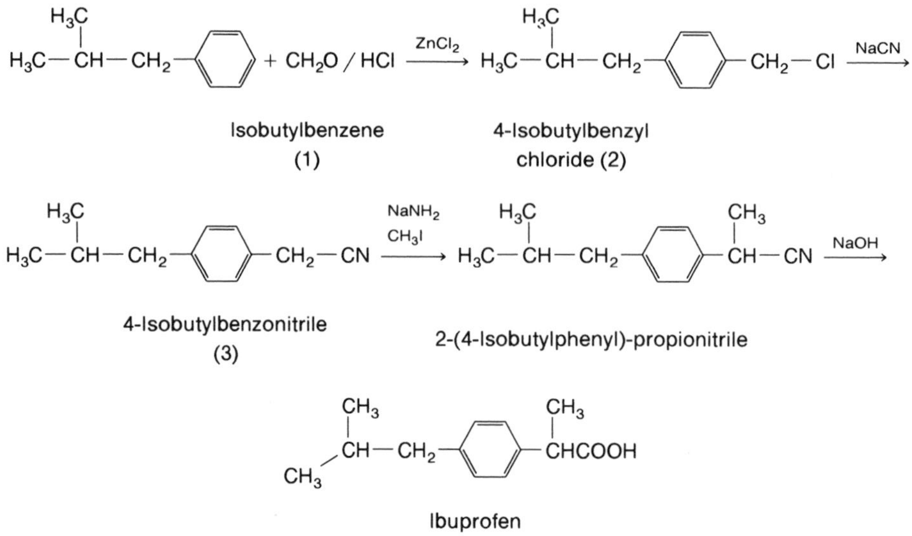

Figure 12.4. A fifth route to ibuprofen.

pyruvate to provide an α-hydroxypropionic acid (3), which can be dehydrated to an acrylic acid (4), which on hydrogenation yields ibuprofen.

The syntheses of ibuprofen are all based on isobutylbenzene, which is also made by a variety of routes. Probably the one used most widely depends on the interaction of toluene with propylene to give not only isobutylbenzene but a number of impurities including *n*-butylbenzene and 2-methylindane, from

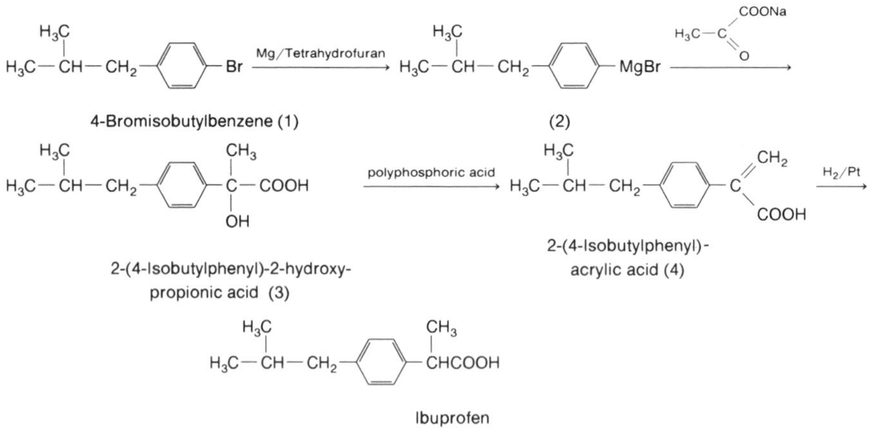

Figure 12.5. A sixth route to ibuprofen.

which the desired compound must be separated:

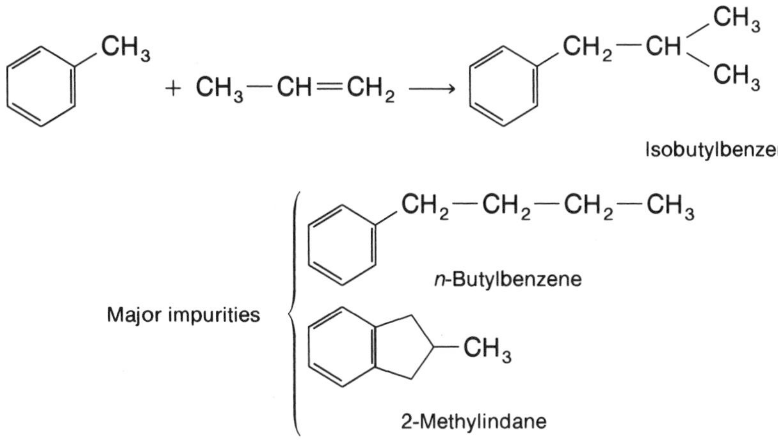

$$C_6H_5CH_3 + CH_3{-}CH{=}CH_2 \longrightarrow C_6H_5CH_2{-}CH(CH_3)_2$$

Isobutylbenzene

Major impurities:

$C_6H_5CH_2{-}CH_2{-}CH_2{-}CH_3$

n-Butylbenzene

2-Methylindane

A conventional Friedel Crafts catalyst would promote alkylation of the ring not the methyl group, and a potassium catalyst is used instead.

A second synthesis involves reaction of methyl benzyl ketone with a methyl Grignard to yield a tertiary carbinol, which on dehydration and hydrogenation gives isobutylbenzene:

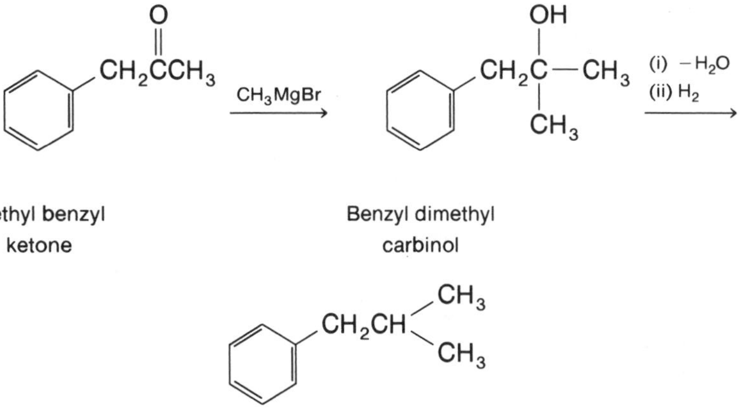

$$C_6H_5CH_2C(=O)CH_3 \xrightarrow{CH_3MgBr} C_6H_5CH_2C(OH)(CH_3){-}CH_3 \xrightarrow[\text{(ii) } H_2]{\text{(i) } -H_2O} C_6H_5CH_2CH(CH_3)_2$$

Methyl benzyl ketone

Benzyl dimethyl carbinol

Isobutylbenzene

In related chemistry, isopropyl phenyl ketone may be hydrogenated to the corresponding alcohol, which on dehydration and hydrogenation yields isobutylbenzene:

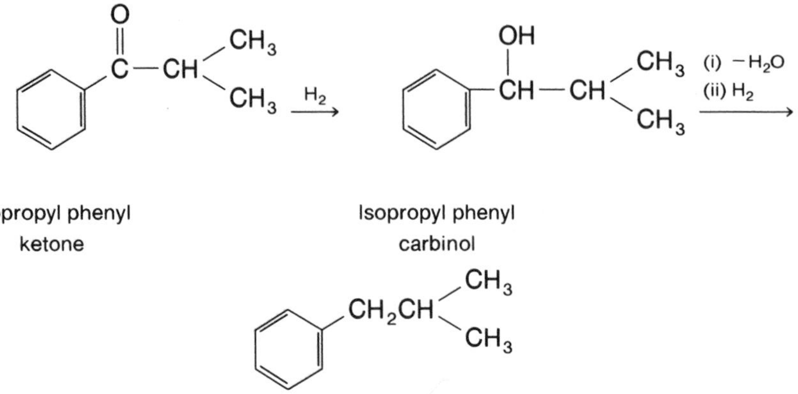

H_2 ; (i) $-H_2O$ (ii) H_2

Isopropyl phenyl ketone

Isopropyl phenyl carbinol

Isobutylbenzene

In yet another synthesis, 1-phenyl-1-propene is carbonylated to the corresponding aldehyde. Hydrogenolysis of the aldehyde group to a methyl group and reduction of the double bond provides isobutylbenzene:

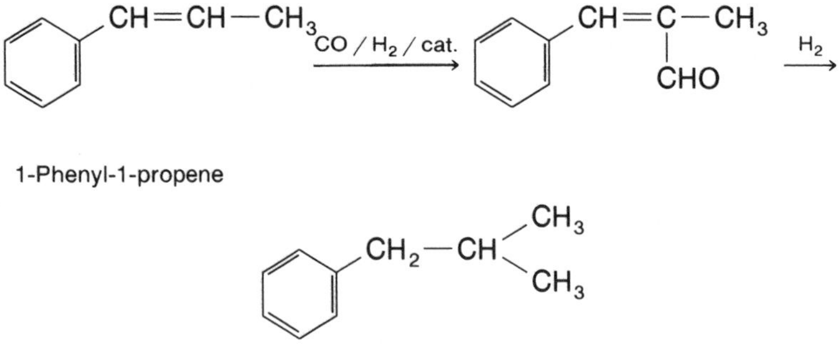

$CO / H_2 / cat.$; H_2

1-Phenyl-1-propene

Isobutylbenzene

12.1.2 Syntheses of Naproxen

Naproxen may also be made by many routes. Indeed, it may be made by most of the methods described for ibuprofen. In one of the earliest, the Willgerodt reaction is used. This synthesis has also been applied to ibuprofen. As shown in Figure 12.6, 2-methoxynaphthalene (1) is acylated to give the methyl ketone (2). Treatment with morpholine and sulfur according to Willgerodt conditions provides the sulfur-containing morpholino compound (3), which, on treatment with acid, gives 6-(methoxy-2-naphthyl)acetic acid (4). Esterification with methanol gives the ester (5). Sodium hydride and methyl iodide insert the

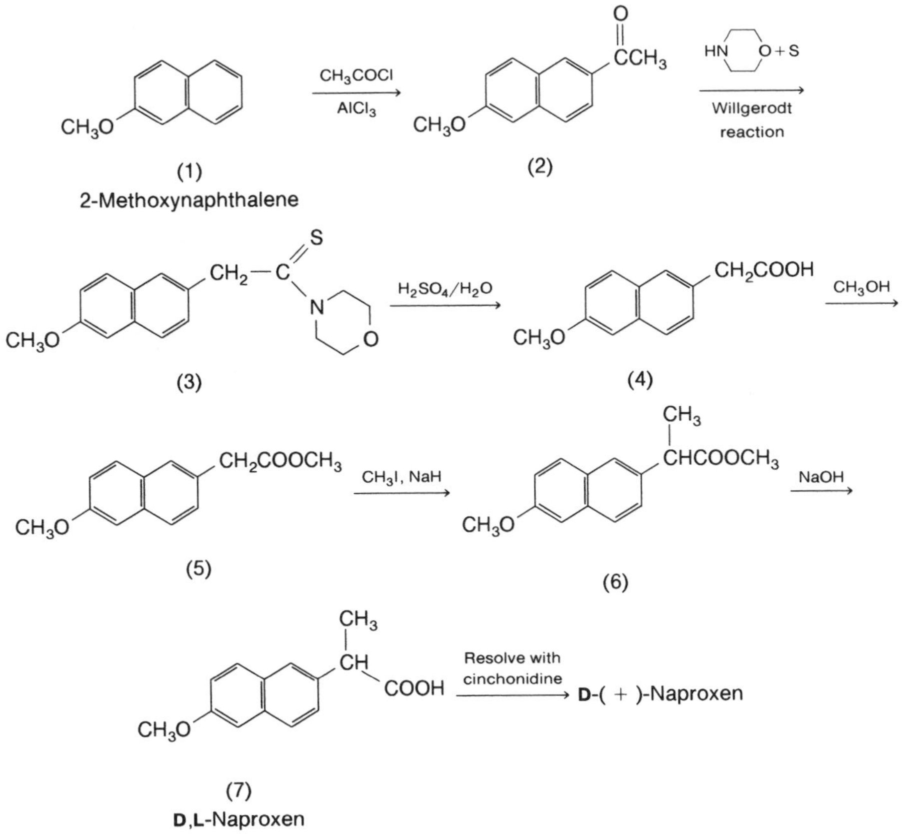

CH_3O (1) 2-Methoxynaphthalene $\xrightarrow[AlCl_3]{CH_3COCl}$ (2) $\xrightarrow[\text{Willgerodt reaction}]{\text{morpholine} + S}$ (3) $\xrightarrow{H_2SO_4/H_2O}$ (4) $\xrightarrow{CH_3OH}$ (5) $\xrightarrow{CH_3I,\ NaH}$ (6) $\xrightarrow{NaOH}$ (7) D,L-Naproxen $\xrightarrow{\text{Resolve with cinchonidine}}$ D-(+)-Naproxen

Figure 12.6. Synthesis of naproxen.

required methyl group to give (6). Saponification gives D,L-naproxen (7), which must be resolved to provide the proper isomer. By contrast, ibuprofen (see above) is used as the racemic mixture.

In another synthesis (Figure 12.7), the methyl ketone (2) from the preceding synthesis, on treatment with HCN, provides a cyanohydrin (8). Yields are higher than in the corresponding route to ibuprofen, shown on the left hand side of Figure 12.2. Dehydration provides an unsaturated nitrile (9), which on hydrolysis gives an unsaturated acid, whose hydrogenation provides D,L-naproxen (7).

Alternatively, (2) may be epoxidized with dimethyl sulfoxide, as shown in the lower part of Figure 12.7. The epoxide (10) isomerizes to an aldehyde (11), which is converted to an oxime (12). This yields the unsaturated nitrile (9) on dehydration.

These syntheses may be carried out with a halogen in the 1-position since this favors β substitution in the initial reaction with acetyl chloride (Figure

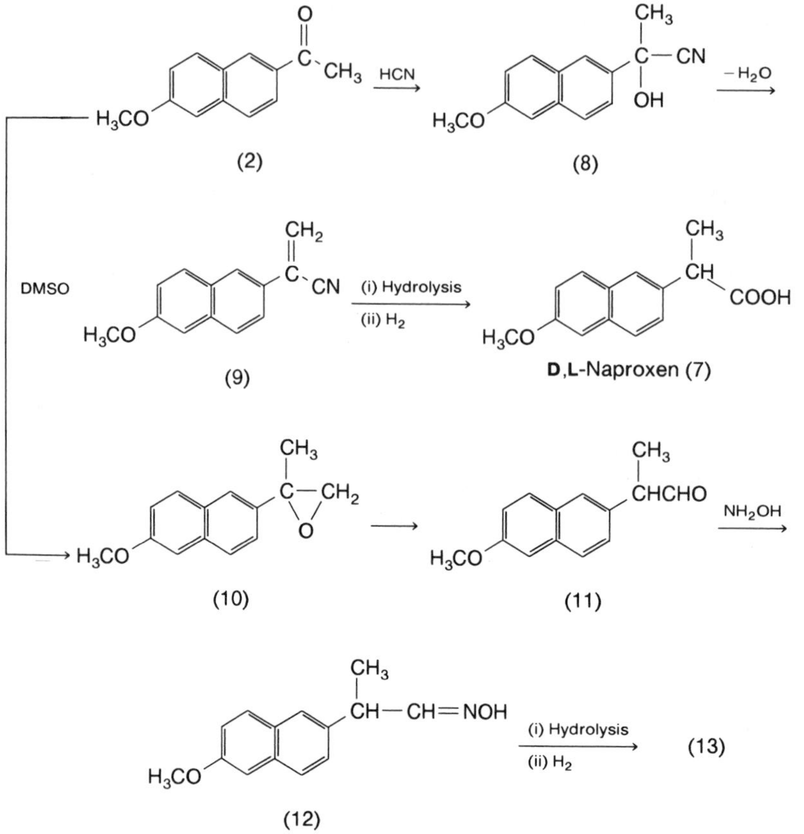

O
C
CH_3
HCN →
H_3CO
(2)
CH_3
C — CN
OH
$-H_2O$ →
H_3CO
(8)

DMSO

CH_2
C
CN
(i) Hydrolysis
(ii) H_2 →
H_3CO
(9)
CH_3
CH
COOH
H_3CO
D,**L**-Naproxen (7)

CH_3
C — CH_2
O
→
H_3CO
(10)
CH_3
CHCHO
NH_2OH →
H_3CO
(11)

CH_3
CH — CH=NOH
(i) Hydrolysis
(ii) H_2 →
(13)
H_3CO
(12)

Figure 12.7. *Second and third routes to naproxen.*

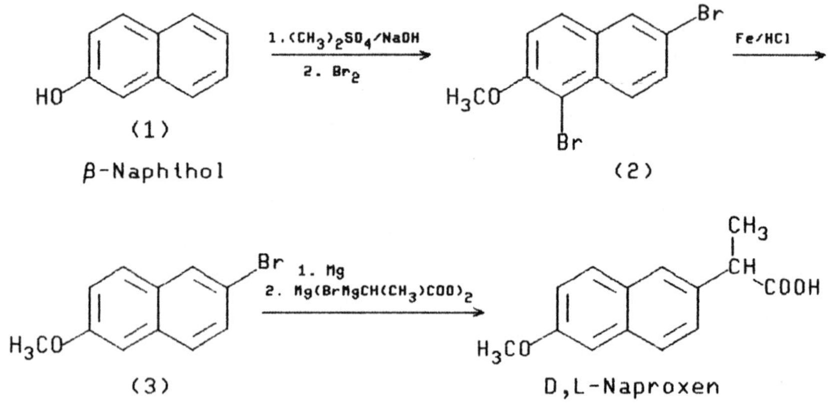

1. $(CH_3)_2SO_4$/NaOH
2. Br_2 →
Br
Fe/HCl →
HO
(1)
β-Naphthol
H_3CO
Br
(2)

Br
1. Mg
2. $Mg(BrMgCH(CH_3)COO)_2$ →
CH_3
CH
COOH
H_3CO
(3)
H_3CO
D,L-Naproxen

Figure 12.8. *A fourth route to naproxen.*

12.6). When the halogen substituent is present, it must be removed by hydrogenolysis, once its directive effect is no longer required.

A fourth synthesis of naproxen is shown in Figure 12.8. β-naphthol (1) is methylated with dimethyl sulfate and then brominated to give (2). Reduction with iron and hydrochloric acid gives 6-methoxy-2-bromonaphthalene (3). This is converted to a Grignard reagent and coupled with the magnesium salt of the Grignard of α-bromopropionic acid in a mixed Grignard coupling. The magnesium salt of naproxen results, which on quenching gives the free acid.

12.1.3 Synthesis of Fenoprofen

A single synthesis of fenoprofen is shown in Figure 12.9. It starts with acetyl substituted diphenyl ether (1). Reduction of the keto group with sodium borohydride gives an alcohol (2), which can be converted to a bromide (3) with phosphorus tribromide. The bromine is in turn replaced with a nitrile group by reaction with sodium cyanide. Hydrolysis of the nitrile (4) gives the carboxyl group required for fenoprofen (5).

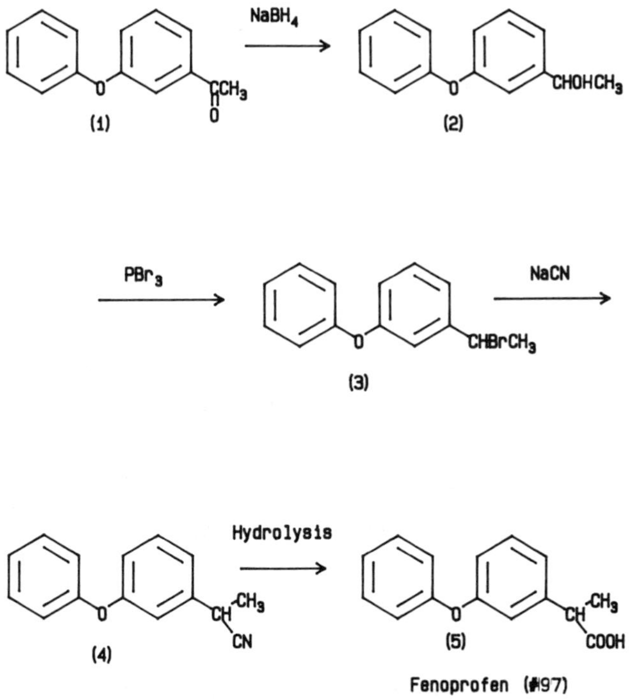

Figure 12.9. Synthesis of fenoprofen.

12.2 SUBSTITUTED ACETIC ACID DERIVATIVES

The next chemical group of anti-inflammatories shown in Figure 12.10 has structures containing a ring attached to an acetic acid residue. Sulindac is #60 and indomethacin #69 in the Top-100; tolmetin is not in the Top-100 but is widely used in the United States. Diclofenac is one of the world's best-selling nonsteroid anti-inflammatories but was only launched in the United States in 1988.

Figure 12.10 shows the synthesis of sulindac. *p*-Fluorobenzyl chloride (1) reacts with diethyl methylmalonate in a conventional malonic ester synthesis to give the acid (2). Polyphosphoric acid catalyzes the cyclization of (2) to the indanone (3). A Reformatskii reaction with zinc amalgam and bromoacetic ester gives the carbinol (4) which is dehydrated with *p*-toluenesulfonic acid to (5). Reaction of (5) with *p*-thiomethylbenzaldehyde (6), which is possible because of the active methylene group in (5), leads, after saponification, to (7). Oxidation of the sulfide to the sulfoxide gives sulindac (8).

Indomethacin is an indole derivative and one of the most potent of the nonsteroidal anti-inflammatories. It has been described as the standard against which the others are measured. Although the drug itself continues to be regarded as safe, a novel dosage form marketed in Britain early in 1983 ran into trouble. The indomethacin was in a slow release form in which water had to diffuse through a semipermeable membrane and expel the drug through a hole drilled by laser in the capsule (Figure 4.8). The preparation was withdrawn after 3 months because of possible toxic effects. In certain people the capsule, having travelled through the stomach, lodged in a fold in the intestine for too long. A high local concentration of drug was reached which perforated the intestine. Here is yet another example of how even a slight change in a drug formulation can alter the action of the drug and possibly render it toxic.

The synthesis of indomethacin is shown in Figure 12.11. It starts with (4-methoxyphenyl)hydrazine hydrochloride (1) and the methyl ester of levulinic acid (2). These condense to give the methyl ester of 2-methyl-5-methoxyindoleacetic acid (3). Saponification gives the free acid (4), which is converted to its tertiary butyl ester (6) by an interesting process that involves the use of *tert*-butanol with dicyclohexylcarbodiimide (5). The amine is then acylated with *p*-chlorobenzoyl chloride (7) to give (8) and the protective tertiary butyl group is removed by pyrolysis to give indomethacin (9). Pyrolysis is used because *N*-acylindoles are unstable to both acids and bases, which might otherwise be used to split off the tertiary butyl group.

12.3 ENOLIC β-DIKETONES

The third group of anti-inflammatories listed in Figure 12.1 is the enolic or potentially enolic *β*-diketones. Phenylbutazone is a pyrazolone and piroxicam is a hydroxybenzothiazine. At first sight, these do not appear to be acids at all,

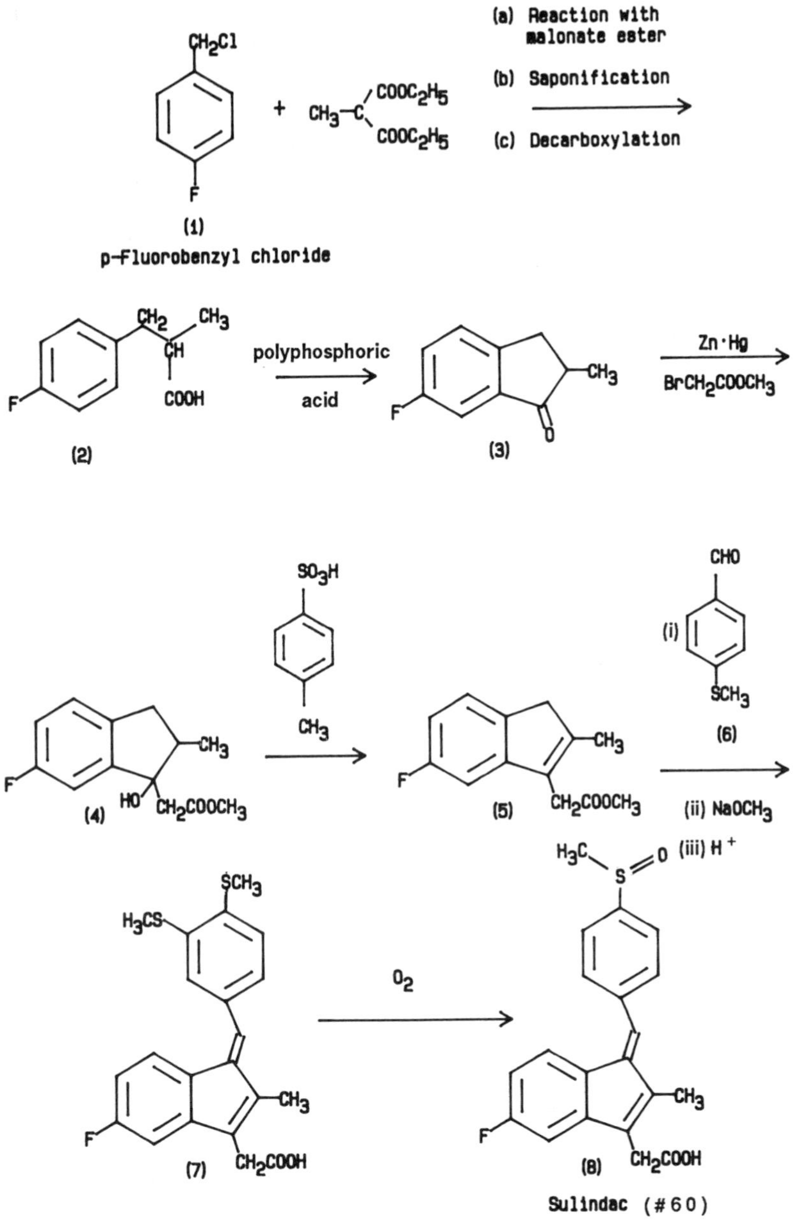

Figure 12.10. Synthesis of sulindac.

but in fact they possess labile protons and have pK_a values between 4 and 7, which makes them comparable with carboxylic acids. The additional aryl groups provide the lipophilicity required for protein binding and tissue distribution. The body eliminates these compounds via the enolic group and the process is slower than with carboxyl groups; hence, they are longer acting. Piroxicam need be given only once daily, the sulfone group providing the additional lipophilicity. The *N*-methyl group is the part of the structure thought to inhibit prostaglandin production and the pyridyl side chain enhances the properties while reducing the acidity of the compound.

The synthesis of piroxicam is shown in Figure 12.12. The sodium salt of saccharin (1), the artificial sweetener, reacts with methyl chloroacetate (2) to give (3). Treatment with sodium methoxide in dimethylsulfoxide expands the

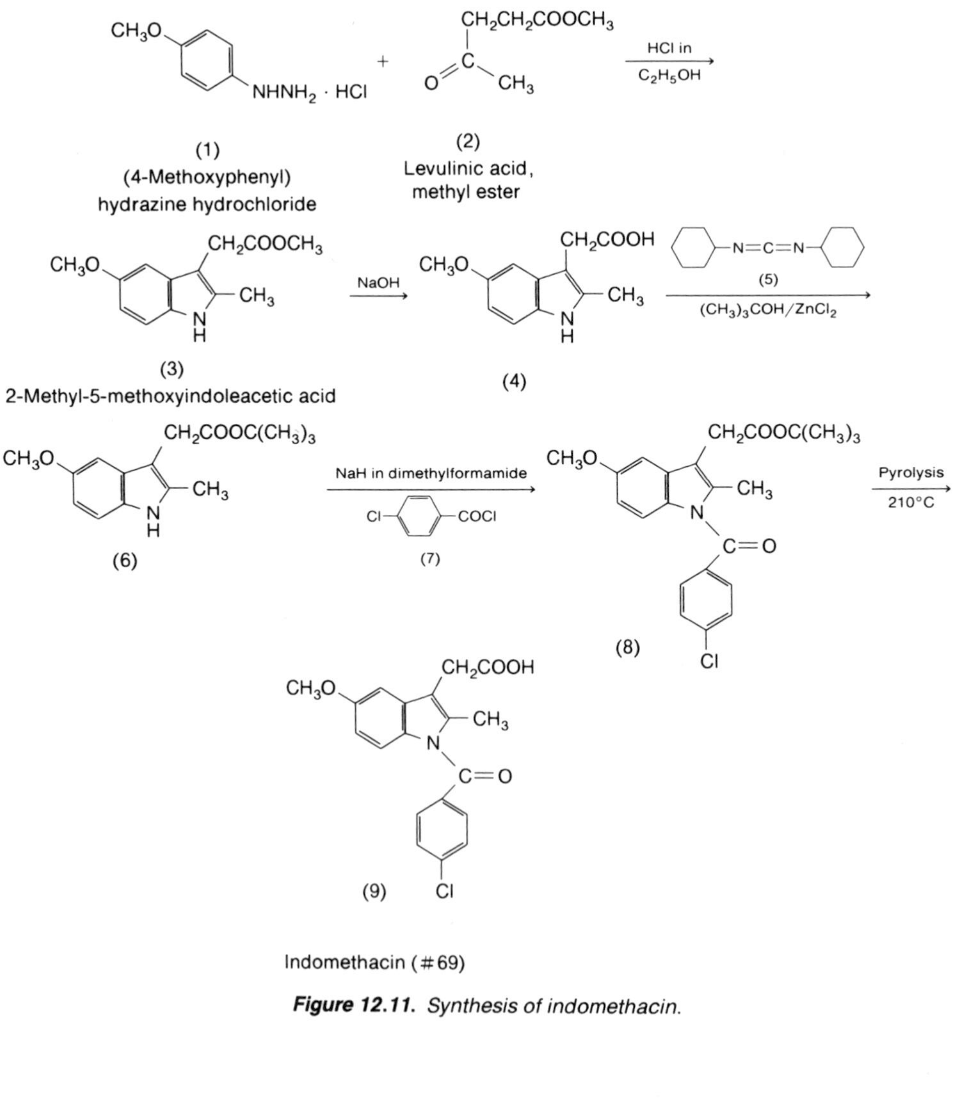

CH_3O NHNH$_2$ · HCl + $CH_2CH_2COOCH_3$ C O CH_3 HCl in C_2H_5OH

(1) (4-Methoxyphenyl) hydrazine hydrochloride

(2) Levulinic acid, methyl ester

CH_3O CH_2COOCH_3 CH_3 N H NaOH CH_3O CH_2COOH CH_3 N H N=C=N (5) $(CH_3)_3COH/ZnCl_2$

(3) 2-Methyl-5-methoxyindoleacetic acid

(4)

CH_3O $CH_2COOC(CH_3)_3$ CH_3 N H NaH in dimethylformamide Cl COCl (7) CH_3O $CH_2COOC(CH_3)_3$ CH_3 N C=O Cl Pyrolysis 210°C

(6)

(8)

CH_3O CH_2COOH CH_3 N C=O Cl

(9)

Indomethacin (#69)

Figure 12.11. *Synthesis of indomethacin.*

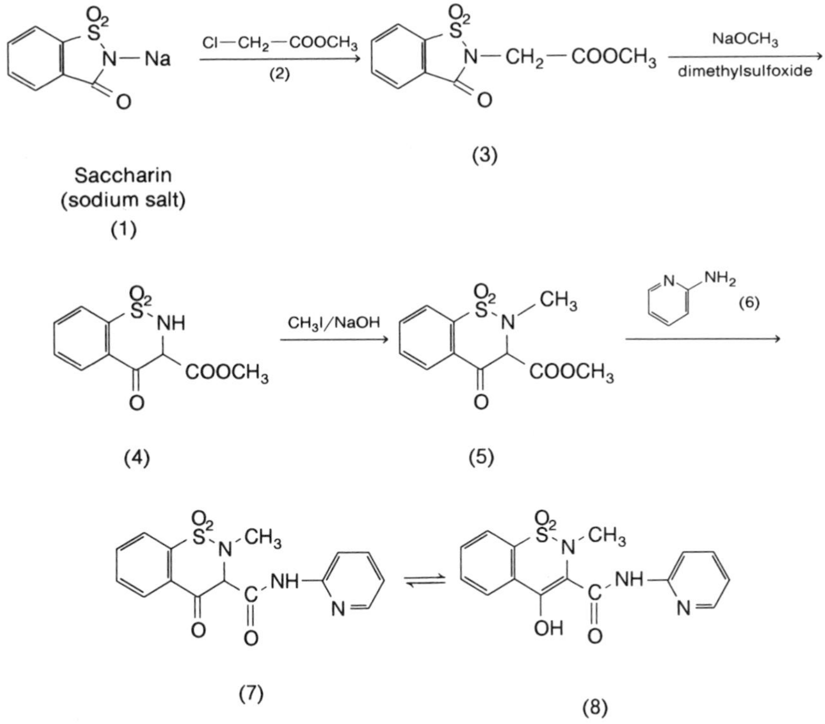

$Cl-CH_2-COOCH_3$ (2)

$NaOCH_3$ / dimethylsulfoxide

Saccharin (sodium salt) (1)

(3)

$CH_3I/NaOH$

(6)

(4)

(5)

(7)

(8)

Piroxicam

Figure 12.12. *Synthesis of piroxicam.*

ring to give the benzothiazine dioxide derivative (4). The nitrogen is methylated with methyl iodide to give (5) and treatment of (5) with 2-aminopyridine (6) gives piroxicam (7). The enolization to (8) is also shown.

Phenylbutazone is potent but often leads to side effects so it is used only when other drugs have failed. It is prepared simply by the condensation of diethyl-*n*-butylmalonate with hydrazobenzene in the presence of a base.

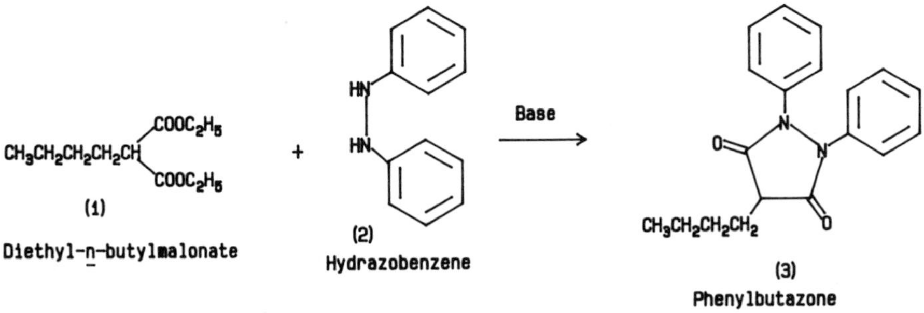

Base

(1) Diethyl-*n*-butylmalonate

(2) Hydrazobenzene

(3) Phenylbutazone

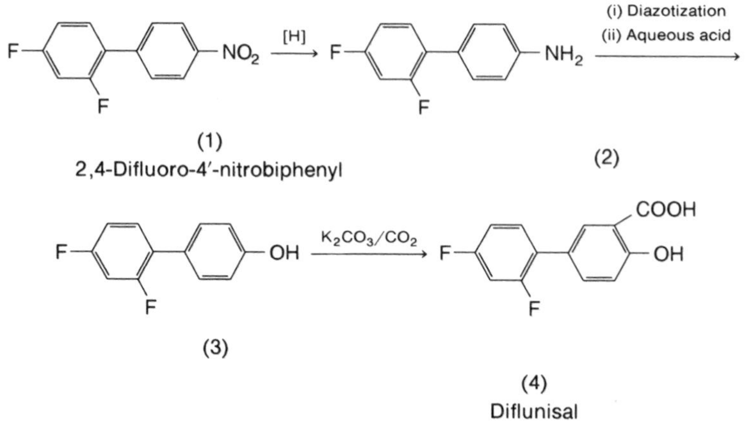

F, F-substituted 4′-nitrobiphenyl (1) $\xrightarrow{[H]}$ amine (2) $\xrightarrow{\text{(i) Diazotization (ii) Aqueous acid}}$ phenol (3) $\xrightarrow{K_2CO_3/CO_2}$ (4)

(1)
2,4-Difluoro-4′-nitrobiphenyl

(2)

(3)

(4)
Diflunisal

Figure 12.13. *Synthesis of Diflunisal.*

12.4 ANTHRANILIC / SALICYLIC ACID DERIVATIVES

The final group of anti-inflammatories in Figure 12.1 has aromatic carboxyl groups. Diflunisal is related to salicylic acid and, hence, to aspirin, and meclofenamic acid to anthranilic acid. The diflunisal synthesis is shown in Figure 12.13. 2,4-Difluoro-4′-nitrobiphenyl (1) is reduced to the amine (2). Diazotization followed by treatment with dilute acid gives the phenol (3). A carboxyl group is inserted with potassium carbonate and carbon dioxide to give diflunisal (4). This last step is reminiscent of the Kolbe reaction used for aspirin (Figure 10.11).

Meclofenamic acid is synthesized by reaction of potassium *o*-bromobenzoate with 2,5-dichloro-3-methylaniline:

Potassium *o*-bromobenzoate + 2,5-Dichloro-3-methylaniline ⟶ Meclofenamic acid

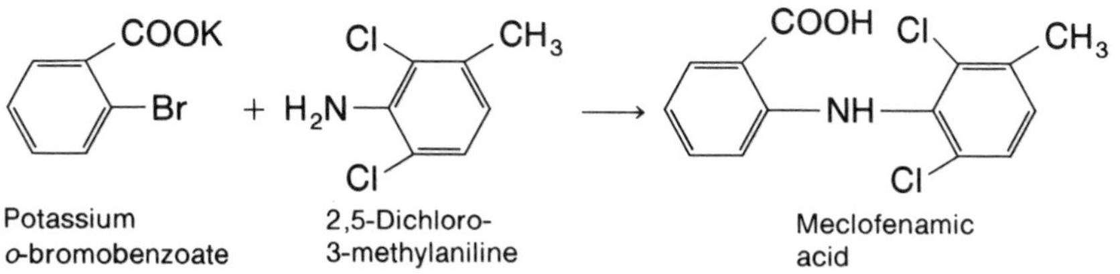

The reaction presents some difficulties because of steric hindrance from the chlorine adjacent to the amine group. It goes in the presence of *N*-ethylmorpholine, dimethylglyoxime, and cupric bromide.

12.5 OTHER ANTI-INFLAMMATORIES

The nonsteroidal anti-inflammatories reduce inflammation from arthritis and diminish pain but cause no remission of the disease. There are a few remission-inducing drugs, all of significant toxicity. Certain antimalarial agents (chloroquine and hydroxychloroquine, Section 18.2.1) are helpful as second-line therapy, and injections of gold compounds have been used for many years. The preferred compounds are aurothioglucose and gold sodium thiomalate. An interesting development is the appearance of auranofin, which is the first oral gold compound. For long-term therapy, the development of a compound that does not need to be injected must be seen as a breakthrough.

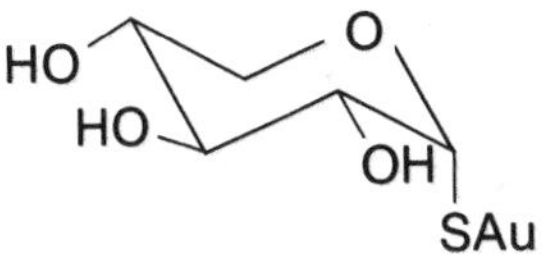

Aurothioglucose

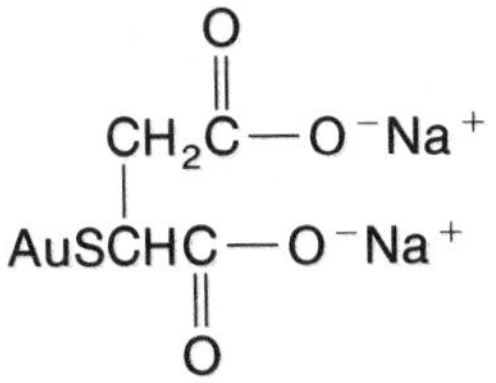

Gold sodium thiomalate

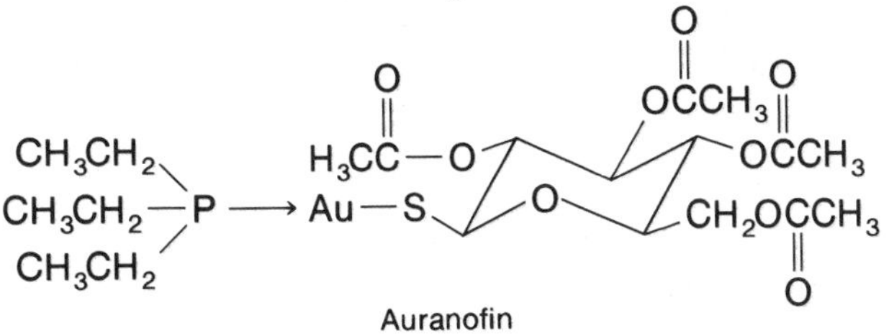

Auranofin

12.6 CONCLUSION

The nonsteroid anti-inflammatories have been tremendously helpful in short-term treatment of illnesses such as bursitis or inflammations of the cartilage, and many of them have been used illegally to prepare slightly injured horses for races. In the long-term treatment of rheumatic illness, however, they have been less successful. They were all developed in the hope that they would cause less stomach irritation than aspirin and to some extent they do. Nonetheless, all of them have side effects usually connected with stomach ulceration and bleeding and, because they are acidic and are taken in high doses over a long period, even minor side effects can be troublesome.

At the last count, there were more than twenty nonsteroid anti-inflammatories on the market for the relief of arthritic pain, and new ones are continually being introduced. Examples are etodolac, a substituted acetic acid, and lobenzarit, an anthranilic acid derivative:

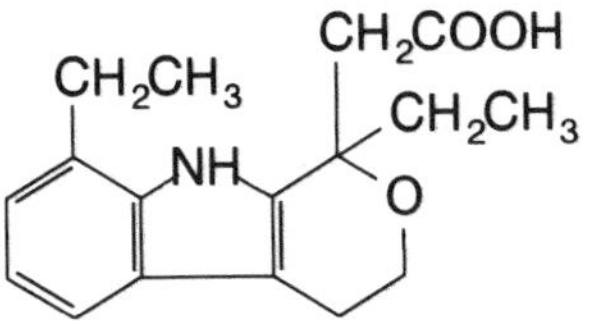

Etodalac

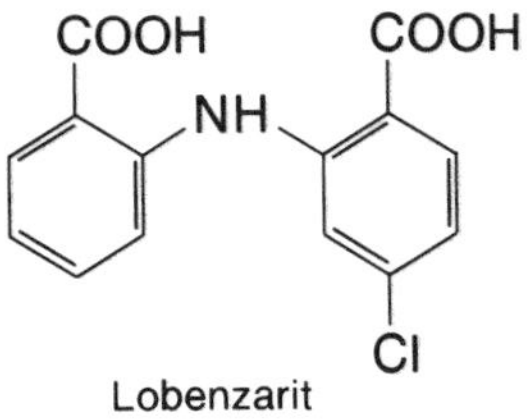

Lobenzarit

Nambumetone is a new pro-drug of some interest. It is metabolized extensively on the first pass through the liver to 6-methoxy-2-naphthylacetic acid:

$$CH_3O\text{-}C_{10}H_6\text{-}CH_2CH_2\overset{O}{\overset{\|}{C}}CH_3 \longrightarrow CH_3O\text{-}C_{10}H_6\text{-}CH_2COOH$$

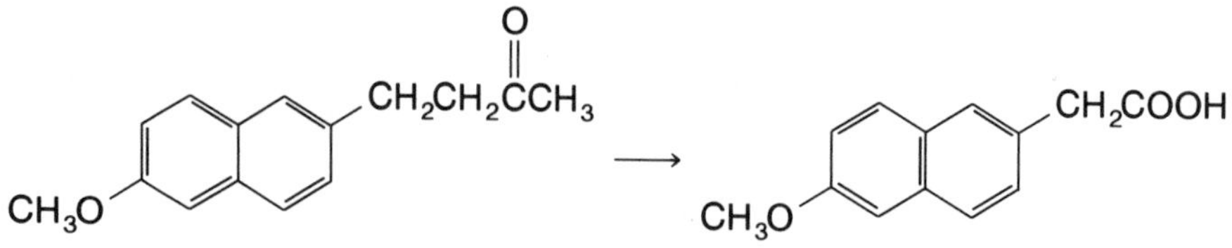

Nambumetone — 6-Methoxy-2-naphthylacetic acid

The latter compound, although closely related to naproxen, is a substituted acetic acid and is the active drug. Nambumetone is only weakly acidic and a weak inhibitor of prostaglandins; hence, it is hoped that it will not cause the stomach problems familiar with the other nonsteroid anti-inflammatories.

The profusion of nonsteroid anti-inflammatories reflects the fact that not one is truly satisfactory, and there is not much difference between any of them, although personal reactions may differ. On the other hand, the manufacturers' point of view is understandable. The U.S. market for nonsteroid anti-inflammatories in 1986 was over a billion dollars. With so many people taking so many pills for so long a time, even a "me-too" drug that can win a 1 or 2% share of the market is economically attractive. There is always the hope, in addition, that the next compound will turn out to be more effective and to have fewer side effects.

13

Hypoglycemics

Hypoglycemics are a group of drugs that lower the level of sugar in the blood. Four such materials appear in the Top-100—insulin (#39), glyburide (#57), chlorpropamide (#71), and glipizide (#98). They are all used in the treatment of diabetes. Diabetes is the common abbreviation for diabetes mellitus, an illness characterized by a deficiency or reduced effectiveness of insulin in the body. Insulin is a hormone produced by glands in the pancreas. It promotes the "burning" of sugar in the tissues to provide the body with energy.

```
                    NH2    ┌──────S-S──────┐                  NH2          NH2          NH2
                     |     |               |                   |            |            |
Gly-Ile-Val-Glu-Glu-Cy-Cy-Ala-Ser-Val-Cy-Ser-Leu-Tyr-Glu-Leu-Glu-Asp-Tyr-Cy-Asp
                         |                                                  |
                         |                                                  S
                         S                                                 /
   NH2 NH2               |                                                S
    |   |                |                                                |
Phe-Val-Asp-Glu-His-Leu-Cy-Gly-Ser-His-Leu-Val-Glu-Ala-Leu-Tyr-Leu-Val-Cy
                                                                          |
                         Ala-Lys-Pro-Thr-Tyr-Phe-Phe-Gly-Arg-Glu-Gly
```

Human insulin

Two forms of diabetes are recognized. Early-onset diabetes occurs before the age of 20, is often accompanied by weight loss and apathy, and is due to failure of insulin production. Nonjuvenile diabetes usually occurs after age 40; the patient suffers from obesity, and the problem is not insulin production but insulin unavailability because it is in some way bound up so that it is not active.

In either case, levels of glucose in the blood rise to cause a condition known as hyperglycemia. The kidneys cannot cope; the urine contains sugar and is produced in larger amounts. Because the diabetic cannot use blood sugar for energy, fat is drawn on instead. It is carried to the liver but, instead of being converted to glucose or glycogen, it is turned into two ketones, acetone and acetoacetic acid, and an acid, β-hydroxybutyric acid. These disrupt the sensitive acid–base balance in the body. The condition is called ketosis and can lead to coma and death if untreated.

Diabetes may result from stresses or infections, particularly if they lead to deterioration of the pancreas. The main predisposing factor, however, is hereditary and ethnic Jews are especially susceptible to the disease.

13.1 REPLACEMENT THERAPY

The original approach to diabetes was replacement therapy. It was discovered by Banting and Best in 1921 that diabetics could be helped by injection of insulin from slaughtered cows and pigs. The insulin had to be injected because it is a protein and is denatured in the digestive tract if taken by mouth. Diabetics had to be trained carefully to give themselves injections several times a day. In spite of the inconvenience of this, it did offer diabetics the chance for a reasonably normal life. The situation could also be helped by restricting intake of sugar, by exercise, and by control of weight and smoking. Insulin therapy is most effective in early-onset diabetes.

Injected insulin is rapidly destroyed by an enzyme, insulase, in the liver and there has been much research into the production of a long-acting insulin. For example, insulin can be combined with a protein such as protamine from fish sperm. This forms an insoluble complex, from which the insulin is slowly released after injection. Addition of a small amount of zinc reduces the number of allergic reactions. Neutralization of the insulin, which is an acid, is said to give a product that takes effect more quickly and acts longer.

A significant recent advance, however, has come about as a result of chromatographic techniques. These have made it possible to remove trace impurities from natural insulin and produce a pure insulin with a greatly reduced chance of antibody formation and allergic reaction.

In spite of the availability of human insulin (see below), the majority of prescriptions are still for cow or pig insulin. In theory, human insulin carries the smallest risk of provoking an immune response, but diabetics who can already control their illness with a particular cow or pig product need not change to something new.

13.2 INSULIN VIA RECOMBINANT DNA

Impressive though the chromatographic work was, it has been overtaken by one of the most dramatic breakthroughs in technology in our generation. Chemical plants now manufacture human insulin by the use of recombinant

DNA. This technique involves the transplantation of genetic material from a higher species, that is, plant or animal cells, into a lower species, such as bacteria or yeasts.

All living organisms contain deoxyribonucleic acid or DNA molecules. These DNA molecules are specific to each species and they carry information that instructs each cell on how and when to produce the substances it needs to live. This is the genetic code of life.

The DNA molecule resembles a spiral ladder, the double helix. The sides are formed of sugar and phosphates. The rungs are formed from pairs of four chemical bases known as nucleotides. The four nucleotides—adenine, cytosine, guanine, and thymine—are arranged in sequences of three along each strand of DNA. Each triplet contains the information needed to produce a given amino acid and the amino acids hook together to form a protein. Thus, the sequence of bases tells the cell what protein to make. Triplet codes exist for each of the 20 amino acids that living cells use to make proteins. This protein-making section of DNA is known as the structural gene.

The genetic information coded into the gene is transcribed onto another similar molecule—messenger ribonucleic acid, known as mRNA. It acts as a template to duplicate the amino acid specified by the structural gene. This is the initiation of protein synthesis. It is a natural process that takes place in every living organism. The aim of genetic manipulation is to alter the natural process of protein synthesis by cells so that they produce a desired protein in larger amounts and at a higher purity.

Early attempts at genetic engineering involved heat treatment or irradiation of genes in the hope that a useful mutant would be produced, but the modern technique involves direct manipulation. The major "tools" are the restriction enzymes, which in effect cut open genetic material, and plasmids, which are small DNA molecules found in lower organisms.

The process is illustrated in Figure 13.1. The first stage in gene splicing, as it is called, is to identify the nucleotide sequence for producing a required protein in the DNA strand of a higher species. Once the appropriate DNA strand has been identified, it is then removed from the DNA by the use of a restriction enzyme. Such enzymes, also known as endonucleases, are specific to the nucleotide sequence required.

The vector DNA molecule of a lower organism, that is, a plasmid, is then cut open, again by restriction enzymes, and the new genetic fragment inserted into it, aided by an annealing enzyme. The new plasmid is the recombinant DNA molecule. It is replaced in the lower organism with the foreign gene included. The lower organism then will produce the required protein and the modified gene will replicate itself during the countless generations of the organism.

A major problem is the selection of the appropriate gene. One way to do this is to use restriction enzymes to produce numerous DNA fragments and then to clone them indiscriminately in bacteria. The clones are then screened for their ability to make the correct protein. The method is time-consuming

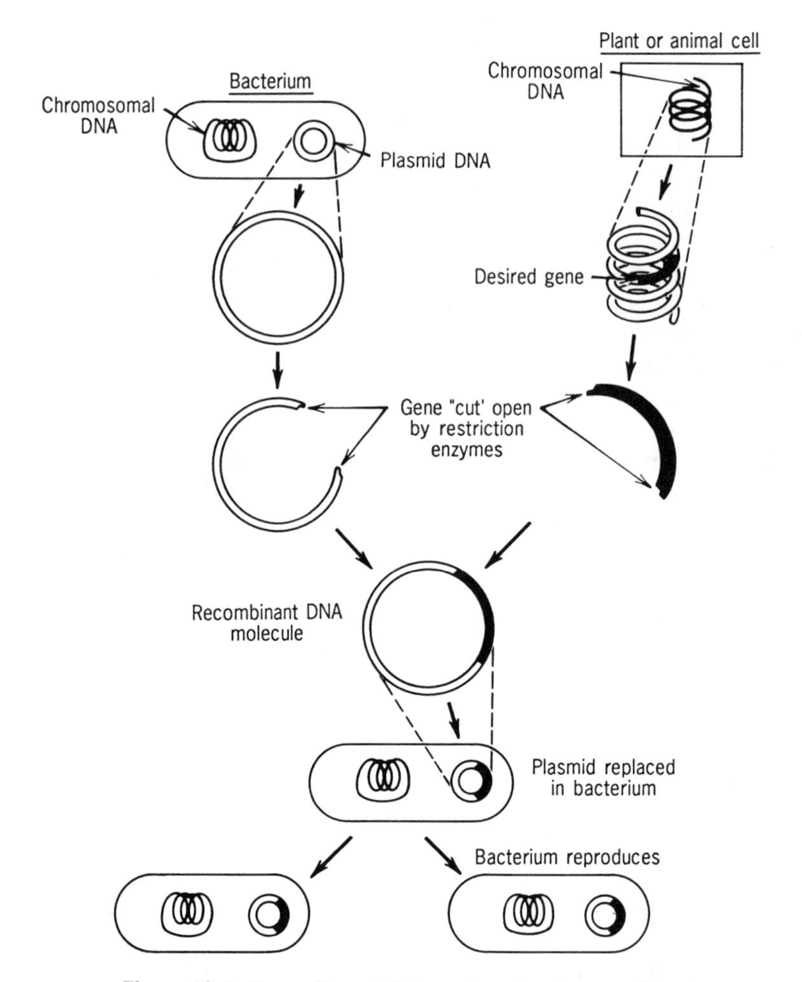

Figure 13.1. *Recombinant DNA engineering (gene splicing).*

and impractical, but was the way in which Charles Weissmann developed a bacterium that would generate human interferon.

The alternative method—the one used for insulin—is the synthesis of DNA by the action of the enzyme reverse transcriptase on isolated mRNA. This exploits the fact that all cells contain genes for the production of all proteins, but only specific cells contain the mRNA required for the production of the desired protein in high yield. The selected mRNA is collected and then exposed to reverse transcriptase, which converts it back to DNA. The resultant DNA is then cloned in bacteria. A regulatory mechanism called the lac operon is also implanted, which enables the insulin genes to be switched on and off.

Most of the work on gene splicing has involved the bacterium *Escherichia coli* because its genetics and metabolism are well known. Early doubts were expressed about the potential hazards if a genetically altered organism were to escape into the environment and the strictest possible safety precautions were enforced. It appears, however, that the strains of modified bacteria produced under such closely controlled conditions are very weak and would not survive in an alien environment.

In the attempts to make insulin, there turned out to be another problem to be faced as well as the production of the correct recombinant DNA. Insulin degrades very rapidly inside a living *E. coli* cell. One solution to this has been the stabilization of the insulin molecule by creation of a recombinant DNA that produces a protein consisting of insulin bound to an *E. coli* enzyme such as β-galactosidase. The relatively stable protein is isolated, purified, and cleaved chemically. The insulin A and B chains are released and they combine chemically to form the intact insulin molecule.

The new biotechnology aroused tremendous excitement and there are prospects for its use for the manufacture of interferon (Chapters 20 and 21), vaccines, single-cell protein, growth hormone, amino acids, and almost every other process where proteins are required (see bibliography).

The technology also created numerous ethical and legal problems. The dangers of the escape of supermicroorganisms now appear to have been overestimated, but there is still the problem of who owns the organisms. Can a life-form be patented? Can the companies who have invested huge sums of money in gene splicing get their money back by patenting the organisms they have developed? The U.S. Supreme Court ruled in 1980 by a narrow 5 to 4 majority that they could, but the decision merely determined what Congress had meant when it passed the patent laws and did not grant any patents or even state which discoveries might be patentable.

In 1987 the Patent and Trademark Office ruled in principle that animals could be patented and gave a favorable ruling on oysters. In April 1988 they granted a patent on a genetically engineered mouse. There are now hundreds of pending applications for patents for life-forms and in addition an application to patent the basic techniques used in gene splicing. Hundreds of companies are entering this expensive and speculative field. Though not a chemical process, the manufacture of drugs by biotechnological processes is of tremendous potential importance to the pharmaceutical industry.

13.3 ORAL HYPOGLYCEMICS

After a story like recombinant DNA, the remaining hypoglycemics must come as an anticlimax. Nonetheless, these drugs make life more pleasant for millions of people. It is unpleasant for diabetics to give themselves regular injections and an oral hypoglycemic was a research aim for many years. At present about 75% of diabetics can manage on such drugs.

Chlorpropamide (#71) appeared in 1958. It is useful for nonjuvenile diabetes. It is a sulfonylurea similar in structure to the sulfonamide antibacterials (Section 6.2), but it has no antibacterial properties.

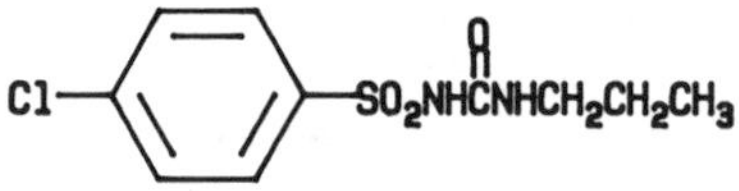

Chlorpropamide

It is believed to act by stimulating the pancreas to produce more insulin. Tolbutamide and tolazamide (#116) are structurally close to chlorpropamide. Tolbutamide appeared in Germany in the mid 1950s and in the United States in 1959:

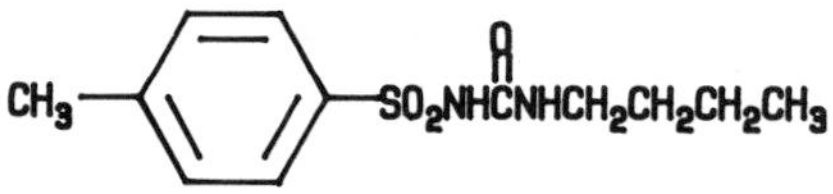

Tolbutamide

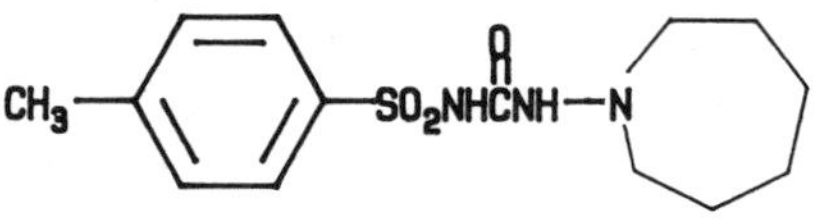

Tolazamide

Of the three first-generation drugs, chlorpropamide had the advantage of requiring only one dose daily. This advantage is shared by the two second-generation sulfonylureas, glyburide (#57) and glipizide (#98):

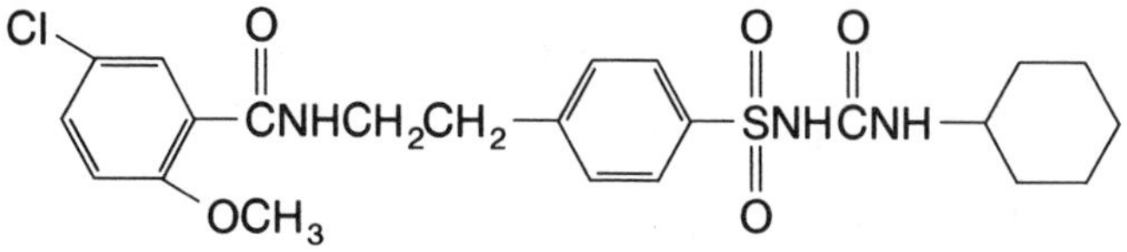

Glyburide

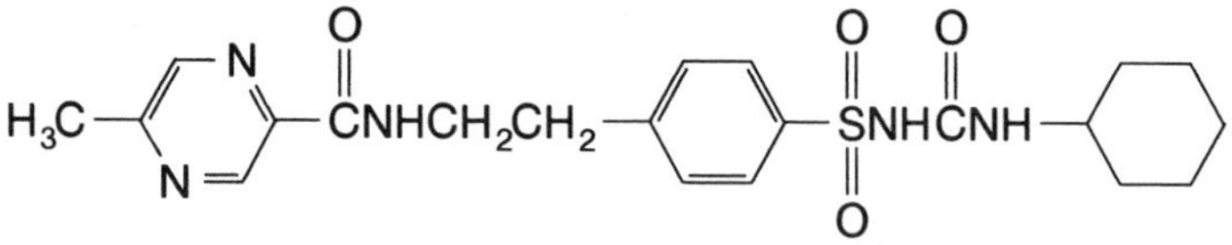

Glipizide

Glipizide has more structural features than most drugs—pyrazine, phenyl, and cyclohexyl rings, together with amide, sulfonamide, and urea linkages.

The new compounds also have the advantage of being more potent; that is, a smaller dose need be taken. This is seen as a benefit, but the difference is not large and chlorpropamide is still a widely used drug.

The oral hypoglycemics are the subject of medical controversy. A major long-term study in the United States has indicated that they are no more effective than careful diet in prolonging the lives of diabetics, and there is some evidence that they lead to heart attacks. It is hard to believe that a group of drugs used so successfully over so many years is without value. While there is some disagreement with these results, meanwhile, there is increased caution in using the drugs and an inclination to restrict them to patients who have problems with insulin, weight reduction, and dietary control.

13.3.1 Syntheses of Oral Hypoglycemics

The synthesis of chlorpropamide is shown in Figure 13.2. Chlorobenzenesulfonamide (1) is reacted with ethyl chloroformate to give a carbamate (2), which, in turn, reacts with propylamine to give chlorpropamide (3). The synthesis of tolbutamide is similar to that of chlorpropamide.

The synthesis of glyburide, shown in Figure 13.3, starts with the acetamide of β-phenylethylamine (1). Chlorosulfonic acid gives the sulfonyl chloride (2), which reacts with ammonia to give the sulfonamide (3). Deacylation gives (4). Treatment with the salicylic acid derivative (5) in the presence of cyanamide gives another amide (6). Finally, condensation with cyclohexyl isocyanate (7) gives glyburide (8).

The synthesis of glipizide, also shown in Figure 13.3, involves similar reagents, but the starting material is 5-methylpyrazine-2-carboxylic acid (9). Treatment with the sulfonamide (4) gives the amide (10), and reaction of this

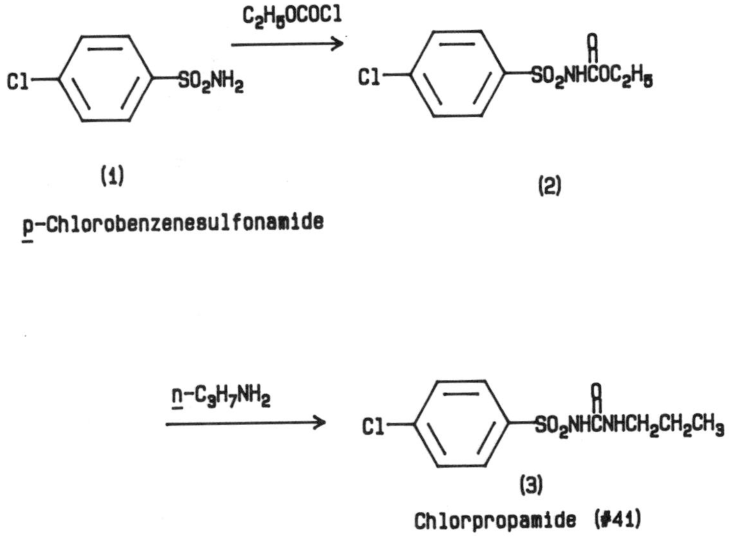

C_2H_5OCOCl

Cl–C_6H_4–SO_2NH_2 (1) p-Chlorobenzenesulfonamide → Cl–C_6H_4–$SO_2NHCOC_2H_5$ (2)

n-$C_3H_7NH_2$ → Cl–C_6H_4–$SO_2NHCNHCH_2CH_2CH_3$ (3) Chlorpropamide (#41)

Figure 13.2. Synthesis of chlorpropamide.

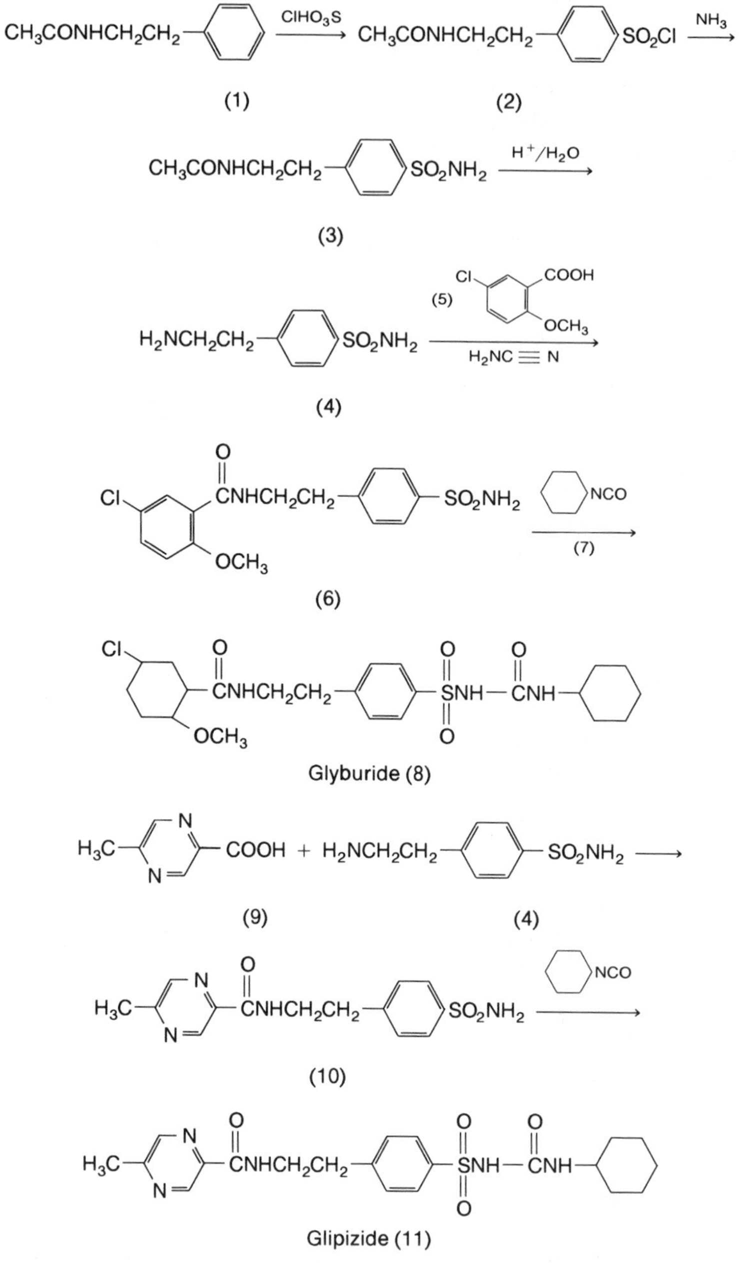

$CH_3CONHCH_2CH_2$–C_6H_5 (1) $\xrightarrow{ClHO_3S}$ $CH_3CONHCH_2CH_2$–C_6H_4–SO_2Cl (2) $\xrightarrow{NH_3}$

$CH_3CONHCH_2CH_2$–C_6H_4–SO_2NH_2 (3) $\xrightarrow{H^+/H_2O}$

$H_2NCH_2CH_2$–C_6H_4–SO_2NH_2 (4) $\xrightarrow[H_2NC\equiv N]{(5)\ Cl, COOH, OCH_3\text{-benzene}}$

(6) Cl, OCH_3–C_6H_3–$\overset{O}{\parallel}{C}NHCH_2CH_2$–$C_6H_4$–$SO_2NH_2$ $\xrightarrow[(7)]{C_6H_{11}NCO}$

Glyburide (8)

H_3C–pyrazine–COOH (9) + $H_2NCH_2CH_2$–C_6H_4–SO_2NH_2 (4) $\longrightarrow$

H_3C–pyrazine–$\overset{O}{\parallel}{C}NHCH_2CH_2$–$C_6H_4$–$SO_2NH_2$ (10) $\xrightarrow{C_6H_{11}NCO}$

Glipizide (11)

Figure 13.3. *Synthesis of glyburide and glipizide.*

with cyclohexyl isocyanate in acetone in the presence of a base gives glipizide (11).

13.4 RECENT DEVELOPMENTS

Much current research is devoted to the development of better dosage forms for insulin. Portable insulin infusion pumps that provide constant, smooth control of glucose levels have been developed and are being evaluated clinically. Patients must be willing and able to monitor their own blood glucose at home. For acute cases, there is a bedside device that automatically monitors glucose levels and injects the appropriate amount of insulin. Infection at the site of needle insertion is the most common problem with these pumps. Mechanical failure of pump, battery, or tubing can also lead to ketosis. The idea of an implantable artificial pancreas has also been studied, but further miniaturization of components is required before it is a possibility.

Attempts are being made to develop an oral insulin in which the dose is covered by an enteric coating that keeps it "safe" until it has passed the digestive enzymes. There are also experiments on an aerosol in which insulin is sprayed up the nose and absorbed by the nasal membranes. A large dose is required and nasal irritation occurred in most patients.

The first truly novel antidiabetic drug to reach the market for some years is acarbose. It is an α-glucosidase inhibitor in the gastrointestinal tract. α-Glucosidase is one of the enzymes that splits sucrose to glucose and fructose. Hence, acarbose inhibits absorption of carbohydrates, reduces peaks of blood sugar levels after meals, and prevents wide fluctuation in those levels. Its obvious use is as an adjunct to correct diet but it may also be a help to formerly obese patients in helping them maintain their new, lower weight.

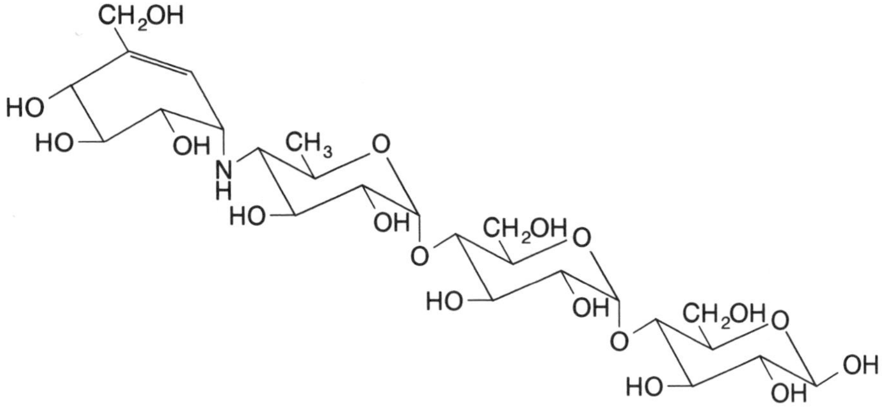

Acarbose

There is also some interesting work on diabetic complications. When the blood sugar concentration in diabetics is high, an enzyme aldose–reductase converts glucose to sorbitol:

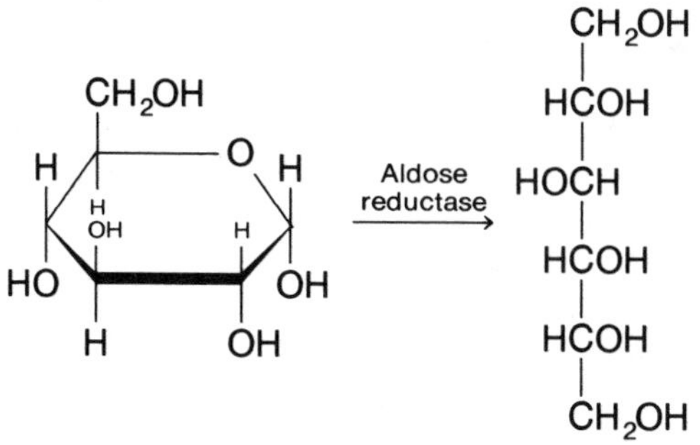

Glucose —Aldose reductase→ Sorbitol

In sufferers, of whom there are about nine million worldwide, sorbitol is linked with neural eye and kidney complications. Statil is one of a number of aldose–reductase inhibitors that will block this reaction and are undergoing trials.

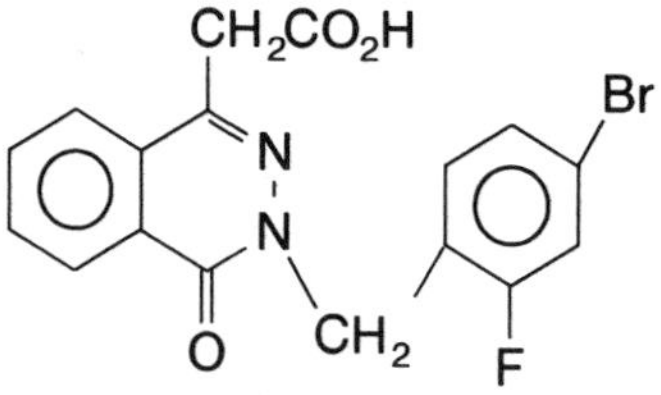

Statil

14

Anticholinergic Drugs

Anticholinergic drugs are generally used for disorders of the bowel. They are often used in drug combinations, which means that their place in the Top-100 is difficult to assess. Atropine occurs at #62 because it is a minor constituent of the antidiarrhea drug Lomotil (Section 10.1.1) and a major constituent of Donnatal. Donnatal contains two other anticholinergic drugs—hyoscyamine sulfate and hyoscine hydrobromide neither of which was listed in Table 5.1A —together with phenobarbital (Section 8.1.1), used here as an anxiolytic. Librax contains the anticholinergic clidinium bromide (#126) with the anxiolytic chlordiazepoxide (Section 8.3.1). Dicyclomine occurs as a single chemical entity at #119. All these drugs are antispasmodics; that is, they counter sudden contractions of the stomach.

They are of particular interest because all the drugs affecting the nervous system discussed so far have affected the adrenergic nervous system. The adrenergic nervous system contains receptors (Section 7.5.4) which are stimulated by the chemical transmitter norepinephrine and can be stimulated or blocked by appropriate drugs. It is normally concerned with the expenditure of energy. The anticholinergic drugs, in contrast, affect the cholinergic nervous system. As Figure 7.12 indicated, this is concerned with the conservation and restoration of energy.

The neurotransmitter is acetylcholine, shown below as the hydrochloride:

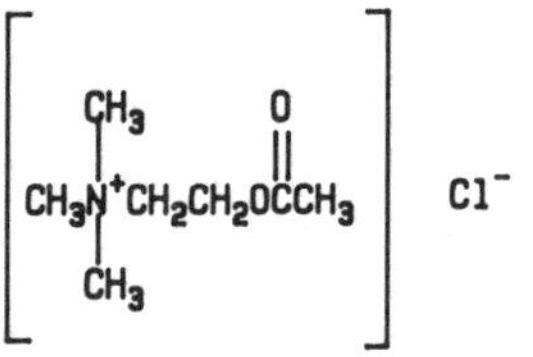

$$\left[CH_3\overset{\overset{\large CH_3}{|}}{\underset{\underset{\large CH_3}{|}}{N^+}}CH_2CH_2O\overset{\overset{\large O}{\|}}{C}CH_3 \right] Cl^-$$

Acetylcholine Chloride

Stimulation of the cholinergic nervous system increases fluid discharge from the saliva, tear, bronchial, and sweat glands. It slows the heart, tightens the chest, increases the movement of the gut, contracts the bladder, and reduces the pupil of the eye to a pinpoint. Conversely, the blocking of the nerve endings that respond to acetylcholine dries the mouth, relaxes the stomach, reduces secretions into the stomach, and dilates the pupil of the eye.

There are two kinds of acetylcholine receptors, which respond to two different plant alkaloids, nicotine and muscarine. They are termed nicotinic and muscarinic receptors, respectively.

N, N, CH_3 — Nicotine

HO, H_3C, O, $CH_2\overset{+}{N}(CH_3)_3$, Cl^- — Muscarine

Atropine blocks the muscarinic receptors, but there is a measure of overlap. The geometry of the muscarinic acetylcholine receptor is well understood, and it is possible that the nicotinic receptor is similar but differs in its surroundings:

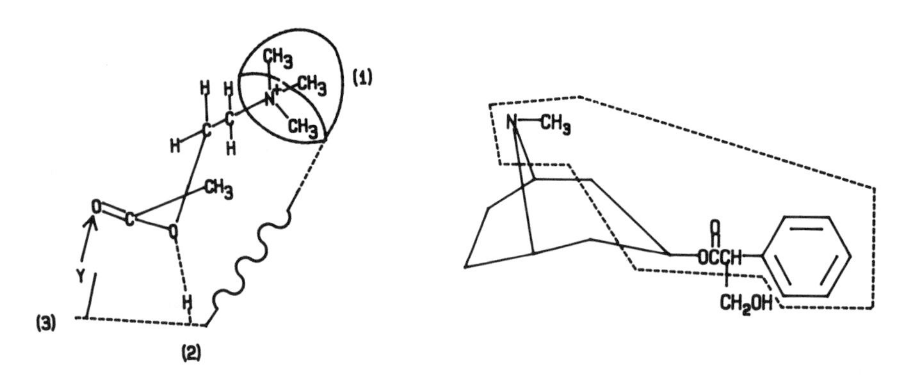

Geometry of the acetylcholine receptor

"Spasmophoric group" of hyoscyamine

The receptor has three main features. The cap (1) contains several phosphate groups with a high electron density which can form an electrostatic bond with the nitrogen of acetylcholine. There is a hydrogen bonding facility at (2) which interacts with the acetyl bridge in the acetylcholine molecule and a further group at (3) which interacts with the carbonyl group of the acetyl. The bond at (2) is thought to be more important for muscarinic action while that at (3) is more important for nicotinic action.

Anticholinergic drugs share structural features believed to be responsible for their ability to block acetylcholine receptors. This is illustrated above by the portion of the hyoscyamine molecule enclosed in the dotted line. The enclosed portion is called the "spasmophoric group" and molecules much simpler than hyoscyamine that contain the group are antispasmodic. With modifications, this spasmophoric group can be found in clidinium bromide and dicyclomine (see below). Comparison of the structures of the spasmophoric group and acetylcholine indicates how blocking can take place.

Anticholinergic drugs are mainly used for treatment of irritable bowel syndrome. They are also used for muscle spasm and travel sickness. At one time, they were widely employed for stomach ulcers. The production of gastric juices including acid was reduced, and so was stomach movement, so there was a reduction in pain. The anxiolytics were added to reduce nervous tension which also plays a part in stomach disorders. The H_2-receptor antagonists (Section 9.6) have largely replaced them in this application. Atropine and other anticholinergics are also used to reduce extrapyramidal symptoms that occur in Parkinson's disease and they will be discussed in Section 16.3. There are other drugs—antihistamines and neuroleptics—that are antispasmodic and counter vertigo and nausea as do the anticholinergics.

A minor but illuminating application of anticholinergics is in the treatment of poisoning by organophosphorus insecticides. Acetylcholine is broken down in the body by the enzyme acetylcholinesterase. The organophosphorus insecticides such as malathion and methyl parathion serve as substrates for acetylcholinesterase and are hydrolysed by it, but much more slowly than acetylcholine itself. Hence, they occupy the acetylcholine sites on the enzyme and inhibit acetylcholine hydrolysis.

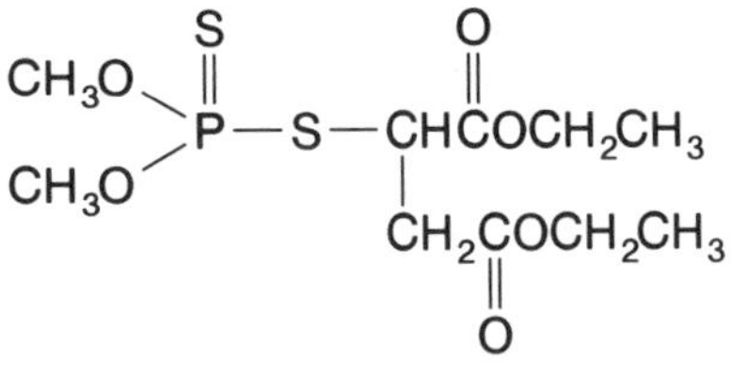

$(CH_3O)_2P(=S)-S-CH(COOCH_2CH_3)CH_2COOCH_2CH_3$

Malathion

$(CH_3O)_2P(=S)-O-C_6H_4-NO_2$

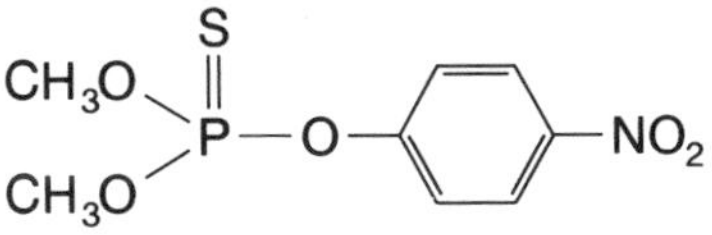

Methyl parathion

In cases of poisoning by organophosphorus compounds, acetylcholine builds up inside the body. A continuous signal is sent along the nerves of the cholinergic nervous system leading to excessive secretion of fluids, asthma, and convulsions. Death is often due to the victim drowning in fluids discharged into the lungs. Part of the antidote is doses of atropine well above the usual toxic limit. These block the acetylcholine receptors and limit the signal from excess acetylcholine. Secretions are diminished and the victim may survive. Because the dose of atropine would normally be fatal in itself, this is a good example of a poison that, in the right circumstances, is a useful drug.

14.1 THE SOLANACEOUS ALKALOIDS

Of the five anticholinergic drugs mentioned above, atropine, hyoscyamine, and hyoscine, otherwise known as scopolamine, are all obtained from natural sources and are called the solanaceous alkaloids. They are extracted from the plant *Atropa belladonna*, or deadly nightshade. They also come from *Hyoscyamus niger*, or black henbane, and *Datura stramonium*, or thorn apple. Laboratory syntheses are known and are excellent examples of the organic chemist's skills but are not used commercially.

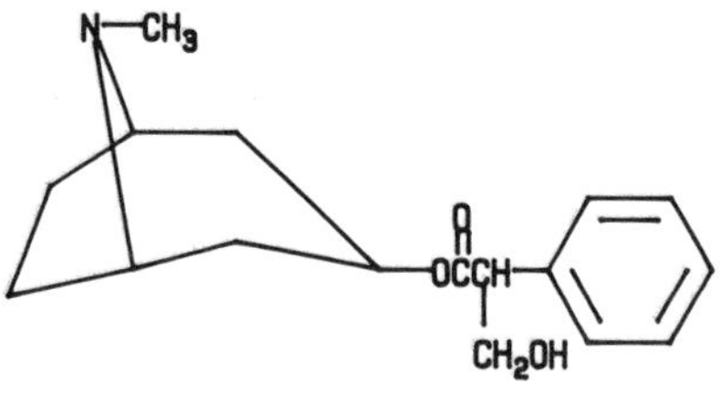

N—CH_3 ... $-\overset{O}{\overset{\|}{C}}CH-$ ... CH_2OH

Atropine;
d,*l*-hyoscyamine

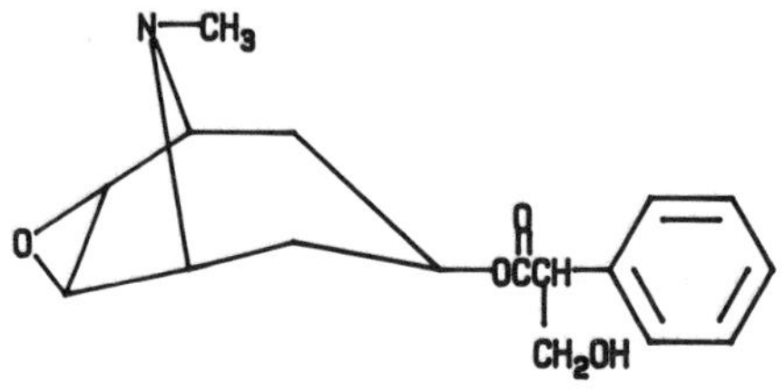

N—CH_3 ... O ... $-\overset{O}{\overset{\|}{C}}CH-$ ... CH_2OH

Hyoscine;
l-scopolamine

The difference between atropine and hyoscyamine is stereochemical. Hyoscyamine is the levo form of the racemic mixture called atropine. The levo isomer, *l*-hyoscyamine, when tested on the cardiac vagus, the iris, or on glands, is much more active than the dextro form and considerably more active than atropine, the *d*,*l* mixture. When tested on the central nervous system, however, all the isomers have the same activity. In the central nervous system, apparently stereochemistry is less important than molecular structure.

14.2 DICYCLOMINE

Dicyclomine is a synthetic anticholinergic. It is a tertiary amine and is said to have a more uniform bioavailability than the natural alkaloids, when taken by mouth.

The synthesis is shown in Figure 14.1. It starts with phenylacetonitrile (1) which, on double alkylation with 1,5-dibromopentane (2), yields the bicyclic compound (3). Hydrolysis of the nitrile group to carbonyl gives (4) and esterification with *N*,*N*-diethylethanolamine gives (5). Catalytic reduction of the benzene ring yields dicyclomine (6).

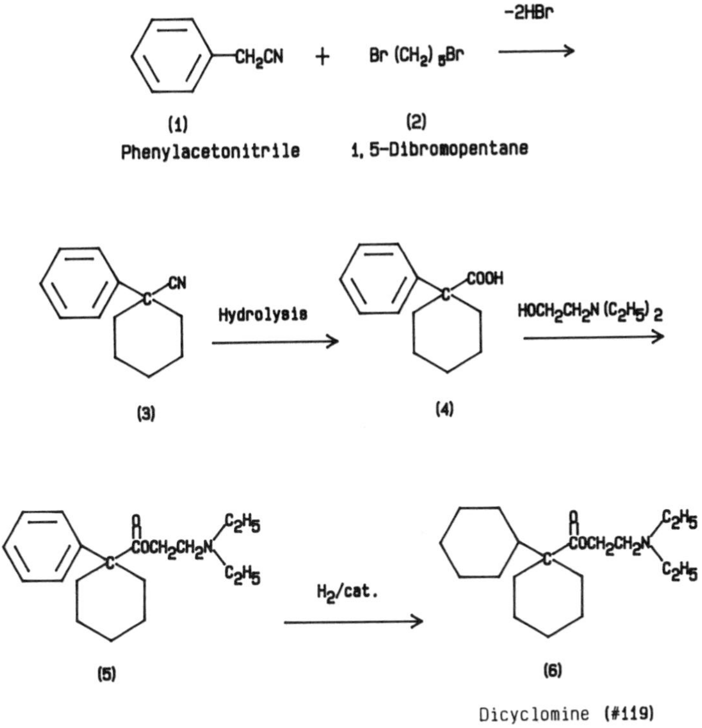

-2HBr

CH_2CN + $Br(CH_2)_5Br$

(1) Phenylacetonitrile

(2) 1,5-Dibromopentane

CN

Hydrolysis

COOH

$HOCH_2CH_2N(C_2H_5)_2$

(3)

(4)

$COCH_2CH_2N(C_2H_5)_2$

H_2/cat.

(5)

(6)

Dicyclomine (#119)

Figure 14.1. Synthesis of dicyclomine.

14.3 CLIDINIUM BROMIDE

Clidinium bromide is a further synthetic anticholinergic and is a quaternary ammonium compound. It is ionized in solution and cannot pass the blood–brain barrier; consequently, it scarcely affects the central nervous system.

The synthesis is shown in Figure 14.2. It starts with the methyl ester of isonicotinic acid (1), which reacts with ethyl bromoacetate (2) to give a quaternary bromide (3). Catalytic reduction removes the ring double bonds to give (4). Successive treatments with potassium and hydrogen chloride lead to the Dieckmann condensation for the formation of a second ring and the hydrochloride (5). Potassium hydroxide liberates the free base (6) with its bridged structure. It is called 3-oxoquinuclidine. Further catalytic reduction gives the alcohol (7). Reaction with the acid chloride (8) gives a quinuclidine ester (9), which is quaternized with methyl bromide to give clidinium bromide (10).

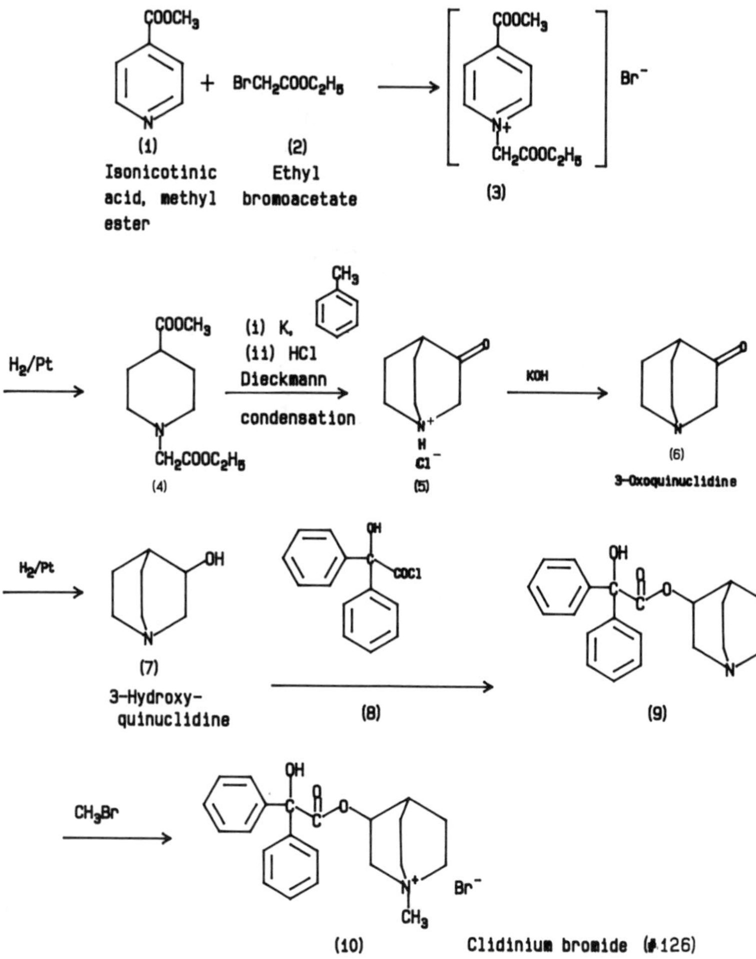

$COOCH_3$ + $BrCH_2COOC_2H_5$ → [$COOCH_3$ / N+ / $CH_2COOC_2H_5$] Br^-

(1) Isonicotinic acid, methyl ester

(2) Ethyl bromoacetate

(3)

H_2/Pt →

$COOCH_3$ / N / $CH_2COOC_2H_5$ (4)

(i) K, CH_3-benzene
(ii) HCl
Dieckmann condensation →

O / N+ / H / Cl^- (5)

KOH →

O / N (6) 3-Oxoquinuclidine

H_2/Pt →

OH / N (7) 3-Hydroxyquinuclidine

OH / C / COCl (8) →

OH / C / C(=O) / O / N (9)

CH_3Br →

OH / C / C(=O) / O / N+ / CH_3 Br^- (10) Clidinium bromide (#126)

Figure 14.2. Synthesis of clidinium bromide.

15

Antiasthma Drugs

Three antiasthma drugs occur in the Top-100, namely theophylline at #26, salbutamol at #33, and metaproterenol at #59. Terbutaline is outside the Top-100 at #107. The number and frequency of prescription of antiasthma drugs in the United States has increased sharply in recent years. The revolution in asthma therapy that occurred in Europe in the early 1970s was blocked in the United States by the Food and Drug Administration and the modern drugs were not licensed until the end of the decade.

Asthma is characterized by attacks of breathlessness caused by overreaction of the airways to certain stimuli. Various stimuli produce asthma but the commonest sources seem to be exercise, emotion, and allergy.

Asthma is more prevalent in children than in adults, and in boys than in girls. Widely different incidences have been recorded in different countries and in different surveys in the same country. A typical incidence is about 6% for boys and 4% for girls. Childhood asthma frequently disappears as the child grows older because the airways get bigger and obstruction becomes less significant.

When a person breathes in, a muscular effort is exerted that enlarges the chest and fills the lungs with air. The muscles are then relaxed. The chest collapses because of its own elasticity and the air is expelled. In asthma, the bronchial smooth muscle, that is, the muscle not subject to voluntary control, contracts and constricts the airways. This bronchospasm is associated with swelling of the lining of the bronchial tubes and increased secretions. Breathing out is a passive process, so there is little a person can do to force air from the lungs. The person starts to wheeze and becomes breathless and distressed.

Until recently most asthma therapy concentrated on relaxing the bronchial muscles. As was pointed out in Section 7.5.4, the bronchial muscles are controlled by β_2 receptors. The various types of adrenergic receptors are listed in Figure 7.12. Stimulation of the β_2 receptors leads to vasodilation and bronchodilation, that is, to the relaxation of the bronchial muscles and the opening up of the airways. Therefore, a drug that stimulates the β_2 receptors, that is, a β_2 agonist, would be expected to relieve asthma.

15.1 β-ADRENERGIC AGONISTS

This is indeed the case. For many years, asthma was treated with epinephrine and ephedrine, shown in Figure 15.1. Unfortunately these stimulate both the α and β receptors. The α stimulation leads to vasoconstriction and the β_1 stimulation to an increased rate of heartbeat, known as tachycardia. There is some evidence that α stimulation increases release of histamine, which is also undesirable for an asthmatic.

In the late 1950s and 1960s, the most widely used drug for asthma was isoproterenol, known in Europe as isoprenaline. It is the isopropyl homologue of epinephrine and is shown in Figure 15.1. An increase in bulk of the substituent on the nitrogen atom from hydrogen (norepinephrine) to isopropyl (isoproterenol) results in the loss of potency of the compound as an α stimulant and an increase in potency as a β stimulant. Hence, although isoproterenol scarcely stimulates the α receptors, it stimulates both the β_1 and β_2 receptors, so in addition to bronchodilation it may still produce tachycardia. It has the advantage when inhaled of acting quickly, but the effect of a single dose wears off after an hour or two. It is inactive when taken by mouth. All the *o*-dihydroxy compounds (catecholamines) are short-acting because they are O-methylated in the body to give inactive ethers:

HO, OH, OH, $CHCH_2NHR$ → (Catechol O-methyl transferase) → HO, OCH_3, OH, $CHCH_2NHR$

In 1961 isoproterenol and related bronchodilators started to be dispensed in pressurized aerosol containers which were much more portable and convenient than the older inhalers. Their use rapidly became widespread but was accompanied by a rise in asthma mortality which almost doubled in England between 1961 and 1965. This is shown in Figure 15.2. Note the rise from about 1200 deaths per year in the early 1960s to a peak of over 2000. Note also that

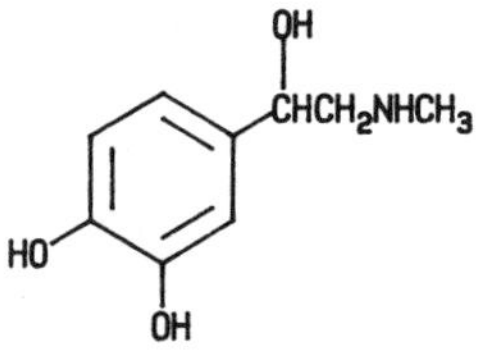

OH
$CHCH_2NHCH_3$
HO
OH

Epinephrine
(Adrenaline)

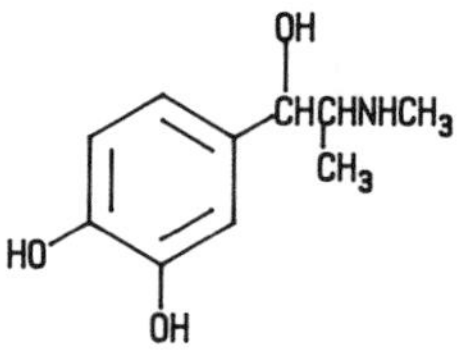

OH
$CHCHNHCH_3$
CH_3
HO
OH

Ephedrine

(Early drugs: stimulate α, β (1) and β (2) receptors.)

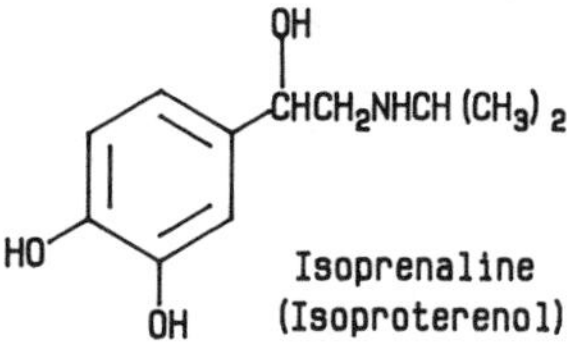

OH
$CHCH_2NHCH(CH_3)_2$
HO
OH

Isoprenaline
(Isoproterenol)

(Stimulates β (1) and β (2) receptors.)

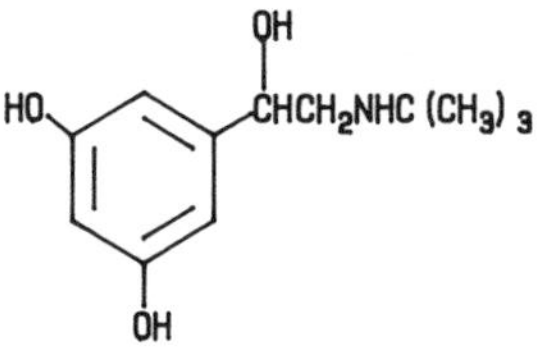

OH
HO
$CHCH_2NHC(CH_3)_3$
OH

Terbutaline (#107)

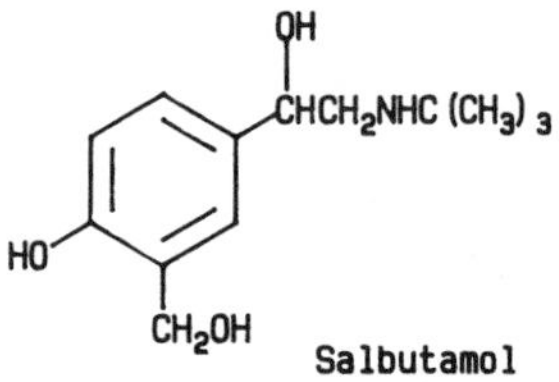

OH
$CHCH_2NHC(CH_3)_3$
HO
CH_2OH

Salbutamol
(Albuterol)

OH
HO
$CHCH_2NHCH(CH_3)_2$
OH

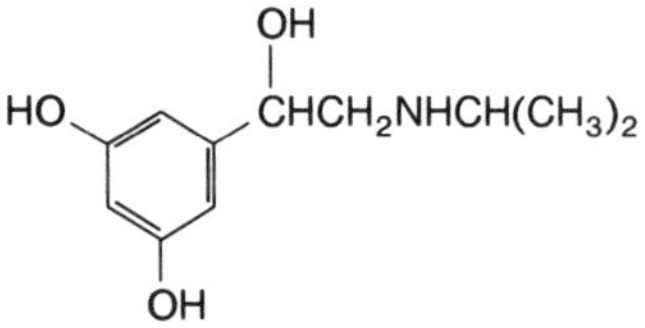

Metaproterenol (#59)

(selective β_2 agonists)

Figure 15.1. *Antiasthma drugs:* β_2 agonists.

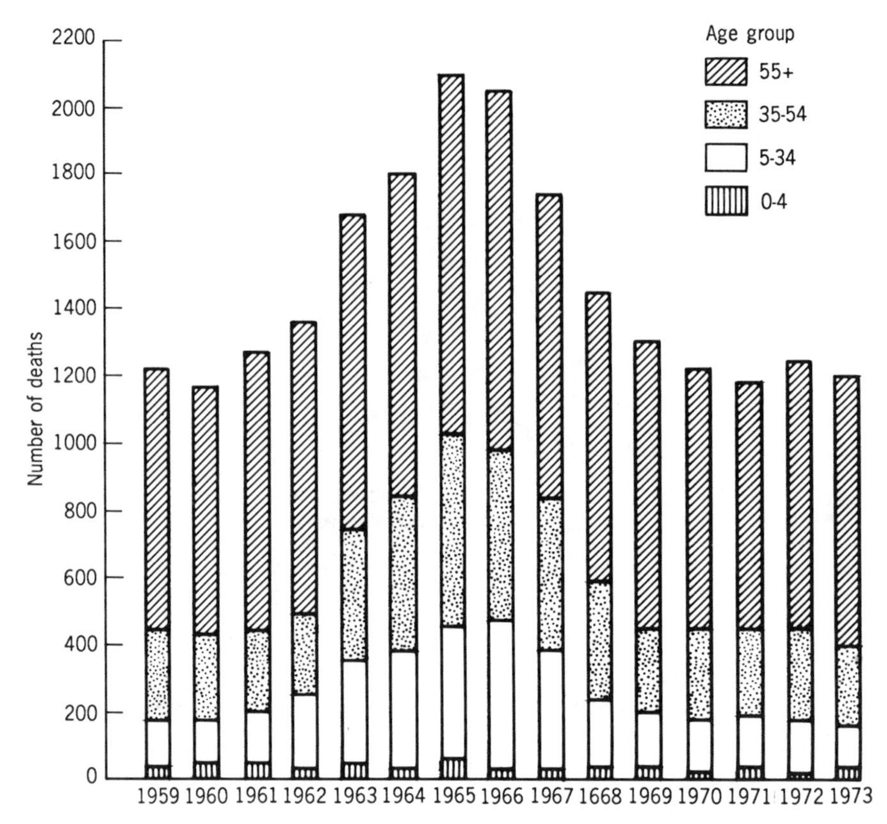

Figure 15.2. Mortality from asthma, by age group (England and Wales 1959–1973). Sources: Medical Tables, Registrar General and Office of Health Economics.

the biggest increase in mortality was in the 5–34 age group where it rose fourfold from about 100 deaths per year to over 400.

After warnings had been published about the indiscriminate use of the aerosols, the mortality fell again and had returned to its original level by the early 1970s. It seems likely that the very effectiveness of the aerosols contributed to their misuse. Instead of going to see a doctor when the condition deteriorated, the asthmatic simply used the aerosol more and more so that it produced a fatal overstimulation of the β_1 receptors in the heart. Similarly, the aerosol often concealed the point at which alternative therapy should have been tried.

This story underlines yet again that, although a drug may have been thoroughly tested in one form, a change in the method of administration may produce a whole new set of safety problems.

The next β_2 agonist after isoprenaline was terbutaline (#107), shown in Figure 15.1. It is much more selective than isoprenaline and has less effect on the β_1 receptors of the heart. Salbutamol (#33) and metaproterenol (#59) are further selective β_2 agonists and these three are the most widely used bronchodilators at present. Salbutamol was licensed much later in the United States, where it is known as albuterol, than in Europe. Because of the delay, most of the literature adheres to the European name. Note the close structural resemblances between the compounds in Figure 15.1. This illustrates that for a drug to be an agonist, its structure must be much closer to the natural agonist than if the drug is intended as an antagonist.

A number of other selective β_2 agonists have been developed in recent years but it is too soon to know if they will displace the above compounds.

15.1.1 Syntheses of β_2 Agonists

Salbutamol is synthesized as shown in Figure 15.3. Methyl salicylate (1) is acetylated to give (2), which reacts with benzyl chloride in the presence of sodium hydroxide to give (3). Bromine in chloroform halogenates the methyl group to give (4) and treatment of (4) with *N*-benzyl-*N*-*tert*-butylamine in methyl ethyl ketone gives (5). Sodium borohydride in ethanol reduces the ketonic oxygen to hydroxyl to give (6). Resolution is performed at this stage by conversion of (6) to the (−)-di-*p*-toluoyltartrate with (−)-di-*p*-toluoyltartaric acid. Only one of the diastereoisomers is soluble in ethyl acetate. The insoluble component is recrystallized, reduced with lithium aluminum hydride to give (7), and then catalytically debenzylated to give *R*-(−)-salbutamol (8). This enantiomer is 68 times as active a β_2 stimulant as the (*S*)-(+)-isomer, and provides another example of the stereospecificity of many drugs and the frequent requirement for optical resolution in pharmaceutical chemistry.

The synthesis of terbutaline is shown in Figure 15.4. The starting material is 3,5-dibenzyloxyacetophenone (1), which is brominated to the bromoketone (2). (2) Is reacted with *N*-benzyl-*N*-*tert*-butylamine to give the ketone (3). Reduction with hydrogen and palladium on charcoal gives terbutaline (4). These three reaction steps and the amine reactant are the same as in the salbutamol synthesis.

The synthesis of metaproterenol is similar to that of terbutaline, the differences being that the starting material is 3,5-dimethoxyacetophenone and the side chain is based on *N*-benzyl-*N*-isobutylamine.

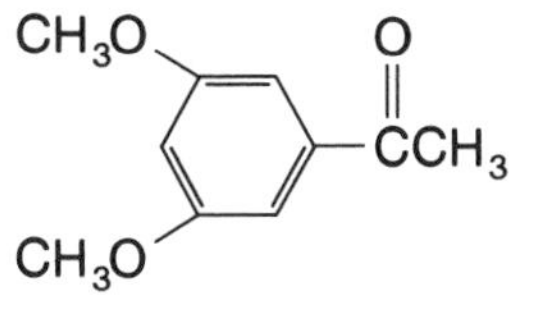

3,5-Dimethoxyacetophenone

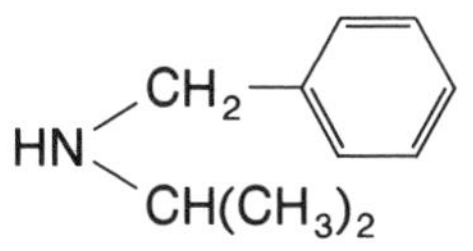

N-Benzyl-*N*-isopropylamine

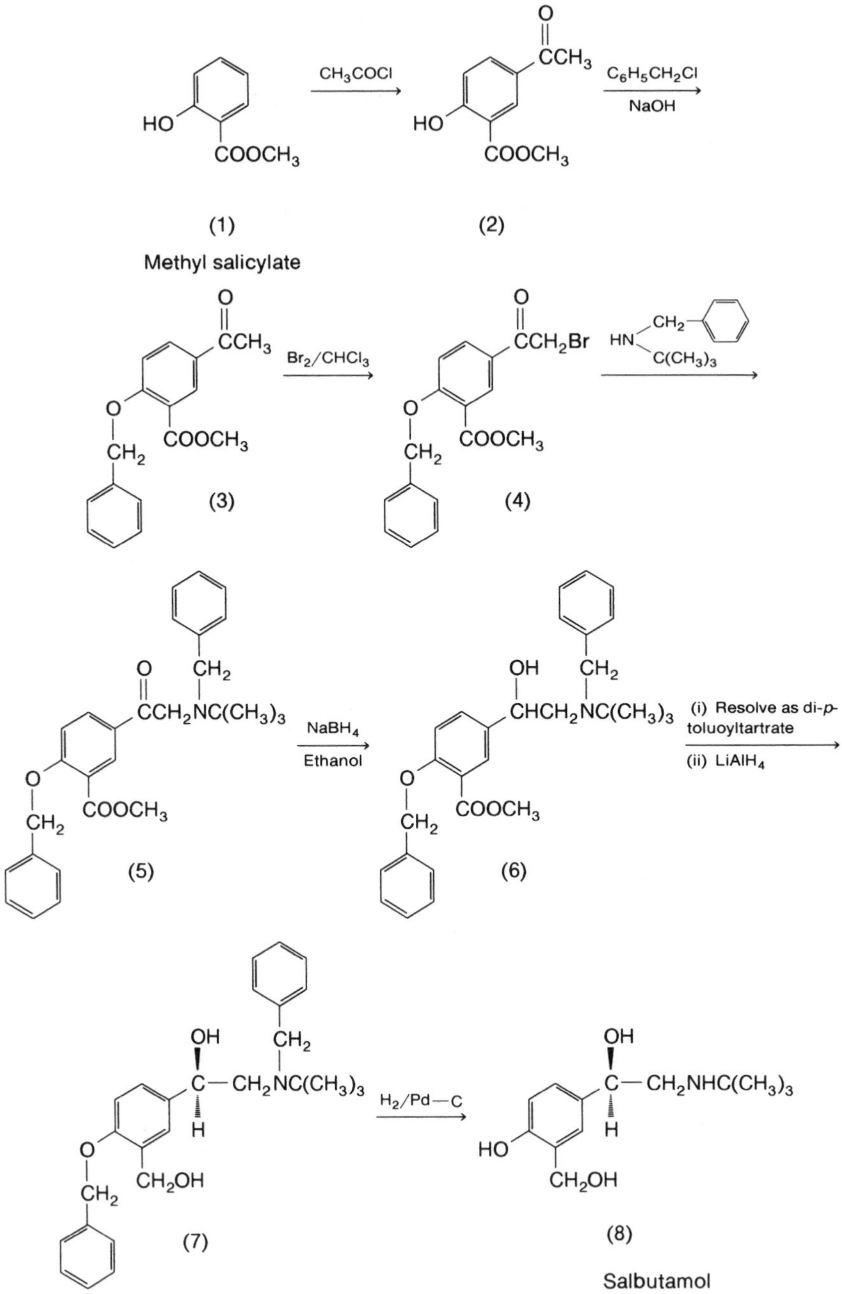
HO
COOCH3
CH3COCl
O
CCH3
HO
COOCH3
C6H5CH2Cl
NaOH
(1)
Methyl salicylate
(2)
O
CCH3
O
CH2
COOCH3
Br2/CHCl3
(3)
O
CCH2Br
O
CH2
COOCH3
(4)
HN
CH2
C(CH3)3
O
CH2
CCH2NC(CH3)3
O
CH2
COOCH3
(5)
NaBH4
Ethanol
OH
CH2
CHCH2NC(CH3)3
O
CH2
COOCH3
(6)
(i) Resolve as di-p-toluoyltartrate
(ii) LiAlH4
OH
CH2
C—CH2NC(CH3)3
H
O
CH2
CH2OH
(7)
H2/Pd—C
OH
C—CH2NHC(CH3)3
H
HO
CH2OH
(8)
Salbutamol

Figure 15.3. *Synthesis of salbutamol (albuterol).*

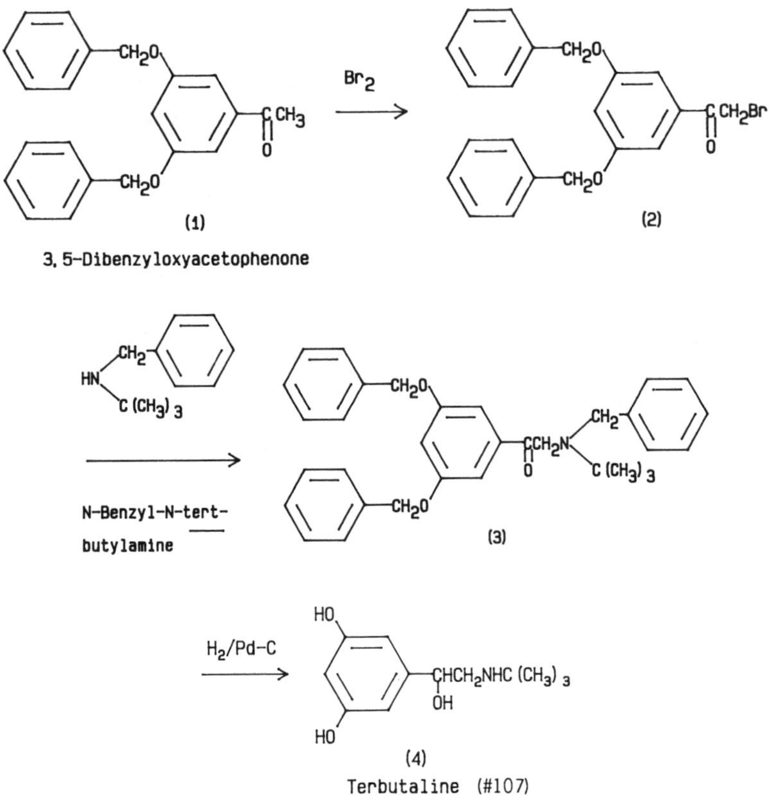

Figure 15.4. Synthesis of terbutaline.

15.2 METHYLXANTHINE BRONCHODILATORS

Methylxanthine bronchodilators may be used instead of β_2 agonists. Theophylline is the preferred drug, but it is absorbed only slowly and is often used as a 2 : 1 complex with ethylene diamine, which is absorbed faster. The complex is known as aminophylline.

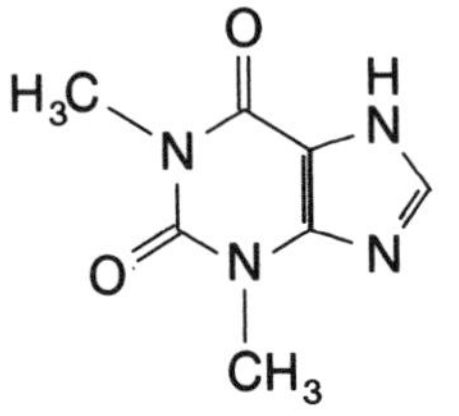

Theophylline

$H_2NCH_2CH_2NH_2$

Ethylene diamine

Theophylline has been known for many years, but its use was limited by its narrow therapeutic index (Section 4.10). Use increased when sustained release preparations and improved methods for determining serum concentrations became available. Theophylline is no more effective than the β_2 agonists, but need be taken only once or twice daily by mouth, so the patient can get a night's sleep. This is important because asthma often increases in severity late at night and early in the morning. Hence, theophylline is useful for maintenance therapy.

Theophylline inhibits an enzyme, phosphodiesterase, which would otherwise destroy a chemical mediator, 3′,5′-cyclic adenosine monophosphate, which limits the release of inflammatory chemicals into the bronchi.

The Traube purine synthesis is one of many routes to theophylline and is shown in Figure 15.5. 1,3-Dimethyl urea (1) amidifies ethyl carboxamidoacetate (2) to give a diamide (3). This cyclizes to a pyrimidone (4) on treatment with a base. Nitrosation with nitrous acid or nitrosyl chloride yields (5) and the nitroso group is then reduced to yield a diamine (6). The two amine groups

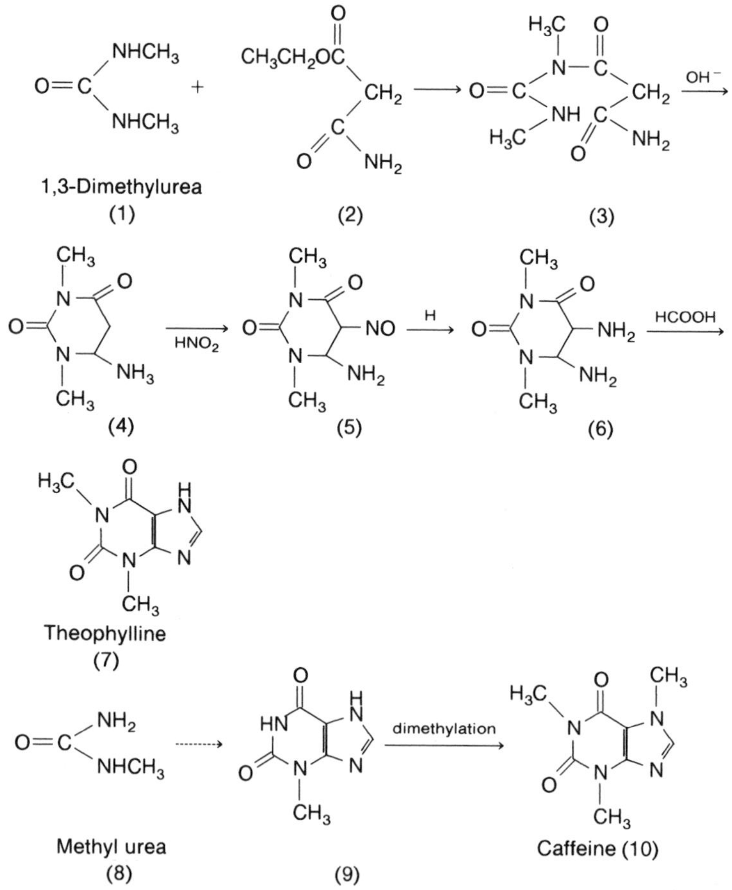

Figure 15.5. Synthesis of theophylline and caffeine.

then form a purine ring with the carbon atom from formic acid, the resulting compound being theophylline.

Theophylline is close in structure to the CNS stimulant caffeine (Section 8.5.2), although its physiological action as a smooth muscle relaxant is totally different. If the Traube synthesis is carried out with methylurea (8), compound (9) results, which on dimethylation yields caffeine (10).

15.3 DISODIUM CROMOGLYCATE

Disodium cromoglycate is a novel compound developed empirically that prevents asthma from developing, but is valueless once the attack has occurred. It acts on the mast cells in the lungs, preventing the release of inflammatory and spasmodic chemicals in response to the presence of allergenic materials. Like theophylline, it is a phosphodiesterase inhibitor, but it acts in other ways too that are not fully understood.

The synthesis of disodium cromoglycate is shown in Figure 15.6. Two molecules of dihydroxyacetophenone (1) condense with one molecule of

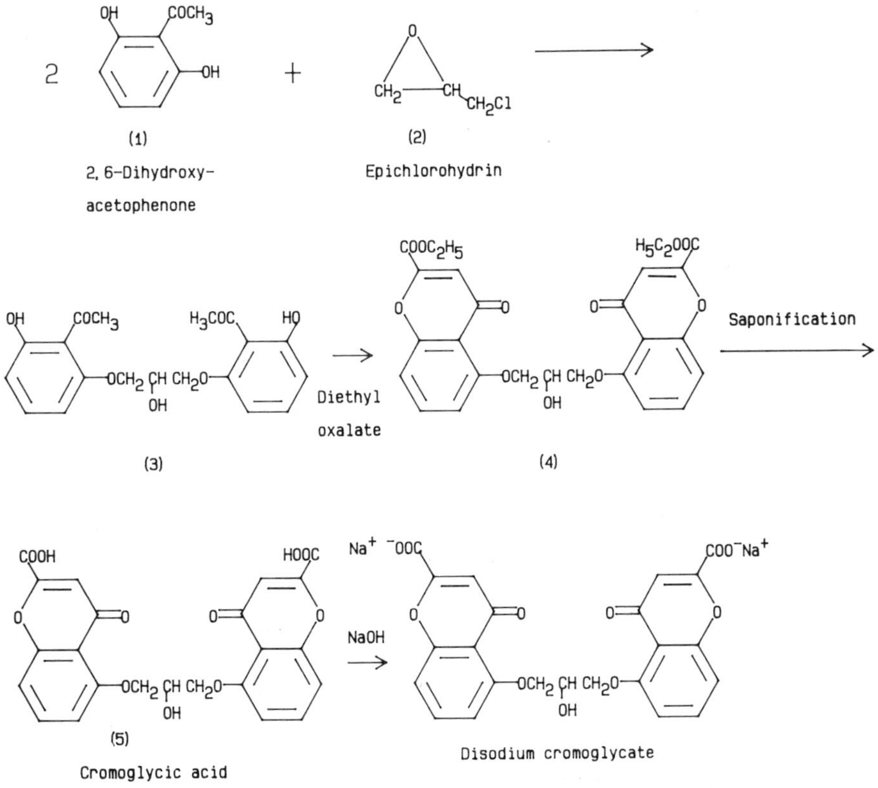

Figure 15.6. Synthesis of disodium cromoglycate.

epichlorohydrin (2) to give the expected intermediate (3). This reacts with diethyl oxalate to give the chromone ester (4). Saponification gives free cromoglycic acid (5) and then the disodium salt, which, unlike most sodium salts, is insoluble.

15.3.1 Dosage Forms

The problems with isoproterenol aerosol inhalers were discussed in Section 15.1. The modern drugs are more selective and, hence, less dangerous. Disodium cromoglycate and the β_2 agonists are all available as metered-dose aerosol inhalers and rigorous instructions for their use are provided. Many of these drugs are also dispensed as solids with an insufflator, a device that allows the solid to be inhaled. The insufflator can be loaded with only a single dose at a time, so the chance of excessive use is diminished. Insufflators are activated by the patient breathing in sharply so these devices are also of value to those whose technique for inhaling aerosols is poor.

The β_2 agonists may be given by injection in cases of severe spasm, or orally, but their action is then much slower. Theophylline and aminophylline are usually taken orally but are occasionally injected. This is not true of disodium cromoglycate. It is ineffective when taken orally, which is a pity because, as a prophylactic, speed of action comes second to convenience. Fisons, the British company that developed it, embarked on a research program to develop an oral modification. The research resulted in proxycromil:

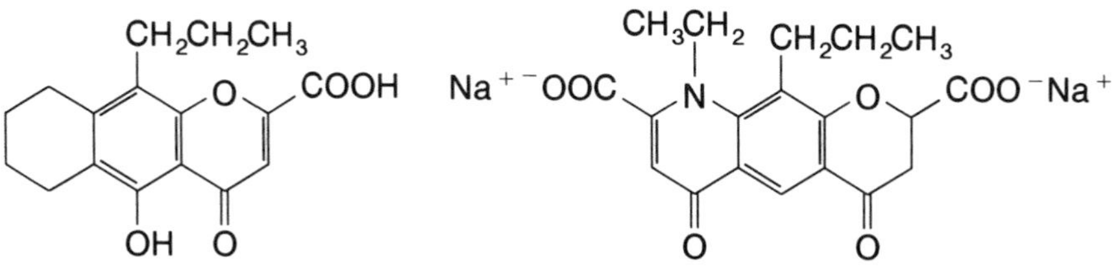

Proxicromil

Nedocromil sodium

They invested tens of millions of dollars in the elaborate test procedures required and were involved in the final clinical trials when it turned out at the beginning of 1981 that proxycromil was potentially carcinogenic in humans. The price of Fisons shares dropped by almost 40%. This story illustrates the huge risk involved in the development of new pharmaceuticals and serves to emphasize that not all stories have as happy an ending as the cimetidine story of SmithKline Corporation. Meanwhile, Fisons have bounced back with the related compound nedocromil sodium (see above), which is said to be similar to disodium cromoglycate but wider ranging and more suitable for adults.

15.4 STEROIDS

Severe bronchospasm may be treated with anti-inflammatory steroids, which reduce the inflammation of the bronchial tubes and are also antiallergic. Beclomethasone dipropionate, an esterified chlorinated analogue of betamethasone (Figure 11.3), is the most widely used. Budesonide has recently come on the market. They are rapidly destroyed in the lungs, hence do not penetrate to other parts of the body. Because of this, patients can tolerate higher doses.

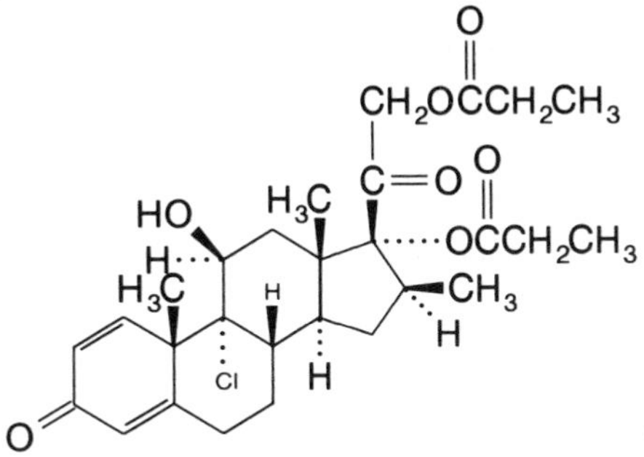

Beclomethasone dipropionate

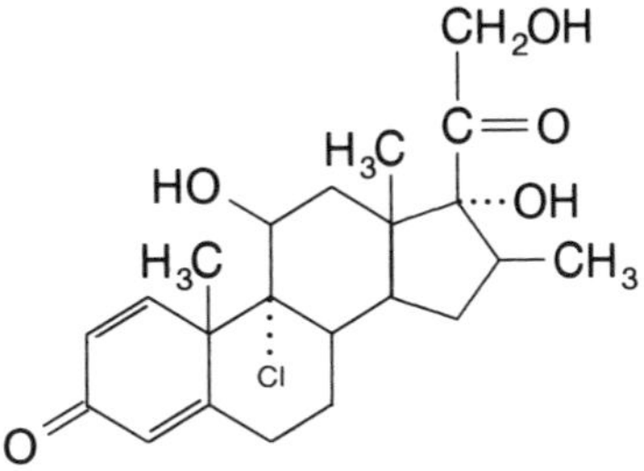

Budesonide

15.5 LEUKOTRIENE INHIBITORS

The most interesting recent development in asthma therapy is the identification of the so-called slow-reacting substance of anaphylaxis, SRS-A. The complex processes that cause bronchospasm were outlined in Chapter 9 and a late stage in the process is the discharge of histamine and SRS-A from the mast cells. These cause immediate constriction of bronchial smooth muscle. Presumably, therefore, it should have been possible to prevent bronchospasm by the use of antihistamines. Unfortunately, their effect was much smaller than hoped. Attention was consequently turned to SRS-A, which is a cysteine hexatriene related to leukotriene C4 (Section 19.1.4):

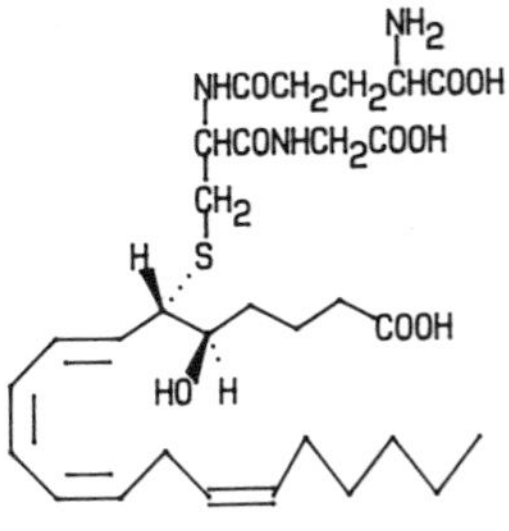

Leucotriene C_4

It is also related to the inflammatory prostaglandins and the race is now on among the pharmaceutical companies to find an SRS-A antagonist, which should be an antiasthma drug.

16

Parkinson's Disease

This chapter deals with Parkinson's disease and its mirror image, Huntingdon's chorea, which are characterized by opposite symptoms. Only one drug to counter parkinsonism appears in the Top-100 because the disease is not widespread. The overall incidence is likely to be about 150 per 100,000 population with a rate above 500 per 100,000 in the over-50 age group. This means between 200,000 and 400,000 sufferers in the United States.

The drugs against Parkinson's disease are not very effective and drugs against Huntingdon's chorea are scarcely effective at all. However, the physicochemical processes involved in these two diseases are interesting. The background to the design of levodopa (L-dopa) as a drug against parkinsonism is also important and the method by which the correct stereoisomer is synthesized is of great chemical significance. Furthermore, the conquest of infectious diseases has led to a marked increase in the prevalence and significance of the chronic disabling diseases of old age, of which parkinsonism is typical.

Parkinson's disease is an illness of the central nervous system in which voluntary movements become slow and shaky (that is, there is a tremor) and the muscles become stiff and the limbs rigid.

In the brain, the adrenergic nervous system is called the dopaminergic system, dopamine being the precursor of norepinephrine (Section 7.5.4). The centers in the brain that control movement do so through the two neurotransmitters, dopamine and acetylcholine, that is, through the dopaminergic (adrenergic) and cholinergic nervous systems.

It was found in the 1960s that South American manganese miners developed symptoms similar to those of Parkinson's disease. There was evidence that this was due to the formation of chelate compounds between manganese and dopamine in the brain, which inhibited the use of dopamine in brain chemistry. It was known that dopamine is converted to norepinephrine, which transmits nervous impulses between certain juxtasposed nerve ends and between some nerve ends and the muscles they control. Interference with norepinephrine formation could perhaps give rise to the tremor and rigidity associated with the manganese miners' disease and with parkinsonism. Further research showed that brain cells of patients with Parkinson's disease were indeed lacking in dopamine. The obvious solution was to dose patients with it.

Dopamine was ineffective, however, because it could not reach the brain. The brain has a specialized means available to prevent access to it of many materials that are in the bloodstream. This protective device is known as the blood–brain barrier. In general it is a nonpolar barrier transmitting lipophilic materials, but it contains water-filled capillaries less than 1 nm wide compared with more than 10 nm for similar capillaries elsewhere in the body. The barrier will thus transmit some inorganic ions, glucose, choline, and some amino acids.

A breakthrough occurred when it was found that the amino acid, 3-(3,4-dihydroxyphenyl)-L-alanine, known as levodopa, could penetrate the blood–brain barrier when dopamine could not. Once there it is decarboxylated to dopamine, at a part of the brain known as the basal ganglia, by the enzyme L-aromatic amino acid decarboxylase.

NH_2 O

CH_2····C····COH

H

HO

OH

L-Aromatic amino acid decarboxylase →

$CH_2CH_2NH_2$

HO

OH

Levodopa

Dopamine

Initial tests suggested the drug might be very effective. It was quickly shown that the D stereoisomer was ineffective and caused a blood disorder in a quarter of the patients. The decarboxylase is specific to the L form; hence, pure levodopa is necessary.

16.1 ASYMMETRIC SYNTHESIS OF LEVODOPA

Levodopa was originally obtained by optical resolution of a racemic mixture of dopa. It also occurs naturally in certain vegetables from which it can be extracted. The need for levodopa, however, led to the development of an

1. Synthesis:

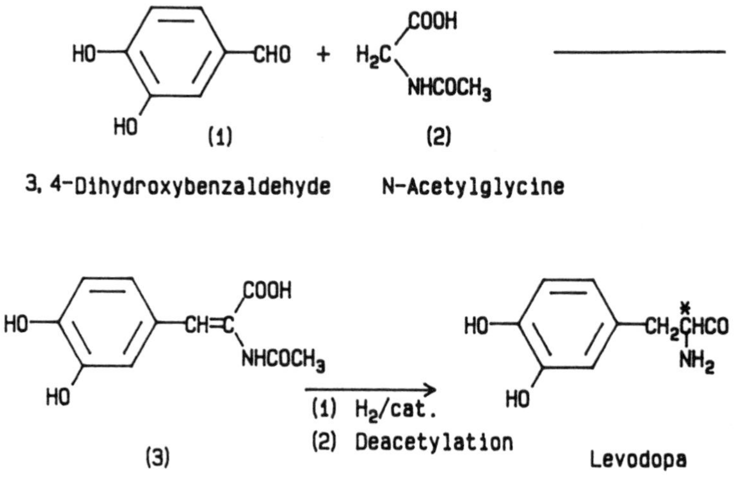

2. Catalyst:

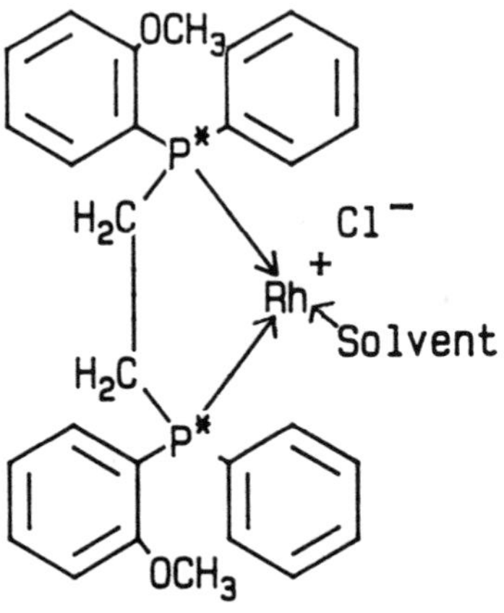

*Figure 16.1. Asymmetric synthesis of L-dopa. *Asymmetric atoms.*

asymmetric synthesis. This was a landmark not only in pharmaceutical chemistry but in organic chemistry generally.

To this day, some organic chemistry textbooks tell us that asymmetric synthesis, the synthesis of pure D or L isomers, is not possible and that laboratory methods give D,L mixtures, which must be tediously resolved.

The stereospecific synthesis of levodopa is shown in Figure 16.1. Condensation of 3,4-dihydroxybenzaldehyde (1) with *N*-acetylglycine (2) gives the unsaturated intermediate (3). Compound (3) is then hydrogenated in the presence of a remarkable catalyst that induces asymmetry. This is shown at the bottom of the figure. It is a rhodium chloride complex in which there is a bidentate (two-pronged) ligand which is both asymmetric and rigidly positioned. This is the key to the synthesis and makes possible the insertion of the hydrogen atoms so as to produce asymmetry.

This synthesis was used commercially by Monsanto. A more recent commercial process is said to be the use of high-pressure liquid chromatography to separate the racemic mixture. An even newer microbiological process employs the action of the enzyme β-tyrosinase on catechol plus an amino acid, such as alanine or serine; or a keto acid, such as pyruvic acid, plus ammonia.

16.2 EFFECTS OF LEVODOPA

The general feeling in the early 1970s was that a drug designed on so clear a theoretical basis and produced by so elegant a synthesis just had to work. Initially it appeared that one-third of the patients were dramatically helped by levodopa, one-third were helped a little, and the remainder not at all. The side effects, however, were unpleasant and included nausea, gastrointestinal disturbance, reduced blood pressure, involuntary grimacing, delusions, and mental disturbance. Because levodopa is rapidly decarboxylated in the body, doses as large as 5–10 g/day are required and costs are high.

The cardiovascular side effects of levodopa seem to be due to its decarboxylation outside the brain. These may be alleviated by the simultaneous use of L-carbidopa:

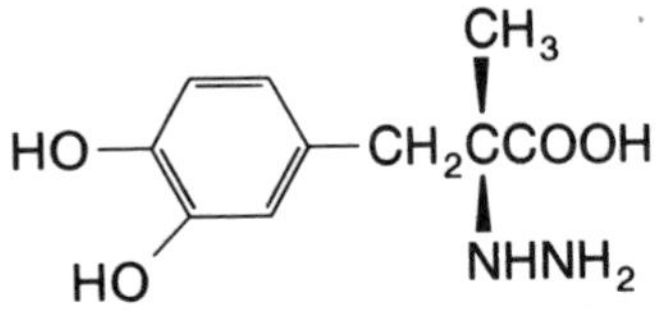

HO, HO — $CH_2C(CH_3)(NHNH_2)COOH$

Carbidopa

HO, HO, HO — $CH_2NHNHC(=O)CH(NH_2)COOH$

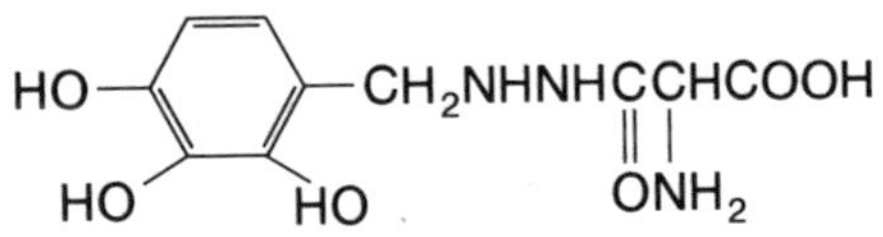

Benserazide

Carbidopa is a decarboxylase inhibitor. It blocks the action of decarboxylases on levodopa and decreases the amount necessary by as much as 75%. The side effects of levodopa inside the brain are not affected by carbidopa, however, presumably because it does not penetrate the blood–brain barrier. The levodopa–carbidopa mixture is #94 in the Top-100. An alternative decarboxylase inhibitor is benserazide (see above). Note the structural similarity between levodopa and these decarboxylase inhibitors, which is the basis of their blocking ability.

A new drug, also intended to be combined with levodopa, is lisuride. It is a serotonin inhibitor and is claimed to reduce involuntary movements in the chronically ill.

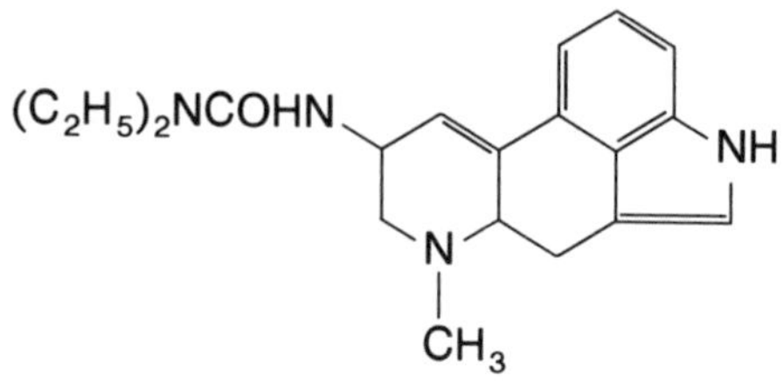

Lisuride

Twenty years after the first clinical trials, levodopa has not lived up to initial expectations in that the proportion of patients helped is lower and the side effects worse than predicted. Even so, it is a worthwhile addition to the handful of anti-Parkinson drugs.

A unique application of L-dopa was in the treatment of *encephalitis lethargica* or sleeping sickness. This was an epidemic illness which swept Europe between 1916 and 1926; it claimed almost 5 million victims of whom a third died. Some of the others recovered but the worst-affected sank into "sleep"–conscious of their surroundings but motionless, speechless and without hope or will, confined to asylums and other institutions. The administration of L-dopa in the early 1970's awakened the survivors after they had spent half a century in a trance. This is not quite a fairy story for the patients had aged by 50 years and there was no handsome prince, but it illustrates graphically the importance of adequate supplies of dopamine in the brain.

16.3 OTHER DRUGS

Because Parkinson's disease results from an inbalance of the dopaminergic and cholinergic nervous systems, in its early stages it can be treated by depression of the cholinergic nervous system by anticholinergic drugs. The antihistamine diphenhydramine (Section 9.2) acts in this way, as does benztropine (#125):

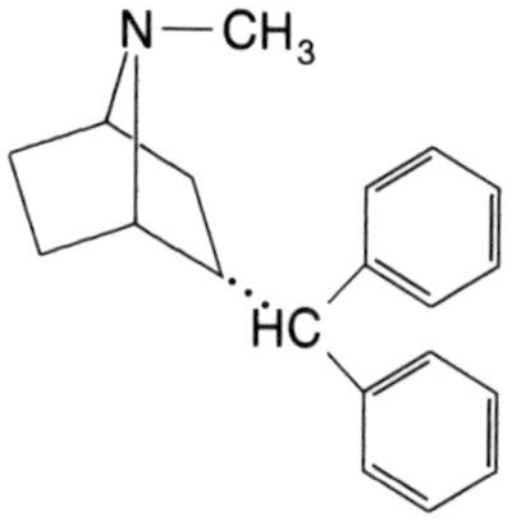

Benztropine

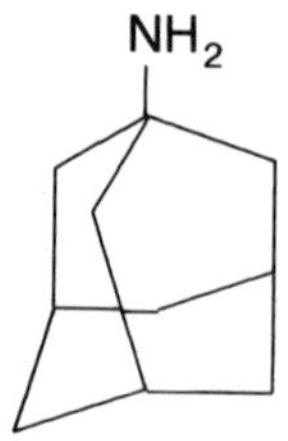

Amantadine

A way to stimulate the release of dopamine in the body is by the use of amantadine (see above). This is thought to act on nerve endings to stimulate the release of dopamine, but is less effective generally than levodopa. It has a range of side effects, many of them similar to those of levodopa. It is also used as an antiviral drug (Section 20.4).

Various other drugs are on the market for Parkinson's disease, most of them anticholinergics, and even atropine, the classic anticholinergic, has been recommended, but these drugs have little effect on the slowness of movement which is one of the most distressing symptoms of parkinsonism. There is still a place for a really effective drug to counter this disease.

A recent idea is the transplantation of dopamine-producing cells from a victim's adrenal gland into the brain. The patient's own tissue is used, so there is no problem with rejection. The question is whether the cells will go on producing dopamine after transplantation. Early trials in Sweden met with little success, but reports are more hopeful on more recent work.

Cells have also been transplanted into patients' brains from aborted fetuses. There appears to be an immediate gain, but it is not clear if this will be preserved in the long term. The problem is that transplanted cells do not "link up" with other brain cells. In addition to medical problems, the technique faces ethical problems concerning the use of fetuses.

Parkinson's disease is not hereditary and the cause of the dopamine deficiency is not known. Drug-induced parkinsonism may result, however, from blockade of dopamine receptors by antipsychotic drugs or metoclopramide, depletion of brain dopamine by rauwolfia alkaloids, or consumption by addicts of meperidine analogues contaminated with 1-methyl-4-phenyl tetrahydropyridine.

16.4 HUNTINGDON'S CHOREA

Huntingdon's chorea (possibly the St. Vitus' dance of olden times) is worthy of mention because it is the mirror image illness of parkinsonism. Instead of rigidity and difficulty with movement, it is characterized by an unsteady walk, and the appearance of restlessness and fidgeting. The involuntary jerks and slurred speech are accompanied by progressive mental deterioration. Like parkinsonism, Huntingdon's chorea is a miserable, frustrating, and tragic illness, but its cause is the opposite. The condition, which is inherited, involves atrophy of the brain, particularly the basal ganglia, which are also concerned with dopamine metabolism. The atrophy, however, seems to spare the dopaminergic nerve pathways and thus they become overactive in relation to other neural systems involving acetylcholine.

The situation is certainly much more complicated than this and effective drugs scarcely exist. Various psychotropic drugs have been tried, especially the major tranquilizers. These block the receptors where the dopamine acts and, hence, decrease its effect. The drug of choice at present is tetrabenazine:

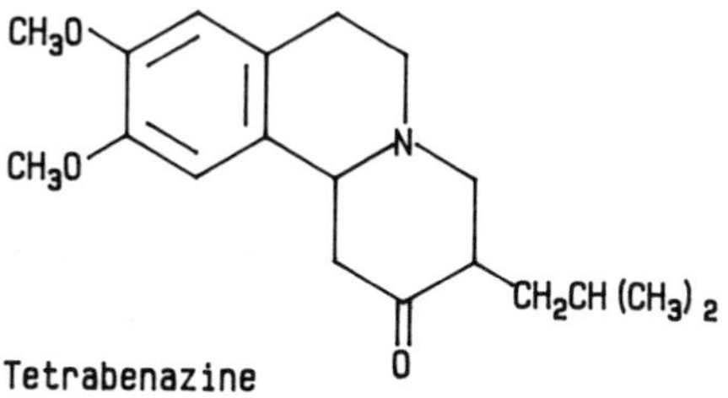

Tetrabenazine

It is believed to act like reserpine, which was discussed as an antihypertensive agent (Section 7.6.1). Isoniazid, an antituberculosis drug (Section 6.7.1), has also been tried, but remains unproved. There is still no effective treatment for this illness. At present, genetic counseling, in which carriers of the defective gene are advised not to marry each other or alternatively not to have children, can do more than pharmaceutical research to reduce the number of sufferers.

17

Miscellaneous Drugs

There remain only seven drugs in the Top-100 that have not been discussed, plus two between #101 and #135. There are two thyroid hormones, two muscle relaxants, an antigout agent, an antismoking drug, a treatment for acne, a stomach drug, and a multivitamin preparation. They will be described in turn.

17.1 THYROID HORMONES

L-Thyroxine and L-triiodothyronine are the primary hormones secreted by the thyroid gland. The latter is different only because one of the iodine atoms next to the hydroxyl has been removed.

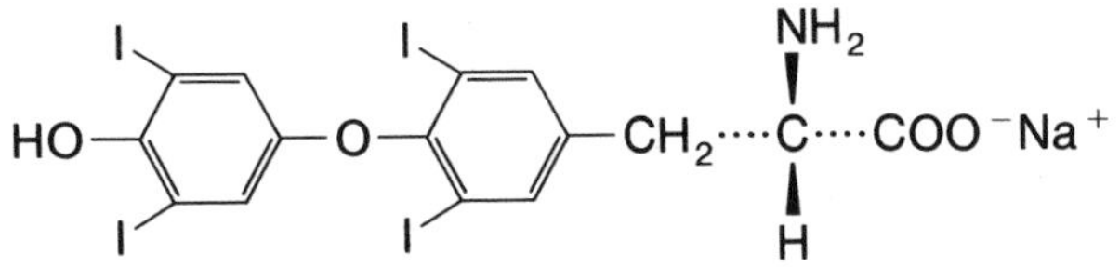

L-Thyroxine, sodium salt

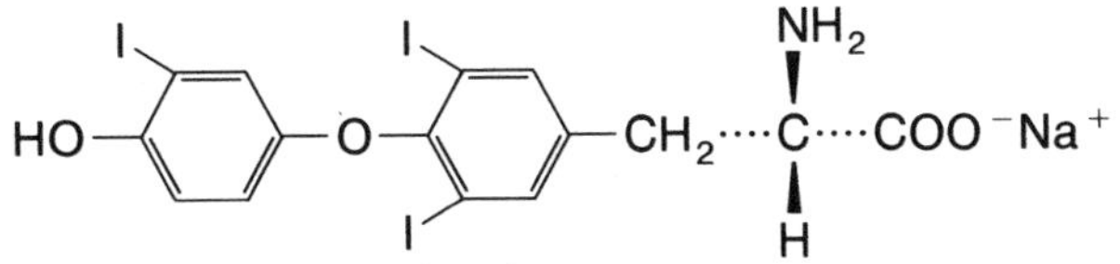

L-Triiodothyronine, sodium salt

They are released into the circulatory system, where the triiodo compound is iodinated to L-thyroxine. There are several illnesses associated with disorders of the thyroid gland, including enlargement of the gland, known as goiter, and underworking and overworking of the gland.

Goiter is due to a deficiency of iodine for the gland to use and was once prevalent in areas where the water supply was exceptionally low in iodine. It has been countered by addition of an iodide to table salt. Overworking of the thyroid gland, called hyperthyroidism, results in a nervous, irritable, overactive patient. It is treated by surgery, part of the gland being cut away.

Underworking of the gland so that it does not produce enough L-thyroxine and L-triiodothyronine is known as hypothyroidism. It results in a slowing down of all the metabolic processes in the body so that the patient is mentally slow, lethargic, and cold. It may be associated with a thick, pale swelling of the skin called myxedema. Hypothyroidism is treated by replacement therapy, that is, by dosing the patient with the hormone that the gland is failing to produce.

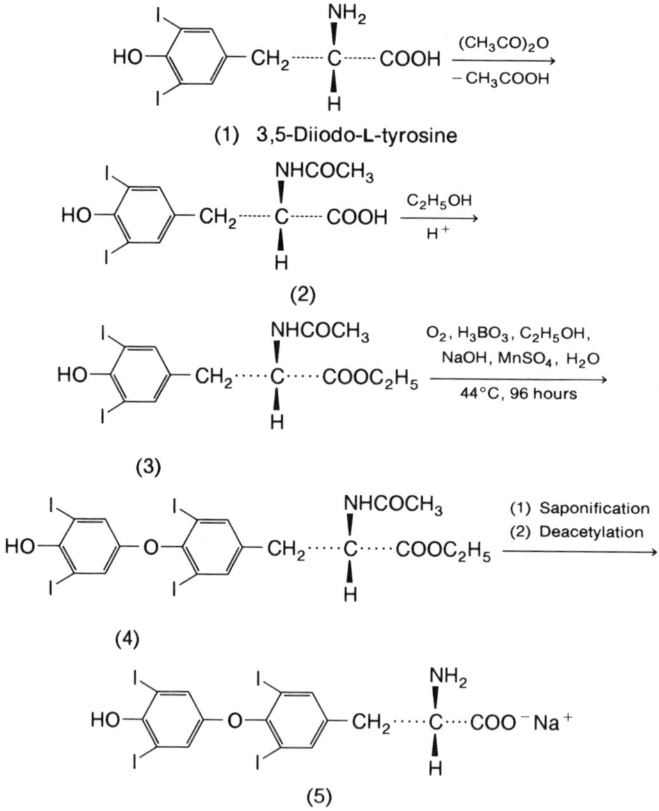

(1) 3,5-Diiodo-L-tyrosine

$(CH_3CO)_2O$, $-CH_3COOH$

(2)

C_2H_5OH, H^+

(3)

O_2, H_3BO_3, C_2H_5OH, NaOH, $MnSO_4$, H_2O; 44°C, 96 hours

(4)

(1) Saponification (2) Deacetylation

(5)

L-Thyroxine, sodium salt (#31)

Figure 17.1. *Synthesis of L-thyroxine, sodium salt*

In the past, mixed thyroid hormones were obtained by desiccation or extraction from pork or beef thyroid glands, but the extract is of variable quality and loses potency on exposure to air. It has largely been replaced by purified extracts which rank #65 in the Top-100 and by synthetic L-thyroxine (#31) synthesized as shown in Figure 17.1. The amino group of L-diiodotyrosine (1) is protected by acetylation to give (2) and the carboxyl group is then esterified giving (3). If this compound is maintained at 44°C for 96 hours in the presence of oxygen, boric acid, ethanol, alkali, and manganese sulfate, it undergoes dimerization yielding an acetylated ethyl ester of L-thyroxine (4). Saponification and deacetylation gives the hormone itself (5).

D-Thyroxine is a drug in its own right and is a hypocholesteremic. It is rare that two stereoisomers have different therapeutic uses. Quinine and quinidine provide another example (Section 7.3).

17.2 MUSCLE RELAXANTS

Chlorzoxazone at #129 and cyclobenzaprine at #73 are skeletal muscle relaxants. They are used in conjunction with rest and physical therapy to relieve muscle spasm, caused perhaps by overextension of a muscle or a bruising blow. The anxiolytic diazepam (#15) is also widely used for this purpose.

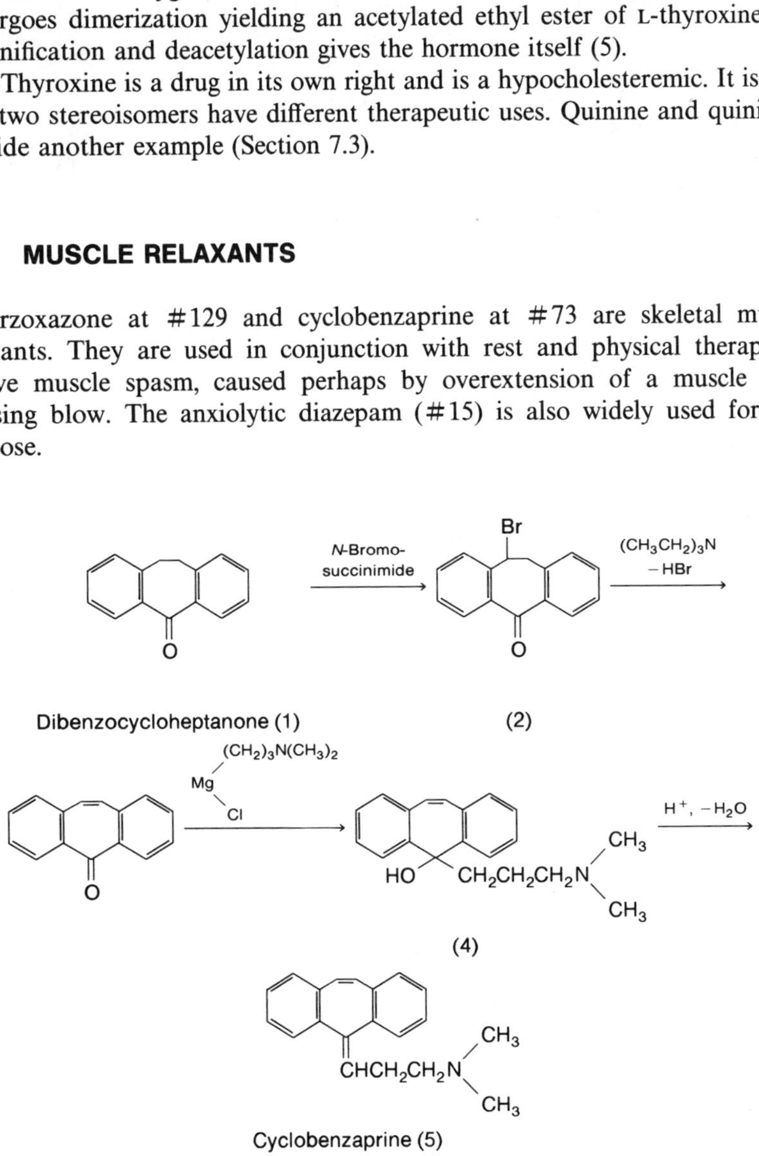

N-Bromo-succinimide

Br

$(CH_3CH_2)_3N$ $-HBr$

O

Dibenzocycloheptanone (1)

(2)

$(CH_2)_3N(CH_3)_2$ Mg Cl

H^+, $-H_2O$

HO $CH_2CH_2CH_2N(CH_3)_2$

(4)

$CHCH_2CH_2N(CH_3)_2$

Cyclobenzaprine (5)

Figure 17.2. Synthesis of cyclobenzaprine.

Chlorzoxazone is synthesized by cyclization of *o*-hydroxy-*m*-chlorobenzoformamide:

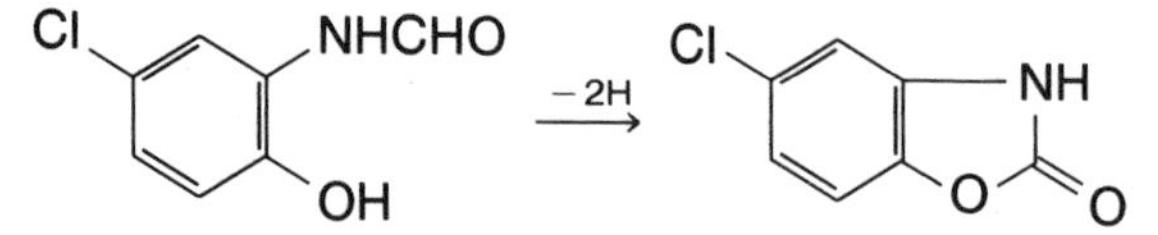

$$\xrightarrow{-2H}$$

o-Hydroxy-*m*-chlorobenzoformamide Chlorzoxazone

Cyclobenzaprine is an analogue of amitriptyline (Section 8.4.2) but contains an additional double bond in the central ring. It is synthesized as shown in Figure 17.2 from dibenzocycloheptanone ((1); also (5) in Figure 8.24). Treatment with *N*-bromosuccinimide inserts a bromine atom to give (2) and dehydrohalogenation with triethylamine eliminates HBr and introduces the extra double bond to give (3). Treatment with a Grignard reagent based on 3-chloropropyl-*N*,*N*-dimethylamine introduces the side chain, and acid-catalyzed dehydration gives cyclobenzaprine (5).

17.3 ALLOPURINOL

Allopurinol (#89) is a drug against gout, a painful disease caused by the accumulation of uric acid in the joints. It was shown in Figure 4.4 as an example of a receptor-blocking drug and it blocks the enzyme xanthine

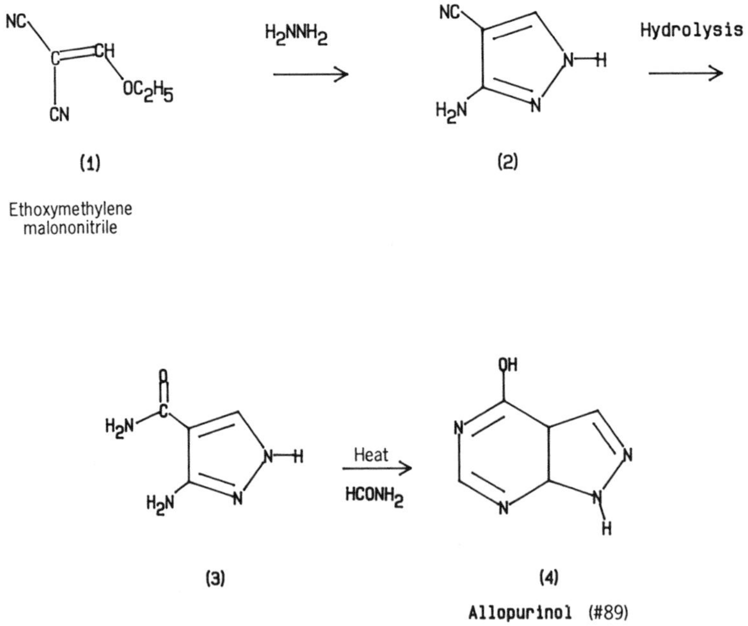

(1) Ethoxymethylene malononitrile $\xrightarrow{H_2NNH_2}$ (2) $\xrightarrow{\text{Hydrolysis}}$

(3) $\xrightarrow[HCONH_2]{\text{Heat}}$ (4) Allopurinol (#89)

Figure 17.3. Synthesis of allopurinol.

oxidase. A synthesis of allopurinol is shown in Figure 17.3. Condensation of ethoxymethylenemalononitrile (1) with hydrazine gives the cyanopyrazole (2). Hydrolysis with sulfuric acid gives the amide (3) and ring closure is brought about by formamide and heat to give allopurinol (4).

17.4 ANTISMOKING DRUG

Fifty-five million Americans smoke cigarettes, in spite of repeated demonstrations of its dangers to health. Sixty percent have tried to quit and 90% say they want to quit. Most smokers are addicted to nicotine and psychologically dependent on smoking behavior. Nicotine polacrilex, an antismoking drug, ranks #92 in the Top-100, and is classified by the AMA under "drugs used in other mental disorders".

Nicotine polacrilex is chewing gum consisting of nicotine bound to an ion exchange resin. It may be represented:

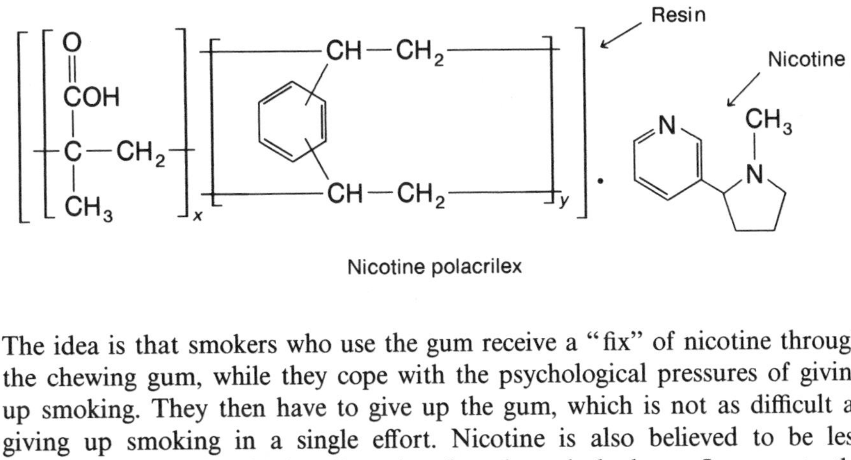

Resin

Nicotine

Nicotine polacrilex

The idea is that smokers who use the gum receive a "fix" of nicotine through the chewing gum, while they cope with the psychological pressures of giving up smoking. They then have to give up the gum, which is not as difficult as giving up smoking in a single effort. Nicotine is also believed to be less dangerous when absorbed orally rather than through the lungs. It passes to the stomach and is inactivated by enzyme systems on its first pass through the liver.

17.5 ACNE TREATMENT

Acne is an embarassing skin disorder principally affecting adolescents. There is probably also a genetic factor. It is caused by an exaggerated response to the increase in androgenic steroids (Section 11.1) accompanying adolescence.

There are a number of therapies available that do not involve drugs. If drug treatment is required, tetracycline antibiotics are often helpful (Section 6.5), and tretinoin appears at #128 in the drug rankings. It is the trans configuration of retinoic acid, that is, vitamin A acid (cf. vitamin A, Section 17.7.2):

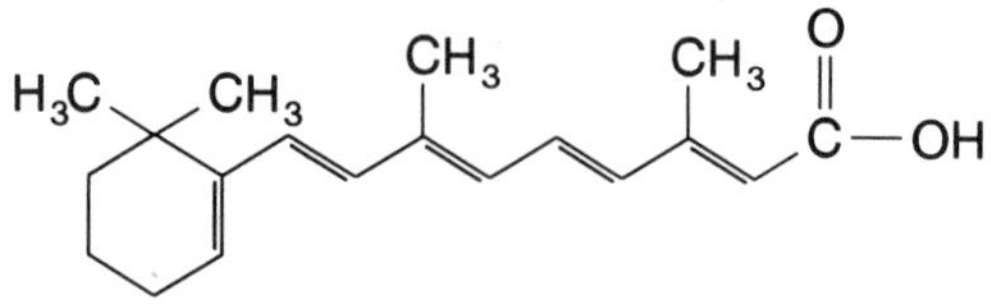

Tretinoin

There is some question of its teratogenicity and it is not recommended for women who are or might become pregnant.

17.6 METOCLOPRAMIDE

Metoclopramide (#90) is a drug used to counter nausea and vomiting, and to increase the rate of emptying of the stomach by normal means. In contrast to the anticholinergics, it *increases* stomach motility. It is a dopamine antagonist but its mode of action is not fully understood.

It is synthesized from 4-acetamido-5-chloro-2-methoxybenzoic acid, which reacts with thionyl chloride to give the acid chloride. This is treated with *N*,*N*-diethylethylenediamine to give metoclopramide:

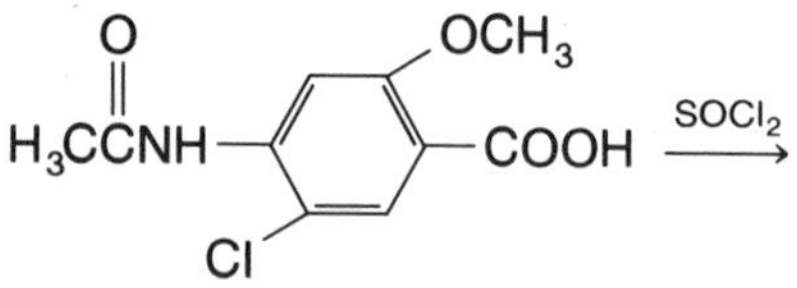

$\xrightarrow{SOCl_2}$

4-Acetamido-5-chloro-2-methoxybenzoic acid

$\xrightarrow{H_2NCH_2CH_2N(CH_2CH_3)_2}$

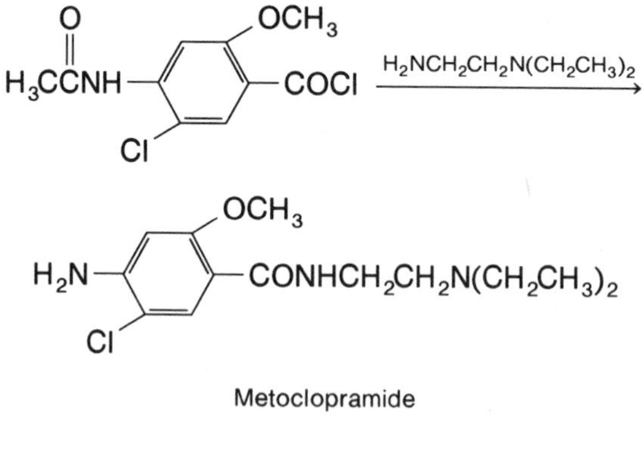

Metoclopramide

17.7 VITAMINS

Number 82 in the Top-100 is a multivitamin/fluoride preparation and two typical formulations for such preparations are shown in Table 17.1. Fluoride for teeth, and minerals such as iron are frequently added. Of course, vitamins are freely available over the counter and do not require a prescription, so that their rank in the Top-100 is unrealistically low in terms of quantities consumed. Vitamins are taken by many people as a diet supplement and in 1979 in the United States about 59 million pounds of vitamins were bought for over $350 million.

Medically established vitamin deficiencies are rare in the United States, but may be associated with gastrointestinal or liver disease. They are usually treated with specific vitamins, not mixtures. It is significant that, of the vitamins in a typical formulation, the minimum daily requirement for two has not been established and another two have not even been shown to be necessary in human nutrition.

The Council on Scientific Affairs of the American Medical Association has taken the position that healthy men and nonpregnant, nonlactating women eating a "usual, varied diet" do not need vitamin supplements. At given times vitamin supplements may be required by infants, pregnant and lactating women, individuals with "unusual life-styles or modified diets," those with deficiency states or pathological conditions that reduce vitamin absorption, and others with nonnutritional disease processes.

Vitamins occur naturally and may be isolated from natural sources. Often the synthesis is more economical and we have reviewed the synthetic methods in some detail (see bibliography). Some of the chemistry is exciting and we shall include three examples here.

17.7.1 Vitamin C

The commercial synthesis of vitamin C is shown in Figure 17.4. The interesting reaction is the second one—the conversion of D-sorbitol to L-sorbose. The problem is to convert one of the four secondary hydroxyl groups to a ketone. This is difficult even with modern techniques and chemists in the 1930s were hard pressed to devise a method. Finally, Haworth and Reichstein reported almost simultaneously that a microbiological process was the answer. A bacterium known as *Acetobacter suboxydans* contained the proper enzyme to bring about the reaction.

Here is yet another example of the importance of microbiological processes in the pharmaceutical industry. Earlier chapters have shown how important they are in the production of antibiotics and in the insertion of a hydroxyl group into the 11-position of progesterone in steroid chemistry. The dramatic possibility of devising "custom-made" bacteria is illustrated by the production

TABLE 17.1. Typical Formulations of Multivitamin Preparations

Vi-Penta Multivitamin Drops

Each 0.6 cc of Vi-Penta multivitamin drops provides:		% Minimum Daily Requirements (MDR)		
		Infants	Children	
		(under 1 year)	(1 – 6 years)	(6 – 12 years)
Vitamin A (as the palmitate)	5000 USP Units	333	166	166
Vitamin D_2	400 USP Units	100	100	100
Vitamin C	50 mg	500	250	250
Vitamin B_1 (as hydrochloride)	1 mg	400	200	133
Vitamin B_2 (as riboflavin-5′-phosphate sodium)	1 mg	166	111	111
Vitamin B_6	1 mg	*	*	*
Vitamin E (as dl-α-tocopheryl acetate)	2 Int. Units	*	*	*
d-Biotin	30 μg	†	†	†
Niacinamide	10 mg	*	200	133
D-Panthenol (equiv. to 11.6 mg Calcium pantothenate)	10 mg	†	†	†

*MDR for these vitamins has not been determined.
†The need for these vitamins in human nutrition has not been established.

Multivitamin capsules with minerals

One tablet daily contains:	
Vitamin A	25,000 USP units
Vitamin D	1,000 USP units
Vitamin C w/rose hips	250 mg
Vitamin E	100 IU
Folic acid	400 μg
Vitamin B_1	80 mg
Vitamin B_2	80 mg
Niacinamide	80 mg
Vitamin B_6	80 mg
Vitamin B_{12}	80 μg
Biotin	80 μg
Pantothenic acid	80 mg
Choline bitartrate	80 mg
Inositol	80 mg
para-Aminobenzoic acid	80 mg
Rutin	30 mg
Citrus bioflavanoids	30 mg
Betaine hydrochloride	30 mg
Glutamic acid	30 mg
Hesperidin complex	5 mg
Iodine (from kelp)	0.15 mg
Calcium gluconate*	50 mg
Zinc gluconate*	25 mg
Potassium gluconate*	10 mg
Ferrous gluconate*	10 mg
Magnesium gluconate*	7 mg
Manganese gluconate*	6 mg
Copper gluconate*	0.5 mg

*Natural mineral chelates in a base containing natural ingredients.

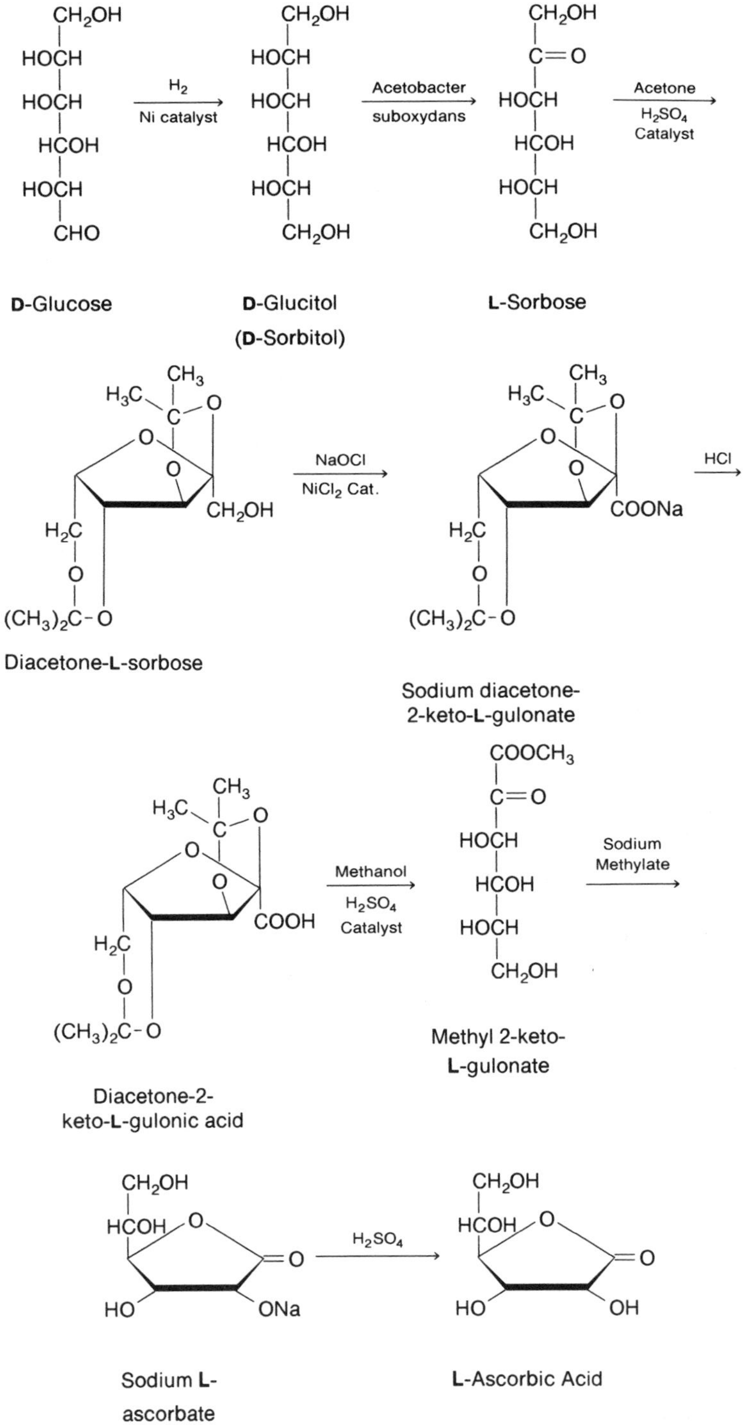

D-Glucose → (H_2, Ni catalyst) → D-Glucitol (D-Sorbitol) → (Acetobacter suboxydans) → L-Sorbose → (Acetone, H_2SO_4 Catalyst) →

Diacetone-L-sorbose → (NaOCl, $NiCl_2$ Cat.) → Sodium diacetone-2-keto-L-gulonate → (HCl) →

Diacetone-2-keto-L-gulonic acid → (Methanol, H_2SO_4 Catalyst) → Methyl 2-keto-L-gulonate → (Sodium Methylate) →

Sodium L-ascorbate → (H_2SO_4) → L-Ascorbic Acid

Figure 17.4. *Synthesis of vitamin C.*

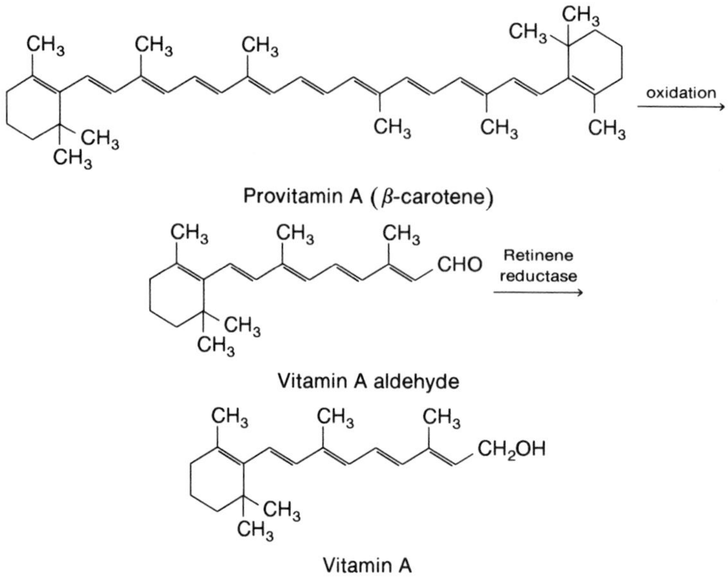

oxidation

Provitamin A (β-carotene)

Retinene reductase

Vitamin A aldehyde

Vitamin A

Figure 17.5. *Metabolization of provitamin A.*

of insulin and interferon via recombinant DNA (Section 13.2). Such processes offer great hope for the future.

17.7.2 Vitamin A

Provitamin A or β-carotene is the orange coloring matter in carrots, tomatoes, and oranges. Figure 17.5 shows how provitamin A is oxidized in the body to vitamin A aldehyde and then reduced to vitamin A by retinene reductase.

Vitamin A used to be made by the traditional Isler synthesis, but this has been overtaken by a route shown in Figure 17.6. Isobutene (1), formaldehyde, and acetone condense to give a ketone (2) that on hydrogenation gives the methylheptenone (3). Reaction of (3) with sodium acetylide gives dehydrolinalool (4), which, in turn, reacts with ethyl acetoacetate to give pseudo-ionone (5). Treatment with acid closes the ring to give β-ionone (6). Reaction of (6) with lithium acetylide followed by hydrogenation gives vinyl β-ionol (7). Treatment of (7) with triphenyl phosphine and hydrogen bromide gives an alkylidene triphenylphosphorane (8), which reacts with β-formylcrotyl acetate (9) to give vitamin A acetate (10).

The preparation of β-formylcrotyl acetate is shown in the lower half of the diagram. Note the hydroformylation reaction with carbon monoxide, hydrogen, and a rhodium catalyst. This reaction is now widely used in industrial chemistry. β-Formylcrotyl acetate and β-ionone (6) are both synthons and are also widely used as intermediates for drugs and perfumes.

1. Vitamin A synthesis:

$H_3C-C(=CH_2)-CH_3$ + $H_2C{=}O$ + $(CH_3)_2C{=}O$ → (2) → Pd catalyzed isomerization → (3)

(1) Isobutene; Formaldehyde; Acetone; (3) Methylheptenone

NaCCH → (4) Dehydrolinalool

$CH_3COCH_2COOC_2H_5$ (Acetoacetic ester) Carroll reaction → (5) Pseudo-ionone

Acid → (6) β-Ionone

(1) LiCCH (2) H_2/catalyst → (7) Vinyl-β-ionol

$(C_6H_5)_3P$ + HBr → (8) $CH_2-P^+(C_6H_5)_3Br^-$ An alkylidene triphenylphosphorane

Base + β-formylcrotyl acetate (see below) Wittig reaction → (10) Vitamin A acetate, CH_2OOCCH_3 (usually converted to palmitate by trans-esterification with methyl palmitate)

2. **β**-Formylcrotyl acetate synthesis:

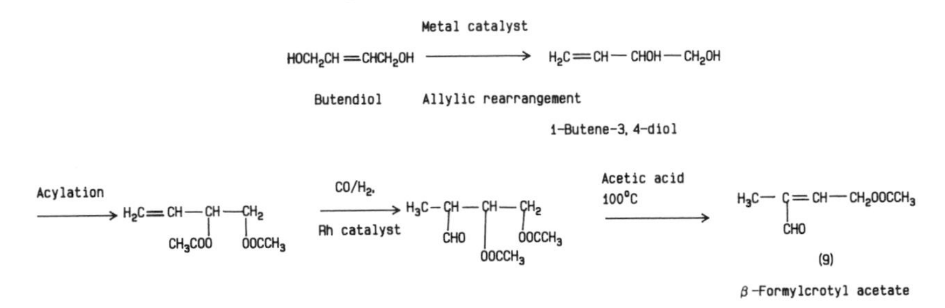

Figure 17.6. *Synthesis of vitamin A.*

17.7.3 Vitamin B_{12}

Vitamin B_{12}, also known as cyanocobalamin, has the complex structure shown in Figure 17.7. Vitamin B_{12} is an organometallic compound, a cyanocobalt complex. The elucidation of its structure was one of the triumphs of x-ray crystallography and won a Nobel prize for Dorothy Hodgkin in 1964. Vitamin B_{12} is made by fermentation with an organism known as *Streptomyces*

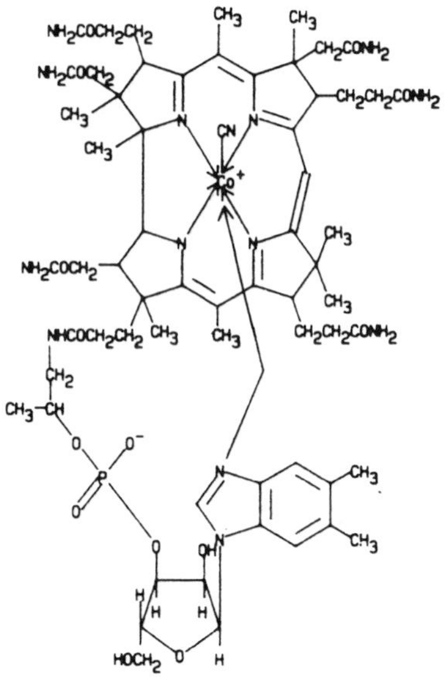

Figure 17.7. Structure of vitamin B_{12}.

olivaceus. The nutrient broth, in addition to carbohydrates and proteins, contains parts per million of a cobalt salt.

Perhaps the greatest achievement in modern synthetic organic chemistry was the synthesis of vitamin B_{12} by Woodward of Harvard and Eschenmosser of the Federal Institute of Technology in Zurich in 1973. The synthesis of such a complex material with its numerous chiral centers is far too complex to include here. But the fact that such a synthesis is possible is a great tribute to the genius inherent in the science of organic chemistry and in the scientists who practice it.

Part III

Drug Groups Outside the Top-100

Chapter 17 concludes our discussion of the 100 most widely prescribed drugs in the United States and their syntheses. We have included various drugs outside the Top-100 to give a well-rounded picture of the pharmaceutical industry. Nonetheless, we have stuck fairly closely to the idea of measuring the importance of a drug by the number of times it is prescribed in the United States. This is not completely realistic, but it has enabled us to impose a pattern on the complexities of the pharmaceutical industry.

There are various reasons why drugs might be important but still not rank high by this criterion. There are the drugs that are bought freely over the counter—nonnarcotic analgesics, cough remedies and "cold cures," hydrocortisone ointments, and vitamin preparations—and we have already mentioned these. We have also discussed some of the drugs that are administered in hospitals but less frequently on prescription. Injectable cephalosporins and penicillins, the more powerful drugs against mental disorders, narcotic analgesics, and certain vasodilators are examples. Other hospital drugs, such as intravenous solutions, sera and gamma globulin, general and local anesthetics, and blood products, fall outside the scope of this book.

Also not represented in the Top-100 are drugs for illnesses that are rare in the United States but widespread in other countries. Drugs for tropical diseases are a good example and they will be discussed in Chapter 18.

Next, there are the potentially important drugs of the future, which are foci of current research. The prostaglandins, and drugs against viruses and cancer are described in Chapters 19, 20 and 21. There are no completely effective drugs in the last two classes, but the ones that exist are of interest. Finally, in Chapter 22, there are some of the drugs for so-called minor diseases where the number of sufferers is not large enough to justify an expensive research program, let alone bring a drug into the Top-100.

18

Tropical Diseases

People who live in tropical countries, particularly those with poor sanitation, suffer from a range of diseases that are rarely found in more affluent countries in more temperate latitudes. Many of the diseases result from infection or infestation by microorganisms other than viruses and bacteria, the two main problems in developed countries. Indeed, there are five types of microorganisms responsible for tropical diseases: viruses, rickettsiae, bacteria, fungi, and protozoa. To these should be added parasitic worms and insects, which, although not strictly microorganisms, cannot be studied without the aid of a microscope.

Viruses are not cellular organisms and do not have a metabolism. They are the source of many diseases all over the world, for example, influenza, measles, and AIDS. They will be discussed in Chapter 20. The most important tropical diseases due to viruses are yellow fever and hepatitis.

Rickettsiae are single-cell organisms falling between viruses and bacteria in size and complexity. They are parasites within cells but otherwise resemble bacteria. They are responsible for Rocky Mountain spotted fever and various forms of typhus. They are countered by tetracyclines (Section 6.5) or chloramphenicol (Section 6.10.1).

Bacterial infections and the drugs to counter them were discussed in Chapter 6. Thirteen percent of prescriptions in the United States are for antibacterials. Bacterial infections particularly prevalent in the tropics are tuberculosis, cholera, typhoid fever, and leprosy. Tuberculosis was discussed in Section 6.7 but has largely been conquered in developed countries.

Fungi are plants that do not contain chlorphyll. The infectious ones are molds and yeasts. They are responsible for diseases such as athlete's foot and

TABLE 18.1. Tropical Diseases

Type of Infection	Worldwide Incidence (million new cases / year)	Worldwide Prevalence (million cases)
Protozoal infections		
Amebiasis	100	450
Malaria	100	
African trypanosomiasis (sleeping sickness)		5
Central and South American trypanosomiasis (Chagas' disease)	1	20
Leishmaniasis		100
Helminth (worm) infections		
Cestodes: tapeworm		150
Nematodes: hookworm		1000
Ascariasis (roundworm)		> 1000
Dracunculiasis (guinea worm)		30
Trichuriasis (whipworm)		> 500
Enteroviasis (pinworm)		
Filariasis	100	
Trichinelliasis	> 1	
Trematodes: Schistosomiasis (bilharzia)	10	200
Bacterial infections		
Tuberculosis	3.2	15–20
Cholera	Epidemic	
Typhoid fever	Epidemic	
Leprosy		12
Viral infections		
Yellow fever		
Hepatitis		
Fungal infections	10–100	

Source: Adapted from reference 3 / 2.

candidiasis. There are two antifungals in the Top-100 and they were described in Section 6.11.

Protozoa are the most complex of the single-celled organisms and are much larger than bacteria. Most are harmless or beneficial, but a number cause serious infections including malaria, amoebiasis, and trypansomiasis.

Parasitic worms and insects are multicellular organisms that live and reproduce on or in a host—sometimes a human one. Most tropical diseases are the result of infection by protozoa and by parasitic worms, also known as helminths. Table 18.1 lists some of the main tropical diseases under the agents causing the infection.

The astonishing feature about tropical diseases is the extraordinary number of people who suffer from them. The estimates given in Table 18.1 are only orders of magnitude figures; indeed, it is hard to imagine how accurate statistics might be collected. About ten million new cases of malaria were reported to the World Health Organization in 1979 but this is perhaps an order of magnitude lower than the actual number of cases. It is apparent that the majority of dwellers in tropical countries, if not most of humanity, suffers from one or more of the diseases listed.

The alleviation of tropical diseases would probably be helped most by effective public health measures, such as the provision of clean water and adequate toilets. After that there is a significant role for immunization. Drugs are probably only third in importance after these. In spite of this, chemotherapy still has an important part to play in alleviating the problems of third world countries.

18.1 INSECTICIDES

Infestation by insects varies in importance depending on whether the insect is a vector for a disease-causing microorganism. Malaria is carried by mosquitoes (Section 18.2.1); rickettsiae by mites, lice, and fleas; and African trypanosomiasis by the tsetse fly. These diseases may be diminished by insect control or the infecting microorganism can be countered.

Sometimes, the insect itself is the problem. Mites (which, strictly speaking, are arachnids not insects) cause scabies, and lice cause pediculosis. Head lice, in particular, are common in developed as well as developing countries. Treatment is with insecticides. Suitable ones include malathion (p. 365), carbaryl, benzyl benzoate, crotamiton, and pyrethrins/piperonyl butoxide.

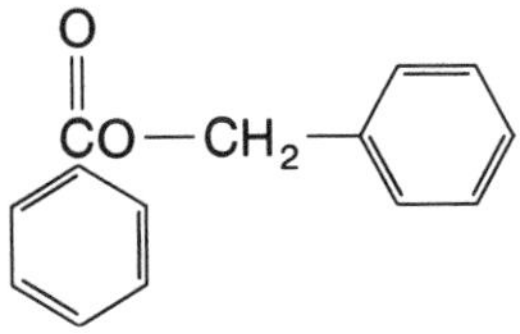

Benzyl benzoate

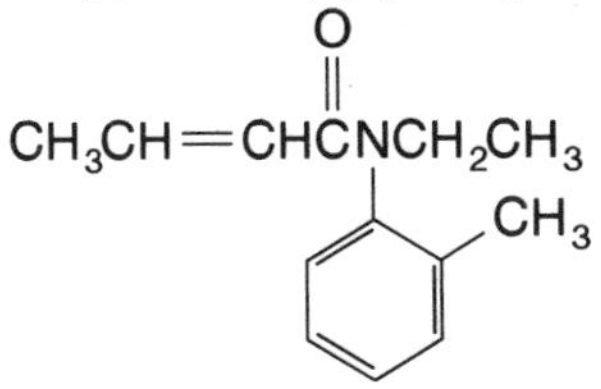

$CH_3CH{=}CHC(O)N(CH_2CH_3)(C_6H_4CH_3)$

Crotamiton

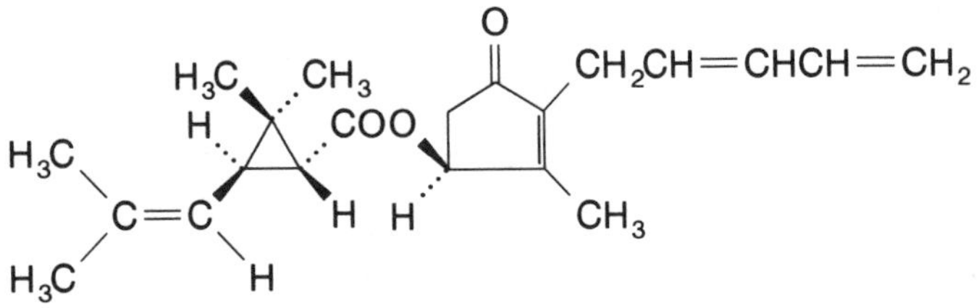

Pyrethrin I

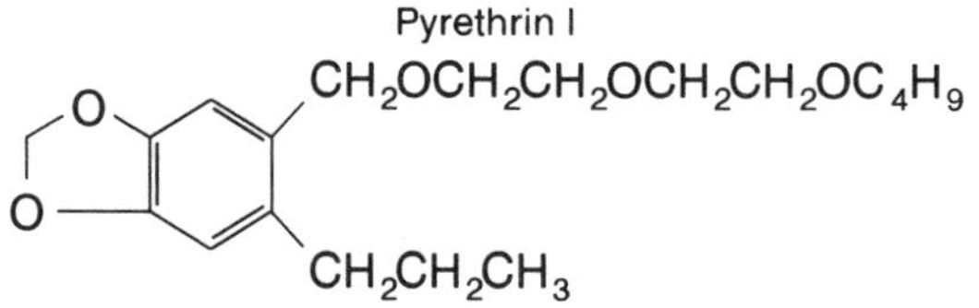

$CH_2OCH_2CH_2OCH_2CH_2OC_4H_9$ / $CH_2CH_2CH_3$

Piperonyl butoxide

$OCONHCH_3$

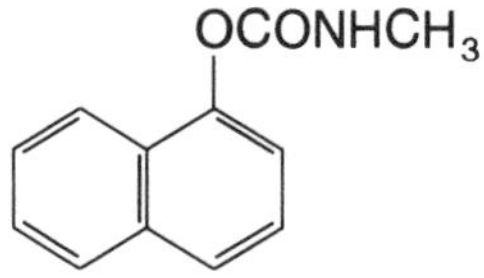

Carbaryl

The preferred agent for many years was lindane. It is a pesticide whose widespread use, like that of DDT, has been prohibited. It is persistent and is readily absorbed through the skin. Furthermore, various resistant strains of lice have emerged. It is still used, however, on a surprisingly large scale. Chemically it is γ-benzene hexachloride and is an alicyclic compound. It should not be confused with its aromatic equivalent hexachlorobenzene. There are eight stereoisomers of benzene hexachloride and lindane is the so-called γ isomer. The 1, 2, 4, and 5 chlorine atoms are cis and the 3 and 6 are trans. This geometric isomer may be separated from a mixture of benzene hexachlorides by crystallization.

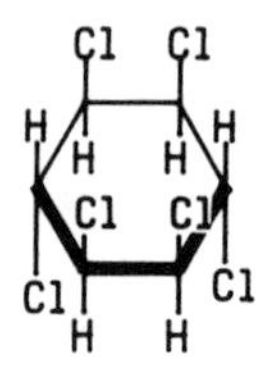

Lindane

18.2 ANTIPROTOZOALS

The main protozoal diseases in tropical countries are carried by insects that are the cause of malaria, trypanosomiasis (African sleeping sickness), Chagas' disease (South American trypanosomiasis), and leishmaniasis. Other protozoal diseases are spread by other routes, especially fecal contamination, and include amebiasis, trichomoniasis, and to some extent Chagas' disease.

18.2.1 Malaria

The world's greatest single health problem is malaria. It is said to be the most common cause of illness and death from infectious disease (see bibliography). It is a threat to half the world's population and kills one in five. It is one of the most debilitating of the diseases listed in Table 18.1. Malaria is spread by the female of a mosquito known as *anopheles*; which harbors the malaria-causing organism. The best way to control it is at its source by the elimination of the mosquitoes. In 1937 the insecticidal properties of DDT were discovered by Mueller and that discovery won him a Nobel prize in 1948. DDT is synthesized from trichloroacetaldehyde and chlorobenzene:

CCl_3CHO + 2 Cl–C_6H_5 ⟶ Cl–C_6H_4–CH(CCl_3)–C_6H_4–Cl

Trichloroacetaldehyde   Chlorobenzene   Dichlorodiphenyltrichloroethane (DDT)

In 1948 the World Health Organization started a program to eradicate malaria, partly by spraying with DDT and other insecticides and partly by antimalarial drugs (see below). The program was a dramatic success and many countries were freed almost entirely from the disease. In Sri Lanka, for example, prevalence declined from 3 million to 20 reported cases by the early 1960s. Unfortunately, the program ran into two problems. First, more and more species of mosquito developed immunity to DDT and other insecticides and, second, DDT was attacked on environmental grounds. DDT is insoluble in water and highly lipophilic. When eaten by animals, birds, or fish, it accumulates in their fatty tissues, and they may ultimately be eaten by humans. The DDT enters the body fat and there is scarcely a person alive today without detectable quantities of DDT in his or her tissues. No one has ever proved this to be harmful but there is an intuitive feeling that people would be better off without such foreign substances trapped in their bodies.

For this reason, DDT was outlawed in the United States and many other countries. There is no evidence that DDT ever killed anyone. Furthermore, there is no doubt that at least half a billion people are alive today because of the elimination of mosquitoes and other pests by DDT. Here is an ethical problem. Should DDT be used to save life now when there are uncertainties about the long-term consequences of its use? The question is a difficult one, but has been made less clear-cut by the development of DDT-resistant mosquitoes. In many parts of the world the traditional methods of mosquito control—the clearing of swamps and covering of infested areas with oil—are being reinstated. And cases of malaria in Sri Lanka once more exceed the million mark.

Quinine was the first antimalarial drug. It is isolated from cinchona bark and has unpleasant side effects. In 1926 Bayer introduced pamaquine which was the first of a series of antimalarials based on the 8-amino-6-methoxyquinoline nucleus.

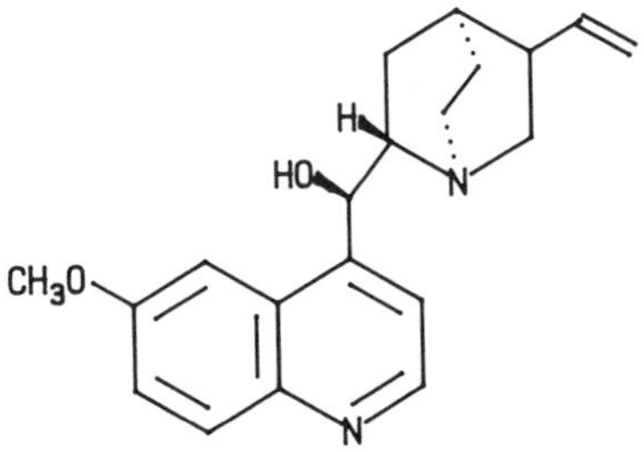

Quinine

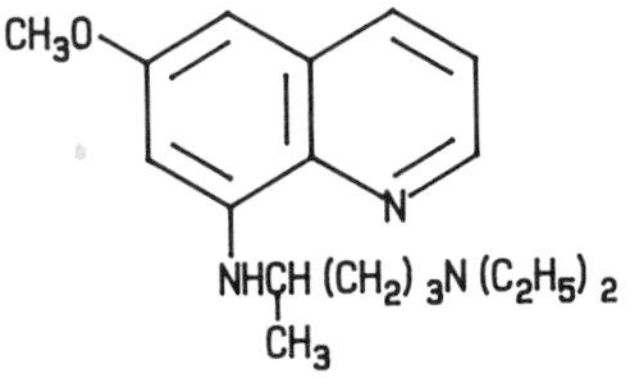

Pamaquine

During World War II, many Allied soldiers were stationed in areas where malaria was endemic and an effective drug to prevent or to cure the disease was of military importance. Two groups of drugs were found. One group related to 8-amino-6-methoxyquinoline, that is, drugs of the pamaquine type of which primaquine is the most widely used today. The other group was based on 4-amino-7-chloroquinoline, and chloroquine is the best example:

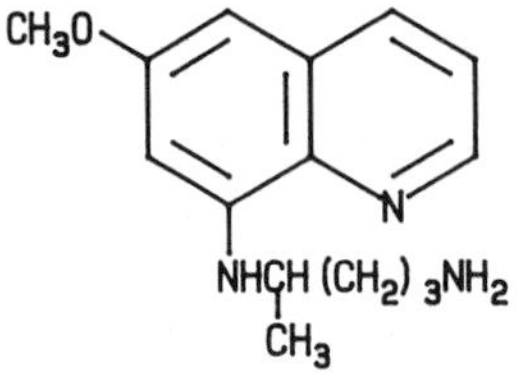

Primaquine

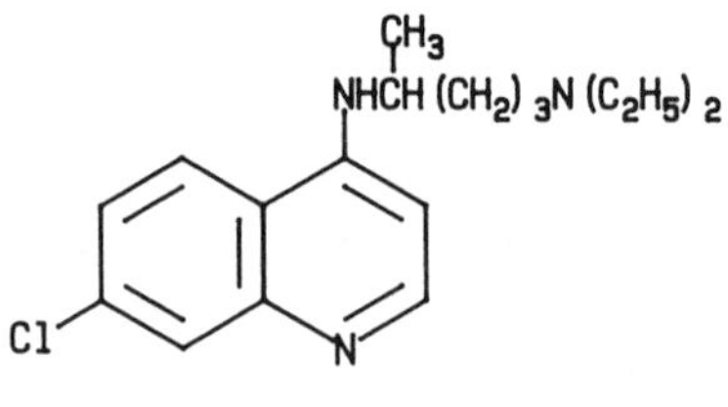

Chloroquine

The chloroquinolines are the more important group and chloroquine is the drug of choice within it. It long ago replaced quinine. Its synthesis is shown in Figure 18.1. *m*-Chloroaniline (1) is condensed with ethyl oxaloacetate (2) to

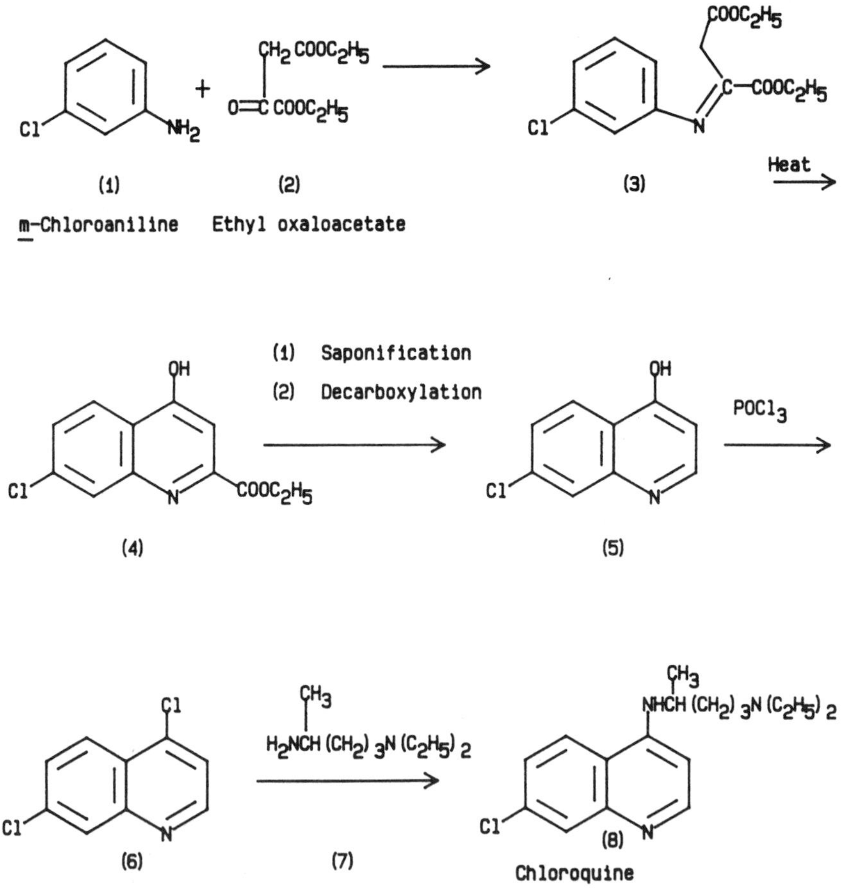

Figure 18.1. *Synthesis of chloroquine.*

1. Pamaquine synthesis:

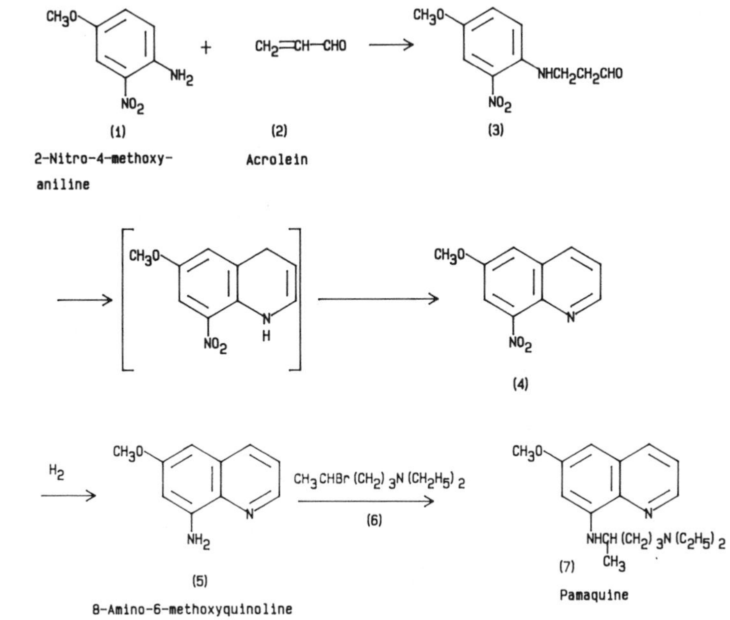

CH_3O NH_2 NO_2 (1) 2-Nitro-4-methoxyaniline + $CH_2{=}CH{-}CHO$ (2) Acrolein → CH_3O $NHCH_2CH_2CHO$ NO_2 (3)

→ [CH_3O NO_2 N H] → CH_3O NO_2 N (4)

H_2 → CH_3O NH_2 N (5) 8-Amino-6-methoxyquinoline — $CH_3CHBr(CH_2)_3N(CH_2H_5)_2$ (6) → CH_3O N $NHCH(CH_2)_3N(C_2H_5)_2$ CH_3 (7) Pamaquine

2. Primaquine synthesis:

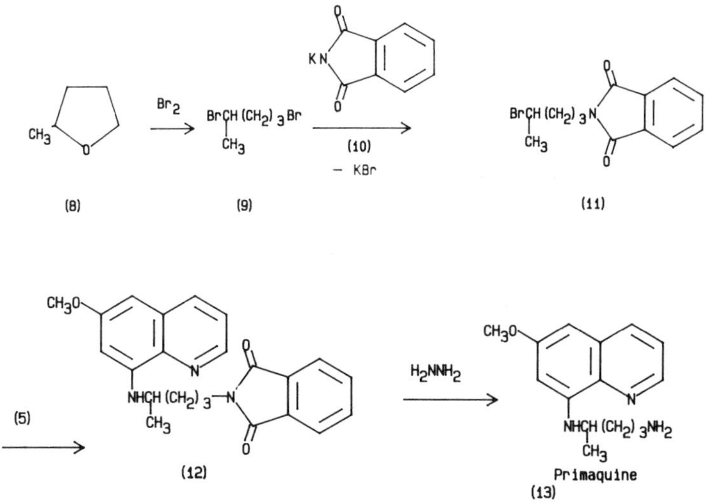

CH_3 O (8) — Br_2 → $BrCH(CH_2)_3Br$ CH_3 (9) — KN O O (10) − KBr → $BrCH(CH_2)_3N$ O O CH_3 (11)

(5) → CH_3O N $NHCH(CH_2)_3{-}N$ O O CH_3 (12) — H_2NNH_2 → CH_3O N $NHCH(CH_2)_3NH_2$ CH_3 Primaquine (13)

Figure 18.2. *Syntheses of pamaquine and primaquine.*

form a Schiff base (3). Heating this compound leads to cyclization and formation of (4), a quinoline nucleus with substituents in three positions. One of these is an ester group whose saponification gives an acid, which is decarboxylated to give a hydroxychloroquinoline (5). Treatment of (5) with phosphorus oxychloride gives dichloroquinoline (6). The chlorine in the 4-position of this molecule is allylic and much more reactive than the chlorine in the 7-position. Accordingly, it will react with the diamine (7) to give chloroquine (8).

The second group of antimalarials is derivatives of 8-amino-6-methoxyquinoline, and it includes pamaquine and primaquine. Their syntheses are shown in Figure 18.2 starting with 2-nitro-4-methoxyaniline (1). This is treated with glycerol in concentrated sulfuric acid. The acid converts the glycerol to acrolein (2) and the amino group of the methoxyaniline adds across the double bond of the acrolein to give (3). Compound (3) rearranges to the

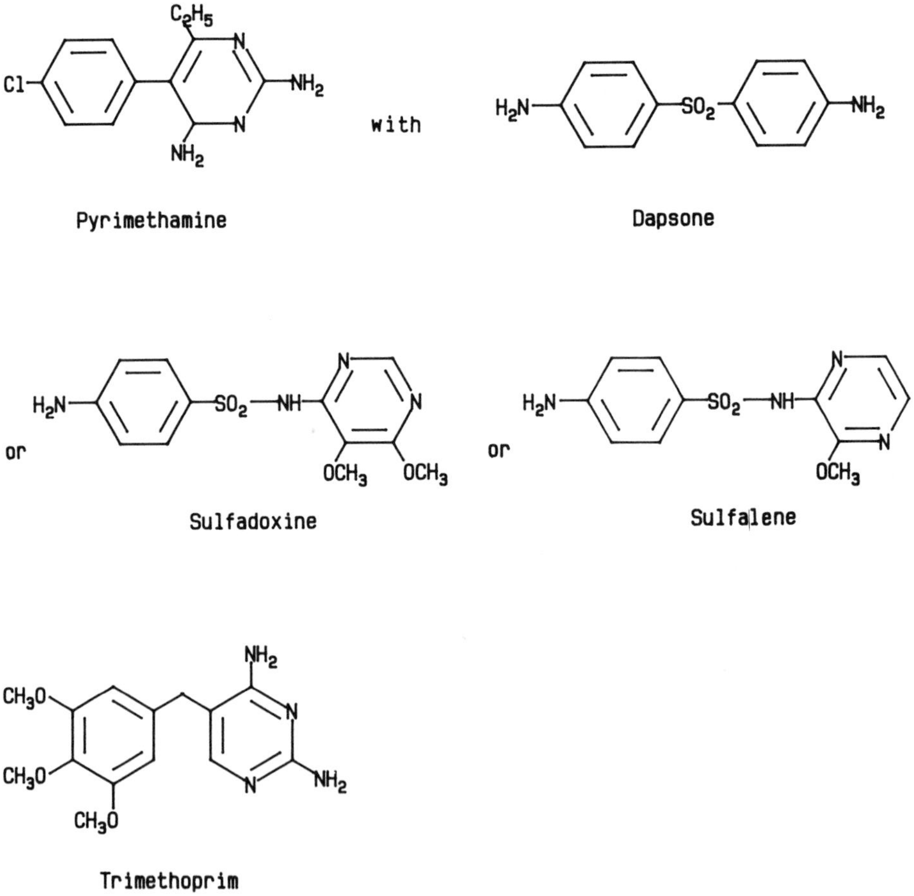

Cl C_2H_5 N NH_2 N NH_2

Pyrimethamine

with

H_2N SO_2 NH_2

Dapsone

or

H_2N SO_2 NH N N OCH_3 OCH_3

Sulfadoxine

or

H_2N SO_2 NH N N OCH_3

Sulfalene

NH_2 CH_3O N CH_3O N NH_2 CH_3O

Trimethoprim

Figure 18.3. *Sulfonamide and sulfone antimalarials, and trimethoprim.*

nitromethoxyquinoline (4). Reduction of the nitro group gives 8-amino-6-methoxyquinoline, the parent compound of the group. Alkylation with the appropriate bromoamine (6) gives pamaquine (7).

Primaquine synthesis starts with 2-methyltetrahydrofuran (8), the ring of which is opened by bromine to give the dibromide (9). The secondary bromine atom is sterically hindered to the extent that treatment with potassium phthalimide (10) gives exclusively compound (11). This reacts with 8-amino-6-methoxy-quinoline (5) to give (12). Removal of the phthalimide group by an exchange reaction with hydrazine gives primaquine (13).

Figure 18.3 shows pyrimethamine, which is used in combination with sulfones, such as dapsone, or sulfonamides, such as sulfadoxine or sulfalene, in the prevention and, in certain cases, the cure of malaria. Trimethoprim, also shown in the figure, is used for the prevention of pyrimethamine-resistant strains. Its major use is as a sulfa drug synergist (Section 6.2).

Unfortunately, chloroquine-resistant strains of the malaria parasite are developing and research is underway to develop drugs to treat them. Amodiaquine, an obsolete drug, was brought back for this reason, but its use was discontinued after it was implicated in four deaths. Proguanil has also been reintroduced as a preventative:

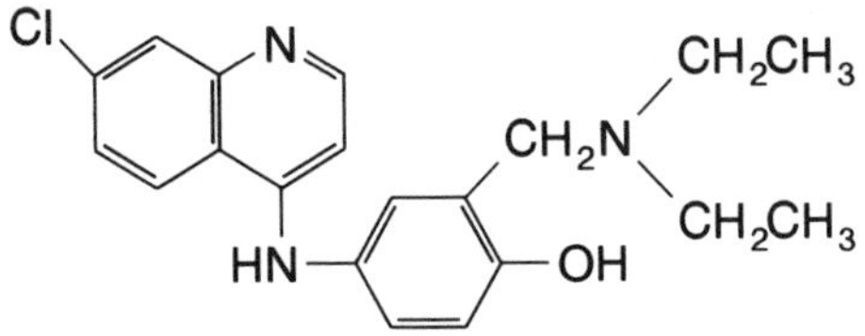

Cl HN N CH_2N CH_2CH_3 CH_2CH_3 OH

Amodiaquine

Cl NH NH NHCNHCNHCH CH_3 CH_3

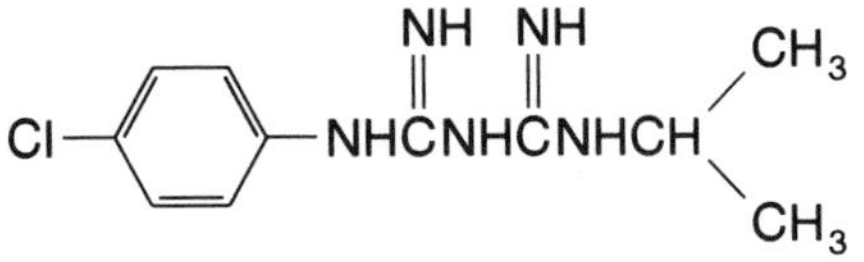

Proguanil

Mefloquine was developed by the U.S. Army medical research in 1963 but has not yet been cleared for civilian use. The Chinese claim a new group of antimalarials developed from old herbal remedies and based on a sesquiterpene lactone called qinghaosu or artemisinin, but there are still questions about its toxicity and effect on fetuses.

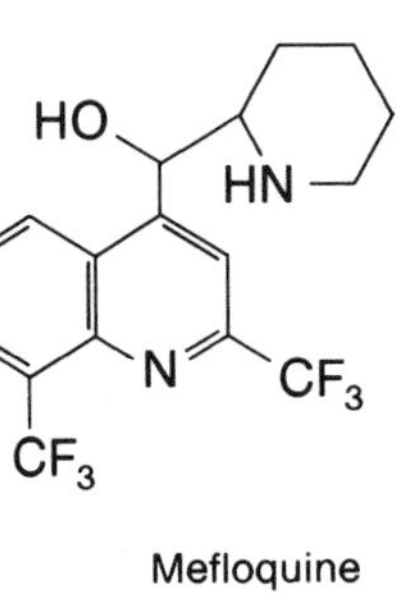

HO HN N CF_3 CF_3

Mefloquine

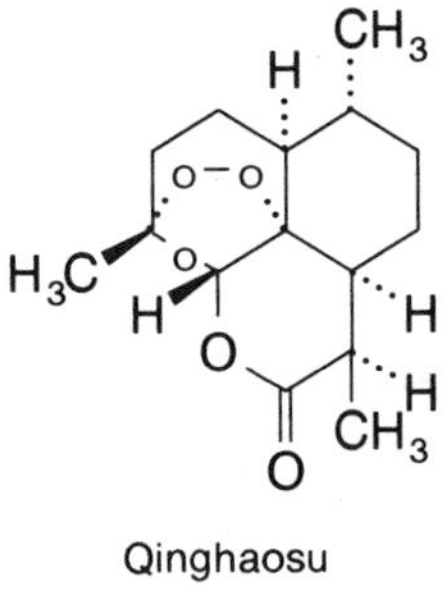

CH_3 H O O O H_3C H O H H O CH_3

Qinghaosu

Current research is directed toward the discovery of a malaria vaccine. The Swiss company Hoffmann-La Roche has announced the isolation of the surface antigen of the sporazoite stages of the parasite causing forms of malaria, and tests of the vaccine have started.

18.2.2 Trypanosomiasis and Leishmaniasis

Two other widespread protozoal diseases are trypanosomiasis and leishmaniasis. Trypanosomiasis occurs in two forms—African trypanosomiasis, or African sleeping sickness; and Central and South American trypanosomiasis, or Chagas' disease. Although seen in a less emotional light than malaria, they nonetheless cause untold misery, death, and economic loss in many parts of the world. The combatting of them, like the eradication of malaria, faces problems not only in the development of suitable drugs, but also in the delivery of such drugs to the population at risk. A depressing sign is the increase in number of cases of leishmaniasis around the Mediterranean basin where the numerous tourists seem particularly at risk and carry the disease home with them.

Two very early drugs, tryparsamide and suramin, were discovered in the 1920s. One is an arsenic compound and the other a dyestuff derivative and a urea, and both arose out of Ehrlich's pioneering work that led to salversan (Chapter 1). Both are still used today and are active against African sleeping sickness.

H_2NCOCH_2NH–C_6H_4–$As(=O)(OH)(ONa)$

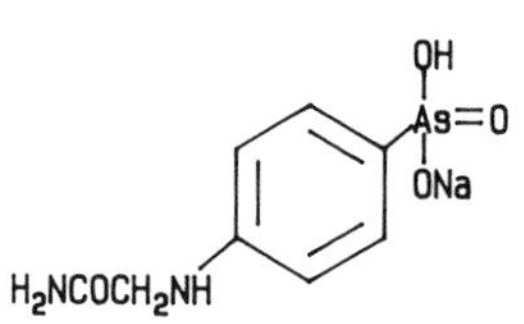

Tryparsamide

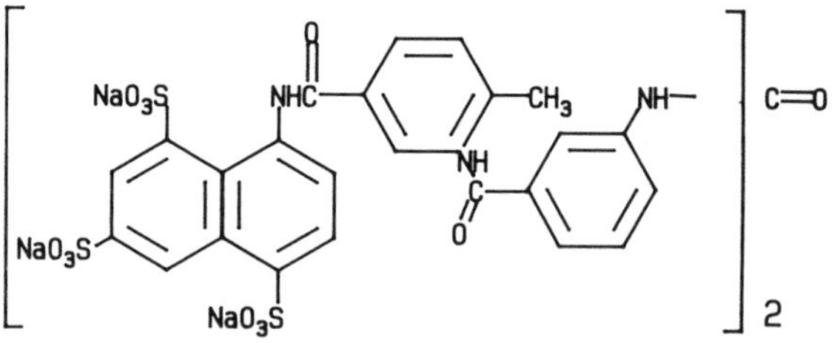

Suramin sodium

Another arsenic compound, melarsoprol, is also active and is the drug of choice to eliminate organisms in the central nervous system in the later stages of the disease.

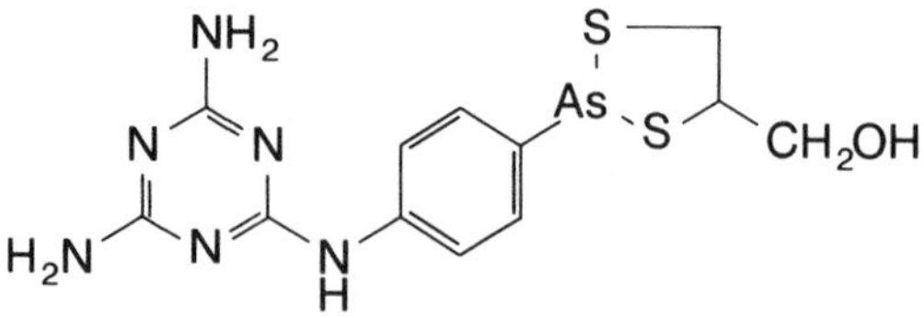

Melarsoprol

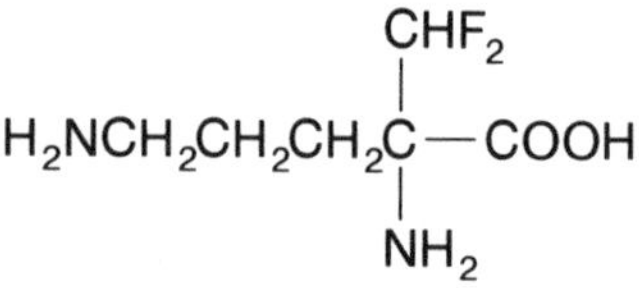

$H_2NCH_2CH_2CH_2C(CHF_2)(NH_2)$—$COOH$

Difluoromethylornithine

A new drug undergoing tests is difluoromethylornithine (see above). It inhibits ornithine decarboxylase, a key enzyme in polyamine synthesis.

Two much newer drugs active against Chagas' disease, which is more difficult to treat than African sleeping sickness, are nitrofurtimox and benznidazole:

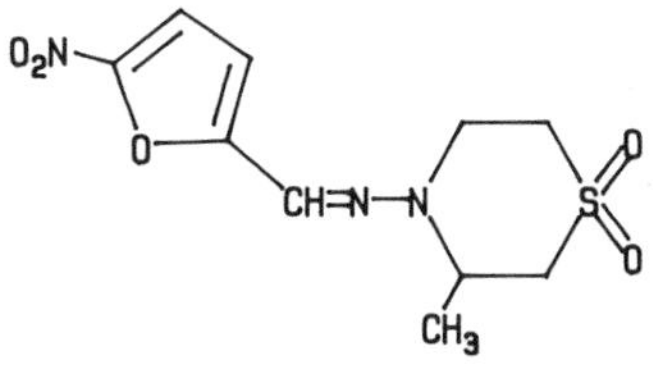

Nifurtimox

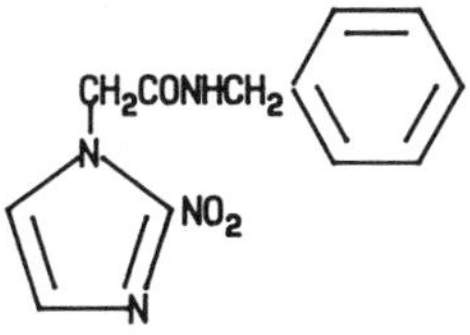

Benznidazole

Nitrofurtimox is a nitrofuran like nitrofurantoin (Section 6.10.1) and seems to be the long-awaited breakthrough in the treatment of the disease.

Leishmaniasis is treated with pentavalent antimony compounds given intravenously. The drug of choice is sodium stibogluconate.

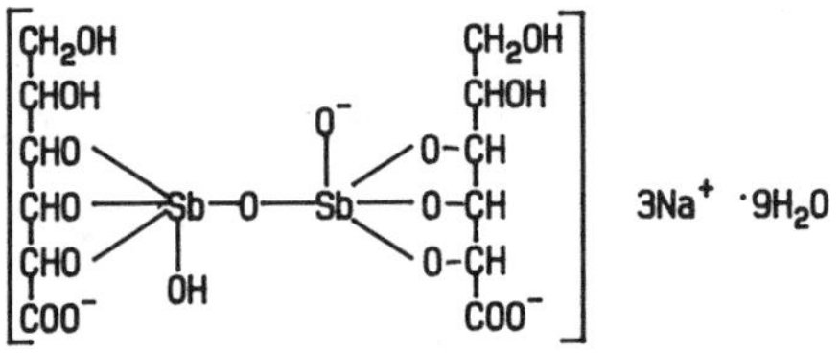

Sodium Stibogluconate

18.2.3 Amebiasis and Trichomoniasis

Amebiasis is the name for infection by the protozoon *Entamoeba histolytica*. About 10% of humanity is afflicted with it and there are 100 million new cases annually. Even in the United States, incidence is 2–5% and among homosexuals it is 25–32%. The main drugs to counter it are iodoquinol and metronidazole:

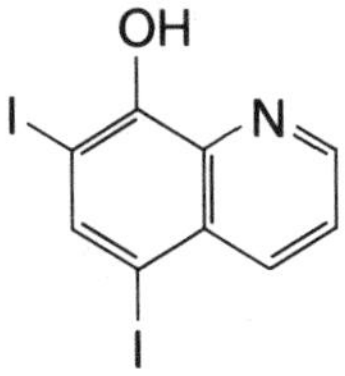

Iodoquinol

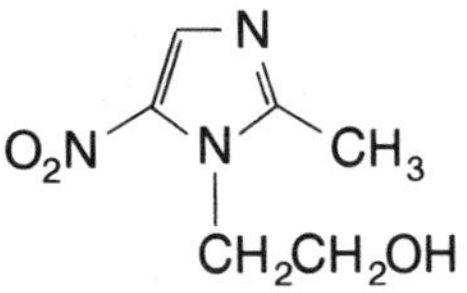

Metronidazole

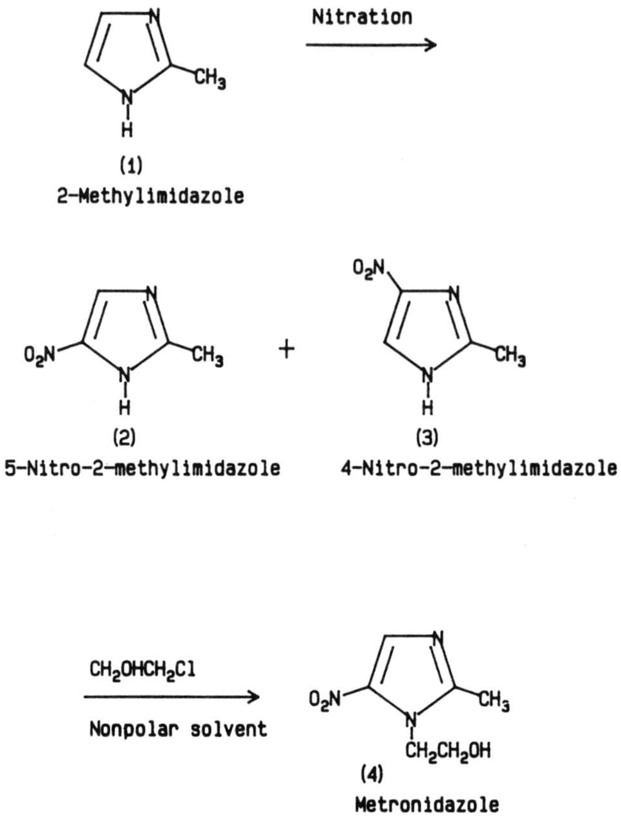

Nitration

(1)
2-Methylimidazole

(2)
5-Nitro-2-methylimidazole

(3)
4-Nitro-2-methylimidazole

CH_2OHCH_2Cl
Nonpolar solvent

CH_2CH_2OH
(4)
Metronidazole

Figure 18.4. Synthesis of metronidazole.

The advantage of metronidazole is that it is effective against amoebas at doses well below those toxic to the host. In general, antibiotics, such as tetracyclines, are given as supportive therapy. Although metronidazole is the drug of choice in amoebic dysentery, chloroquine, a quinoline derivative used in malaria, is sometimes an alternative.

Trichomoniasis is an infection of the vagina, or of the male prostate or sexual organs by *Trichomonas vaginalis*. Metronidazole is the drug of choice. Like miconazole and clotrimazole, it is an imidazole. In addition to the above applications, it is used in hospitals for preoperative preparation of the bowels, to prevent fecal contamination or infection by anaerobic bacteria, against which it is also active. It has a veterinary application against treponema, a parasite in pigs related to those causing syphilis and yaws in humans.

The synthesis of metronidazole is shown in Figure 18.4 Nitration of 2-methylimidazole (1) gives a mixture of 4- and 5-nitroimidazoles (2, 3). It has been found that alkylation of this mixture in a nonpolar solvent brings about a shift from the 4- to 5-isomer so that the 5-nitro isomer predominates. On the other hand, alkylation in an aprotic solvent such as dimethylformamide brings about a shift from the 5- to the 4-nitro isomer. The mixture containing the

5-nitro compound is treated with ethylene chlorohydrin in a nonpolar solvent. Metronidazole (4) results.

18.3 HELMINTH INFECTIONS

In addition to protozoal diseases, the populations of tropical areas suffer severely from the so-called helminth diseases in which worms infest the body. The three main families of worms that infest humans are called cestodes, nematodes, and trematodes.

Cestodes, usually known as tapeworms, infect about 150 million people. Unpleasant though they are, they do not usually lead to serious illness, and medically and economically tapeworms are less significant than the other helminth infections. They may be treated with a single dose of niclosamide or praziquantel (Section 18.3.2):

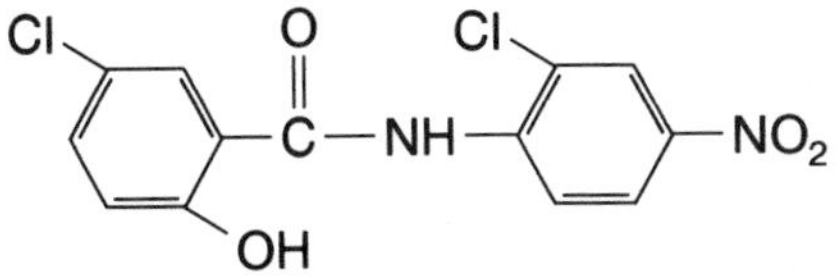

Niclosamide

Of the nematode infections, that by the hookworm is the most dangerous. The mature parasites attach themselves to the wall of the intestine and suck blood for the rest of their lifetime, estimated at about 5 years. There are thought to be about one billion hookworm sufferers, and the worm is thought to consume blood each year equivalent to the blood content of 200 million people.

The roundworm *Ascaris lumbricoides* accounts for another serious nematode infection. The statistics associated with this disease are astounding. The one billion people afflicted with roundworm excrete each year a total of 58,000 tons of *Ascaris* eggs.

There are also a number of other nematode infections shown in Table 18.1 —Guinea worm, whipworm, pinworm, filariasis, and trichinella. Trichinosis, that is infection by trichinella, is the classic illness from eating undercooked pork. These infections are less life-shortening than roundworm and hookworm but nevertheless cause suffering to a large proportion of the human race.

The most important of the trematode diseases is bilharzia or schistosomiasis, which afflicts some two to three hundred million people in the tropics. Infection occurs by penetration of the unbroken skin by species carried by various aquatic snails. It eventually leads to half-inch worms inside the veins of the infected person and slow destruction of the liver and kidneys. It is not in itself life-threatening, but is said to be a contributory factor to half of all deaths in Egypt.

It is sad that progress in areas such as controlled irrigation has led to an increase in bilharzia. The Aswan high dam, for example, has made possible extensive land reclamation by controlling the annual flooding of the Nile. It has also provided electricity for much of Egypt. The irrigation canals are filled with controlled amounts of water all the year round. This is excellent except that, in the past, when the irrigation canals dried up, they were dredged to clear them of silt. The dredging reduced the number of aquatic snails and, hence, reduced bilharzia. Controlled irrigation has led to its increase. Similarly, the irrigation of the Gezira area of the Sudan has led to an increase in infection to the point at which 60% of the seven-year-olds have bilharzia.

Unlike malaria, the diseases caused by protozoa and helminths are debilitating rather than fatal. Sufferers may live for many years in a weakened state in which they are susceptible to other diseases. Their economic potential is low and they are a burden on the economies in which they live. In Brazil, for example, 80–90% of the rural population of about 43 million are believed to suffer from infestation. Ten million have schistosomiasis, 8 million Chagas' disease, and the remainder hookworm, tapeworm, and similar infections, not to mention leprosy and tuberculosis. Only a small fraction of the population is healthy.

The great European epidemics such as the Black Death in the mid-fourteenth century carried off perhaps a half or two-thirds of the population, but they left healthy survivors to rebuild society. These tropical diseases help tie societies to a standard of living always on the border of subsistence.

18.3.1 Drugs against Nematodes

There are several drugs that have been used for many years as anthelmintics. Some of them are shown in Figure 18.5. They start with carbon tetrachloride, which was introduced in 1921 against hookworm, even though at that time it was well known that it could cause severe liver damage. Carbon tetrachloride was superseded by tetrachloroethylene, which is still used against hookworm. Phenothiazine, now better known as a source of the major tranquilizers, was used in large doses against helminth infections in the 1930s. During the 1960s two new groups of drugs against nematodes were developed which could be taken in far smaller doses. The first group was the benzimidazoles, exemplified by thiabendazole, parbendazole, and mebendazole, shown in the figure. Thiabendazole is effective against threadworm and pinworm, and moderately effective against roundworm and hookworm. It is widely used for livestock. Parbendazole was the first anthelmintic to contain a methyl carbamate group in the 2-position but it was displaced by mebendazole, which is active against threadworm, roundworm, and hookworm. It is believed to act by irreversibly blocking glucose uptake by nematodes in the colon.

The other group was imidazole–thiazoles and the figure shows levamisole, which has remained the preferred compound and is active against roundworm and hookworm in humans and animals. Other valuable drugs against nematodes include piperazine and its salts such as the citrate or tartrate, and pyrantel pamoate. Piperazine induces narcosis in the worms and is the drug of choice for pinworm infections.

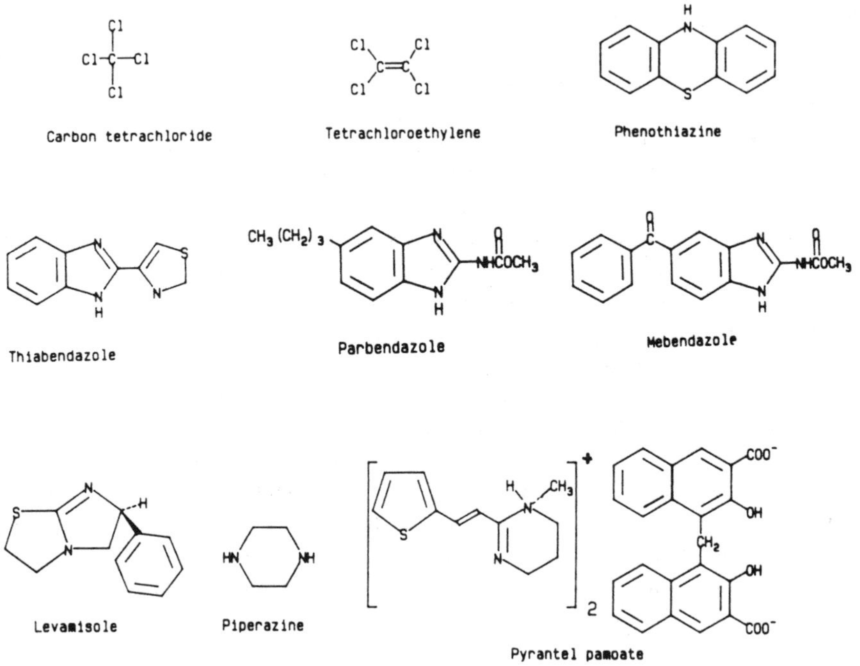

Carbon tetrachloride

Tetrachloroethylene

Phenothiazine

Thiabendazole

Parbendazole

Mebendazole

Levamisole

Piperazine

Pyrantel pamoate

Figure 18.5. *Drugs against nematodes.*

The above drugs can be used at a level of 5% of the dosage required for phenothiazine and the chlorinated hydrocarbons. A new series of anthelmintics has recently been isolated that appears to be effective at 3% of the dosage level of even the newer drugs. They are the fermentation products of a bacterium *Streptomyces avermitilis* and are called avermectins. A typical structure is

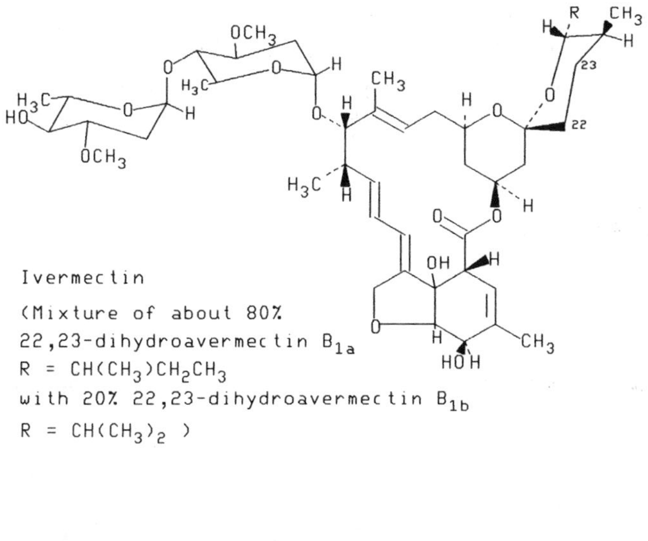

Ivermectin

(Mixture of about 80% 22,23-dihydroavermectin B_{1a} R = $CH(CH_3)CH_2CH_3$ with 20% 22,23-dihydroavermectin B_{1b} R = $CH(CH_3)_2$)

The avermectins appear to be more potent and to have a broader spectrum of activity than any previous compounds and may well be the anthelmintics of the future. Note how the newest drugs require the smallest doses. This is important because people are more likely to comply with instructions to take a single small dose of medicine each day than several large doses.

Clinical trials are currently in progress for the use of ivermectin (the name for a mixture of avermectins) against river blindness. This is a disease caused by the parasitic worm *Onchocerca volvulus* and transmitted by blackfly found on the shores of fast-flowing rivers. An immature form of the parasite, introduced by the bite of the fly, grows to adulthood in the human host and, in a lifespan of ten years, continually produces microscopic immature worms. In a severely infected person, they may number 50–200 million. These cause lesions and scar tissue, which is particularly serious in the eye. In some African villages, 15% of the population is blind, and a recent study of three villages in Ghana showed 12% of the inhabitants blind by age 30, and 60% by age 55.

Ivermectin disrupts the parasite's nervous system but has no effect on humans. It need only be taken once or twice per year. Thus, the logistical problems of supplying it to third world countries are much reduced. It is said to have fewer side effects and is much easier to administer than the present drug, diethylcarbamazine:

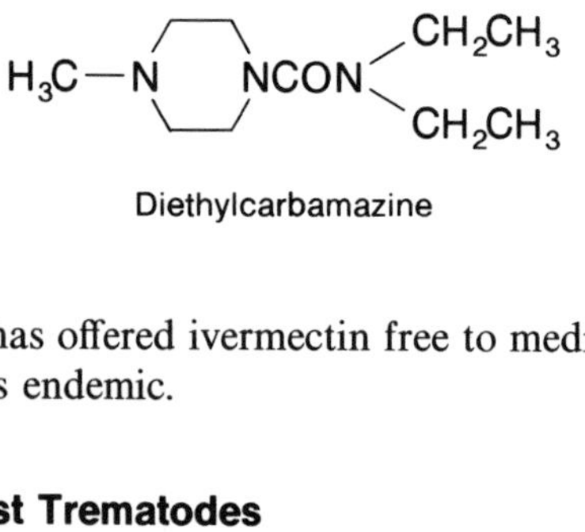

Diethylcarbamazine

Merck and Company has offered ivermectin free to medical programs in areas where onchocerciasis is endemic.

18.3.2 Drugs against Trematodes

Trematode worms responsible for schistosomiasis or bilharzia differ from roundworms and are sensitive to different drugs. The earliest drugs were trivalent antimony compounds such as potassium antimony tartrate intro-

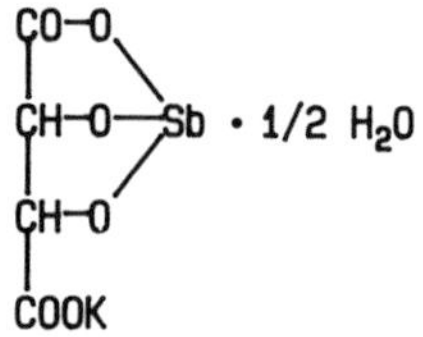

Potassium antimony tartrate

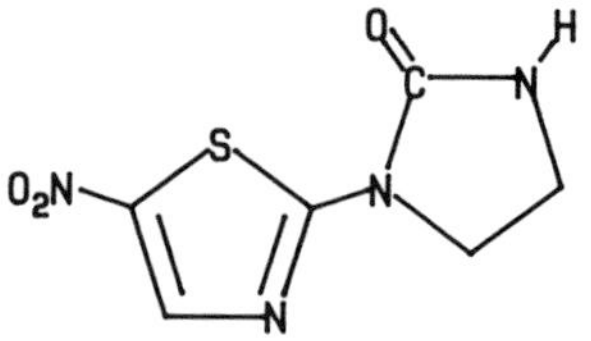

Niridazole

duced in 1918. Although the discovery was made by Christopherson, a British scientist, the pattern of thinking clearly is derived from Ehrlich's work with arsenicals. Antimonials have toxic side effects and prolonged treatment is required, so while they may still be of value in individual cases, they are of no use for mass chemotherapy.

Lucanthone appeared in the 1960s but had the drawbacks that multiple doses were required and there were toxic side effects. Niridazole, which appeared subsequently, suffered from the same problems. It is active against *Schistoma hematobium* worms and, even though it has been used in the United States for many years, it still only has investigational status.

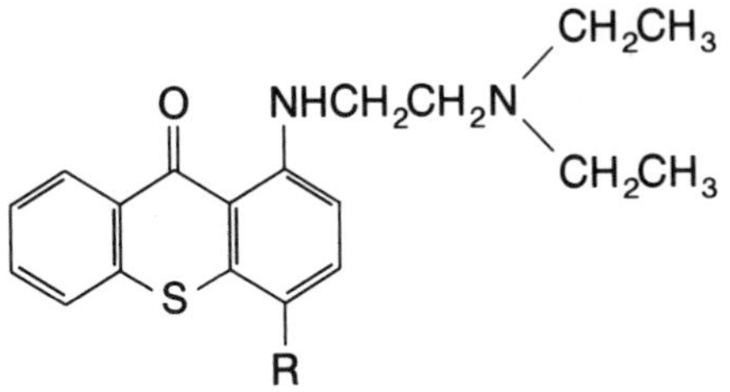

Lucanthone (R = CH_3) Hycanthone (R = CH_2OH)

Three drugs developed more recently are praziquantel, metrifonate, and oxamniquine:

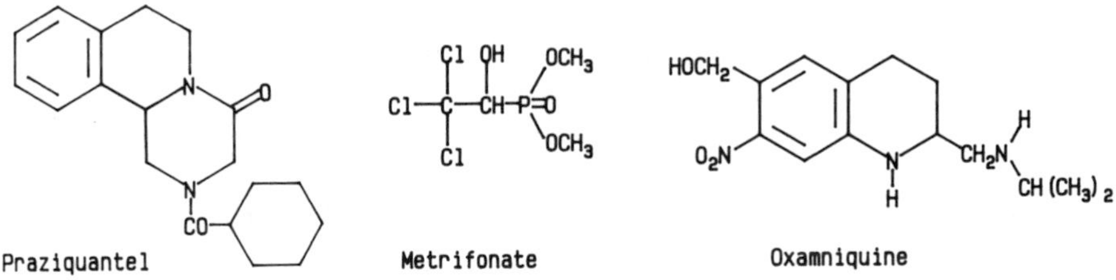

Praziquantel Metrifonate Oxamniquine

Praziquantel can be given in relatively low doses, although not nearly as low as the avermectins. Its long-term side effects are not yet known, but it appears to have the unique distinction of activity against all forms of schistosomiasis. Metrifonate is related to the organophosphorus insecticides and like them is a cholinesterase inhibitor. It is, unfortunately, active only against *Schistoma hematobium*, but is well tolerated by humans, although it lowers their plasma cholinesterase levels. Unlike the earlier compounds, which had to be taken daily, it need be given only three times at three-week intervals. Oxamniquine is the drug of choice against *Schistoma mansoni* worms and is used in most areas where bilharzia is endemic, but it is not available in the United States.

The development of oxamniquine has been reviewed by Richards (see bibliography). Research started from the lucanthone structure and led to the basic compound mirasan, which killed *Schistoma* in mice:

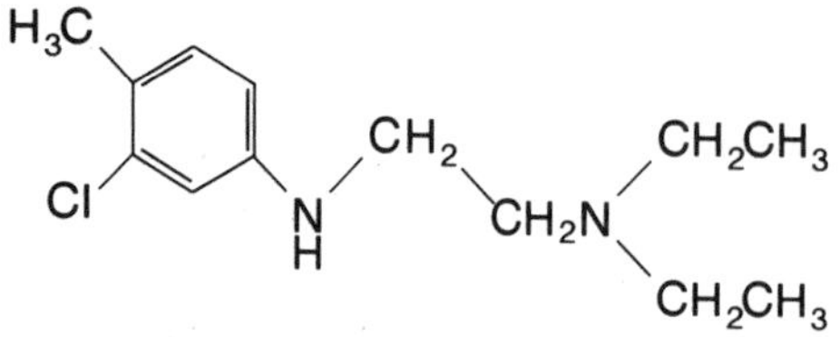

Mirasan

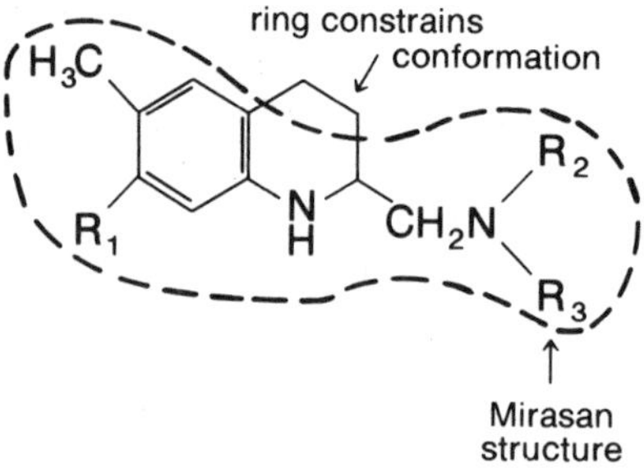

Tetrahydroquinolines

The important structural features of mirasan were the *β*-diethylaminoethyl side chain, the methyl group para to it and an electronegative group ortho to the methyl. It turned out that conformational constraint on the side chain was necessary if the compound was to lock efficiently to the receptor, and the incorporation of the first side chain carbon into a ring achieved this.

A range of tetrahydroquinolines of the type above was evaluated. The breakthrough for Pfizer, who was performing this research, came when a rival company, Winthrop, showed that lucanthone (above) was not active itself but was metabolized to hycanthone. In hycanthone, the para-methyl group has been converted to hydroxymethyl. Hycanthone was launched by Winthrop as a drug against schistosomiasis that could be administered in a single intramuscular injection, but its success was limited by toxicity problems. Meanwhile, Pfizer replaced their para-methyl by hydroxymethyl and came up with oxamniquine. Oxamniquine can be given either orally or in a single intramuscular injection. Even better, simultaneous administration of oxamniquine and praziquantel produces a synergistic effect and lower doses of both drugs can be used.

Recent field trials of the three above drugs by the World Health Organization have been remarkably successful and the Organization has described them as the most encouraging development since the eradication of smallpox.

18.4 TUBERCULOSIS, CHOLERA, TYPHOID FEVER, LEPROSY

A number of bacterial infections are particularly prevalent in tropical zones, among them tuberculosis, cholera, typhoid fever, and leprosy. Tuberculosis was discussed in Section 6.7.1. Effective drugs are available, yet half a million people a year worldwide still die of tuberculosis. The problem is one of drug delivery and compliance, not of chemotherapy.

Cholera is not endemic and occurs in epidemics. The last great cholera epidemic in Britain was in 1866, after which the provision of sewers and clean water countered the disease. Cholera vaccination has been available since 1914. The reason cholera may prove fatal is because it causes dehydration. Once the disease has been contracted, the consequences can be avoided by an intravenous saline drip which replaces the fluids in the body. Recent research on glucose-mediated electrolyte absorption in the small intestine has led to a technique for oral rehydration with solutions of sodium chloride, glucose, sodium bicarbonate, and potassium chloride. This is certainly effective for acute diarrhea and is advocated by the World Health Organization to counter cholera. Cholera is still a threat, especially in India, but the disease is best controlled environmentally and by the above techniques.

In countries where it is endemic, typhoid fever is best avoided by personal care—bottled or sterilized drinking water, avoidance of raw salads and vegetables, and consumption only of fruit that can be peeled. Monovalent typhoid vaccine is not as unpleasant as the traditional antityphoid–paratyphoid vaccine, known as TAB, but is still nasty. If the disease is contracted, chloramphenicol, which was mentioned in Section 6.10.1, is the drug of choice. Although chloramphenicol is of value for both diseases, typhoid fever, which is due to *Salmonella typhosa*, is not the same as typhus fever, which is a rickettsial infection.

Leprosy remains a tragic and widespread disease. Although it is well known to be scarcely infectious, its victims are cast out from society. The bacterium causing leprosy is related to that causing tuberculosis and some antituberculosis drugs have an effect on leprosy. Also like tuberculosis, treatment of leprosy is prolonged and in many parts of the world it is difficult to maintain continuity of treatment. The preferred drug is dapsone, but it is sometimes administered in a depot form as acedapsone:

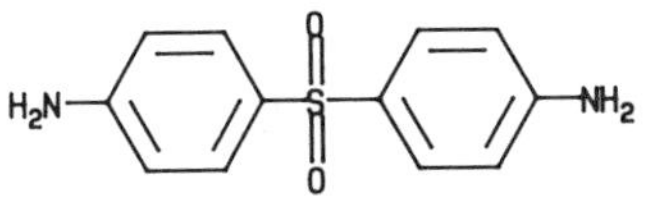

Dapsone

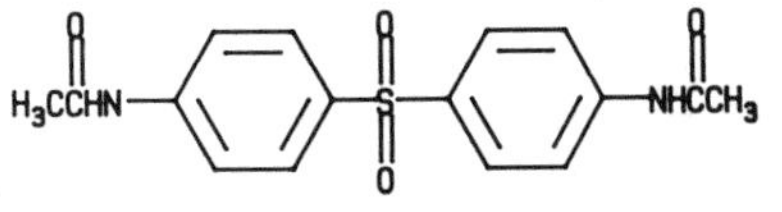

Acedapsone

This need only be given every 15 days and hydrolyzes to dapsone in the body. Curiously enough, dapsone is also a curing agent for epoxy resins. Dapsone and acedapsone are inexpensive, readily synthesized drugs, an important point in the economically deprived areas where leprosy is prevalent.

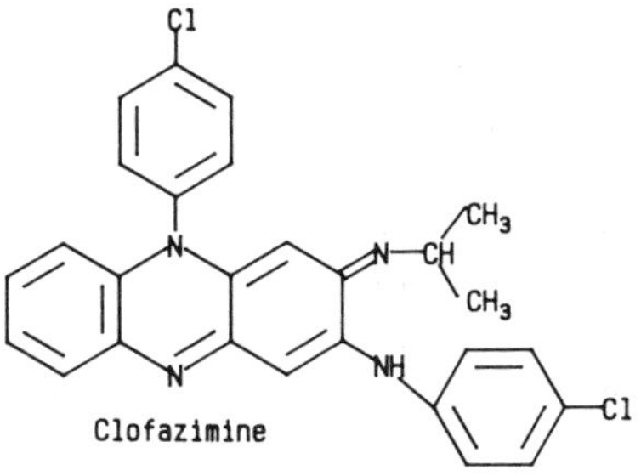

Clofazimine

Dapsone-resistant leprosy is usually treated with mixtures of clofazimine with the antituberculosis drug rifampin. Concurrent treatment with all three is seen as a potential route to the elimination of leprosy.

18.5 YELLOW FEVER AND HEPATITIS

Yellow fever is caused by a virus transmitted by a mosquito. Like typhoid fever it may be controlled by vaccination. The first yellow fever vaccine was produced in 1928, and currently the freeze-dried living attenuated 17D vaccine virus strain is the preferred prophylactic and lasts for at least 10 years. The mosquito vector lives in the thatched roofs of huts and the incidence of the disease may be decreased by the use of other roofing materials.

Hepatitis is a virus infection which attacks the liver. It has only recently been realized that it is the most common virus infection worldwide. Up to 90% of the population of some Chinese villages is infected and more than 20% in some African villages. It is linked with liver cancer, which is particularly prevalent in these areas. It is of current research interest and various vaccines are being developed to combat it. It is also a shared target for anti-AIDS drugs.

18.6 CONCLUSION

As should be apparent from the above, the range of drugs available to third world sufferers from tropical diseases is small compared with that available to sick people in developed countries. In some cases the drugs are not particularly effective and in others effective treatment would depend on a change in the way of life of the sufferers. Few effective new drugs have been developed in recent years compared with the dramatic innovations in, for example, heart drugs. With the increase in costs of developing new drugs, many firms have withdrawn from the potentially less profitable area of tropical diseases to concentrate on those offering a worthwhile economic return. Like most of the problems afflicting the third world, there is no obvious solution. The World Health Organization is doing its best with its Essential Drugs Program and its Tropical Disease Research and Training Program, but the organizational and political aspects of health care are among the most intractable of problems.

Writing in 1980, Burger (see bibliography) said it was disturbing that the drug firms of the developed countries were withdrawing from research on tropical diseases. At that time, the international pharmaceutical industry spent over $50 million/year on third world drug research out of a total R & D budget of $5000 million. The breakthroughs with drugs for schistosomiasis and river blindness indicate that the picture has perhaps improved since then.

What is perhaps more disturbing is that such a high proportion of humanity lacks access to clean water, adequate sanitation, and the drugs that are already available. The World Bank gave as an example that cholera vaccination costs 15 cents and gives only 50% protection for 6 months. In the Phillipines, privies could be built with "self-help" labor for $1 and would cut cholera rates by 60%. The latter are clearly the best buy and third world countries would be foolish in such situations to spend their foreign currency on sophisticated drugs.

19

Prostaglandins

The prostaglandins do not occur in the Top-100 and are not sold in large quantities since they are only now becoming available. They are potentially of great importance in the cure of diseases and may be among the important drugs of the future. They have certainly provided one of the most dramatic areas of chemistry of our generation. Like the steroids, the prostaglandins are a class of compounds of great therapeutic interest. Furthermore, this interest is not confined to the relief of a single illness. Rather, these materials may alleviate the symptoms of a variety of illnesses.

The prostaglandins received their name because they were first detected in human seminal fluid from the prostate gland. Excellent reviews have been published by Burger and by Nelson et al. (see Bibliography). Prostaglandins occur widely in animal tissues such as lungs, eyes, stomach, intestinal mucosa, brain, thymus, heart, liver, and kidneys. They were not detected for many years because they occur in such low concentrations and have such short half-lives. Their discovery and that of their precursors was made possible by modern analytical techniques.

The prostaglandins are named as derivatives of prostanoic acid:

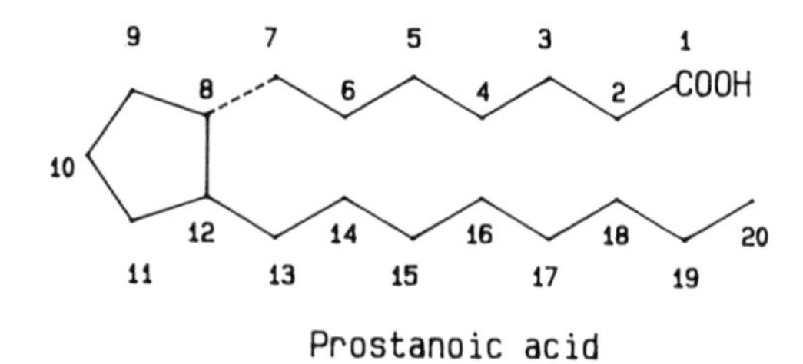

Prostanoic acid

They are designated by the letters PG (prostaglandin), followed by another letter depending on the oxidation state of the cyclopentane ring:

A = 9-keto-10-ene
B = 9-keto-8-12-ene
C = 9-keto-11-ene
D = 9-hydroxy-11-keto
E = 9-keto-11-hydroxy
F = 9,11-dihydroxy
G and H = cyclic endoperoxides

Numerical subscripts refer to the number of double bonds in the side chain. Alpha or beta subscripts refer to the stereochemical configuration of the 9-hydroxy substituent on the ring. The substituent is called alpha if it projects downward from the plane of the molecule when the molecule is shown with the ring to the left and beta if it projects above the plane. None of the naturally occurring prostaglandins has the beta configuration. As an example of the nomenclature, consider the structure:

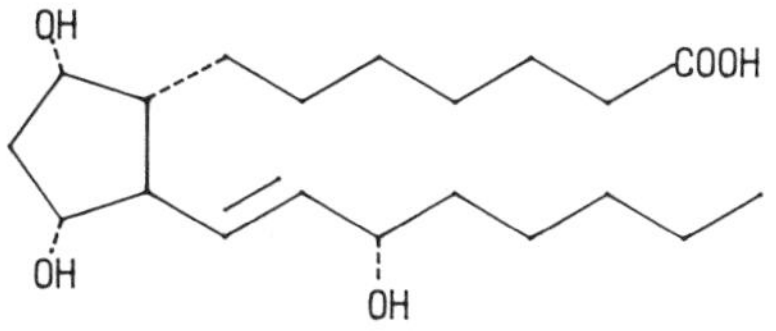

OH COOH OH OH

$PGF_{1\alpha}$

It is a dihydroxy compound; hence, the ring letter is F. The 9-hydroxy substituent is below the plane; hence, α. And there is one double bond in the side chain; hence, the compound is $PGF_{1\alpha}$.

19.1 THE ARACHIDONIC ACID CASCADE

The prostaglandins are actually a subgroup of a much wider class known as the eicosanoids and it is this class rather than the prostaglandins themselves that is of such importance. The eicosanoids are the products of enzymic oxidation of arachidonic acid, which is one of the essential fatty acids. These are the fatty acids that the body needs but cannot itself synthesize and thus must be obtained from diet. Arachidonic acid is a scarce, highly unsaturated C_{20} acid derived from cell membrane phospholipids by the action of phospholipase A_2. Its enzymic oxidation is represented (perhaps oversimplified!) in Figure 19.1. The overall scheme is described vividly as the arachidonic acid cascade. The subgroups into which the products fall are:

1. The hydroxyeicosatetraenoic acids (HETE) and the hydroperoxyeicosatetraenoic acids (HPETE), at the top of the diagram.
2. The prostaglandins themselves (PG), at the bottom left of the diagram.

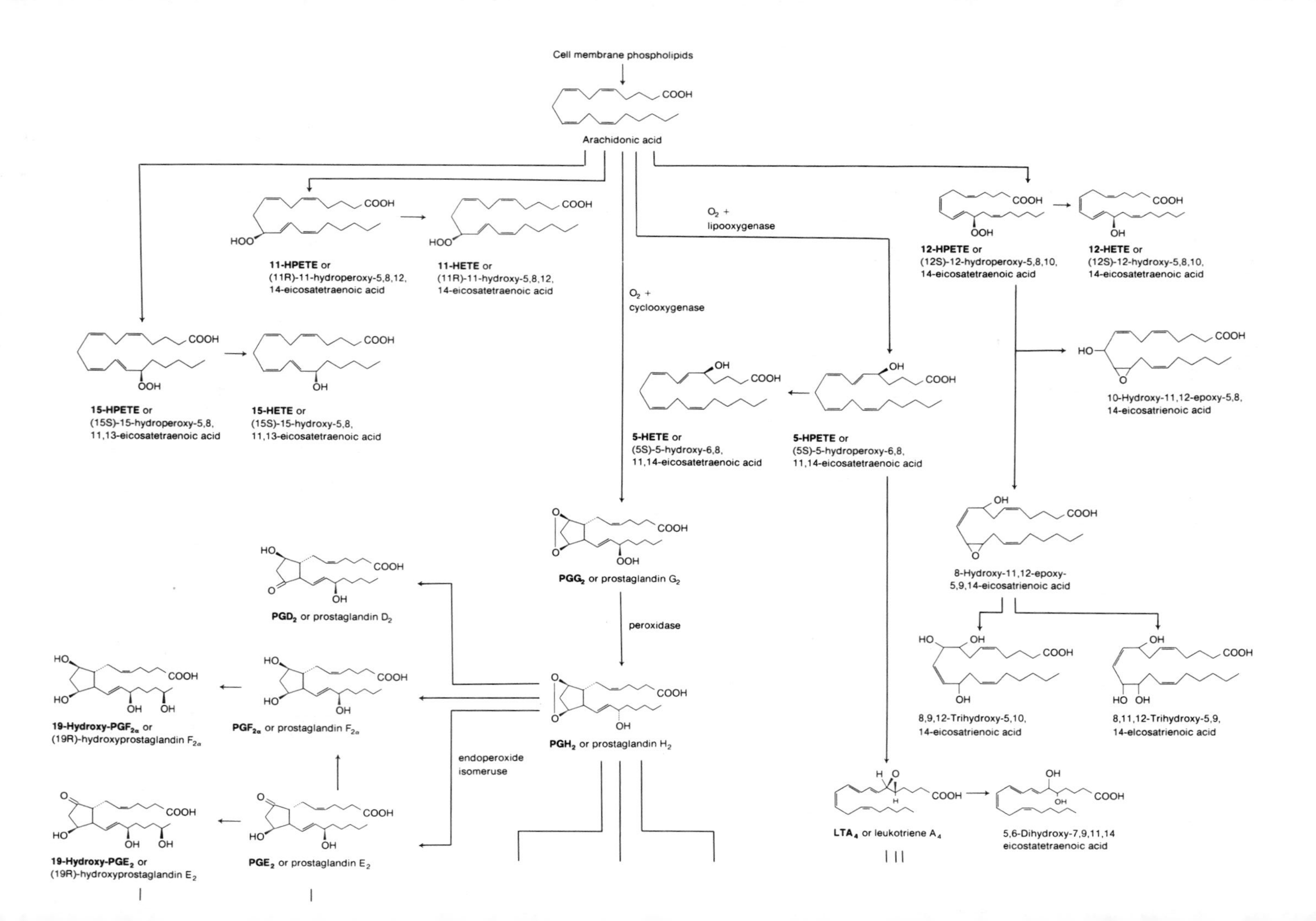

Cell membrane phospholipids
COOH
Arachidonic acid
O_2 +
lipooxygenase
O_2 +
cyclooxygenase
HOO
11-HPETE or
(11R)-11-hydroperoxy-5,8,12,
14-eicosatetraenoic acid
11-HETE or
(11R)-11-hydroxy-5,8,12,
14-eicosatetraenoic acid
OOH
15-HPETE or
(15S)-15-hydroperoxy-5,8,
11,13-eicosatetraenoic acid
OH
15-HETE or
(15S)-15-hydroxy-5,8,
11,13-eicosatetraenoic acid
12-HPETE or
(12S)-12-hydroperoxy-5,8,10,
14-eicosatetraenoic acid
12-HETE or
(12S)-12-hydroxy-5,8,10,
14-eicosatetraenoic acid
HO
O
10-Hydroxy-11,12-epoxy-5,8,
14-eicosatrienoic acid
5-HETE or
(5S)-5-hydroxy-6,8,
11,14-eicosatetraenoic acid
5-HPETE or
(5S)-5-hydroperoxy-6,8,
11,14-eicosatetraenoic acid
8-Hydroxy-11,12-epoxy-
5,9,14-eicosatrienoic acid
8,9,12-Trihydroxy-5,10,
14-eicosatrienoic acid
8,11,12-Trihydroxy-5,9,
14-eicosatrienoic acid
H
LTA4 or leukotriene A_4
5,6-Dihydroxy-7,9,11,14
eicostatetraenoic acid
PGG2 or prostaglandin G_2
peroxidase
PGH2 or prostaglandin H_2
endoperoxide
isomeruse
PGD2 or prostaglandin D_2
PGF2α or prostaglandin $F_{2\alpha}$
PGE2 or prostaglandin E_2
19-Hydroxy-PGF2α or
(19R)-hydroxyprostaglandin $F_{2\alpha}$
19-Hydroxy-PGE2 or
(19R)-hydroxyprostaglandin E_2

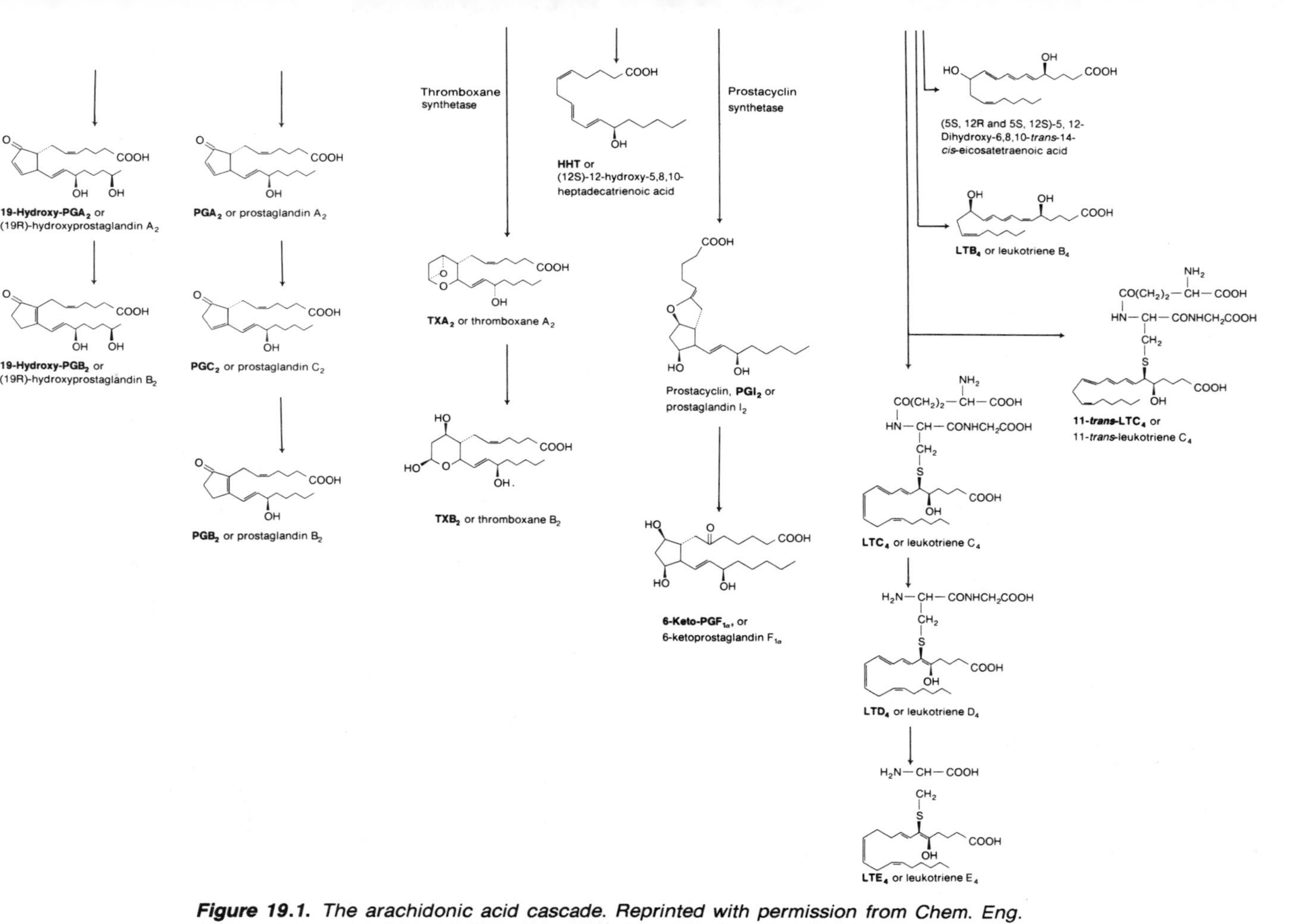

Figure 19.1. *The arachidonic acid cascade. Reprinted with permission from Chem. Eng. News, 16 August 1982 p. 30, Copyright 1982, American Chemical Society.*

3. The thromboxanes (TX), at the bottom center of the diagram.
4. The important prostaglandin I_2, known as prostacyclin, in the bottom center.
5. The leukotrienes (LT), at the bottom right of the diagram.

Many of these transformations can be brought about by enzymes *in vitro*. All of them have been demonstrated in the body. The pathways discussed below are those of particular significance in drug development.

19.1.1 Prostaglandin Endoperoxides and Prostaglandins

The key to the biosynthesis of prostaglandins was the identification of the prostaglandin endoperoxides in 1973 by Hamberg and Samuellson at Karlinska Institutet, and Nugteren and Hazelhof at Unilever. Oxygen plus a cyclooxygenase converts arachidonic acid to an endoperoxide, PGG_2. PGG_2 can, in turn, be converted by a peroxidase to PGH_2. Under the conditions of the laboratory biosynthesis, the half-lives of the endoperoxides were about 5 minutes, but they were isolated by the inhibition of further enzymic conversions and the use of rapid isolation techniques. PGH_2 is a crucial intermediate in prostaglandin, thromboxane, and prostacyclin synthesis.

PGH_2 undergoes numerous enzymic and chemical reactions. Reduction by an endoperoxide reductase converts it to $PGF_{2\alpha}$; under nonreductive conditions, such as silica gel or an endoperoxide isomerase, it rearranges to PGE_2 and PGD_2. It is also a prostacyclin and thromboxane precursor, and enzymic or nonenzymic cleavage gives 12-hydroxy-5-*cis*, 8-*trans*, 10-*trans*-heptadecatrienoic acid (HHT).

Some of the prostaglandins have adverse effects on the body. It has been suggested that schizophrenia may be caused by an excess of prostaglandin and there is evidence that intravenous sodium salicylate, which is known to be a prostaglandin inhibitor, improves glucose disposal in diabetic patients.

It is well established that when body cells are injured an enzyme makes prostaglandins, which aggravate inflammation, fever, and pain. Aspirin and nonsteroid anti-inflammatories, such as indomethacin, act by blocking the active site on the enzyme cyclooxygenase and preventing synthesis of the inflammatory prostaglandins from arachidonic acid.

Figure 19.2 illustrates this process for aspirin. In the first frame the arachidonic acid and oxygen enter the active site of the cyclooxygenase. In the second frame the molecules are aligned by the enzyme and in the third they join to form PGG_2. When aspirin is present, as in frames 4 and 5, it blocks the active site and prevents prostaglandin formation.

In Section 10.2.1, discussion of why aspirin and other nonsteroid anti-inflammatories irritate the stomach was deferred. The reason is that aspirin

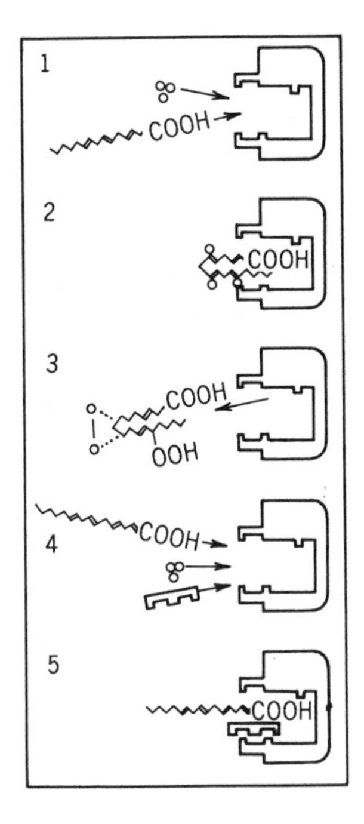

Figure 19.2. *Mode of action of aspirin.*

and the nonsteroid anti-inflammatories block prostaglandin formation in the stomach. Prostaglandin PGE_2 is not only inflammatory, so that its blockage inhibits pain, it also protects the cells of the stomach wall by facilitating formation of a protective layer of mucus.

Prostaglandin E_2
(Dinoprost)

Thus, less mucus is formed and the cells of the stomach wall are more likely to be damaged and are less able to repair themselves. Lack of PGE_2 also depresses levels of cyclic adenosine monophosphate, a compound that mediates the action of certain hormones. In this case, the consequence is that more hydrochloric acid is formed in the stomach.

The elucidation of the mechanisms by which aspirin works and by which primary prostaglandins are biosynthesized from arachidonic acid won the 1982

Nobel prize for medicine for John Vane of Britain and Sune Bergstrom and Bengt Samuellson of Sweden.

Prostaglandins $F_{2\alpha}$ (below) and E_2 (above) cause contractions in the smooth muscle of the uterus and are used both for abortions and to induce labor. One of the problems with prostaglandins is their instability and this can be diminished by structural modification. Addition of a methyl group at C_{15} and esterification gives carboprost, a drug with similar properties but a longer lifetime in the body. These drugs have important veterinary as well as human applications.

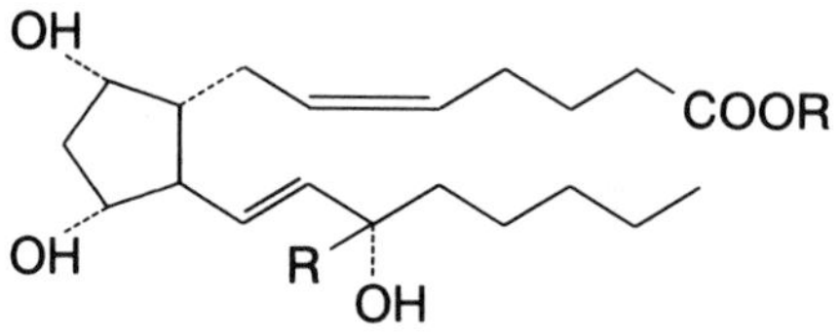

Prostaglandin $F_{2\alpha}$ (R = H)

Carboprost (R = CH_3)

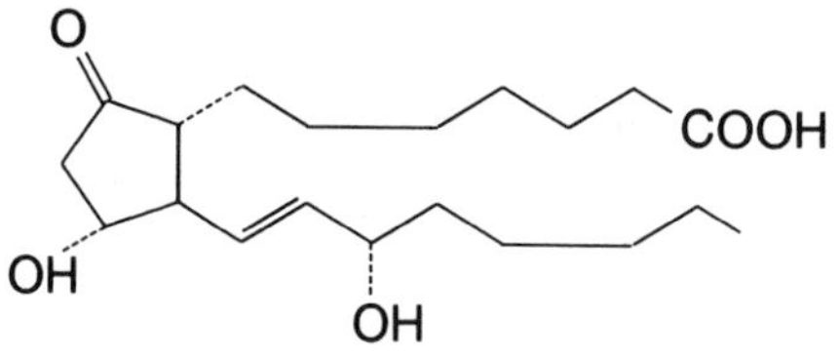

Prostaglandin E_1

Prostaglandin E_1 is a powerful vasodilator. It may be used to treat vascular disease in the leg of the kind that leads to gangrene. It may also be of help in keeping open the blood vessels near the heart in cases of angina and stroke. There is some evidence of its being of value in asthma but the development of leukotriene inhibitors (see below) seems a better bet.

Commercially, the most interesting prostaglandin drugs are those intended for the treatment of stomach ulcers. The two major H_2 antagonists, cimetidine and ranitidine, were enormously profitable (Section 9.6) and the fact that prostaglandins also inhibit secretion of stomach acid provided a great stimulus for drug development. In addition, they are claimed to facilitate the production of mucus, that is, they are cytoprotective. In 1984, at least 24 compounds were under test and a number of them have reached the market. Misoprostol is PGE_1 methyl ester, ornoprostil is 17,20-dimethyl-6-oxo-PGE_1 methyl ester, and Enprostil is (+)-4,5-didehydro-16-phenoxy-ω-tetranor-PGE_2 methyl ester:

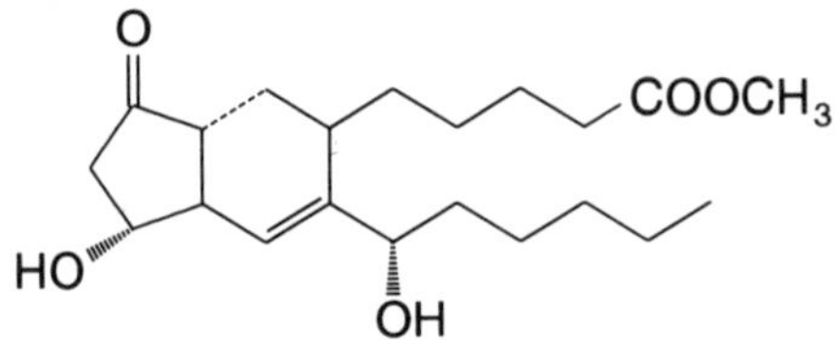

Misoprostol

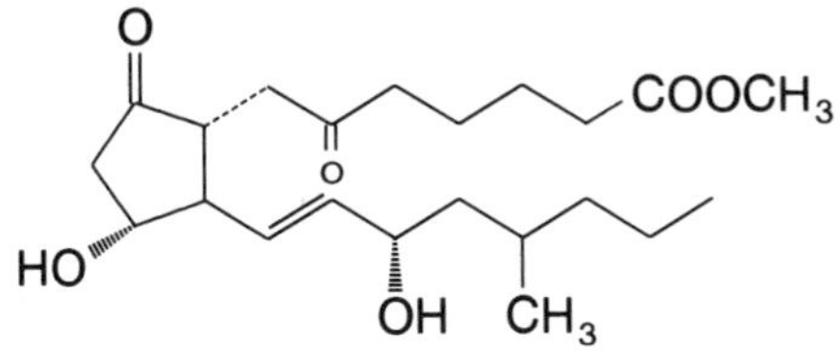

Ornoprostil

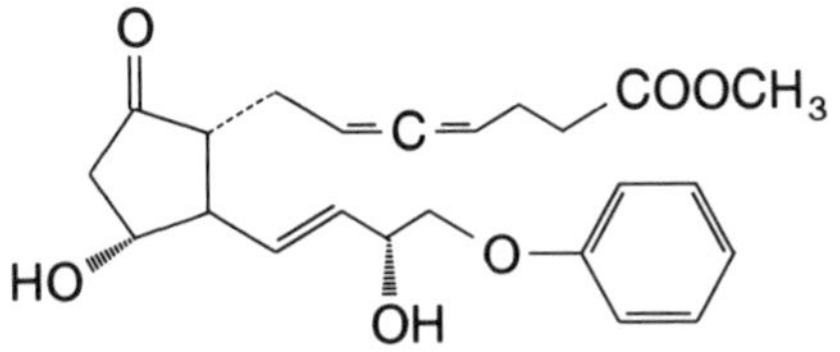

Enprostil

The cytoprotective action has been challenged by other workers (see bibliography) and it is still too early to decide whether these new compounds will represent a major challenge to the H_2 antagonists.

19.1.2 Thromboxanes

The action of blood platelets on arachidonic acid was found to give two unsaturated fatty acids, 12-HETE and HHT, together with two compounds that were named thromboxanes (Figure 19.1). The first (TXA_2) was unstable and rapidly gave the second (TXB_2). TXA_2 is a vigorous contractor of smooth muscle and a potent aggregator of platelets in the blood. In spite of these unattractive qualities, TXA_2 appears to function together with prostacyclin to maintain a healthy system of blood circulation.

TXA_2 proved difficult to synthesize, but TXB_2 could be made simply from $PGF_{2\alpha}$. It was converted to its 11,15-diacetate methyl ester (1). Lead tetraacetate opened the ring to give the 11,12-ketoaldehyde (2). Masking of the aldehyde as its dimethyl acetal, followed by saponification of the ester groups and subsequent acid hydrolysis gave TXB_2.

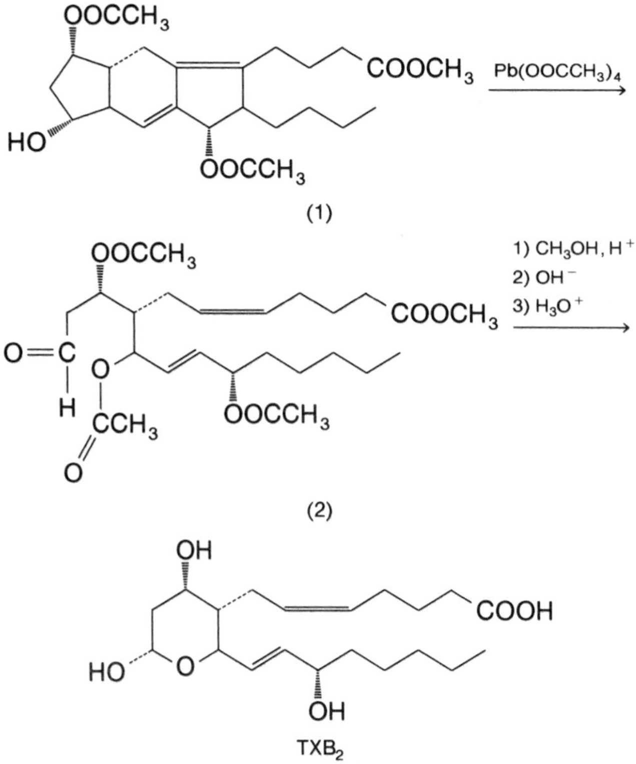

(1)

(2)

TXB_2

19.1.3 Prostacyclin

Prostacyclin was discovered by John Vane's group at the Wellcome laboratories working with another group at Upjohn. They were screening various biological tissues for the ability to produce TXA_2 from PGH_2 and found, instead, a substance that had the opposite properties to the thromboxane. Instead of contracting smooth muscle and aggregating platelets, it relaxed smooth muscle and inhibited platelet aggregation. The new substance was named prostacyclin and designated PGI_2 (bottom centre Figure 19.1).

In the cardiovascular system, it appears that biosynthesis of thromboxane A_2 is concentrated in the platelets, whereas prostacyclin is produced in the walls of blood vessels. It is astonishing that two similar molecules with an identical precursor should have such potent yet opposite properties. The cardiovascular system is clearly controlled by the balance between the two. Both are necessary yet too much thromboxane A_2 or too little prostacyclin could lead to heart disease.

The instability of prostacyclin appears to rule out its usefulness as a drug and, in fact, many more stable derivatives have been made in the hope that they would retain the biological properties of prostacyclin. None of these has yet appeared on the market but prostacyclin itself has been used in cases where surgery requires circulation of blood outside the body, for example, in coronary bypass operations. The inhibition of platelet aggregation is beneficial and the short half-life is advantageous in that the patient's blood returns rapidly to normal after the operation.

Various thromboxane antagonists are said to be under development. In principle, they should prevent further clot formation in patients suffering a heart attack. Typical antagonists are the 1,3-dioxanes.

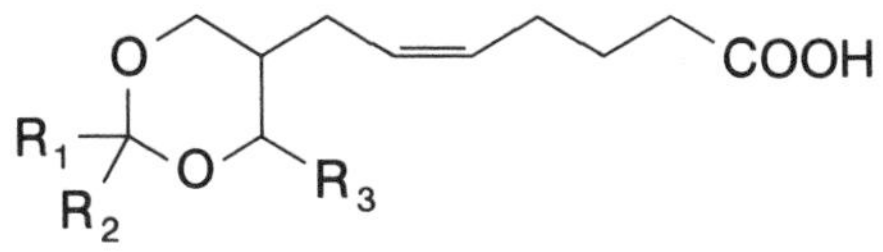

O COOH
R_1 O R_3
R_2

1,3-Dioxanes

19.1.4 Leukotrienes

Having completed their work on the effect of red blood cells on arachidonic acid, Samuellson and his colleagues decided to examine the white cells. The blood contains a species of white cell called a polymorphonuclear leukocyte. Incubation of these with arachidonic acid yielded 5-HETE and another product designated leukotriene (because of the leukocytes) B_4. An enzyme, lipooxygenase, was involved (Figure 19.1). Further work identified a series of leukotrienes and especially a metastable precursor, leukotriene A_4. It was hoped that these substances would prove identical with a material generated from an anaphylactic event and released from mast cells during experiments with allergy-producing substances. This was known as the slow-reacting substance of anaphylaxis (SRS or SRS-A).

There were initial disappointments but eventually SRS was shown to be a mixture of three other leukotrienes—LTC_4, LTD_4, and small amounts of LTE_4. Various threads could then be drawn together. SRS has long been implicated in the occurrence of asthma and is one of the inflammatory substances produced along with histamines when mast cells are damaged. Corticosteroids are effective against asthma because they block release of arachidonic acid from phospholipids, thus closing down the whole cascade. Nonsteroid anti-inflammatory agents, however, block only the cyclooxygenase branch of the cascade. Hence, leukotrienes are still formed and nonsteroid anti-inflammatory agents are inactive against asthma. As noted at the end of Chapter 15, a leukotriene C_4 antagonist should be active against asthma. A number of companies have compounds under development, and Ly 171883

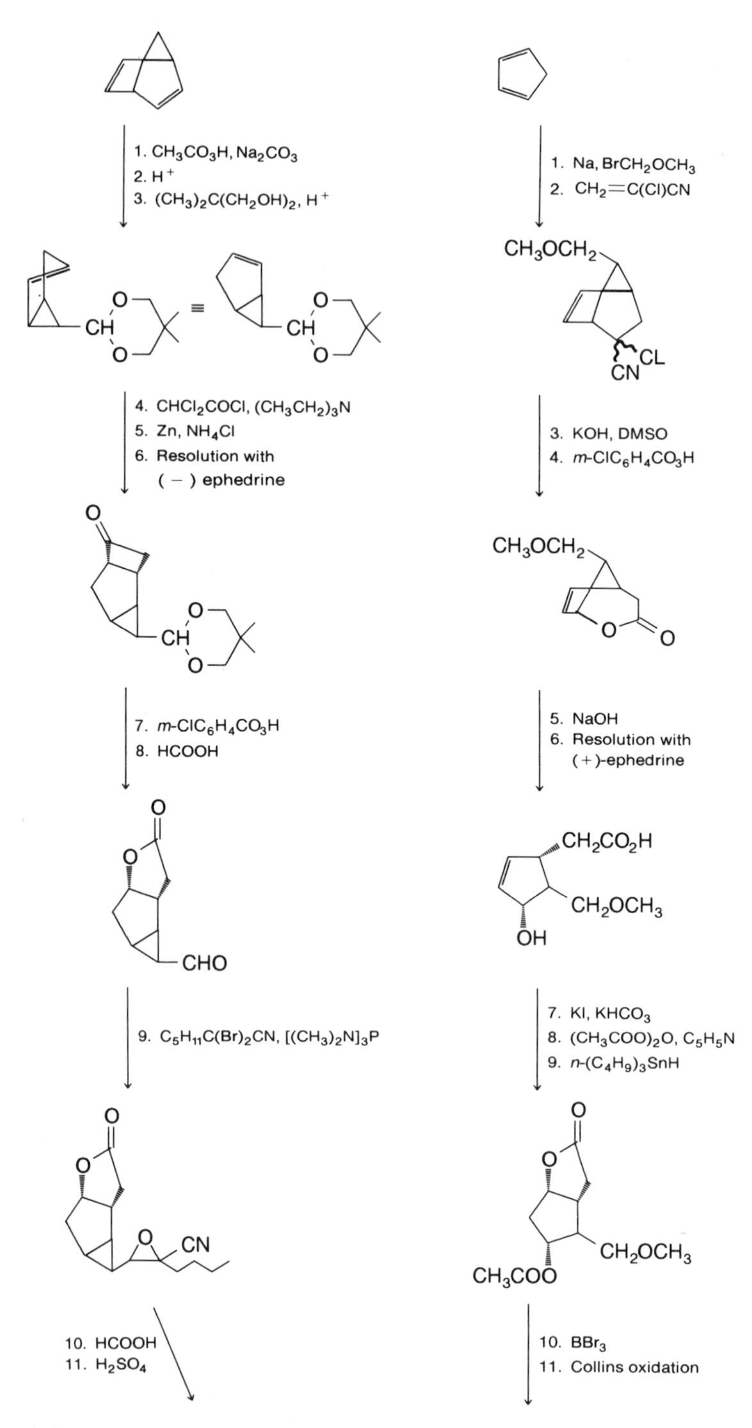

Figure 19.3. *Two commercial prostaglandin syntheses. Process on right was developed by Elias J. Corey at Harvard; that on left was developed at Upjohn. DMSO is dimethyl sulfoxide; THP is tetrahydropyran. Adapted with permission from Chem. Eng. News, 16 August 1972, p. 30, Copyright 1982, American Chemical Society.*

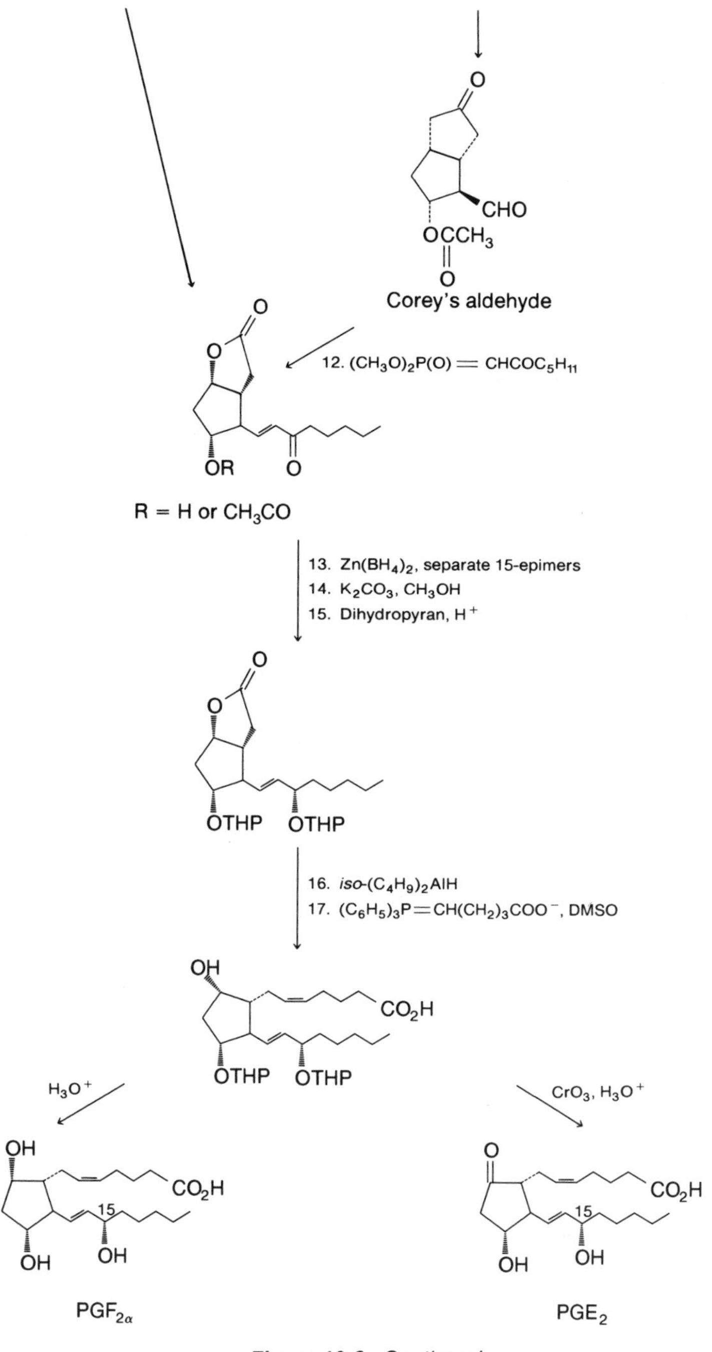

Figure 19.3. *Continued.*

and 163443 are undergoing phase II trials:

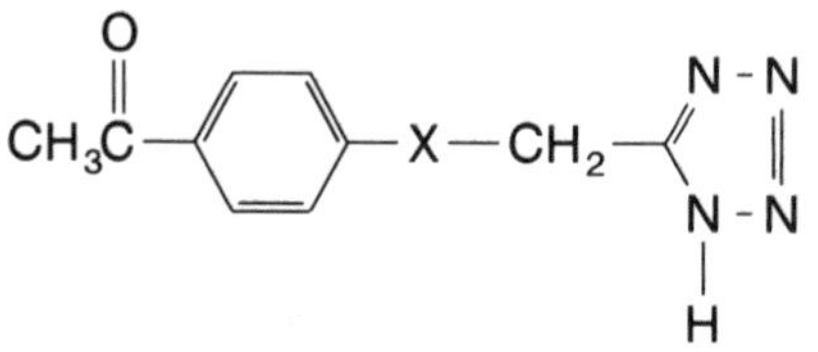

$CH_3C(=O)-C_6H_4-X-CH_2-(CN_4H)$

Ly 171883 X = $-O-CH_2CH_2CH_2-$

Ly 163443 X = $-CH_2-O-C_6H_4-$

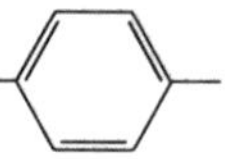

19.2 PROSTAGLANDIN SYNTHESES

Prostaglandins occur in such minute concentrations that extraction of commercial quantities from natural sources is not practical, thus they have to be synthesized. Many syntheses have been developed and these are summarized by Axen, Pike, and Schneider (see bibliography).

Two of the processes used commercially were developed by Corey and his group at Harvard and by another group at Upjohn. They are shown in an abridged form in Figure 19.3. Stereochemistry is controlled at almost every stage and good yields are obtained. The chemical syntheses are not as simple as the biochemical syntheses. It is not yet given to organic chemists to match nature's elegance, nor will it be until they learn to synthesize enzymes. On the other hand, the chemical method produces prostaglandins at 1% of the cost of the biosynthetic method.

19.3 NOVEL NONNARCOTIC ANALGESICS

The understanding of prostaglandin biochemistry and the mode of action of aspirin has stimulated research into the mechanism of pain and the design of nonnarcotic analgesics. Damage to the body initiates pain by releasing chemicals called algogens within tissue or internal organs. Some algogens have already been mentioned in other connections, for example, histamine, acetylcholine, and 5-hydroxytryptamine. These stimulate the small sensory nerves so that pain is felt. Prostaglandins are also released into damaged tissue and potentiate the action of the algogens by sensitizing the receptors. Pain relief could be brought about by prostaglandin blockers more effective than aspirin.

One material currently awaiting U.S. registration is ketorolac, which is said to be as effective as morphine in treating postoperative pain.

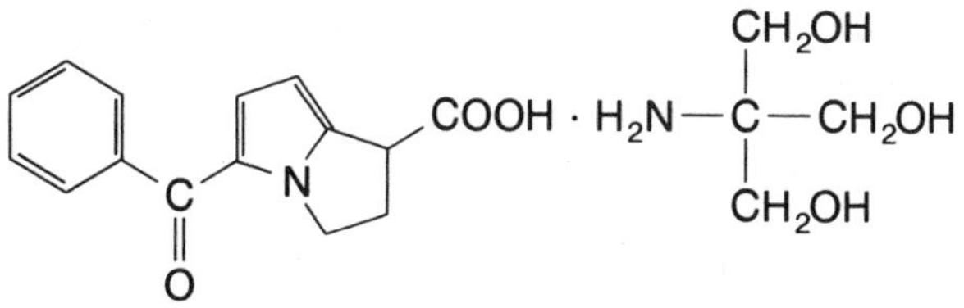

Ketorolac

The drawback of prostaglandin blockers, apart from the stomach irritation they cause, is that they are most effective when prostaglandin production is high and sensitization of the receptors is important. In severe injuries, the quantities of algogens are so high that they override any effect of prostaglandins. In these circumstances, drugs are needed that will block the algogens directly. The snag is that a whole cocktail of drugs will presumably be needed, one to block each algogen.

Nonetheless, a range of 5-hydroxytryptamine (serotonin) antagonists is being evaluated, including ICS 205-930 which contains the 5-hydroxytryptamine ring structure. Like so many of the narcotic analgesics, it is an *N*-methylpiperidine.

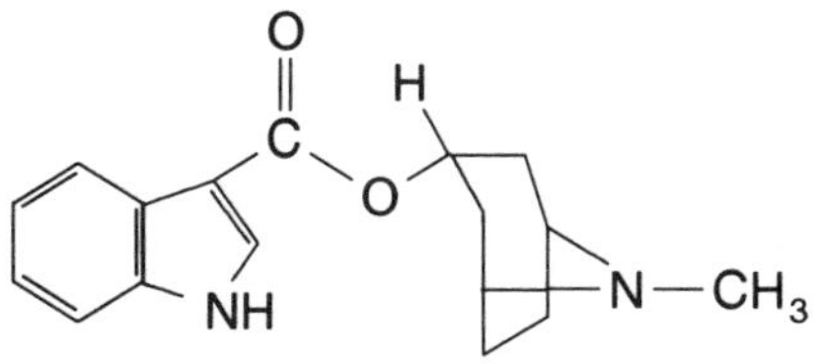

ICS 205-930

Bradykinin, a peptide, is the most potent algogen known in humans. Although its receptor at the ends of the neurons has not yet been characterized, a number of antagonists have already been prepared. In the one shown below, three of the bradykinin residues have been replaced:

Arg-Pro-Pro-Gly-Phe-Ser-Pro-Phe-Arg

Bradykinin

Arg-Pro-Pro-Gly-*Thi*-Ser-*Phe*-*Thi*-Arg

Bradykinin antagonist (Thi = β-(2-thienyl)alanine)

Bradykinin has been found at high levels in nasal secretions associated with the common cold and, early in 1988, Nova Pharmaceutical filed a new drug

TABLE 19.1. Time Required for Development of Steroids and Prostaglandins

Steroids		Prostaglandins	
Late 18th century	Crystalline cholesterol isolated	1930	First detected in human semen
1932	Structure determined (Windaus and Wieland)	1956	Serious research starts
1932 – 1942	All primary steroid hormones except aldosterone isolated and structures determined	1957	First crystalline prostaglandin (Bergstrom)
1949	First commercial production of cortisone at $200 / gram	1962 – 1966	Structures of entire family determined
Mid 1960s	Oral contraceptives introduced	1960s	All naturally occurring prostaglandins synthesized (Corey)

application for a bradykinin blocker to be used in a nasal spray against the common cold.

These compounds are all at the development stage but provide an idea of the type of sophisticated research that is being pursued in this area.

19.4 STEROIDS AND PROSTAGLANDINS

The American Chemical Society has compared the time required for the development of steroids and prostaglandins. The data are summarized in Table 19.1. The point is that steroid research was carried out much earlier than prostaglandin research and prostaglandins had the benefit of modern instrumental techniques that were not available in the heyday of steroid research. Because of this lack, although cholesterol was isolated in crystalline form in the late eighteenth century, its structure remained a mystery until 1932. In the 10 years that followed, all the primary steroid hormones except aldosterone were isolated. Their structures were determined and partial laboratory syntheses were carried out with the aid of the classical techniques of organic chemistry. So important and difficult was this work that no fewer than six scientists received Nobel Prizes for steroid research.

In the 1940s, research started to focus on cortisone and it was produced commercially in 1949 at $200 a gram. Oral contraceptives were not introduced until the mid 1960s.

In contrast, the prostaglandins were first detected in human semen in 1930. Serious work was not started until 1956, but progress was rapid from then on. The first crystalline prostaglandin was isolated in 1957, the first structure was determined in 1962 and the structures of the entire family of prostaglandins were known by 1966 after a mere ten years of serious effort. During the 1960s optically active prostaglandins were synthesized that duplicated the natural materials, and the synthetic methods made possible the preparation of several hundred analogues, many of them pharmacologically active. The explosive growth of research in the area is indicated by the size of the literature. Patents grew from 21 in 1971 to 226 in 1975. Between 1971 and 1976, 723 patents were granted and 5113 papers and monographs were published.

Thus, a century and a half elapsed between the first isolation of crystalline cholesterol and the determination of its structure, and another 20 years before commercial quantities of steroid drugs were available. Less than ten years elapsed between the first crystalline prostaglandin and determination of its structure and complete synthesis.

It is still not clear whether prostaglandins will be as important medically as steroids, but the speed with which they have been developed is a tribute to the power of the techniques of modern chemistry.

20

Antiviral Drugs

Virus infections are the commonest type of infection. It has been estimated that 60% of all human illnesses result from them, compared with a mere 15% from bacteria. In the 1918/19 influenza epidemic there were 500 million cases worldwide and 20 million deaths. In 1968/69 in a United States epidemic there were more than 51 million reported cases of influenza and more than 80,000 deaths.

Diseases caused by bacteria were largely conquered during the 1940s and 1950s by the development of antibiotics. Virus diseases have proved less tractable. There have been some notable successes with immunization. For example, smallpox, yellow fever, poliomyelitis, measles, and mumps may be controlled in this way. Influenza vaccines have been prepared with more limited success. A drawback is that new influenza vaccines must be prepared every year to cope with changes in the virus.

Viruses are not cellular organisms and do not have a metabolism with which it is possible to interfere by means of antibiotics. There was dispute for many years about whether they should be regarded as living systems. On the one hand, they invade body cells, take over their metabolism, and use it to reproduce themselves. On the other hand, they can be obtained in crystalline form. Thus, they occupy a twilight zone of ambiguous life. Virus infections are the source of many diseases, for example, influenza, measles, and AIDS, but viruses are so intractable that drugs to counter them are not yet highly effective nor well established and none of them finds a place in the Top-100.

Viruses are about one-thousandth the length of bacteria. They are nucleic acids—ribonucleic acid (RNA) or deoxyribonucleic acid (DNA)—packed into a protective protein coating. Many small viruses are icosahedral with 20

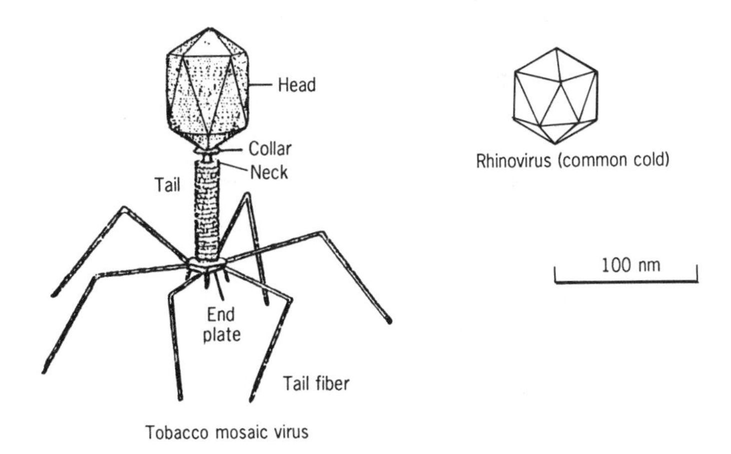

Figure 20.1. *Tobacco mosaic virus and rhinovirus.*

triangular faces. Larger viruses look like space modules, as shown in Figure 20.1. The T4 bacterial virus has a protein membrane head packed with DNA. It is attached via a neck to a tail consisting of a hollow core surrounded by a contractile sheath and based on a spiked end plate to which six fibers are attached. The spikes and fibers attach the virus to a cell wall. The spikes also help a virus recognize a cell and the virus' specificity lies in its ability to attach to a particular cell wall receptor site. The polio virus locks only to a certain subset of cells in the spinal cord; the hepatitis virus goes for the liver.

20.1 REPLICATION OF VIRUSES

By itself a virus cannot reproduce; hence, the question as to whether viruses are living systems. To reproduce, a virus affixes itself to a receptor on a mammalian cell wall (Figure 20.2). It is then enveloped by a section of the cell's membrane and drawn inside. Cell-to-cell transfer is another way in which viruses spread and is a good way for them to avoid the host's immune system. After the virus has entered the cell, the virus' protein coat is removed and the virus DNA—its genetic material—is released. The cell DNA is disrupted. The viral DNA takes over the enzyme system of the cell and programs it to replicate the virus DNA instead of its own cell nucleic acids, and also to produce the enzymes and proteins required for the new virus coatings. These viral structural proteins are synthesized from cell material and, together with the replicated DNA, are assembled into virus.

The viruses can escape from the cell in several ways. Sometimes the cell may burst after a few minutes, spreading identical new viruses throughout the host. Sometimes the viruses may "bud" out of the cell, leaving it intact. When the cell "explodes," cellular protein and debris from its breakdown are

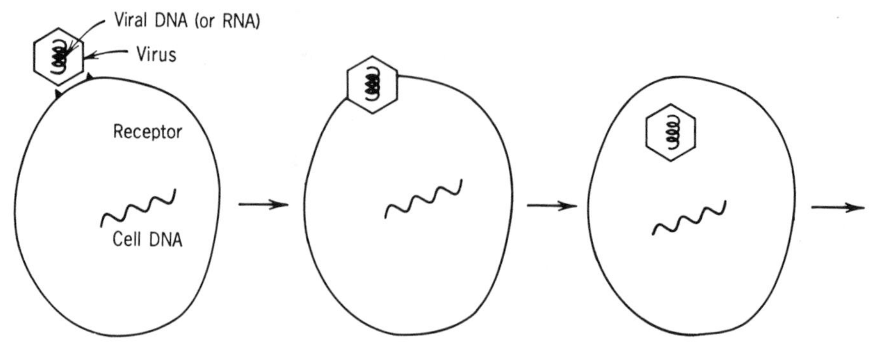

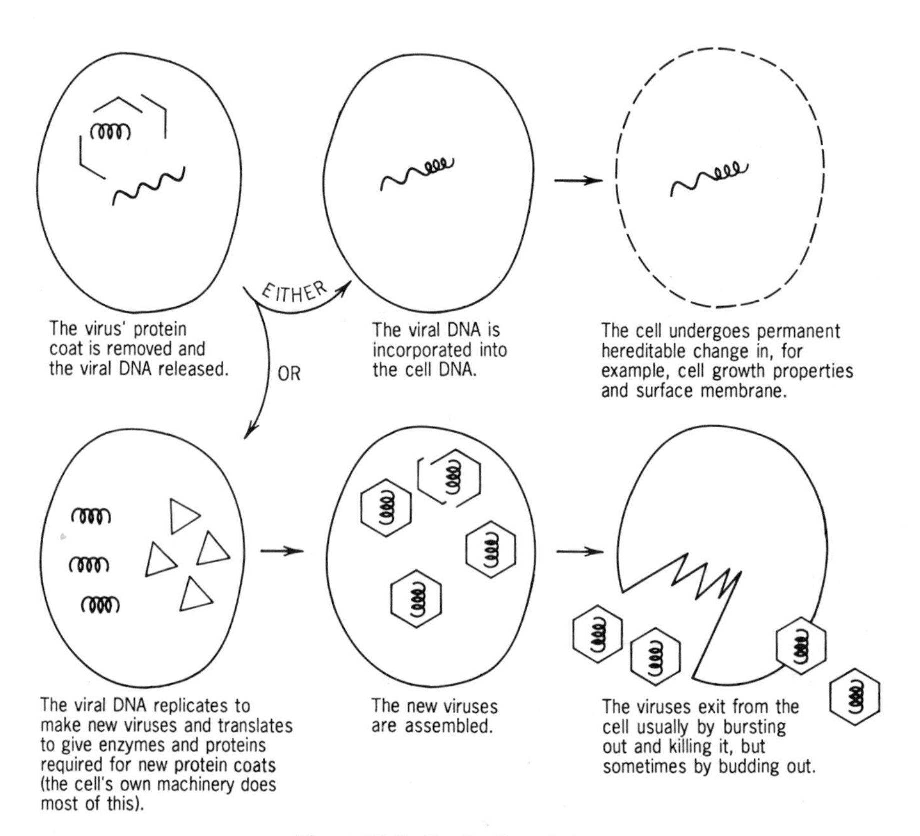

EITHER
OR
The virus' protein coat is removed and the viral DNA released.
The viral DNA is incorporated into the cell DNA.
The cell undergoes permanent hereditable change in, for example, cell growth properties and surface membrane.
The viral DNA replicates to make new viruses and translates to give enzymes and proteins required for new protein coats (the cell's own machinery does most of this).
The new viruses are assembled.
The viruses exit from the cell usually by bursting out and killing it, but sometimes by budding out.

Figure 20.2. *Replication of viruses.*

released into the blood, and these toxins cause fever and the other symptoms of infection. Viruses are thus parasites that change a cell's metabolism to suit their own purposes.

In some instances the virus does not destroy a cell but transforms it. The viral DNA is incorporated into cellular DNA to give a new cell with permanent inheritable changes in cell growth properties and surface membrane. This transformation is characteristic of viruses that cause cancer.

20.2 VIRUS INFECTIONS

Some viral infections are chronic, for example, chronic mononucleosis and hepatitis B. Others terminate when the body's immune response develops sufficiently, as in influenza and the common cold. The immunity to some viruses, for example, smallpox and chickenpox, lasts for life. Many virus infections recur, however, possibly because the virus persists in a latent state, as in herpes, and possibly because many closely related viruses exist for which there is no cross immunity, as in the common cold.

The difficulty with treating virus disease is that, even more than with bacteria, a chemical that attacks a virus is also likely to attack the living cell. Five possible ways around this dilemma may be envisaged:

1. Drugs might be found that would prevent the virus from attaching itself to the cell wall.
2. Drugs might prevent the genetic material from being injected through the cell wall and taking control of the cell.
3. Drugs might interfere with the correct replication of the virus genetic material once it is inside the cell.
4. There are drugs that stimulate the natural defense system of the body, the so-called pro-host drugs.
5. There are drugs that mimic the body's defense system, such as the monoclonal antibodies.

The two illnesses for which most of the present antiviral drugs have been developed are influenza and herpes. Influenza has been well known for centuries, as have the cold sores on the mouth and lips caused by the herpes simplex type 1 virus. Quite recently there has been a spectacular increase in infection by the herpes simplex type 2 virus, which is known as genital herpes and which is sexually transmitted. In 1981, 11,000 new cases were reported in the United Kingdom and the number is increasing at 13% per year. In the United States, in 1982, an estimated 20 million adults had the disease. This puts genital herpes in the same category as syphilis and gonorrhea. Female sufferers have a high risk of cervical cancer and it may be that the virus is itself a cause of the cancer. There is also the risk that babies born to infected women will have infections of the eyes and mouth and perhaps even be blind.

If the spread of herpes 2 caused concern, the appearance of acquired immune deficiency syndrome (AIDS) a few years later has caused panic. The AIDS virus finds its receptor on the surface of a T lymphocyte (Section 9.1). These are the cells that trigger the B lymphocytes to produce antibodies. Deprived of a crucial number of these T cells, the immune system breaks down irreversibly and the person becomes susceptible to infections and cancers that, in the normal way, could be overcome. About 14% of AIDS patients contract a cancer, Kaposi's syndrome, while two-thirds of surviving AIDS sufferers have a specific pneumonia, *Pneumocystitis carinii*.

By the beginning of November 1986, there had been 15,000 deaths from AIDS in the United States. Another 11,500 people showed symptoms, while 1–2 million were believed to harbor the virus. The expectation of life of a person showing symptoms of AIDS is about 2 years. The development of anti-AIDS drugs is seen as a national, indeed a world priority. It has encouraged an unusual level of cooperation between drug companies and has also drawn much money away from cancer research.

20.3 PYRIMIDINE AND PURINE ANTIMETABOLITES

The main antiviral drugs are antimetabolites that inhibit nucleic acid synthesis. In principle, they should selectively block viral rather than host DNA synthesis. A number are shown in Figure 20.3. The first widely used antiviral drug was idoxuridine (1), which was introduced in 1962; it was followed in

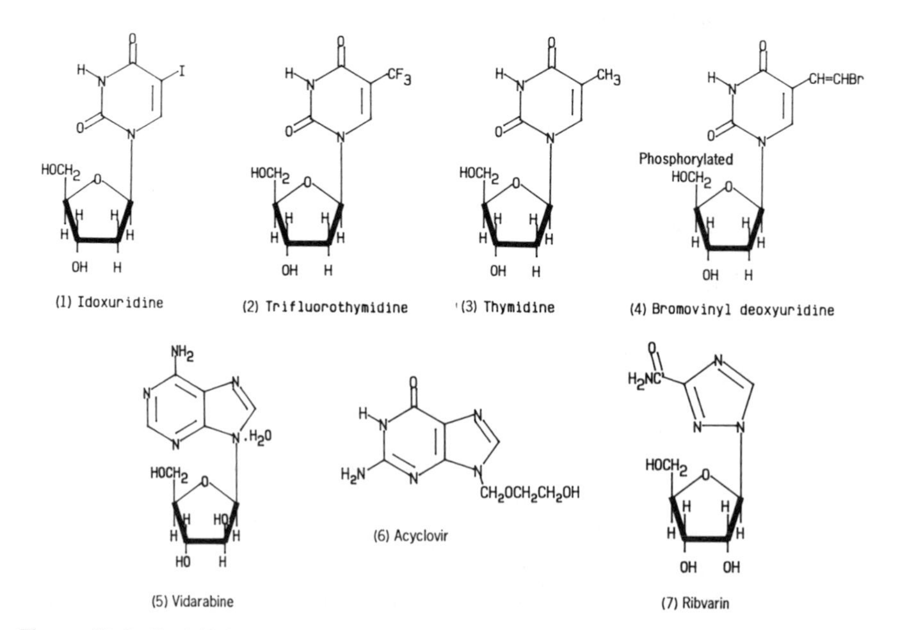

Figure 20.3. *Antiviral drugs: pyrimidines and purines. *Thymidine is not a drug but a building block in DNA synthesis.*

1964 by the preparation of trifluridine (2). Both compounds mimic thymidine (3), a building block in DNA synthesis, and are said to be thymidine kinase inhibitors. If given systemically, they are phosphorylated by thymidine kinases to give triphosphates which are incorporated into DNA instead of thymidine. This false DNA is more liable to breakage, and altered viral proteins result from faulty transcription. Thus, the virus cannot replicate itself. Unfortunately, the drugs do not distinguish between human and viral thymidine kinase; hence, they inhibit cellular as well as viral DNA. They are toxic, but idoxuridine is effective in ointments for treatment of herpes simplex infection of the cornea of the eye, which previously had to be treated surgically. Trifluridine has limited uses against human cytomegalovirus. Idoxuridine has recently been replaced by vidarabine, its adenine analogue (see below).

Bromovinyl deoxyuridine (BVDU) (4) is a selective antiviral agent in that it is phosphorylated in the 5′-position specifically by the herpes simplex virus type-1-induced thymidine kinase. This enzyme also gives the diphosphate, and cellular enzymes then give the triphosphate. The DNA from infected cells incorporates the BVDU and is, hence, defective, but normal cells are unaffected. It showed an unprecedentedly high activity for viral enzymes, but, unfortunately, prolonged dosing of animals with BDVU caused tumors and work on it was stopped in 1984.

Two antiviral drugs based on the purine nucleus are vidarabine (5) and acyclovir (6). Vidarabine is also called adenine arabinoside. It was originally isolated from a Caribbean sponge but is now made by culture of a strain of *Streptomyces antibioticus*. It was tried originally as an anticancer drug and is selectively active against some viruses. It is phosphorylated by host kinases to the triphosphate, and the triphosphate inhibits virus DNA polymerases about 40 times as much as host polymerases. It also "chain terminates" both cellular and herpes virus DNA, which increases its activity but also its toxicity. It may

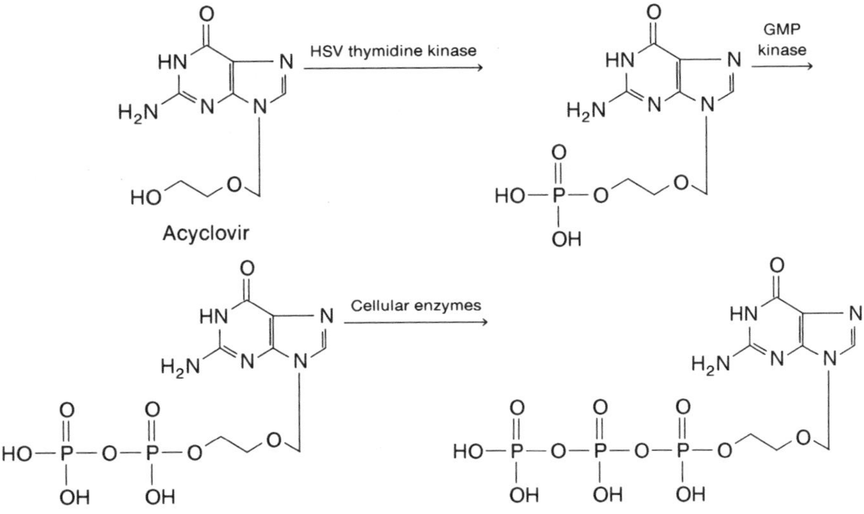

Acyclovir

HSV thymidine kinase

GMP kinase

Cellular enzymes

Figure 20.4. *Mode of action of acyclovir.*

be given intravenously, but its limited solubility means that a large volume of fluid is required. It is active against a number of viruses including the herpes simplex type-1, hepatitis B, and encephalitis viruses. A drawback is that as soon as treatment is stopped the infection recurs. At present it is used to counter serious herpes infections in patients who are taking immunosuppressive drugs and who are therefore susceptible to infections.

Acyclovir is more important than most of the above drugs in that it is already widely used. It appears to be the least toxic and most promising drug against the herpes type-2 virus, although there is evidence that the viruses can develop resistance to it. It is the only drug licensed for use against herpes-2 by the FDA. It may be given orally, intravenously, or as an ointment.

Its mode of action is shown in Figure 20.4. In an uninfected cell, acyclovir is inactive. In infected cells, acyclovir is converted to its monophosphate by viral-induced thymidine kinase. Acyclovir monophosphate is converted to the diphosphate by guanosine monophosphate kinase from the host cell. Unidentified enzymes then give the triphosphate. This inhibits viral DNA polymerase but not normal cell polymerase. The mechanism is that acyclovir triphosphate has no 3′-hydroxyl functional group on which the chain can continue to build. (Cf dideoxycytidine, Section 20.5.)

Ribavirin (7) has a single triazole ring. Its spectrum of activity is broader than any other antiviral drug and includes hepatitis A, measles, and influenza A and B viruses. It copes with the respiratory syncytial virus responsible for 10% of the lower respiratory tract infections that attack children each winter, and has been used against Lassa fever. It may be active against AIDS. The drug has recently been licensed for use as an aerosol. Nonetheless, it is thought to cause anemia of the blood cells and deterioration of lung function and is likely to be used only in life-threatening illnesses.

20.4 AMANTADINE

When it was licensed in 1966, amantadine hydrochloride became the first systemic antiviral. It is structurally totally different from the compounds discussed so far, being tricyclic. It is active against the influenza A virus and is said to reduce illness by 60% if administered as a prophylactic in the early stages of an epidemic. It has no action on the influenza B virus. It is also used in the treatment of shingles.

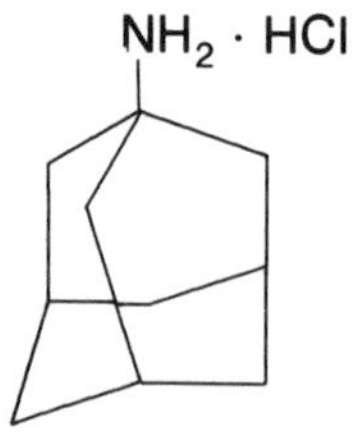

$NH_2 \cdot HCl$ CH_3 $HC—NH_2$

Amantadine hydrochloride 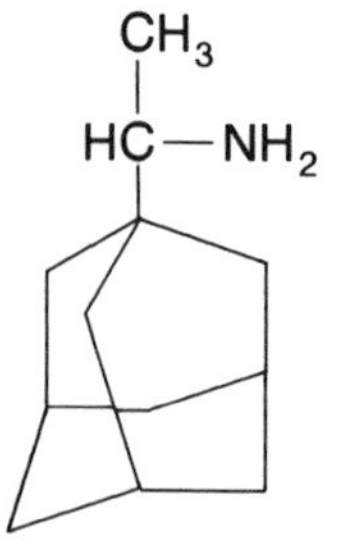 Rimantadine

It was thought to act by preventing the penetration of the cell wall by the virus or by inhibiting removal of the protein coating of the virus and, hence, preventing takeover of the cell by the virus DNA. More recent studies suggest it may inhibit virus assembly. The drawback of amantadine and similar materials is that infected people do not know that a virus is developing inside their bodies and, by the time they do, it is too late to take the drug. As noted in Section 16.3, amantadine is also used in Parkinson's disease, an illness that has nothing to do with viruses.

Rimantadine is an analogue of amantadine that still has only investigational status. Its spectrum is similar to amantadine but it may have fewer side effects.

20.5 DRUGS AGAINST AIDS

The AIDS virus differs from those discussed above in that it is a retrovirus. It contains RNA, not DNA, and, to replicate, it requires an enzyme called reverse transcriptase to convert its RNA to DNA within the cell. Most of the AIDS drugs under development aim to block the enzyme without disrupting the host T cell.

The only drug licensed at the time of writing is azidothymidine:

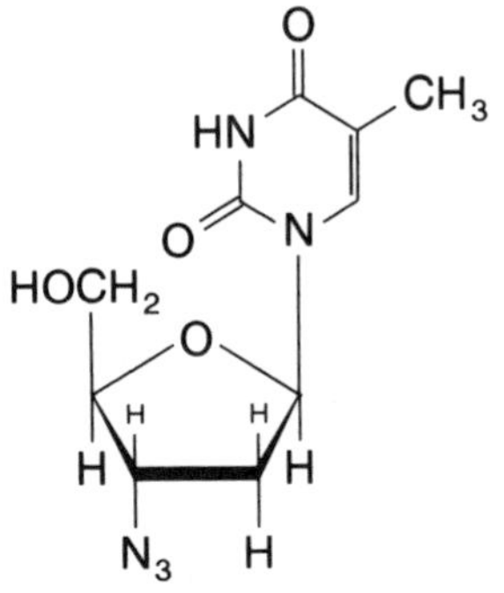

3′-Azidothymidine

It is not a cure for the disease but a palliative and its side effects include anemia, headache, confusion, anxiety, nausea, and insomnia. As with the thymidine kinase inhibitors, it mimics thymidine and is incorporated in the DNA strand being built by the reverse transcriptase. Because there is an azido group in the 3′-position, instead of a hydroxyl, the chain is broken and the process is inhibited.

At least 80 other anti-AIDS drugs are under development and it is far from clear which, if any, will prove more effective than azidothymidine. Front runners include 2′,3′-dideoxycytidine and castanospermine:

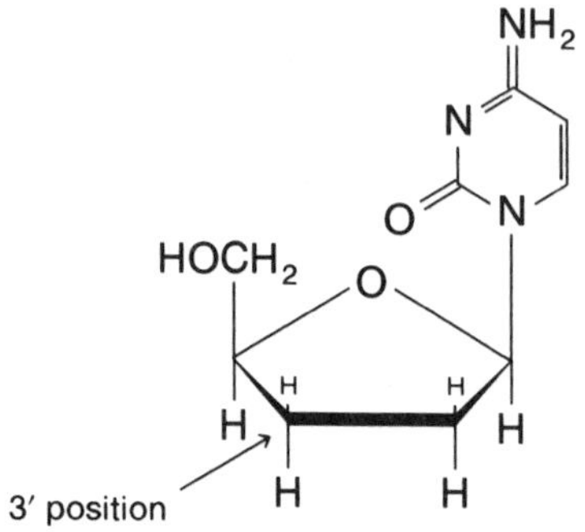

2′,3′-Dideoxycytidine

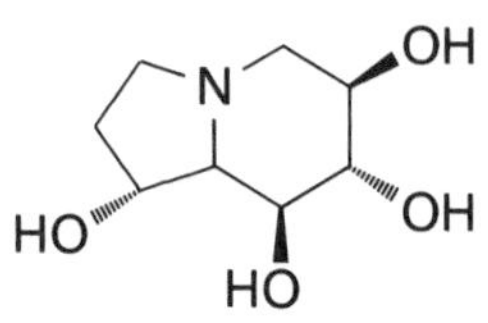

Castanospermine

2′,3′-Dideoxycytidine has no functional group in the 3′-position and presumably acts in much the same way as azidothymidine. Castanospermine, a plant alkaloid, acts on the cell membrane. It does not prevent the virus locking to the membrane but in some way interferes with its entry into the cell.

20.6 OTHER ANTIVIRAL DRUGS

In the list of ways of interfering with virus action (Section 20.2) the two final groups were drugs that stimulate or mimic the natural defense systems of the body. They are outside the scope of this book because they are biological rather than chemical products. Such drugs might be expected to be less toxic than present drugs. Drugs that stimulate defense systems are called "prohost" drugs, but as yet there are no examples apart from conventional vaccines. Vaccines are still very important, of course. New hepatitis B vaccines were introduced in 1986 and an AIDS vaccine is eagerly sought.

A final area where a breakthrough might occur is the development of drugs that would mimic the body's defense system, which means that pure human antibodies would be produced in sufficient quantities to counteract disease. Such antibodies could be made from white blood cells from a person who had just recovered from hepatitis. These could be fused with a special human cancer cell that would grow indefinitely in a test tube. Accordingly, the new cell would both grow indefinitely and produce antibodies, in this case against hepatitis. Such antibodies are already under investigation and are called monoclonal antibodies.

Cell fusion, the technique by which monoclonal antibodies are produced, is clearly related to the recombinant DNA techniques for insulin. The problem is that there is no way one can be sure that the fused cell will combine the good rather than the bad properties of the individual cells. The important part of the technique is the softening of the cell membranes to permit fusion. Cell fusion is a much older technique than recombinant DNA techniques. Frequently it will accomplish the same end and is free of the legal restrictions that in some countries are associated with gene splicing.

Interferon is a potential antiviral drug. Interferon is a general term applied to a family of glycoproteins and there are several subtypes. It is mentioned in Section 21.8 as a possible anticancer drug and is one of several antiviral protein molecules formed when mammalian cells are invaded by viruses. It is an investigational drug and has a very broad antiviral spectrum. For many years, research was inhibited by unavailability of pure product in sufficient quantity, but it is now made by recombinant DNA technology. In the beginning it appeared to be a miracle drug, but, so far, the trials have been disappointing. Systemic interferon has not prevented disease in experimental animals nor has it been successful in isolated trials on terminal patients. A nasal spray is said to have reduced the incidence and severity of colds. Nonetheless, if interferon could be inserted into cells, if cells could be persuaded to make more of it, or if its action could be mimicked by other compounds, then this would be a breakthrough in the treatment of virus diseases.

The development of antibodies, however, would not make chemotherapy unnecessary. Cellular immunity, called cell mediated immunity, is just as important as antibodies in the bloodstream in combatting viral infection.

Although many of the above drugs carry hope for the future, there are still very few effective antiviral drugs. Until quite recently, however, the view was widely held that there was no possibility of chemotherapeutic agents against viruses. That some have emerged means that antiviral drugs are now at the stage antibiotics had reached by the early 1930s. That, of course, is no guarantee that the search will be successful, but it provides one of the most interesting and important areas of current research.

21

Anticancer Drugs

Cancer is second only to heart disease as a cause of death in the United States and there were over 400,000 deaths from cancer in 1981. Trends in cancer deaths in the 55–64 age group are shown in Figures 21.1 and 21.2. Figure 21.1 shows that the total death rate from cancer in men has risen substantially since 1940, but that this rise is due entirely to the rise in lung cancer. The death rate among women dropped until 1960 and has since climbed slowly. Women smoke less than men and cigarette smoking by women became fashionable later, so female lung cancer deaths are low but climbing, whereas among men the death rate seems to be approaching a plateau. If lung cancer is excluded, therefore, the pattern is of a steady or slightly declining mortality.

Figure 21.2 subdivides the cancers by type. Deaths from cancers of the digestive and genital organs have declined for both men and women. Deaths from breast cancer have remained static. There was a sharp rise between 1940 and 1970 in "other cancers," which is now leveling off. Cancers of the urinary organs, mouth, throat, and pharynx, together with Hodgkin's disease and leukemia, seem to show fairly steady death rates over this period and the nature of these miscellaneous cancers is not clear from the published statistics.

21.1 ORIGINS OF CANCER

Cancer is the name given to a dreaded group of about 200 different diseases in which various types of body cells in some way alter. Normal body cells reproduce and replace each other up to the point at which the necessary tissue has been replaced. Cancer cells multiply in an abnormal and uncontrolled

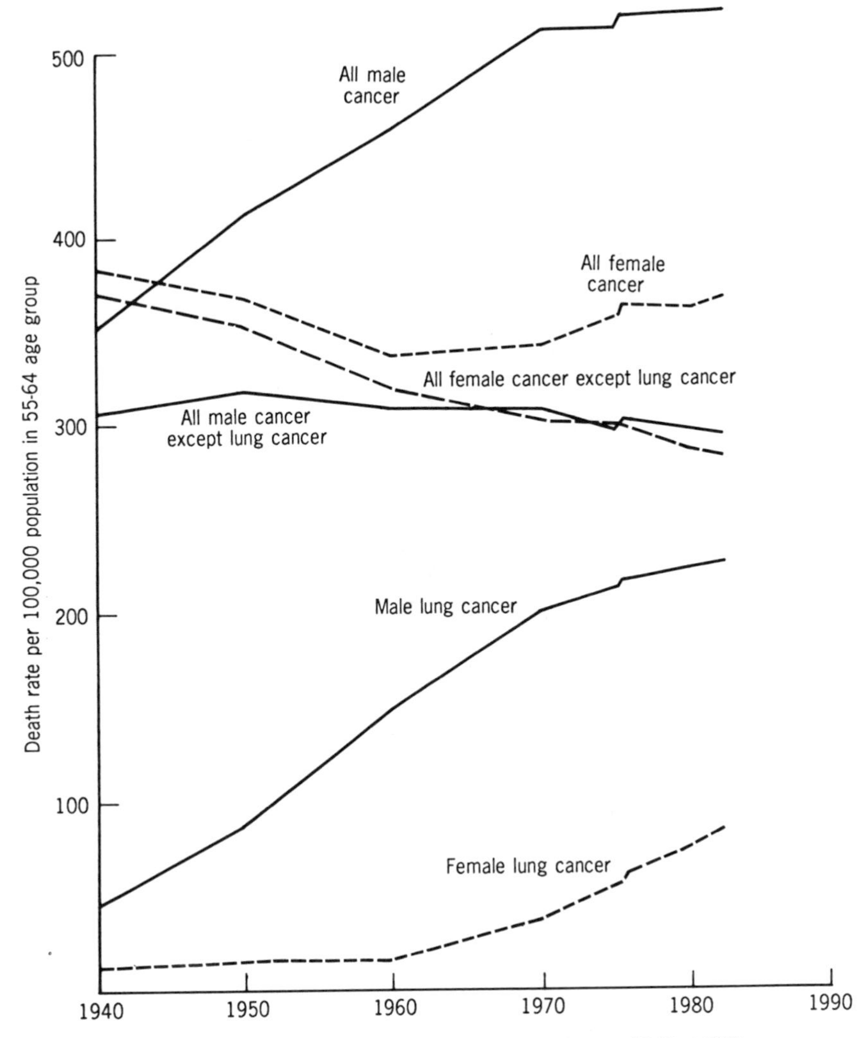

Figure 21.1. *Death rate from cancer, (United States: 1940–1982).*

fashion at a rate greater than that of normal body cells. They crowd out the normal body cells and eventually kill the victim.

There are a number of known and possible causes of cancer. Cellular mutations due to exposure to high-energy radiation is a well-established cause, but many cancers seem to result from cellular mutations of unknown origin. These seem to occur most readily late in life, so old age is a predisposing factor. There may well be a hereditary factor disposing certain individuals to cancer. Some cancers are due to virus infections and others may be related to contact with carcinogenic materials in the environment. Contact with polycyclic aromatic hydrocarbons or nitrosamines is known to lead to cancer and

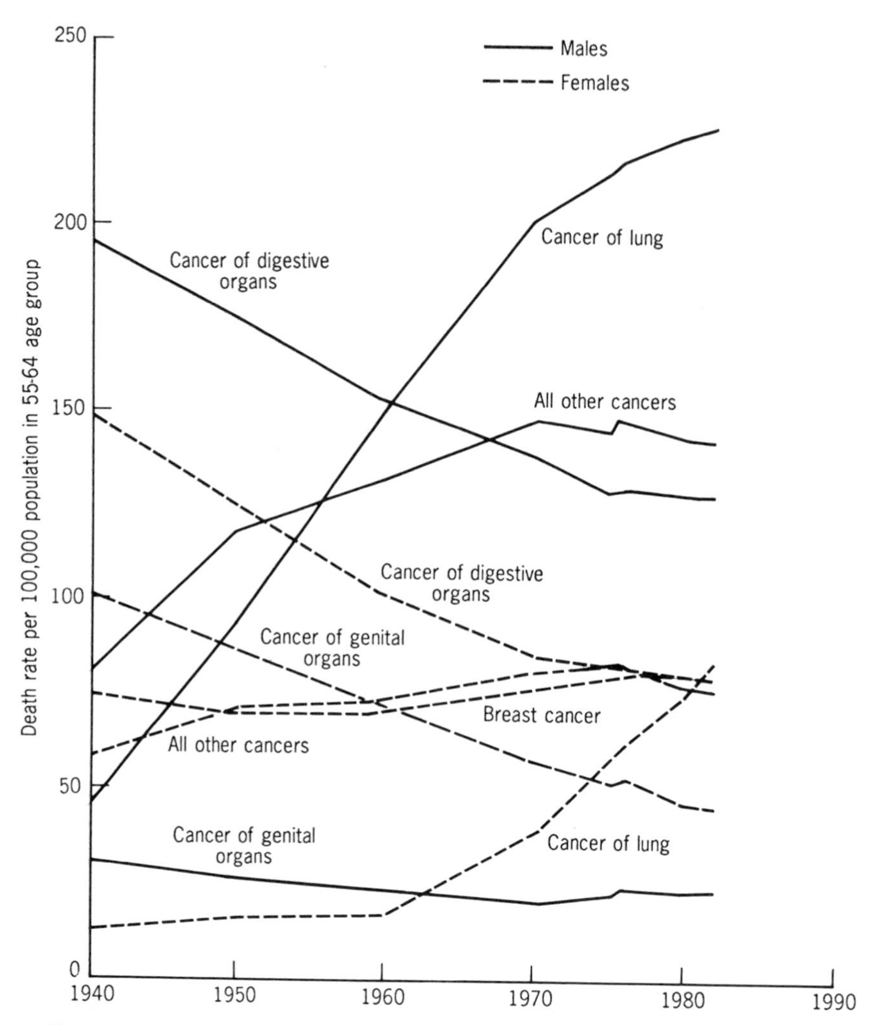

Figure 21.2. *Death rate from different types of cancer (United States: 1940–1982).*

there are many other materials from which the FDA does its best to protect the population. There is, of course, no reasonable doubt of the link between cigarette smoking and lung cancer.

It was thought at one stage that there were ethnic patterns of cancer. The Japanese, for example, have a high rate of cancer of the stomach, while North Americans get cancer of the breast or bowel. Studies of Japanese communities in the United States have shown, however, that after a few years' exposure to United States culture, the Japanese pattern moves close to the American. It is now suggested that the Japanese in Japan eat raw and dried fish, which may predispose them to stomach cancer, while North Americans have a diet that emphasizes broiled foods, such as hamburgers, which may predispose them to bowel cancer. This is not proved and other dietary or cultural differences may be responsible.

21.2 PREVENTION AND TREATMENT

An estimate by Richard Doll, who first demonstrated the smoking–lung cancer link, suggested that 80% of all cancers are environmentally produced. Thirty percent of cancer deaths are attributable to smoking, including passive smoking—the breathing of cigarette smoke by nonsmokers. Thirty-five percent are attributable to diet, 3% to alcohol, and 1% to sexual behavior. What preventative action people should take is less clear. Thirty percent of deaths could be avoided if people stopped smoking, but abstinence from alcohol would only save 3%. Two percent could be eliminated by avoiding obesity and 1% by cervical screening and genital hygiene. The numbers of deaths attributable to occupational exposure to carcinogens, hormone treatment, radiology, and overexposure to sunlight are all very small. It is difficult to identify the factors in diet that cause cancer. There is no evidence that food additives are to blame and the suggestions above are far from proved. There is a tentative conclusion that diets high in fiber, β-carotene, and vitamin A may help.

Until 1970 the only means of treatment for cancer were radiotherapy, in which the abnormal cells were destroyed by high-energy atomic beams, and surgery, in which the cancerous tumor was removed. These treatments are limited to accessible tumors and, moreover, to cancers where the malignant cells have not metastasized; that is, they have not broken away from the main tumor and spread to other parts of the body.

Chemotherapy is an obvious way to attack widespread or inaccessible cancers since drugs are carried throughout the body in the bloodstream. The difficulty is to find effective drugs, and good ones are sorely needed. Most of the available drugs interfere with the DNA of cancerous cells to prevent reproduction. The problem is to find drugs that attack cancerous cells without attacking healthy cells. A number of partially effective drugs has been developed in the past decade and their use has grown, though not to the point at which they appear in the Top-100. In any case, most cancer chemotherapy is carried out in hospitals, so a large number of prescriptions for such drugs would not be expected in the list.

The empirical search for anticancer drugs has been generously funded but has had only limited success because the origins of cancer are not understood. It is also important for cancer to be diagnosed early, a problem that is compounded by public fear of the disease.

The drugs that are available today may prevent metastasis and even cause temporary remission of tumors. They also relieve pain and increase survival time. A problem is that the effective dose is frequently close to the toxic dose

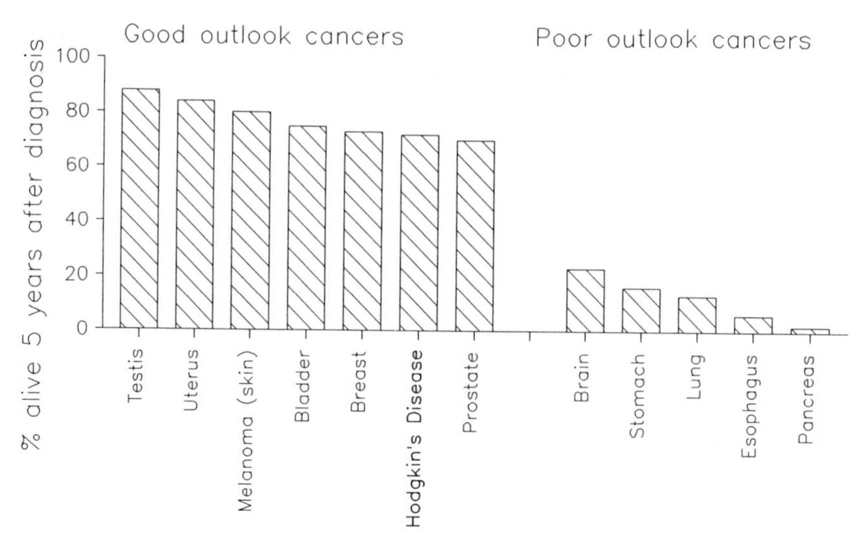

Figure 21.3. *Five-year cancer survival rates. Source: Mayo Clinic Health Letters. March 1986.*

and side effects are unpleasant. Nonetheless, they have led to a slow but significant reduction of mortality from most cancers. Cancer patients are often considered cured if they survive five years after treatment without a recurrence of the cancer, and the number of such survivals has risen as indicated in Figure 21.3. For seven cancers, the patients have at least a 70% chance of being alive after five years. On the other hand, lung and stomach cancer and adult leukemia are the least tractable of the more common cancers, and the prognosis for cancer of the brain, esophagus, and pancreas is also poor. Thus, progress has been made in the fight against cancer even though this success is masked in overall cancer statistics by the steady rise in lung cancer.

21.3 STEROIDS

The main groups of anticancer (antineoplastic) drugs are steroids, alkylating agents, antimetabolites, antibiotics, platinum complexes, and a miscellaneous category. The first anticancer drugs were the female hormones, that is, the estrogens. They were shown during World War II to slow the growth of prostate cancer in men. Subsequently, other steroids were found to have similar effects. Today, estrogens, especially ethynylestradiol, are still used for breast cancer, and diethylstilbestrol is the drug of choice for prostate cancer. Prednisone and prednisolone are used in certain forms of leukemia. It is not known how these steroids work but they are believed to inhibit production of proteins or key enzymes.

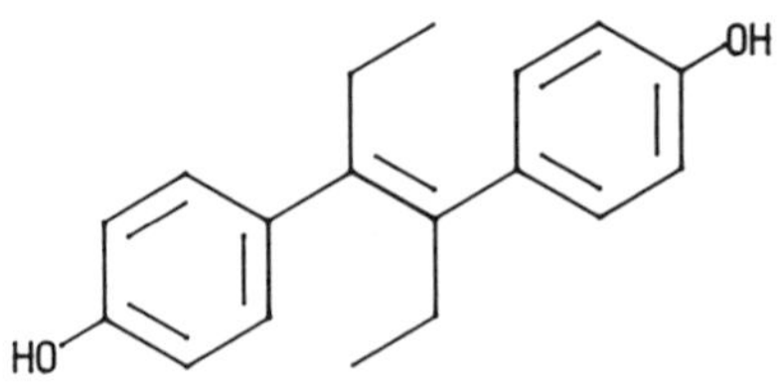

Diethylstilbestrol (Estrogen)

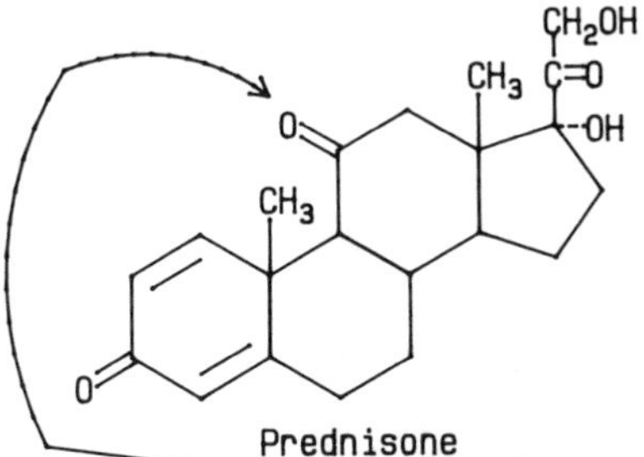

Prednisone
Prednisolone has an -OH instead of =O in #11 position.

Tamoxifen citrate (below) is a nonsteroidal antiestrogenic agent. It forms a stable complex with estrogen receptors, for example, in breast cancer cells, but does not stimulate RNA and protein synthesis. It has become the drug of choice for breast cancer in postmenopausal women.

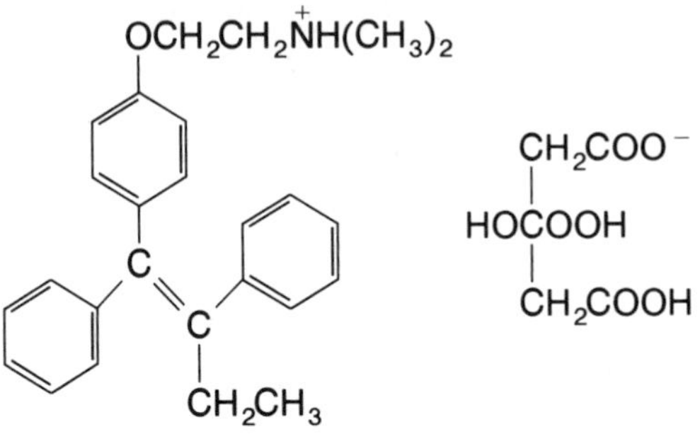

Tamoxifen citrate

21.4 ALKYLATING AGENTS

The second group of anticancer drugs is alkylating agents, of which a selection is shown in Figure 21.4. The genetic material DNA is made up of molecules that must be duplicated and precisely paired when the cell divides. Alkylating

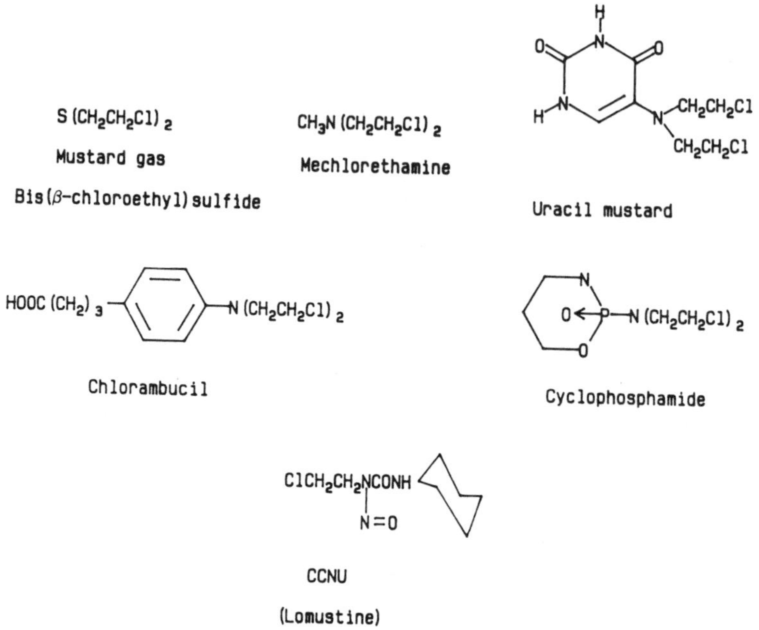

$S(CH_2CH_2Cl)_2$
Mustard gas
Bis(β-chloroethyl)sulfide

$CH_3N(CH_2CH_2Cl)_2$
Mechlorethamine

Uracil mustard

$HOOC(CH_2)_3$–C_6H_4–$N(CH_2CH_2Cl)_2$
Chlorambucil

Cyclophosphamide

$ClCH_2CH_2N(N{=}O)CONH$–
CCNU
(Lomustine)

***Figure 21.4**. Alkylating agents.*

agents covalently modify DNA strands and this interferes with orderly pairing in the helix. Thus, the cell cannot divide successfully. An early drug was mustard gas, that is, bis(β-chloroethyl)sulfide. Its high toxicity led to the synthesis of the nitrogen analogues, mechlorethamine, uracil mustard, chlorambucil, and cyclophosphamide. Especially promising are the nitrosoureas, such as CCNU (lomustine), which are less closely related to the nitrogen mustards. Nitrogen mustards attack lymph tissues and are still drugs of choice in Hodgkin's disease.

21.5 ANTIMETABOLITES

The third group of anticancer drugs is the antimetabolites shown in Figure 21.5. Some of these resemble vitamins or other nutrients and so are absorbed by cells. Once inside, however, they disrupt the cell's metabolic machinery. Aminopterin and methotrexate interfere with folic acid metabolism, that is, they are folic acid antagonists. Compare their structures with that of folic acid in the figure. The slight structural differences have been marked by arrows.

The remaining antimetabolites are pyrimidine and purine derivatives. They resemble the antiviral drugs (Section 20.3) that inhibit DNA formation in virus-infected cells. 5-Fluorouracil resembles uracil (shown for comparison), a

1. Folic acid antagonists and folic acid: (Structural differences from folic acid are shown by arrows.)

Aminopterin

Methotrexate

Folic acid

2. Pyrimidine antagonists and uracil:

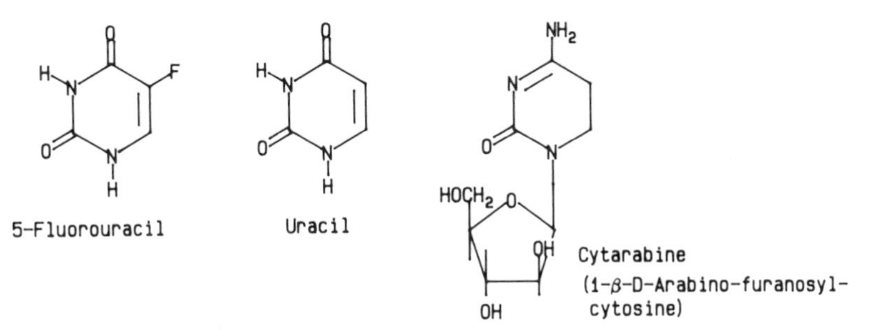

3. Purine antagonists and purine:

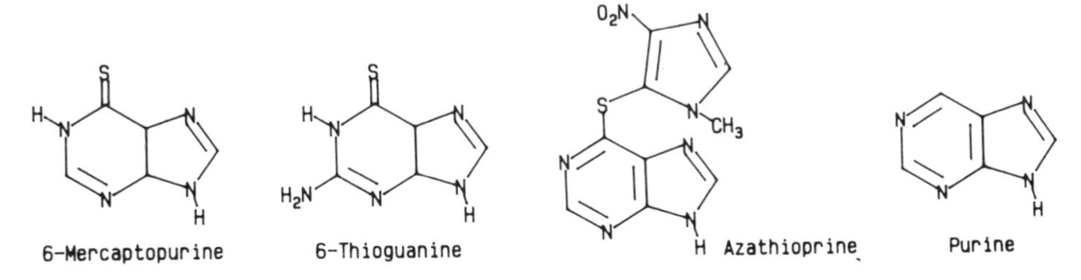

Figure 21.5. *Antimetabolites.*

substance the cell needs to make pyrimidine nucleotides, but it is not an effective substitute and synthesis is blocked. It is said to be a pyrimidine antagonist. Cytarabine is another pyrimidine antagonist in clinical use.

The cell also uses purine to make purine nucleotides and three purine antagonists in clinical use are shown at the bottom of Figure 21.5 together with the purine structure. They are 6-mercaptopurine, 6-thioguanine, and azathioprine. Many of these antimetabolites inhibit the body's immune response and, consequently, are of value as immunosuppressive drugs, because they inhibit the rejection of organ transplants. Azathioprine is currently the drug of choice in this area especially for kidney transplants.

21.6 ANTIBIOTICS

The fourth group of anticancer drugs is related to antibiotics. They disrupt the synthesis of RNA, a substance the cell needs to make essential proteins. They include daunorubicin, adriamycin, and bleomycin, which is a complex mixture of fermentation products.

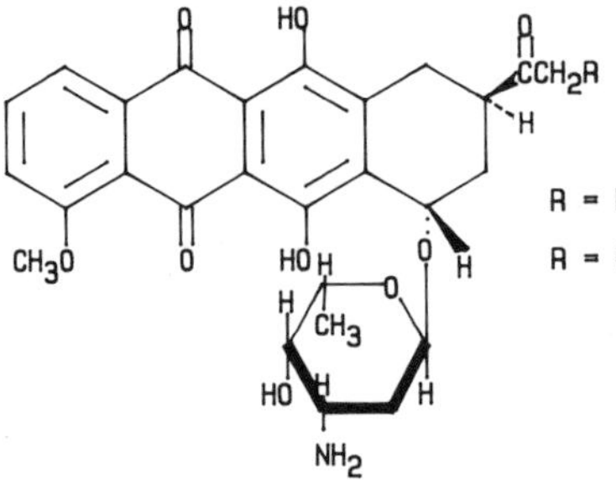

R = H, Daunorubicin
R = OH, Adriamycin (doxorubicin)

Recent work has focused on the mitomycins, a group of compounds proliferated by *Streptomyces caespitosus*. Their mode of action is shown in Figure 21.6. The quinone form of the mitomycin C (1) is inactive to nucleophiles. In an oxygen-poor environment, such as a cancer tumor cell, there is one electron reduction to a semiquinone (2), which loses methanol to give an active intermediate (3). The intermediate undergoes nucleophilic attack by DNA and, hence, cross-links DNA strands through guanine groups (4). The cross-linking interferes with pairing and the cell cannot reproduce.

21.7 PLATINUM COMPLEXES

A fifth group of anticancer drugs is based on platinum. They are particularly interesting because drugs based on inorganics are rare. Examples mentioned so far are antimony compounds for leishmaniasis, potassium chloride as a potassium replenisher, cobalt in vitamin B_{12}, and gold compounds in severe arthritis.

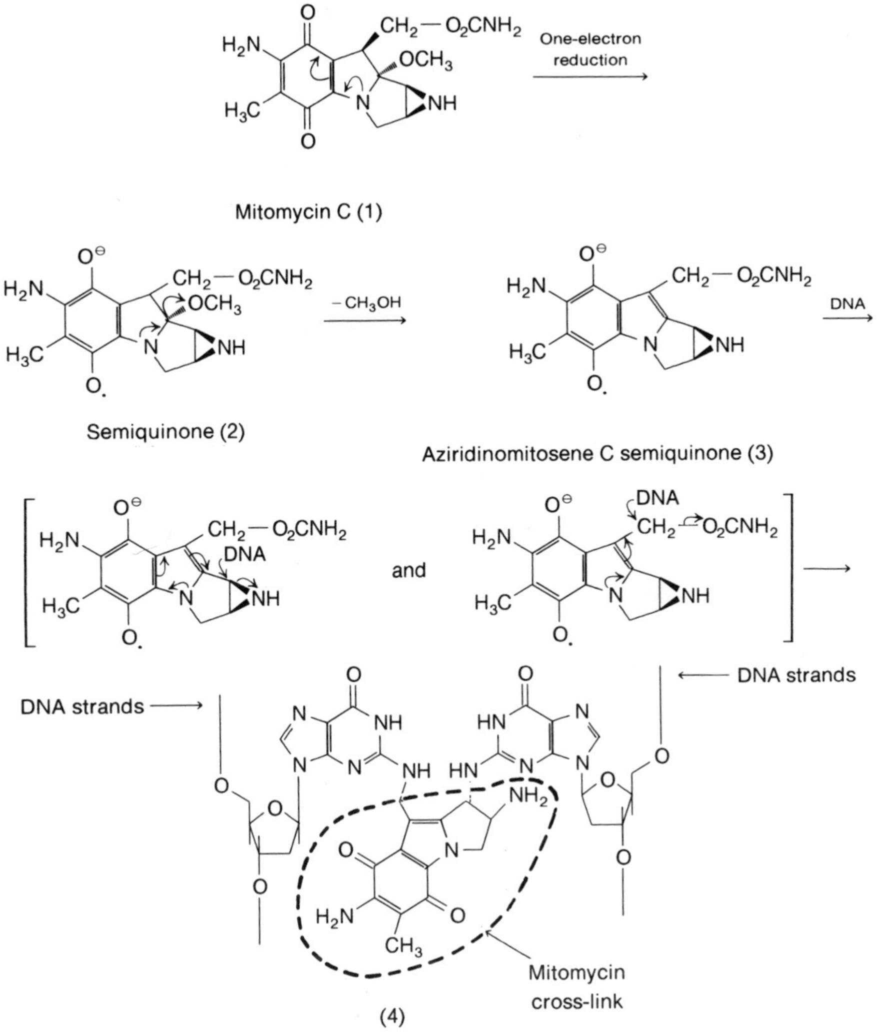

Figure 21.6. Mode of action of mitomycin C. Source: Chemical & Engineering News, 1 September 1986, p. 26.

Cisplatin has comparable standing to adriamycin and fluorouracil. It seems to be hydrolyzed in the body to the compound shown below, and this complex damages DNA in a way that healthy cells can repair but cancerous cells cannot. Anticancer activity is confined to square-planar Pt(II) complexes with two cis nitrogen-containing ligands and two other groups that are sufficiently labile. Pt(IV) analogues are usually also active. A second-generation platinum complex licensed in 1986 in the United Kingdom is carboplatin. It appears to have fewer side effects than cisplatin.

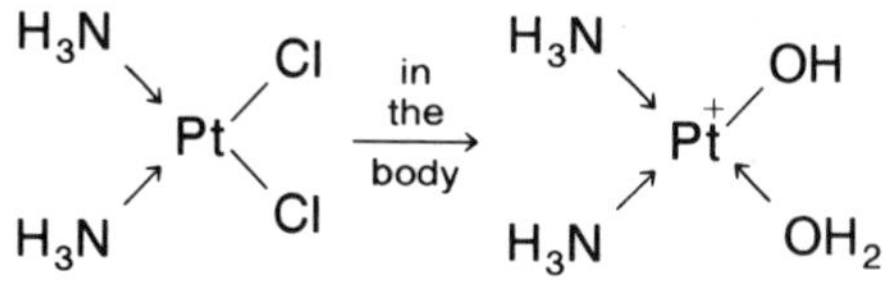

Cisplatin

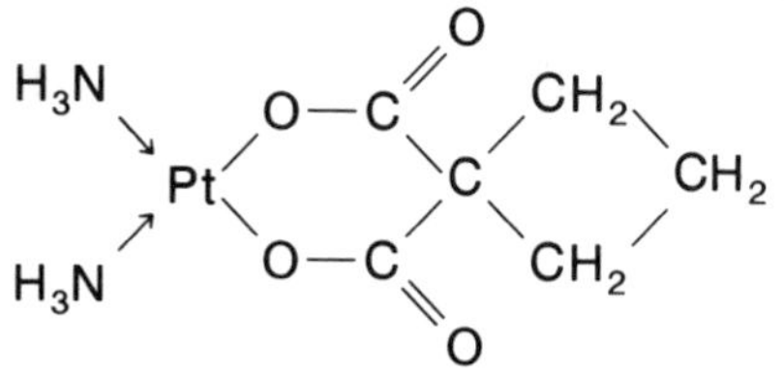

Carboplatin

21.8 MISCELLANEOUS DRUGS

Finally there are some drugs that do not fit into general categories. The alkaloids, vinblastine and vincristine, obtained from the periwinkle plant, interfere with DNA synthesis and prevent cell division.

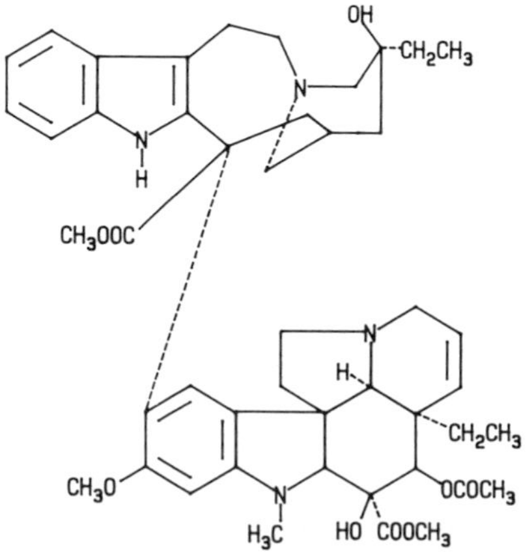

Vinblastine

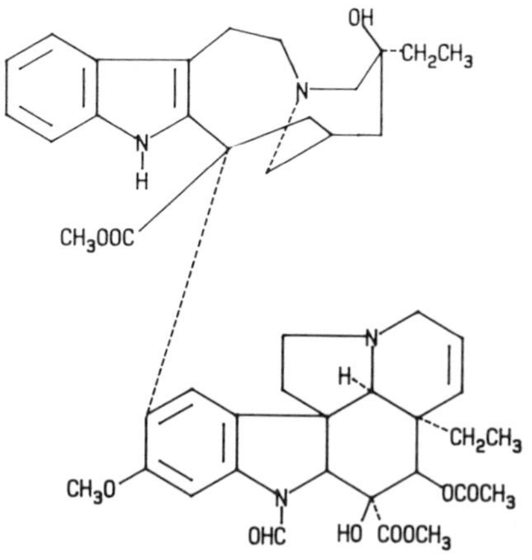

Vincristine

L-Asparaginase, another drug, is an enzyme that deaminates asparagine to L-aspartate and ammonia. Asparagine is an amino acid that some cancer cells cannot make for themselves and must draw from the bloodstream. Hence,

L-asparaginase prevents their multiplication. Normal cells have no difficulty in synthesizing asparagine and are unaffected by the drug.

A number of anthracycline analogues are undergoing clinical evaluation, of which the most advanced is mitozantrone:

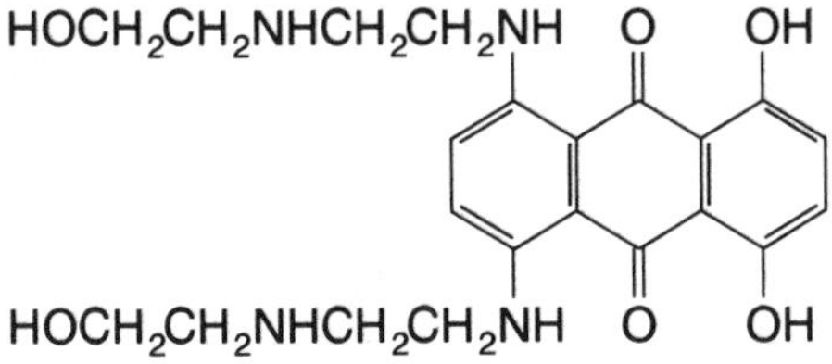

$HOCH_2CH_2NHCH_2CH_2NH$ O OH

$HOCH_2CH_2NHCH_2CH_2NH$ O OH

Mitozantrone

All aminoanthraquinones bind strongly to DNA and inhibit RNA and DNA synthesis, presumably by intercalating the base pairs of the DNA double helix (see also proflavine, Figure 4.1). Mitozantrone seems particularly effective in breast cancer.

Interferon (Section 20.6) was thought to show promise as an anticancer drug. It appears to fasten onto cells and cause the release of enzymes that inhibit growth. Because it is a natural substance it should have no side effects, but unfortunately, in the high doses required for therapy, these appear. It has, however, been used with some success against hairy cell leukemia.

The large number of cancer drugs available is partly a measure of the number of different types of cancer, partly due to the fact that no single drug is truly satisfactory, and partly because the regulatory authorities are inclined to be generous to drugs intended for so intractable a disease. The grave drawback of all anticancer drugs is the severe side effects, and these are among the most unpleasant of any group of drugs.

22

Orphan Drugs

This chapter deals with drugs that are aimed at rare diseases. They are not consumed in large quantities, they are not prescribed frequently, and they offer little prospect of substantial profits. Nonetheless, there are about 4000 rare diseases detailed in the literature, which might or do respond to chemotherapy. Many are known by the names of those who first described them—Huntingdon's chorea, Paget's disease, Tourette's syndrome, and so on. It is in the public interest that these diseases be treated, but the problem is who should pay.

The cost of developing a new drug was estimated in Table 2.8 as \$93–129 million. If a ten-year patent life is assumed, then to cover these costs alone a revenue of \$12–21 million/year worldwide (about a quarter of this in the United States) is required from a new drug. There are very few drugs in the United States that cost more than \$60 for a month's supply; hence, the new drug requires 17,000–30,000 people taking the drug at any one time even if all other costs are ignored. If they are included, an order of magnitude estimate is that 100,000–250,000 U.S. sufferers are needed if a drug is to have any hope of commercial viability. Compare this with the estimate of 16,000 victims of Huntingdon's chorea.

Few people would wish to return to the lax regulation of the drug industry that prevailed before 1960. Nonetheless, there can be no question that expensive regulation inhibits speculative research and causes drug companies to be more cautious and less venturesome. It may put them in a position where they do not carry out research on certain diseases because, even if they find a drug to cure them, the diseases are so rare that they cannot hope to recoup their outlay. Alternatively, a drug may be available but no one wants to produce it.

This may be because the drug is not patentable or because the cost of testing is high or both.

Such drugs have been named "orphan drugs." The problem emerged with the rising cost of testing in the mid 1970s. Until 1983 there was little progress. In that year it became possible for the U.S. Secretary of State to hold the product licence and for the compound to be produced by a supplier who was not required to take responsibility. The British equivalent was a clinical trials certificate, which allowed patients under close supervision to be given a drug that had not been fully tested.

A typical case is Wilson's disease. It is an abnormality of the copper metabolism in which there is a deficiency of copper-carrying protein. The body is unable to excrete traces of copper, and these deposit in tissues and cause damage, especially to the brain and liver. In the 1950s, penicillamine was found to chelate the copper and carry it into the urine to be excreted.

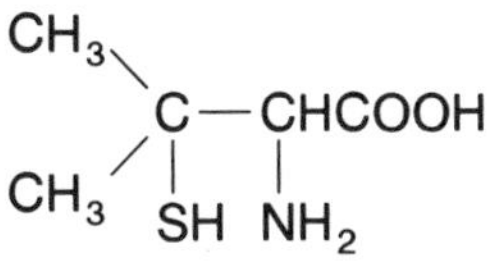

$$(CH_3)_2C(SH)-CH(NH_2)COOH$$

Penicillamine

$$[H_3\overset{+}{N}CH_2CH_2NHCH_2CH_2NHCH_2CH_2\overset{+}{N}H_3]\,Cl_2^-$$

Trientine

In the relaxed climate of the 1950s, there were few problems in having the drug accepted and licensed, particularly as it also turned out to be of value in acute rheumatoid arthritis.

Penicillamine, however, was not tolerated by all patients. There are about 1000 victims of Wilson's disease in the United States and about 100 do not tolerate penicillamine. During the 1970s it was found that triethylene tetramine dihydrochloride (trientine, see above) was effective in the patients who did not tolerate penicillamine and, indeed, that it might be superior to penicillamine for other patients too. John Walshe of Cambridge University, England, who had discovered both drugs, tried to find a drug company who would adopt his product. The conclusion of all of them was the same: The possible rewards from the tiny number of victims of Wilson's disease in no way matched the risks involved. The companies feared the expense of clinical trials, feared litigation, feared damages, and feared the loss of their good name if things went wrong. And that all happened in Britain, which is less strict than the United States about the introduction of new drugs.

It is unfair to assume that the drug firms themselves are not prepared to have anything to do with drugs that do not bring them a substantial profit. Kreitzman (see bibliography) lists many uneconomic drugs produced as a public service by the drug companies. The point is that there was no active search for drugs to combat rare diseases.

Because of this, there was a vigorous campaign in the United States to alter the law. One of the principal spokesmen was David Abelow, who suffered from neurofibromatosis, which became widely known as a result of the film,

The Elephant Man. John Merrick, the subject of the film, had it in a severe form. As a result of the campaign, the so-called Orphan Drugs Act was passed in January 1983 which allows for grants, provides tax credits, eases regulatory pressures, and permits seven years' exclusive marketing rights for non-patentable drugs to companies making them.

By the middle of 1986, the FDA had approved 12 orphan drugs for the treatment of a total of 400,000 people, and another 93 were still waiting testing and approval. Among the drugs approved were trientine (above), acetohydroxamic acid, and sodium cellulose phosphate for the treatment of kidney stones, and ethanolamine oleate for swollen blood vessels in the esophagus.

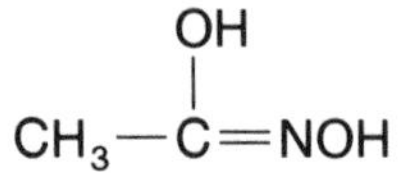

$$CH_3-C(OH)=NOH$$

Acetohydroxamic acid

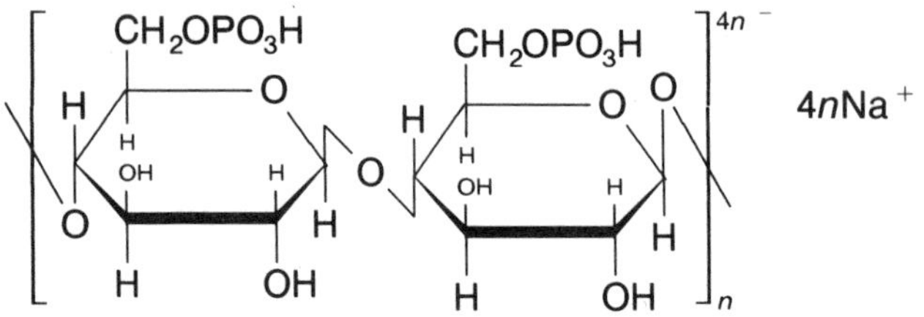

$$\left[(C_6H_7O_2(OH)_2CH_2OPO_3H)_2O \right]_n^{4n-} \quad 4nNa^+$$

Sodium cellulose phosphate

$$CH_3(CH_2)_7CH=CH(CH_2)_7COO^-H_3\overset{+}{N}CH_2CH_2OH$$

Ethanolamine oleate

Another orphan drug of interest is pimozide (Figure 8.7), a neuroleptic of value against Tourette's syndrome. This syndrome develops in children of about ten years of age. It causes involuntary muscle movements and vocal sounds, making victims so disruptive that they cannot gain social acceptance. There are about 100,000 U.S. sufferers, many of whom respond to haloperidol (Section 8.2). Pimozide is of value in those who fail to respond. It appears to act on the dopaminergic nervous system in the brain and to block the postsynaptic dopamine receptors.

The Orphan Drugs Act evolved as a result of the interaction of pressure groups characteristic of the democratic system. It can be accounted a success. Indeed, the legislation of the 1980s concerning the pharmaceutical industry appears more carefully thought out and more helpful than the apparent overregulation of the 1970s.

23

Innovations and Issues

The pharmaceutical industry lives by innovation. It is committed to the discovery, testing, and marketing of new chemical entities. Of course, there are the less successful companies, which did not spend enough on research or whose research produced little of value, and there are other companies—the generic companies—which earn what is generally a poor living by manufacturing successful drugs once they are free of patent protection. They are not the mainstream of the industry, however. About 50 new chemical entities are licensed each year worldwide and about 25 per year reach the U.S. market. This represents a research investment of about $5 billion—the sort of sum which might be the national income of a small country.

23.1 NEW CHEMICAL ENTITIES: 1986 – 1988

It is a formidable undertaking even to keep track of such a level of innovation, but to give some idea of its scope we have listed in Table 23.1 the new drugs that have been licensed between 1986 (the baseline year for this book) and 1988. We have concentrated on chemical entities and omitted some vaccines and biotechnological products. The entities are classified according to the chapter in which they would have been included earlier in the book. Of the approximately 150 products listed, 25 are antibiotics or antibacterials; 31 are cardiovasculars; 16 are central nervous system drugs; 16 are antihistamines, "cold cures," and antiulcer drugs; and 13 are antineoplastic drugs. Antiasthma drugs number 7, nonsteroid anti-inflammatories 6, steroids 6, and there is a substantial miscellaneous category.

TABLE 23.1 New Chemical Entities (World Market 1986 – 1988)

		Antibacterials and Antibiotics (I = injectable; O = Oral)			
1988	Isepamicin	Aminoglycoside	1987	Cefixime	Cephalosporin (O)
1987	Rifaximine	Antibacterial	1987	Cefteram pivoxyl	Cephalosporin (O)
1987	Roxithromycin	Antibacterial	1987	Cefuroxime axetil	Cephalosporin (O)
1988	Azithromycin	Antibacterial	1986	Rokitamycin	Macrolide
1988	Rifapentine	Antibacterial	1986	Butoconazole	Miconazole antifungal
1988	Teicoplanin	Antibacterial	1988	Fluconazole	Miconazole antifungal
			1988	Itraconazole	Miconazole antifungal
1986	Sultamicillin	Antibiotic	1987	Fenticonazole	Miconazole antifungal
1988	Flomoxef	Antibiotic	1986	Cloconazole	Miconazole antifungal
1986	Sulbactam	β-Lactamase inhibitor	1988	Carumonam	Monobactam
1987	Cefminox Na	Cephalosporin (I)	1986	Lenampicillin	Penicillin
1987	Cefpimazole	Cephalosporin (I)	1987	Aspoxicillin	Penicillin (I)
1987	Cefuzonam Na	Cephalosporin (I)	1986	Enoxacin	Quinolone
		Cardiovascular Drugs			
1988	Alacepril	ACE inhibitor	1987	Tertatolol	β-Blocker
1988	Lisinopril	ACE inhibitor	1988	Nipradilol	β-Blocker
1988	Perindopril	ACE inhibitor	1988	Felodipine	Calcium blocker
1987	Enalaprilat	ACE-inhibitor	1988	Denopamine	Cardiac stimulant
1988	Amosulalol	α, β-Blocker	1988	Xamoterol	Cardiac stimulant
1987	Encainide	Antiarrhythmic	1986	Beclobrate	Hypolipemic
1987	Enoxaparin	Anticoagulant	1986	Binifibrate	Hypolipemic
1987	Picotamide	Anticoagulant	1987	Lovastatin	Hypolipemic
1988	Cicletanine	Antihypertensive	1988	Beclobrate	Hypolipemic
1988	Doxazosin	Antihypertensive	1988	Simvastatin	Hypolipemic
1988	Rilmenidine	Antihypertensive	1987	Alteplase	Thrombolytic
1988	Tiamenidine	Antihypertensive	1987	Eminase	Thrombolytic
1986	Defibrotide	Antithrombotic	1986	Brovincamine	Vasodilator
1986	Bisoprolol	β-Blocker	1987	Pinacidil	Vasodilator
1987	Bevantolol	β-Blocker	1988	Cadralazine	Vasodilator
1987	Esmolol	β-Blocker			
		Drugs Affecting the Central Nervous System			
1988	Tianeptine	Antidepressant	1987	Zuclopenthixol	Neuroleptic
1986	Flutoprazepam	Anxiolytic	1986	Idebenone	Nootropic
1987	Metaclazepam	Anxiolytic	1987	Bifemelane	Nootropic
1987	Zopiclone	Anxiolytic	1987	Oxiracetam	Nootropic
1987	Flumazenil	Benzodiazepine antagonist	1988	Exifone	Nootropic
1988	Zolpidem	Hypnotic	1988	Indeloxazine	Nootropic
1986	Amisulpride	Neuroleptic	1988	Nizofenone fumarate	Nootropic
1987	Biriperone	Neuroleptic	1986	Adrafinil	Psychostimulant
		Antihistamines, "Cold Cures," and Antiulcer Drugs			
1986	Azelastine	Antiallergic	1987	Benexate	Antiulcer
1987	Cetirizine	Antiallergic	1987	Spizofurone	Antiulcer
1987	Repirinast	Antiallergic	1987	Ornoprostil	Antiulcer prostaglandin

TABLE 23.1 Continued.

		Antihistamines, "Cold Cures," and Antiulcer Drugs			
1987	Setastine	Antiallergic	1988	Levodropropizine	Antitussive
1988	Acrivastine	Antiallergic	1988	Tinazoline	Decongestant
1988	Loratadine	Antiallergic	1986	Roxatidine	H_2-Antagonist
1986	Troxipide	Antiulcer	1987	Nizatidine	H_2-Antagonist
1986	Plaunotol	Antiulcer	1988	Omeprazole	Proton pump inhibitor
		Analgesics			
1986	Flupirtine	Analgesic	1987	Eptazocine	Analgesic (I)
		Steroid Drugs			
1986	Deflazacort	Anti-inflammatory steroid	1987	Gestodene	Progestogen
1987	Gestrinone	Endometriosis	1986	Prednicarbate	Topical steroid
1986	Norgestimate	Progestogen	1987	Mometasone furoate	Topical steroid
		Nonsteroid Anti-Inflammatory Agents and Antirheumatic Drugs			
1987	Bucillamine	Antirheumatic	1986	Loxoprofen	NSAI
1986	Lobenzarit	Antirheumatic immunodilator	1987	Tenoxicam	NSAI
1986	Amfenac	NSAI	1986	Felbinac	Topical anti-inflammatory
		Antiasthma Drugs			
1987	Amlexanox	Antiasthmatic	1988	Flutropium Br	Bronchodilator
1986	Formoterol	Bronchodilator	1986	Nedocromil	Prophylactic
1986	Mabuterol	Bronchodilator	1987	Surfactant bovine	Respiratory distress
1988	Doxofylline	Bronchodilator			
		Miscellaneous and Orphan Drugs			
1988	Octreotide	Acromegaly GI tumors	1987	Dapiprazole	Glaucoma
1988	α_1-proteinase inhibitor	α_1-antitrypsin deficiency	1986	rDNA hepatitis-B	Hepatitis-B
1988	Erythropoietin	Anemia	1987	Somatropin	HGH-deficiency
1986	Propofol	Anesthetic	1988	Alpiropride	Migraine
1986	Orthoclone OKT3	Antigraft rejection	1986	Nafamostat	Pancreatitis
1986	H-Influenzae vaccine	Bacterial meningitis	1988	Cilostazol	Thromboangiitis obliterans
1986	Clodronate	Bone resorb inhibitor	1988	Limaprost	Thromboangiitis obliterans
1988	Ozagrel	Cerebral ischemia	1988	Olsalazine	Ulcerative colitis
1986	Trientine	Chelating agent	1988	Alfuzosin	Urinary incontinence
1988	Apraclonidine vaccine	Eye surgery	1988	Clobenoside	Venous insufficiency
1988	Cisapride	Gastroprokinetic			

TABLE 23.1 Continued.

		Tropical Diseases			
1987	Artemisinin	Antimalarial	1988	Halofantrine	Antimalarial
1988	Artesunate	Antimalarial	1987	Ivermectin	Onchocerciasis
		Antiviral Drugs			
1987	Rimantadine	Antiviral	1988	Ganciclovir	Antiviral
1987	Zidovudine	Antiviral			
		Antineoplastic Drugs			
1986	Decapeptyl	Anticancer	1987	Ranimustine	Anticancer
1986	Lentinan	Anticancer	1987	Ubenimex	Anticancer
1986	Carbaplatin	Anticancer	1988	Lonidamine	Anticancer
1986	Interferon-α-2A	Anticancer	1988	Pirarubicin	Anticancer
1986	Schizophyllan	Anticancer	1987	Nilutamide	Anticancer (prostate)
1986	Interferon-α-N1	Anticancer	1987	Gugulipid	Anticancer (prostate)
1987	Doxifluridine	Anticancer			

Study of this list gives an idea of the current areas of interest to the research community. Among antibiotics and antibacterials, emphasis is on quinolones, monobactams, β-lactamase inhibitors, and yet more third generation injectable cephalosporins with, it is hoped, a slightly wider spectrum than current materials. Cardiovasculars include various materials with a "traditional" mode of action, but interest is concentrated on thrombolytics, angiotensin-converting enzyme inhibitors, and hypolipemics.

There has been a general decrease of interest in central nervous system drugs of the mood-altering type and almost half the new entities in this class are intended to counter senility. Some of them are intended to increase neuron activity in the brain while others are supposed to increase alertness and protect the cerebral cortex against hypoxia, that is, a shortage of oxygen.

In the area of antihistamines, there are a number of nonsoporific antiallergic compounds. Antiulcer drugs are still an area of excitement. A single prostaglandin has been licensed, plus various H_2-antagonists and a proton pump inhibitor. The elucidation of the role of prostaglandins in asthma has increased interest in this disease and a number of new preparations have come on the world market. All the same, no leukotriene antagonist has yet been licensed.

There is still a modest level of activity in steroids and nonsteroid anti-inflammatories, but apart from mifepristone (Section 11.3.1) there seems little that is truly novel. The large number of cancer drugs reflects the high incidence of the disease but also the ease of registering a drug for the treatment of patients who would certainly die without it.

The miscellaneous category includes a number of biotechnological products, including vaccines against hepatitis-B and bacterial meningitis, and a product for human growth-hormone deficiency. Most interesting is ozagrel, which is the first thromboxane synthetase inhibitor (Section 19.1.3) to be licensed. It inhibits thromboxane production and thereby reduces blood clotting. It is used in cerebral vasospasm and ischemia following subarachnoid hemorrhage.

Finally, mention should be made of the handful of antivirals. The absence of large numbers of such compounds is certainly not due to lack of research interest.

23.2 THE ACHIEVEMENT OF THE PHARMACEUTICAL INDUSTRY

The past few years have seen the consolidation of the second chemotherapeutic revolution. Few companies would now begin the search for a new drug without an understanding of the biochemical nature of the disease it was intended to ameliorate and the enzyme systems it was hoped to modify. Screening has its place, certainly, and receptor technology which accelerates screening is itself a breakthrough. Nonetheless, the days of molecular roulette have passed and drug development rests on a solid theoretical basis.

The understanding of the biochemical mechanisms of the body is a great achievement of the past generation and the pharmaceutical industry has played a seminal role. The steam engine was invented in the eighteenth century, long before Josiah Willard Gibbs developed the science of thermodynamics to explain how it worked. In the same way, the often accidental discovery of new drugs by the pharmaceutical industry has prompted biochemists to discover the body mechanisms on which they act. This, in turn, has fed back into the industry and suggested the design of possible new drugs. Indeed, much of the basic biochemistry was performed in industrial laboratories and the pharmaceutical industry can claim credit for many of the breakthroughs.

Nor were the biochemists acting in isolation. The millionfold increase in sensitivity of analytical chemical techniques was a prerequisite for the kind of research that led to the elucidation of the arachidonic acid cascade. Sophisticated chemical techniques were required for the determination of the structures of biochemical materials and the syntheses both of them and of related compounds. All these, together with the contributions of other disciplines, are part of the scientific romance of our times.

Where is it leading? On a practical level, it should lead to better drugs, which should lead to greater life expectancy and a higher quality of life. Many diseases remain unconquered, viral infections and cancer being the two most noteworthy.

On the intellectual side, too, the outlook is exciting. Science is approaching an understanding of brain chemistry. This is a much bigger problem than understanding the cosmos and is the last frontier of human knowledge. Most of our knowledge of the chemistry of the brain's action has been developed in the last 30 years and progress is rapid. Is it possible for the human brain to understand how the human brain understands? Whatever the philosophic doubts, there are no obvious boundaries to the scientific work. It is a privilege even to be a bystander at such an explosion of the human intellect.

23.3 THE PROBLEMS OF THE PHARMACEUTICAL INDUSTRY

By any standards, the achievements of the pharmaceutical industry since the early 1930s have been remarkable and one might reasonably expect it to occupy a secure place in public esteem. To paraphrase Jonathan Swift, anyone who could cure a sore throat, let alone alleviate an angina attack, would deserve better of mankind than the whole race of politicians put together.

In contrast to such expectations, the pharmaceutical industry is widely unpopular. It is frequently attacked in the media and is a particular bugbear of "green" politicians who, in the naive view, might be expected to be particularly enthusiastic about the curing of illness. The industry allegedly markets dangerous products, encourages overprescribing of expensive drugs, makes excessive profits, and spends unnecessarily on marketing and administration. All these criticisms are denied by the industry but have wide circulation and it is not a matter of indifference where governments and legislators stand. The future prosperity of the industry depends crucially on its public image.

23.3.1 Adverse Drug Reactions

The question of drug testing was reviewed in Section 2.10. The testing procedure usually takes a long time and is expensive. This does not guarantee its efficacy but appears to be the best that anyone can do, bearing in mind that some adverse effects will appear only after the drug has been used by millions of patients. An official "Phase IV" involving postapproval testing might be a good idea, particularly if physicians were more punctilious about reporting side effects.

The search for a 100% safe drug, however, is a chimera. Drugs are selective poisons and there are bound to be sensitive individuals. The snag is that the more careful the monitoring, the more alerts will be put out, and the more the public will believe that the drugs are in fact dangerous. Furthermore, the pressure on the industry to bring new products to market has led to occasional falsification of test records (see bibliography).

In the end, there are what appear to be miscarriages of justice on both sides. For example, in the benoxaprofen case (Section 12.1), the U.K. victims received far less compensation than the U.S. victims. On the the other hand, the antinausea compound debendox had to be removed from the market even though the evidence that it was teratogenic was rejected by a number of courts.

A factor which makes judgement more difficult is that people's attitude to drugs depends on whether they are sick or well. Invalids are desperately anxious to obtain drugs that might alleviate their symptoms and, if the drugs are not legal in their home country, they may well have them smuggled in from abroad. Healthy people, on the other hand, are more cautious about drug safety and less sympathetic about the occasional person who dies when an unlicensed drug might have saved him. Furthermore, while statistics on people who die from adverse drug reactions are readily available, no one really knows how many lives might have been saved had a drug been permitted or even licensed more quickly. It has been calculated that if reserpine, the first modern drug for high blood pressure, had been delayed for five years by the regulations in force now, another 52,000 Americans would have died of heart disease.

23.3.2 Overprescribing

Rates of prescribing vary among different countries (Section 3.3.4) and different physicians. The lowest prescribing physician in the U.K. National Health Service prescribes on one first consultation out of four, compared with some colleagues who have virtually a 100% record. As there are no gross differences in health between different developed countries or patients of different physicians, there is an implication that many prescriptions are unnecessary.

Physicians accused of overprescribing blame the drug industry for high-pressure sales techniques and patients for expecting a prescription every time they come to the office. Patients comment sourly that they do not demand prescriptions; the physician hands them out because it is quicker than talking to them to find out what the trouble really is. The drug industry complains that their sales techniques are necessary because physicians do not read medical journals sufficiently carefully to get news of new pharmaceuticals. Furthermore, the public demand for elaborate testing means that the cost of developing new drugs is high (Section 2.5) and takes such a large proportion of the patent life (Section 2.7) that without vigorous promotion to build up sales of a new drug rapidly the companies would never recover their investment. This is undoubtedly true in some cases but in others the public (and often the physicians and the professional associations) have justifiably looked askance at the "freebies"—heavily subsidized conferences in attractive places and so on —that are used to induce physicians to prescribe one brand name drug rather than another brand name drug or a generic equivalent.

23.3.3 Are Profits Excessive?

Attitudes to the pharmaceutical industry are influenced by a rarely stated but profound belief that it is immoral to make profits out of the illness of others. The donation of the original penicillin to the world was praiseworthy, as was Merck's more recent gift of ivermectin (Section 18.3.1). The high cost of cimetidine in relation to other drugs is widely seen as reprehensible, in spite of the much greater saving in costs of surgery (Figure 9.7).

The usual economic constraints of supply and demand which keep profits at an acceptable level—neither too high nor too low—do not apply to the drug industry. On the one hand, drug companies are monopoly sellers of patent-protected products; on the other hand, a high proportion of the drug bill, varying from country to country, is paid directly or indirectly by governments, so that they are to some extent monopoly buyers. In countries with innovative pharmaceutical industries which contribute to the balance of payments, governments have to set the cost of pharmaceuticals against the need to keep their industries alive; countries without innovative pharmaceutical industries have an interest in driving prices to a minimum. Faced with a decision on allocation of limited resources, a government will frequently prefer to fund a nurses' pay claim than to increase drug companies' remuneration. Nurses are politically more visible than research scientists and the effects of a cut in research expenditure will, in any case, have no effect for about 10 years and even then no one will know which disease would have been curable had the cuts not been made.

Prices are thus the result of a battle of wills between multinational companies and governments. The fact that the drug industry is still so profitable is a measure of the extent to which the industry has in fact managed to stand up to such pressures. Whether this will remain true in the future (and especially when the unified European market is established in 1992) remains to be seen. One danger is that the industry will become a victim of its own success. New drugs are emerging at an incredible rate. Several calcium antagonists and alpha- and beta-blockers appear each year. Four angiotensin-converting-enzyme inhibitors joined captopril and enalapril in 1987–1988. At least 20 cepahlosporins and 20 nonsteroid anti-inflammatories are marketed internationally and more are sold in a limited number of countries. The Dutch government has refused to license any more NSAIs unless something completely new is offered. Fourteen years after the introduction of cimetidine, there are at least 20 antiulcer compounds on the market working by four quite different and well understood mechanisms. Are there enough ulcers to support so many research programs?

The dominant characteristic of the pharmaceutical industry at present is risk. Hundreds of millions of dollars are invested with the hope of a return in many years' time. A minor adverse drug reaction at the last minute can cause tens of millions of dollars to be lost. Bureaucratic delays in registration can destroy the profitability of a drug but, equally, the pressure on companies to be first on the market may cause them to skimp on testing. It is this high level of risk which should be set against the high profitability of the industry. It would take a talented forecaster to predict the financial position of any individual pharmaceutical company in ten or even five years' time.

The political battles over the pharmaceutical industry will continue and there will no doubt be drug disasters in the future, along with the same accusations of dishonesty, greed and malice. Like so many aspects of democratic life, the present system is demonstrably inadequate. The problem is to find anything that is not worse. Any other century and most other countries would regard even the present imperfect situation as a golden age.

23.4 DRUGS OF THE FUTURE

Government policies towards the pharmaceutical industry are likely only to slow down or speed up the development of new drugs. It is therefore legitimate to look forward to the type of new drugs to be expected in the future without setting a time scale on their arrival. We have mentioned at the end of each section specific examples of new drugs in the pipeline and the following paragraphs are devoted to general trends.

First, there will be an enlargement of present areas of activity. The search will continue for receptor agonists or antagonists of the types which are now familiar. Table 23.1 provides a sample of what has come on the market recently and such innovations will be the staple of the industry for years to come. The two breakthroughs in technique—receptor technology (Section 4.2) and the use of computer graphics for modelling new drugs (Section 7.6.4)—will speed up the screening process for new compounds.

Second, there will be development of many new dosage forms which will localize drug action in time and space. Examples of time localization are the many forms of controlled release tablets already on the market and such things as insulin pumps which release drugs when appropriate.

Space localization is exemplified by the antiasthma steroids mentioned in Section 15.4, by topical drugs generally, and by antibiotics that are not absorbed in the gut and are therefore used against urinary infections. Methods of directing drugs are becoming more sophisticated, however, and there are likely to be more drugs like the ADEPT system and acyclovir (Sections 4.5.5 and 20.3) where the disease-curing system is in some way "switched on" when it reaches its point of action. Rather than "magic bullets," such drugs are more like "magic guided missiles."

Finally, we foresee developments arising out of the current interest in biotechnology and in proteins and small peptides. We have not dealt in detail with biotechnology in this book because our objective was to concentrate on the chemical side. Nonetheless, the advent of biotechnology has altered how biomedical research is conducted and what products humanity can expect from such research. Many physiological processes are controlled by peptides and proteins, and these are the products of most biotechnological processes. We have already mentioned bradykinin (Section 19.3), angiotensin II (Section 7.6.4), endorphins and enkephalins (Section 10.3), insulin (Section 13.2), and various others.

The first generation of biotechnological products (e.g., insulin and interferon) are identical with the natural products. These are sometimes effective but many peptides are ineffective as drugs because they are destroyed by peptidases before they have a useful effect. The second generation will be chemically modified peptides or proteins. One example is Eminase, the anisoylated enzyme used as a thrombolytic agent (Section 7.8). Proteins and peptides may also be made synthetically and an example is ICI's Zoladex, a complex peptide active against prostate cancer, and an analogue of the naturally occurring luteinizing-hormone-releasing hormone.

The third generation of products will be the peptides which are not peptides! Chemists will try to develop small nonpeptide molecules which have the same effect as biologically active peptides but which are easier to manufacture and more convenient to administer to patients. The problem may be stated as follows: if one knew the structures of endorphins and enkephalins, would it be possible to invent morphine? If the poppy did not provide it, how easily could it be made? As the structures of more biologically active peptides are elucidated, we may expect a range of synthetic "morphine-equivalents."

23.5 CONCLUSION

In this book we have reviewed the chemistry, technology, and a little of the pharmacology of the most widely prescribed drugs. We have placed the social and economic background of the industry in context and commented also on physicians and patients—the end-users of pharmaceuticals. We have also mentioned some of the drugs that are likely to be important in the future. We have indicated the tremendous benefits that the world obtains from the pharmaceutical industry: the increased life expectancy, the reduced infant mortality, the conquest of infectious diseases, and so on. We have also pointed out various political problems faced by the pharmaceutical industry, including legislation, profitability, patent term erosion, increasing costs of testing, pricing, drug safety, overprescribing, and high-pressure selling techniques. In the final analysis, how does society balance the benefits brought by the pharmaceutical industry against the risks that arise from it?

These are complicated issues that merit extensive treatment. Indeed, there is no chapter in this book which could not be dealt with in a complete book, if not an encyclopedia. Our objective, however, has been to provide a perspective and an overview. The pharmaceutical industry *as a whole* is a fascinating topic and we hope that we have managed to communicate some of our enthusiasm.

Appendixes

Appendix 1

The Most Widely Prescribed Drugs by Chemical Entity

Table 5.1 purports to list the hundred most widely prescribed ethical drugs in the United States by chemical entity. Such a list could most easily be obtained from the National Prescription Audit conducted each year by IMS America, Ltd., Ambler, PA. The audit, however, is not in the public domain and, even if we had access to it, it would not be legitimate for us to reproduce it.

On the other hand, the *Pharmacy Times* publishes each April a review of the audit and we have devised a method of calculating the most widely prescribed drugs by chemical entity based on these published data. The method has certain limitations and sources of error and is described in this appendix.

The review lists the Top-200 prescription drugs in the United States for the previous year in rank order. The drugs are categorized not by chemical entity but by brand name, hence drugs such as penicillin V, which are produced by a number of manufacturers, do not figure prominently in the list. In addition, the top 30 generic drugs are listed together with the number of first (as opposed to repeat) prescriptions for each.

The final piece of information is the percentage of prescriptions for the top-50 drugs, the second-50, the third-50, and the fourth-50, plus the total of all prescriptions. The figures for 1986 are given in the second column of Table A.1.

It seemed likely that the distribution of prescriptions with rank order would follow a Pareto distribution. This distribution was devised to classify the spread of incomes in different countries (V. Pareto, *Cours d'Economie Politique*, 2 Vols., Rouge, Lausanne, 1896/7) but has since been shown to fit many other statistical patterns, including the borrowing of journals from the British

TABLE A.1. Reported and Calculated Percentages of Prescriptions in the Top-200 Drugs Arranged in Groups of 50 (United States, 1986)

Percentage of all Prescriptions			
	Reported	Calculated for $B = 0.635$	Calculated for $B =$ 0.738 Plus *Scrip* Data
Top-50	31.4	31.4	31.4
Second-50	11.4	11.1	11.5
Third-50	8.0	7.9	7.8
Fourth-50	6.0	6.4	6.1
RMS deviation		0.252	0.125

library (see D. de Solla Price, *Big Science*, *Little Science*, Columbia University Press, NY, 1963). The distribution in the case here gives the number of prescriptions of the Nth ranking drug as

$$\ln(\text{no. of prescriptions}) = A - B \ln N$$

where B is a constant representing the rapidity with which numbers of prescriptions drop off with rank order and A is a constant related to B and the total prescriptions. The percentages of prescriptions falling into the first-, second-, third-, and fourth-50s depend entirely on B and are independent of A.

A computer program was written which summed the terms of the above equation for different values of B. The optimum least squares fit occurred at $B = 0.635$. The extent of agreement was excellent and is shown in the third column of Table A.1. The total number of prescriptions in 1986 was 1557 million and the constant A therefore becomes 17.777.

In principle, the above values can be used to calculate the numbers of prescriptions for each of the Top-200 prescription drugs by trade name. The chief and serious source of error is that the Pareto distribution is recognized to break down for the richest members of society in Pareto's work and for the drugs prescribed most frequently in this.

Fortunately, the journal *Scrip* publishes each year the numbers of prescriptions dispensed of each of the top ten drugs and such data are available for 1986. These numbers can be installed in the computer program and the Pareto distribution continued from the point where the known data end. The fit with the *Pharmacy Times* figures is shown in the fourth column of Table A.1 and it is even better than the original fit.

The number of prescriptions plotted against rank number is shown in Figure A.1 for the two methods of calculation. The error for the highest ranked drugs is clearly visible but the change in numbers at low ranks is surprisingly small considering the large change in the exponent B.

The numbers of prescriptions for each of the brand name drugs were calculated on the above basis and the prescription numbers for drugs contain-

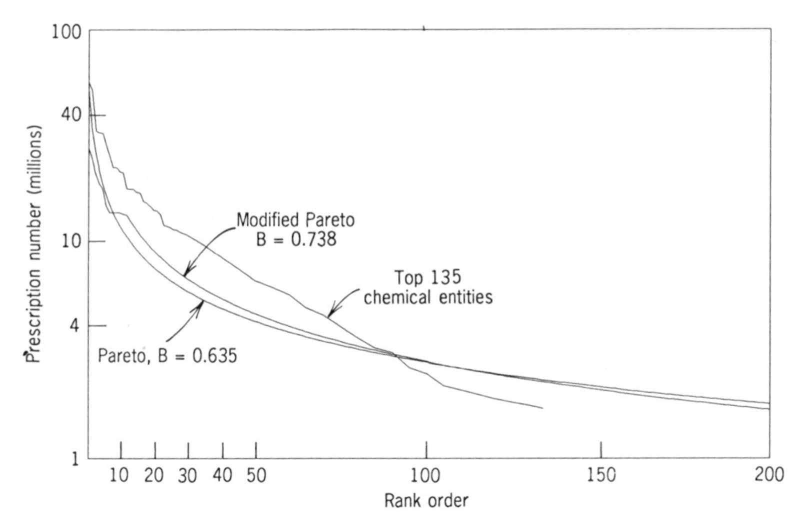

Figure A.1. *Number of prescriptions versus rank order for the United States, 1986.*

ing the same chemical entities were aggregated. Drugs containing two ingredients which occurred in other mixtures were counted separately under each ingredient. For example, Dyazide contains triamterene and hydrochlorothiazide and Dyazide prescriptions were counted under each heading. Trimethoprim and sulfamethoxazole are always prescribed together (at least in the U.S. Top-200) and hence were counted as a single entity. Drugs containing the over-the-counter components aspirin and acetaminophen were counted only under the ethical drug present in the formulation. Generic drug prescription numbers were added in. The Top-200 branded drugs contained 135 chemical entities. These were ranked to give Table 5.1 and the third line on Figure A.1.

There are various errors in the method. In particular, brand name drugs falling outside the Top-200 are ignored. Inclusion of, say, a million prescriptions of a chemical entity otherwise ranked, say, number 100, would send it 40 places up the rankings. Nonetheless, we believe the list to provide a broadly accurate picture of which drugs are important and it is more suitable for our purposes than one based on brand names.

Appendix 2

International Classification of Diseases

The World Health Organization (WHO) requests member countries who report to it to classify diseases according to the scheme laid down in the *Manual of the International Statistical Classification of Diseases*, 9th revision (1975), WHO, Geneva, 1977.

There are 17 main categories shown on the following pages and we have also inserted some of the subdivisions. Individual diseases are represented by a single three digit number. Under some headings we have given in parentheses examples of the diseases falling into the category, but there are usually other diseases that we have not mentioned. For example, at the top of the first column of the table we list 001–009, intestinal infectious diseases. Cholera, typhoid, and food poisoning, which we quote, are 001, 002, and 005, respectively, but the category also contains 006, amebiasis, which we have not mentioned.

In addition to the 17 categories listed, there are two supplementary classifications. One of them, bearing the numbers E800–899, classifies detailed external causes of injury and poisoning and the other, V01–V82 classifies factors such as pregnancy, which are not illnesses but influence health status and contact with health services.

Countries lacking the facilities to gather detailed statistical information use less detailed lists comprising significant and easily diagnosed diseases.

The names in the list are carefully defined by the medical profession but are not the common names. For example, malignant neoplasms are better known as cancer. In general we have used the common names of illnesses, hoping to gain in comprehensibility more than we lose in precision.

I. Infectious and Parasitic Diseases

001 – 009	Intestinal infectious diseases (cholera, typhoid, food poisoning)
010 – 018	Tuberculosis
020 – 027	Zoonotic bacterial diseases (anthrax, plague)
030 – 041	Other bacterial diseases (leprosy, diphtheria, whooping cough)
045 – 049	Poliomyelitis and other non-arthropod-borne viral diseases of central nervous system
050 – 057	Viral diseases accompanied by exanthem (smallpox, herpes, measles, rubella, chicken pox)
060 – 066	Arthropod-borne viral diseases (yellow fever, encephalitis)
070 – 079	Other diseases due to viruses and chlamydia (viral hepatitis, rabies, mumps, trachoma)
080 – 088	Rickettsioses and other arthropod-borne diseases (typhus, malaria, leishmaniasis, trypansomiasis)
090 – 099	Syphilis and other venereal diseases
100 – 104	Other spirochetal diseases
110 – 118	Mycoses
120 – 129	Helminthiases
130 – 136	Other infectious and parasitic diseases
137 – 139	Late effects of infectious and parasitic diseases

II. Neoplasms (Cancers)

140 – 149	Malignant neoplasm of lip, oral cavity, and pharynx
150 – 159	Malignant neoplasm of digestive organs and peritoneum
160 – 165	Malignant neoplasm of respiratory and intrathoracic organs
170 – 175	Malignant neoplasm of bone, connective tissue, skin and breast
179 – 189	Malignant neoplasm of genitourinary organs
190 – 199	Malignant neoplasm of other and unspecified sites
200 – 206	Malignant neoplasm of lymphatic and hematopoietic tissue
210 – 229	Benign neoplasms
230 – 234	Carcinoma *in situ*
235 – 238	Neoplasms of uncertain behavior
239	Neoplasms of unspecified nature

III. Endocrine, Nutritional and Metabolic Diseases, and Immunity Disorders

240 – 246	Disorders of thyroid gland
250 – 259	Diseases of other endocrine glands
260 – 269	Nutritional deficiencies (vitamin deficiencies, kwashiorkor)
270 – 279	Other metabolic disorders and immunity disorders

IV. Diseases of Blood and Blood-Forming Organs

280 – 289	Diseases of blood and bloodforming organs (anemia)

V. Mental Disorders

290 – 294	Organic psychotic conditions (alcohol and drug psychoses)
295 – 299	Other psychoses (schizophrenic psychoses)
300 – 316	Neurotic disorders, personality disorders, and other non-psychotic mental disorders
317 – 319	Mental retardation

VI. Diseases of the Nervous System and Sense Organs

320–326 Inflammatory disease of the central nervous system
330–337 Hereditary and degenerative diseases of the central nervous system (cerebral degeneration, Parkinson's disease)
340–349 Other disorders of the central nervous system (multiple sclerosis, epilepsy, migraine)
350–359 Disorders of the peripheral nervous system
360–379 Disorders of the eye and adnexa (blindness, glaucoma, cataract)
380–389 Diseases of the ear and mastoid process (deafness)

VII. Diseases of the Circulatory System

390–392 Acute rheumatic fever
393–398 Chronic rheumatic heart disease
401–405 Hypertensive disease
410–414 Ischemic heart disease (angina pectoris, myocardial infarction)
415–417 Diseases of pulmonary circulation
420–429 Other forms of heart disease
430–438 Cerebrovascular disease
440–448 Diseases of arteries, arterioles and capillaries (atherosclerosis, aneurysm)
451–459 Diseases of veins and lymphatics and other diseases of circulatory system

VIII. Diseases of the Respiratory System

460–466 Acute respiratory infections (common cold, acute tonsilitis, laryngitis and bronchitis)
470–478 Other diseases of upper respiratory tract
480–487 Pneumonia and influenza
490–496 Chronic obstructive pulmonary disease and allied conditions (asthma, chronic bronchictisis)
500–508 Pneumoconioses and other lung diseases due to external agents
510–519 Other diseases of respiratory system

IX. Diseases of the Digestive System

520–529 Diseases of oral cavity, salivary glands, and jaws
530–537 Diseases of esophagus, stomach, and duodenum
540–543 Appendicitis
550–553 Hernia of abdominal cavity
555–558 Noninfective enteritis and colitis
560–569 Other diseases of intestines and peritoneum
570–579 Other diseases of digestive system (cirrhosis of the liver)

X. Diseases of the Genitourinary System

580–589 Nephritis, nephrotic syndrome, and nephrosis
590–599 Other diseases of urinary system
600–608 Diseases of male genital organs
610–611 Disorders of breast
614–616 Inflammatory disease of female pelvic organs
617–629 Other disorders of female genital tract

XI. Complications of Pregnancy, Childbirth, and the Puerperium

630–639 Pregnancy with abortive outcome
640–648 Complications mainly related to pregnancy
650–659 Normal delivery and other indications for care in pregnancy, labor, and delivery

XI. Complications of Pregnancy, Childbirth, and the Puerperium

660–669 Complications occurring mainly in the course of labor and delivery

670–676 Complications of the puerperium (puerperal fever)

XII. Diseases of the Skin and Subcutaneous Tissue

680–686 Infections of skin and subcutaneous tissue

690–698 Other inflammatory conditions of skin and subcutaneous tissue

700–709 Other diseases of skin and subcutaneous tissue

XIII. Diseases of the Musculoskeletal System and Connective Tissue

710–719 Arthropathies and related disorders

720–724 Dorsopathies

725–729 Rheumatism, excluding the back

730–739 Osteopathies, chondropathies, and acquired musculoskeletal deformities

XIV. Congential Anomalies

740–759 Congenital anomalies (spinabifida, cleft palate and lip)

XV. Conditions Originating in the Prenatal Period

760–779 Certain conditions originating in the prenatal period

XVI. Symptoms, Signs, and Ill-Defined Conditions

780–789 Symptoms, signs and ill-defined conditions

790–796 Nonspecific abnormal findings

797–799 Ill-defined and unknown causes of morbidity and mortality (senility)

XVII. Injury and Poisoning

800–804 Fracture of skull

805–809 Fracture of spine and trunk

810–819 Fracture of upper limb

820–829 Fracture of lower limb

830–839 Dislocation

840–848 Sprains and strains of joints and adjacent muscles

850–854 Intracranial injury, excluding those with skull fractures (concussion)

860–869 Internal injury of chest, abdomen, and pelvis

870–879 Open wound of head, neck and trunk

880–887 Open wound of upper limb

890–897 Open wound of lower limb

900–904 Injury to blood vessels

905–909 Late effects of injuries, poisonings, toxic effects, and other external causes

910–919 Superficial injury

920–924 Contusion with intact skin surface

925–929 Crushing injury

930–939 Effects of foreign body entering through orifice

940–949 Burns

950–957 Injury to nerves and spinal cord

958–959 Certain traumatic complications and unspecified injuries

960–979 Poisoning by drugs, medicaments, and biological substances

980–989 Toxic effects of substances chiefly nonmedicinal as to source

990–995 Other and unspecified effects of external causes

996–999 Complications of surgical and medical care not elsewhere specified

Appendix 3

Bibliography and Notes

The first part of this bibliography lists standard works on the broad topics we have covered. The second part lists, section by section, the sources of our information when they are neither personal nor included in the standard works. Numbers in parentheses in the second section refer to books in the first section. The bibliography makes no attempt to be comprehensive but simply lists the items we have found useful.

LISTS OF DRUGS, PHARMACOPEIAS, AND SO ON

1. *The Merck Index*, 10th ed., Merck & Co., Inc., Rahway, NJ, 1976, is a useful reference work for the chemist interested in pharmaceuticals. Along with many other chemicals, it lists almost 10,000 drugs by generic and chemical names and with cross references to trade names.
2. *The Physician's Desk Reference*, B. B. Huff, Ed., Medical Econ. Co., Oradell, NJ, published annually, is a collection of data on the various drugs that can be prescribed.
3. *The ABPI Data Sheet Compendium*, A. J. M. Bailey and M. Murray, Eds., Pharmind Publications Ltd, London, 1988, is the British equivalent of the *Physician's Desk Reference*.
4. *The British National Formulary*, published by the British Medical Association (Tavistock Square, London WC1H 9JP) and the Pharmaceutical

Society of Great Britain was completely revised in 1981. Its scope was broadened and it now not only lists drugs and drug formulations, but also provides guidance on drugs of choice in particular situations. It appears twice a year. The U.S. equivalent is the *National Formulary*, American Pharmaceutical Association, Washington DC.

5. *Drug Evaluations*, American Medical Association, 6th ed., 1986, distributed by W. B. Saunders, Philadelphia, is a seminal volume, listing all the major pharmaceuticals and commenting in detail on the new ones. It gives chemical structures and modes of action, together with the usual information on dosages.
6. Useful pharmacopeias include *Pharmacopeia of the USA*, U.S. Pharmaceutical Convention, New York; *Homeopathic Pharmacopeia of the USA*, Boericke and Tafel, Philadelphia; *The British Pharmacopeia*, General Medical Council, London; *British Pharmaceutical Codex*, 11th ed., Pharmaceutical Society of Great Britain, London, 1979; *The International Pharmacopeia*, 3d ed., World Health Organization, Geneva, 1981; and *Martindale—The Extra Pharmacopeia*, 28th ed., J. E. F. Reynolds, Ed., The Pharmaceutical Press, London, 1982. A new edition of *Martindale*, an exceptionally useful source of pharmaceutical information, appeared early in 1989.
7. *The Drugs Handbook*, 3d ed., P. Turner and G. Volans, Macmillan, London, 1983, is a slim paperback containing an alphabetical listing of the main drugs with fairly elementary text describing them.
8. *British Approved Names*, British Pharmacopeia Commission, London, 1981, lists the generic and chemical names of drugs together with their (English) pronunciation. Updates appear twice yearly.
9. *Drugs of Choice, 1980–81*, W. Modell, Ed., Mosby, St. Louis, 1980, also provides guidance on the relative merits of various drugs.
10. *Drug Facts and Comparisons*, E. K. Kastrup and B. Olan, Eds., Lippincott, Philadelphia, is a looseleaf guide with monthly updates. It is also available on microfiche.
11. *Avery's Drug Treatment*, 3d ed., T. M. Speight, Ed., ADIS, Auckland, NZ, 1987, contains an extensive discussion of clinical pharmacology and the therapeutics required for the treatment of many categories of diseases.
12. *The Red Book*, Medical Econ. Co., Oradell, NJ, appears annually and lists the prices of all the drugs on the market and their manufacturers. The *Chemist and Druggist Price List*, Benn, Tonbridge, UK, does the same for the United Kingdom. It appears monthly with weekly updates.
13. *Approved Drug Products, with Therapeutic Equivalence Evaluations*, U.S. Department of Health and Human Services, Rockville, MD, known as "The Orange Book," has appeared annually since 1985 with quarterly updates. It lists drugs with their therapeutic equivalents approved by the FDA on the basis of safety and effectiveness. This is to aid the prescribing of generic rather than branded products, but also contains useful information on patent expiry dates and exclusivity arrangements.

MEDICINAL CHEMISTRY AND PHARMACOLOGY

20. Burger's *The Basis of Medicinal Chemistry* is the authoritative book dealing with the theories and principles that guide the medicinal chemist. The fourth edition, edited by M. E. Wolff, Wiley, New York, 1980, runs to three volumes and over 3000 pages and provides an unparallelled overview of the field.
21. *Introductory Medicinal Chemistry*, J. B. Taylor and P. D. Kennewell, Ellis Horwood/Wiley, Chichester, 1981, is an excellent paperback with a mere 200 pages, which, unlike Burger, can practically be read in a single sitting.
22. *The Pharmacological Basis of Therapeutics*, A. G. Gilman, L. S. Goodman, T. W. Rall, and F. Murad, Eds., 7th ed., Macmillan, New York, 1985, is a pharmacological classic.
23. *Textbook of Organic Medicinal and Pharmaceutical Chemistry*, C. O. Wilson, O. Giswold, and R. F. Doerge, 6th ed., Lippincott, PA, 1971, is another classic.
24. *Lecture Notes on Pharmacology*, M. F. Grundy, Blackwell, Oxford, 1986., is intended as a college-level text. It contains precise descriptions of many pharmacological phenomena and attempts to provide fundamentals on which a body of knowledge can be built.
25. *Clinical Pharmacology*, 6th ed., D. R. Laurence and P. N. Bennett, Churchill/Livingstone, 1987, is a standard text and has a chapter by Sir James Black, the discoverer of propranolol and cimetidine.
26. Useful books on pharmacology include *Textbook of Pharmacology*, W. C. Bowman and M. J. Rand, 2d ed., Blackwell, Oxford, 1980; *Remington's Pharmaceutical Sciences*, 16th ed., Mack Publishing Co, Easton, PA, 1980; *Essentials of Molecular Pharmacology*, A. Korolkoras, Wiley, New York, 1970; *Quantum Pharmacology*, W. G. Richards, Butterworths, London, 1972; *Natural and Synthetic Organic Medicinal Compounds*, L. C. Salerni, Mosby, St. Louis, 1976; *Medicinal Chemistry: A Biochemical Approach*, T. Norgrady, OUP, Oxford, 1985; *Principles of Drug Action: The Basis of Pharmacology*, A. Goldstein, Aronow & Kalman, Wiley, New York, 1974. *The Physicochemical Principles of Pharmacology*, A. T. Florence and D. Attwood, Macmillan, New York, 1981, provides valuable physical chemistry for pharmacologists.

DRUG SYNTHESIS

There are not many books that describe drug syntheses and most of them are concerned inevitably with the syntheses reported in the literature. Pharmaceutical firms are understandably reluctant to divulge which routes they actually use.

30. *The Organic Chemistry of Drug Synthesis*, D. Lednicer and L. A. Mitscher, Wiley, New York, 1977, 1980, 1984, is a comprehensive three-volume work

which categorizes drugs by their structure. A cross-reference system classifies drugs by their activity. The books are poorly produced and contain a depressing number of errors (Volume 2 carries 14 pages of errata to Volume 1 and the list is by no means comprehensive). Used with caution, however, these books are an invaluable source of information and reference and we have drawn heavily on them.

31. *Pharmazeutische Wirkstoffe: Synthese, Patente, Anwendungen*, A. Kleemann and J. Engel, Thieme, Stuttgart, 1984, is a more attractive and formal presentation of a smaller number of drug syntheses. There is little text and the book is written in the "international" language of chemical formulae. Drugs are listed by their German generic names and comprehensive patent and literature references are given. We suspect the routes described are closer to those used in industry.
32. *Encyclopedia of Chemical Technology*, R. E. Kirk and D. F. Othmer, 3d. ed., Wiley–Interscience, New York, 1983 deals with every aspect of the chemical industry. The third edition is particularly strong on pharmaceuticals.
33. Part 2 of our own book, *Industrial Organic Chemicals in Perspective*, H. A. Wittcoff and B. G. Reuben, Wiley, New York, 1980, contains a chapter on the syntheses of the Top-50 drugs in 1976. *The Chemical Economy*, B. G. Reuben and M. L. Burstall, Longman, London, 1974, contains a shorter chapter, now somewhat dated. There is also a chapter on "Pharmaceuticals" by B. G. Reuben in the *Encyclopedia of Physical Science and Technology*, Vol. 10, Academic Press, San Diego, 1987.
34. The forerunner of this current book was an American Chemical Society Audiotape, *The Pharmaceutical Industry—Chemistry and Concepts*, H. A. Wittcoff and B. G. Reuben, Washington, DC, 1986.
35. A fascinating book that contains much chemistry is *Medicinal Chemistry—The Role of Organic Chemistry in Drug Research*, S. M. Roberts and B. J. Price, Eds., Academic Press, London, 1985. It has three chapters on receptors and enzymes and eleven on case studies of the development of specific drugs.
36. *Chronicles of Drug Discovery*, Vol. 1, J. S. Bindra and D. Lednicer, Wiley, New York, 1982, covers much of the same ground and even some of the same drugs as the previous reference.
37. Noyes Publications, Park Ridge, NJ, produces a two-volume *Pharmaceutical Manufacturing Encyclopedia* with manufacturers' names and details of syntheses. The second edition was published in 1988.

"POPULAR" BOOKS ON PHARMACEUTICALS

The following books are generally paperbacks and are intended for the educated layman. We have found many of them compulsive reading. In some

cases they can be accused of oversimplification, but they provide valuable insights in medical areas, although less so in the field of chemistry.
40. *Medicines—A Guide for Everybody*, P. Parish, 2d ed., Penguin, Harmondsworth, England, 1984, is an outstanding paperback which describes various diseases and the drugs used to treat them. It is intended to help patients to assess the risks and benefits of any drug they may take.
41. An American book on similar lines is *Common Medicines—An Introduction for Consumers*, D. J. George, W. H. Freeman, San Francisco, 1979.
42. *A Dictionary of Drugs*, R. B. Fisher and G. A. Christie, Paladin Books, Frogmore, St. Albans, England, 1975, is a paperback that provides a popular introduction to drugs and their mode of action. Although laid out as a dictionary, it is ideal for browsing.
43. A useful although brief and dated overview of the drug industry is provided in *Pharmaceuticals*, J. N. T. Gilbert and L. K. Sharp, Butterworths, London, 1971.

BIBLIOGRAPHY TO INDIVIDUAL SECTIONS

1 History

1/1 Two fascinating books published just prior to the chemotherapeutic revolution are *The Microbe Hunters*, Paul de Kruif, Jonathan Cape, London, 1927, and *Devils, Drugs and Doctors*, H. W. Haggard, Harper & Bros., New York, 1929. We have drawn heavily on *Medicines, 50 Years of Progress 1930–1980*, N. Wells, Office of Health Economics, London, 1980. We have also drawn on *The Search for the Magic Bullet*, E. Baumler, Econ-Verlag, Dusseldorf, 1963 (English edition: Thames and Hudson, London, 1965); and *Alive and Well: Medicine and Public Health 1830 to the Present Day*, N. Longmate, Penguin, Harmondsworth, 1970.

1/2 We consulted the *British Pharmacopeia* of the appropriate date for details of early drugs and also *Materia Medica and Therapeutics*, W. J. Dilling, 14th ed., Cassell, London, 1933.

1/3 Sources of statistical data are cited beneath figures listing them. We should mention in particular the series of publications by the Office of Health Economics, 12 Whitehall, London SW1A 2DY; the *Statistical Abstract of the United States*, U.S. Dept of Commerce, Bureau of the Census, Washington, DC, published annually and *Historical Statistics of the United States*, U.S. Dept of Commerce, Bureau of the Census, Washington, DC, 1975 (2 Vols.).

2 The Characteristics of the Pharmaceutical Industry

2/1 The statistics in this section were assembled from a number of sources, including *U.S. Industrial Outlook 1983*, U.S. Department of Commerce, Washington, DC, *Synthetic Organic Chemicals*, *U.S. Production of Sales*, U.S. International Trade Commission publication 776, U.S. Government Printing

Office, Washington, DC, 1982; *Census of Manufacturers and Annual Survey of Manufacturers*, U.S. Dept of Commerce, Bureau of the Census, Washington, DC. The Pharmaceutical Manufacturers' Association (1155 Fifteenth Street NW, Washington, DC 20005) produces the useful *Pharmaceutical Industry Fact Book, 1986*, and their Annual Report. The Office of Health Economics in the United Kingdom produces a *Compendium of Health Statistics*, London, 1987.

2/2 Especially valuable was "R & D Scoreboard," *Business Week*, 20 June 1983, p. 126.

2/3 The Japanese effort to become a major pharmaceutical producer is described in *Business Week*, 10 May 1982, p. 150.

2/4 The role of the multinationals in the pharmaceutical industry is described in *Multinational Enterprises, Governments and Technology—The Pharmaceutical Industry*, M. L. Burstall, J. H. Dunning, and A. Lake, OECD, Paris, 1981.

2/5 Competition in the drug industry is discussed in numerous books including *The Economics of the Pharmaceutical Industry*, W. D. Reekie, Macmillan, London, 1975; *The Canberra Hypothesis*, G. Teeling Smith, Office of Health Economics, London, 1975; *Innovation in the Pharmaceutical Industry*, D. Schwartzmann, Johns Hopkins University Press, Baltimore, 1976; and *Drug Regulation and Innovation*, H. Grabowski, American Enterprise, 1976.

2/6 Figure 2.3 is taken from *Chemistry in Medicine*, American Chemical Society, Washington, DC, 1977. This book is the result of a study by a distinguished task force headed by Professor Marshall Gates and deals with the way in which chemistry impinges on health care. The report on the Massengill sulfanilamide elixir appeared in the *Report of the Secretary of Agriculture*, 109, 1985 (1937).

2/7 Problems with present methods of toxicity testing are outlined in "Principles, Practice, Problems and Priorities in Toxicology", J. V. Bridges and S. A. Hubbard, in *Current Approaches in Occupational Health*, Vol. 2, Ward Gardener, Ed., Wright, Bristol, 1982. See also *Newsweek*, 14 July 1980, p. 60, and N. Henson, *New Scientist*, 5 May 1983, p. 275. The trend away from animal testing is described in *Chemical Week*, 26 May 1982, p. 26.

2/8 The use of generic drug names has been described by R. L. Rawls, *Chem. Eng. News*, 11 Nov. 1974, p. 7.

2/9 One of many attacks on brand names in prescribing appears in *Chemistry in the Market Place*, B. Selinger, Australian National University Press, Canberra, 1978. A defence appears in *Brand Names in Prescribing*, Office of Health Economics, London, 1976. An editorial, "Generic Prescribing and the Drug Industry", in the *British Medical Journal*, 284, 919 (1982) gives several references.

2/10 The scandal of substrength tetracyclines was hushed up quite effectively in the United Kingdom, but details may be found in the *Sunday Mercury*, 13 June 1965, *Hansard* 14 & 21 June 1965, and *Medical News*, 28 May 1965, p. 12.

2/11 Drug names and pronunciation are listed in reference (8).

2/12 There have been many highly priced reports on generic pharmaceuticals from consultancy firms. One of the more reasonable is *Generic Pharmaceuticals in Europe—Blessing or Threat*, M. L. Burstall, Economists' Advisory Group, 35 Albemarle Street, London W1X 3LB, 1986. Also recommended is B. G. Reuben and M. L. Burstall, *Generic Pharmaceuticals—The Threat: Products and Companies at Risk*, Economists Advisory Group, London, 1989.

2/13 Figures of the British generics market come from *Structure of British Industry*, P. Johnson, Ed., Unwin Hyman, London, 1988.

2/14 The Waxman–Hatch Act is formally described as *Drug Price Competition and Patent Term Restoration Act of 1984*, S1538/98th Congress, 2d session. Therapeutically equivalent drugs are listed in the Orange Book (13).

2/15 An article on "The Drug Industry: Harder Going" featuring generics especially was reprinted from the *London Economist* in *CHEMTECH*, January and February 1988.

2/16 An attack on the pharmaceutical industry for its strategy on generics was published under the title "The Big Lie about Generic Drugs", *Consumer Reports* 52, 480 (1987).

2/17 The costs and benefits of the removal of non-tariff barriers in the European Community in 1992 are assessed in *The Cost of "non-Europe" in the Pharmaceutical Industry*, M. L. Burstall and B. G. Reuben, Directorate-General III of the European Commission, Luxembourg, 1988.

3 Patterns of Illness and Health Care

3/1 Details of the nonprescription drug market appeared in *Chemical Marketing Reporter*, 5 July 1982, pp. 5, 26, which referred to a less accessible survey by C. H. Kline.

3/2 A valuable discussion of the problems of health care in developing countries is to be found in *Medicines, Health, and the Poor World*, D. Taylor, Office of Health Economics, London, 1982. See also *Pharmaceuticals in Developing Countries 1981–82*, Office of Health Economics, London, 1982, and *The World Drug Situation*, World Health Organization, Geneva, 1988.

3/3 Worldwide mortality figures may be obtained from the *United Nations Demographic Yearbook* or, in more detail, from the *World Health Statistics Annual*, WHO, Geneva.

3/4 Disability and physical impairment in Britain were discussed in an Office of Health Economics briefing, No. 15, July 1981. The most up-to-date survey is J. Martin, H. Meltzer, and D. Elliot, *OPCS Surveys of Disability in Great Britain: Report 1—The Prevalence of Disability among Adults*, HMSO, London, 1988.

3/5 Statistics about the illnesses for which people visit their physician were published in *Morbidity Statistics from General Practice, 1981–2, Studies on Medical and Population Subjects No. 36*, HMSO, London, 1986.

3/6 Medical sociology is an important topic which we have dealt with only cursorily. A valuable introduction to the subject is *Health, Illness and Medicine*, G. L. Albrecht and P. C. Higgins, Rand McNally, Chicago, 1979.

3/7 Some patterns of U.K. prescribing are discussed in *Prescribing in General Practice*, P. A. Parish et al., Royal College of General Practitioners, 1976. Differences between countries are fascinatingly reviewed in *Patterns of European Diagnoses and Prescribing*, B. O'Brien, Office of Health Economics, London, 1984.

4 Pharmacological Concepts

4/1 We have drawn especially on Burger (20), Taylor and Kennewell, (21) and "The Molecular Basis of Drug Action," A. Albert, *Chemistry International* No. 3 (1980) p. 19. We have dealt briefly with the topics of drug design and structure–activity relations, but those who are interested might like to consult the above references followed by *Drug Design*, Vols. 9 and 10, E. J. Ariens, Ed., Academic Press, New York, 1980, and *Medical Research Series* Vol. 10, *Physical Chemical Properties of Drugs*, S. H. Yalkovssky et al., Dekker, New York, 1980. In addition to the *Drug Design* series, there are also the *Annual Reports on Medicinal Chemistry*, Academic Press, New York. The photographs in Figure 4.3 were taken by Mike Hann of Glaxo Pharmaceuticals.

4/2 The digoxin affair was reported by T. R. D. Shaw, *Postgraduate Medical Journal* 50, 98 (1974). The general problems raised by it were discussed in *Brand Names in Prescribing* (2/9) and by A. H. Beckett, *Postgraduate Medical Journal* 50, 125 (1974), and in *The Pharmaceutical Industry and Society*, G. Teeling-Smith, Ed., Office of Health Economics, London, 1972.

4/3 The properties of an ideal drug are summarized from Giswold et al. (23).

4/4 The relation of drug design to drug structures is discussed in *Introduction to the Principles of Drug Design*, J. Smith and H. Williams, Wright, London, 1983.

4/5 Dosage forms are described in Burger (20). *Chemical Engineering*, May 1982, has a number of articles on the equipment for making pills, and the problems associated with pill manufacture. Sustained release formulations are discussed in *Chemical Week*, 11 May 1983, p. 46. We have drawn heavily on H. J. Sanders, *Chem. Eng. News*, 1 April 1985. Drugs incorporated in polymers are described in two papers presented to the ACS 179th Annual Meeting, Houston, Texas, March 1980, namely "Antimicrobial and Antiparasitic Polymers," J. A. Brierley et al., (preprints p. 432), and "Organometallic Polymers as Chemotherapeutic Drug Delivery Agents," C. E. Carraher, Jr. (preprints p. 427). Polymeric drugs are also described by C. A. Samour, *CHEMTECH* 8, 494 (1978). "Delivery Drugs" are described in *CHEMTECH* 10, 82 (1980). A comprehensive work is *Pharmaceutical Technology*, M. Rubinstein, Ed., Ellis Horwood, London, 1987. Volume 1 is said to deal with tableting technology, but also covers controlled release, stability, and targeting. The ADEPT system was described in *Chem. Ind.* 2 January 1989, p. 2.

4/6 Two books with information on the newer aspects of controlled release are *Rate Control in Drug Therapy*, L. F. Prescott and W. S. Nimmo, Butler & Tanner, London, 1985, and *Controlled Drug Delivery*, J. R. Robinson and V. H. L. Lee, Dekker, New York, 1987.

4/7 The formulation question is also dealt with in *Materials Used in Pharmaceutical Formulation*, A. T. Florence and D. Attwood, Blackwell, Oxford, 1984. *Pharmaceutical Dosage Forms: Disperse Systems*, Dekker, New York, 1988 discusses the latest information on dosage forms, including the many auxiliary materials required. It is Volume 1; Volume 2 is in preparation.

4/8 The design of pro-drugs is the subject of a whole edition of *Chemistry and Industry*, 7 June 1980, pp. 433–461.

4/9 A useful book on drug absorption and metabolization is *An Introduction to Pharmacokinetics*, B. Clark and D. A. Smith, Blackwell, Oxford, 1981.

5 The Manufacture of Pharmaceuticals

5/1 Statistics on the top prescription drugs are recorded by the National Prescription audit, conducted annually by IMS America Ltd, Ambler, PA, and published in the April issue of *Pharmacy Times*.

5/2 Statistics about pharmaceuticals and their sales by companies are of great importance to the companies themselves and to investment analysts and so on. Many details can be found in company reports and other data are published by firms of stockbrokers and other interested groups. Much useful information is published in the British journal, *Scrip*, and in the American *IMS Market Letter*. See also B. G. Reuben and M. L. Burstall (2/12).

6 Antibacterials and Antibiotics

6/1 *Encyclopedia of Antibiotics* J. S. Glasby, Wiley, London, 1976, is a useful if slightly dated reference book. More recent is *The Biochemistry and Pharmacology of Antibacterial Agents*, R. A. D. Williams and Z. L. Kruk, Croom Helm, London, 1981.

6/2 *Bergey's Manual of Determinative Bacteriology*, R. E. Buchanan and N. E. Gibbons 8th ed. Williams & Wilkins, Baltimore, 1974, is an excellent reference on bacterial classification.

6/3 The detailed techniques of antibiotic production are beyond the scope of this book but are dealt with comprehensively in *Biotechnology of Industrial Antibiotics*, E. J. Vandamme, Ed., Dekker, New York, 1984. Figure 6.7 comes from an article in this book, "The Penicillins—Properties, Biosynthesis and Fermentation," G. J. M. Hersbach, C. P. Van der Beek, and P. W. M. van Dijck.

6/4 Another solid work is *The Antimicrobial Drugs*, W. B. Pratt and R. Fekety, OUP, Oxford, 1986. In spite of its title, it covers antifungals and antiparasitics too.

6/5 The photographs in Figure 6.11 were taken by G. R. Parish of Beecham Pharmaceuticals.

6/6 The history of penicillin is presented by G. B. Kauffman in *J. Chem. Ed.*, July 1979, p. 454 in an article entitled "The Discovery of Penicillin, 20th Century Wonder Drug." See also *The Life of Alexander Fleming*, A. Maurois, Jonathan Cape, London, 1959, and *Penicillin in Perspective*, D. Wilson, Faber, London, 1976. More recently there was "Penicillin, the First Half Century," P. Craig, *Chem. Brit.*, 15, 392 (1979). Sheehan's version of the G-APA story is told in *The Enchanted Ring—The Untold Story of Penicillin*, J. C. Sheehan, MIT Press, Cambridge, MASS, 1983, and is criticized by the Chairman of Beecham Pharmaceuticals in *Chem. Ind.*, 4 April 1983, p. 275.

6/7 The stereospecific synthesis of penicillins was reported by J. E. Baldwin et al., *J. Am. Chem. Soc.* 98, 3085 (1976). Other interesting articles include J. H. C. Nayler, "Structure–Activity Relationships in Semi-Synthetic Penicillins," *Proc. R. Soc. London Ser. B* 179, 357 (1971), and "Advances in Penicillin Research," in *Advances in Drug Research*, A. B. Simmonds, Ed. 7, 1 (1973). The penems are described in *Chem. Eng. News*, 5 Nov. 1979, p. 19.

6/8 B. G. Reuben and K. Sjoberg discuss ion pair extraction of penicillin and phase transfer catalysis in the production of penicillin esters in *CHEMTECH* 11, 315 (1981).

6/9 We have drawn heavily on "Ceftazidime, an Injectable Antibiotic," C. E. Newall, *Chem. Brit.*, 23, 876 (1987). Moxalactam and cefotaxime were described in *Chem. Brit.* 17, 318 (1981).

6/10 There is an excellent article on "Antibiotic Resistance" by R. C. Moellering, Jr., and B. E. Murray in *CHEMTECH* 11, 280 (1981), and another by R. E. Cape on "Microbial Genetics" in *CHEMTECH* 9, 638 (1979).

6/11 The authoritative article on rifampin is by W. Lester, *Ann. Rev. Microbiol.* 26, 85 (1972).

6/12 Table 6.3 is taken from the British National Formulary (4).

6/13 The increase in penicillin-resistant gonorrhea is described by J. A. McCutchan et al., *Brit. Med. J.* 285, 337 (1982). Resistance to *Hemophilus influenzae* is reported by J. Philpott-Howard and J. D. Williams, *Brit. Med. J.* 284, 1597 (1982).

6/14 The newer antibacterials are dealt with by S. C. Stinson, *Chem. Eng. News*, 29 Sept. 1986, p. 33.

7 Cardiovascular Drugs

7/1 The medical and socio-economic aspects of heart disease are discussed succinctly in *Coronary Heart Disease*, Office of Health Economics, London, 1982. A more recent booklet deals with *Stroke*, London, 1988.

7/2 An excellent updating report on heart drugs is given by S. C. Stinson, *Chem. Eng. News*, 3 Oct. 1988, p. 35.

7/3 The Nobel prize-winning synthesis of quinine was reported by R. B. Woodward and W. E. Doering, *J. Am. Chem. Soc.* 67, 860 (1945).

7/4 The history of digoxin is told in "Foxglove as a Therapeutic Agent", J. Aronson, *Chem. Brit.*, January 1987, p. 33.

7/5 The diuretics section was based mainly on Burger, Part 3, p. 147–224 (20) and "Diuretic Research Saves Lives," D. Bormann, *Chem. Brit.* 15, 72 (1979).

7/6 Developments in calcium antagonists and β blockers are summarized in "New Drugs for Combatting Heart Disease", H. W. Sanders, *Chem. Eng. News*, 12 July 1982, p. 26.

7/7 We have mentioned neurotransmitters rather late in this book although they are basic to modern pharmacology. A useful paperback is *Neurotransmitters and Drugs*, Z. L. Kruk and P. J. Pycock, Croom-Helm, London, 1987.

7/8 We have grossly oversimplified our description of β blockers by failing to refer to their antihypertensive action. Relevant articles include B. N. C. Prichard, *Cardiology* 64 (Suppl. 1), 44 (1979), D. G. McDevitt, *Drugs* 17, 267 (1979), and M. Meier et al. in *Pharmacology of Antihypertensive Drugs*, A. Scriabine, Ed., Raven, New York, 1980. For details of the differences between the various β blockers, see "Beta-Adrenergic Blockade in Clinical Practice," W. H. Frishman, *Hospital Practice*, Sept. 1982, p. 57.

7/9 Avoidance of coronary disease is discussed by J. T. Hart, *Brit. Med. J.* 285, 347 (1982).

7/10 For details of β-adrenergic agents we have drawn on E. E. Smissman in Wilson et al. (23).

7/11 Anticholesterol drugs are reviewed in *Chem. Week*, 11 Feb. 1987, p. 72. A graphic description of the role of lipoproteins in cholesterol metabolism appeared in *Time Magazine*, 12 December 1988, p. 66.

7/12 The design of ACE inhibitors is covered in "Computer Graphics as an Aid to Drug Design", C. H. Hassall, *Chem. Brit.*, Jan. 1985, p. 7.

8 Drugs Affecting the Central Nervous System

8/1 As usual, there is an Office of Health Economics booklet giving an overview, in this case *Medicines Which Affect the Mind*, London, 1975.

8/2 Suicides due to barbiturates are reported in the *Pharmaceutical Journal*, 12 June 1971. There have been many more anticonvulsant drugs recently than we have been able to mention. They are reviewed by H. Kohn and J. D. Conley, *Chem. Brit.*, Mar. 1988, p. 231.

8/3 The behavior of the drug companies after the thalidomide tragedy was strongly criticized by the *London Sunday Times* whose campaign was believed to have increased the compensation eventually paid by an order of magnitude. It was reported in "The Thalidomide Children and the Law," Andre Deutsch, London, 1973. The view that Grunenthal knew of the dangers of thalidomide to the central nervous system before they marketed it is advanced in H.

Sjostrom and R. Nilsson, *Thalidomide and the Power of the Drug Companies*, Penguin, Harmondsworth, 1972.

8/4 The synthesis of diazepam used in industry is something of a mystery and is probably not the one given here.

8/5 The conversion of amphetamine to a mescaline-like substance is discussed in *Nature* 237, 454 (1972).

8/6 The Quaalude case was around for some time. Early reports appeared in *Newsweek*, 28 Sept. 1981, p. 93, and more recent ones in *Medical World*, 15 Mar. 1983, p. 46.

8/7 The β-carboline esters may be anxiogenic or anxiolytic and are described in *Science* 218, 645 (1982), and the *Lancet*, 6 Nov. 1982, p. 1030. Three papers dealing with benzodiazepine antagonists are R. Amrein et al., *Medical Toxicology*, 2, 411–429, (1987); R. N. Brogden and K. L. Goa, *Drugs* 35, 448–467 (1988); and U. Klotz and J. Kanto, *Clinical Pharmacokinetics* 14, 1–12, (1988).

8/8 The serotonin reuptake inhibitors are reviewed in "What's New in Antidepressants," *Drug Topics*, 3 Aug. 1987, p. 32.

8/9 The effect of anorectics is discussed in *Drugs and Appetite*, T. Silverstone, Academic Press, New York, 1982.

8/10 Anxiety about the long-term side effects of diazepam is reported in the *London Observer*, 24 Feb. 1980; see also "The Storm before the Calm," *CHEMTECH* 9, 686 (1979). The related problem of alcoholism is reviewed in an Office of Health Economics booklet, *Alcohol*, London, 1981.

8/11 Questions about drugs and drug-related problems are discussed in detail in *Encyclopedia of Psychoactive Drugs*, S. H. Snyder and M. H. Lader, Eds., Burke, London, 1983. There are 25 volumes, the one we have seen deals with tranquilizers.

8/12 Figures 8.10 and 8.11 are taken from C. Baum, D. L. Kennedy, D. E. Knapp and G. M. Faich, *Drug Utilization in the US–1985*, FDA, MD, 1986.

9 Antihistamines

9/1 We are indebted to P. Goddard of SmithKline (UK) for much of the cimetidine story, and to Roberts and Price (35) for more of it. The firm itself published a booklet *The Discovery of H_2-Receptor Antagonists*, which is more medically than chemically oriented. The two kinds of histamine receptors were proposed by A. S. F. Ash and H. O. Schild, *Br. J. Pharmacol.* 27, 427 (1966), and the first H_2 blocker (called burimamide) was reported by J. W. Black et al., *Nature* 236, 385 (1972). The advantages of cimetidine were reported by R. W. Brimblecombe et al., *J. Int. Med. Res.* 3, 86 (1975). The range of compounds aiming to displace cimetidine was described in *Chem. Eng. News*, 12 Apr. 1982, p. 24. The press release launching ranitidine in the United Kingdom was published by Glaxo, 13 Oct. 1981 and the prospects for the drug in the United States were reviewed in *Business Week*, 22 Nov. 1982, p. 36.

9/2 Recent work on antiallergy drugs is reviewed in Lunt (15/2).

10 Analgesics

10/1 Useful references include *Chemistry in Medicine* (2/6), Gilbert and Sharp (43) and Burger (20) Vol. 3, Chap. 52, pp. 645–696, Wilson et al. (23) Chap. 24, pp. 699–754, together with J. L. Fox, *Chem. Eng. News*, 22 Nov. 1982, p. 29, and S. H. Snyder, *Chem. Eng. News*, 28 Nov. 1977, p. 26. There is also an excellent case study in Lednicer and Mitscher, Vol. 1, Chap. 2 (30).

10/2 For background on metabolization and elimination of acetaminophen, see J. R. Mitchell et al., *Drug Metabolism Reviews* 13, 539 (1982), and J. A. Hinson et al., *Life Science* 29, 107 (1981).

10/3 J. S. Morley has written an excellent review of structure–activity relation of enkephalin-like peptides in *Ann. Rev. Pharmacol. Toxicol.* 20, 31 (1980).

10/4 The literature on enkephalins and endorphins is legion. The following sources were useful: R. C. A. Frederickson et al., *Science* 211, 603 (1981); G. A. Stacher et al., *Pain* 7, 159 (1979); D. Roehmer et al., *Nature* 268, 547 (1977); Anon., *Chem. Eng. News*, 19 Apr. 1982, p. 41; ibid., 21 Sep. 1981, p. 23; R. C. A. Frederickson et al., *Dev. Neurosci.* (*Amsterdam*) 4, 215 (1978); J. F. Calimlin et al., *Lancet* 1, 1374 (1982); P. D. Gesellchen and R. T. Shuman, U.S. Patent 4,322,339 to Eli Lilly, 20 Aug. 1981. Endogenous morphine was reported by H. W. Kosterlitz, *Nature*, 317, 671 (1985).

10/5 Algogens and new ideas for nonnarcotic analgesics are discussed in "New Ideas for Pain Relief," *Chem. Brit.*, Aug. 1987, p. 758.

11 Steroid Drugs

11/1 The safety or otherwise of the contraceptive pill is a long-running epic. The evidence is summarized in reference (2). The suggestion that it may protect against certain ailments was written up in the *New York Times*, 13 July 1982. An attack on the regulatory procedures is made by C. Djerassi, one of the discoverers of oral contraceptives, in *The Politics of Contraception*, Norton, New York, 1980. See also *Chem. Ind.*, 4 Oct. 1980, p. 773.

11/2 There are articles on contraceptive and anti-inflammatory steroids in Roberts and Price (35), and various syntheses are given.

11/3 Two excellent reviews on the attachment of side chains at the 17-position in 17-ketosteroids have been written by D. M. Piatak and J. Wicha, *Chemical Reviews* 78, 199–241 (1978), and by J. Redpath and F. J. Zeelen, *Chem. Soc. Rev.* 12, 75 (1983). Typical of the many articles in this field is D. van Leusen and A. M. van Leusen, *Tetrahedron Letters* 2581 (1984).

12 Nonsteroid Anti-Inflammatory Agents

12/1 The benoxaprofen disaster was discussed in *Forbes*, 25 Oct. 1982, p. 41, and in many other places.

13 Hypoglycemics

13/1 A worthwhile account of the application of gene splicing is to be found in "The Genetic Programming of Industrial Microorganisms," D. A. Hopwood, *Scientific American* 245, 91 (1981).

13/2 A survey of the growth of the biochemical engineering industry appeared in the *New York Times*, 29 June 1980. A more sophisticated article appeared in *Newsweek*, 24 Mar. 1980, p. 44. A survey by R. M. Baum of the whole area of biotechnology with emphasis on new pharmaceuticals on the verge of commercialization appeared in *Chem. Eng. News*, 20 July 1987, p. 11.

13/3 The patented mouse was reported in *Chem. Ind.*, 16 May 1988, p. 310.

14 Anticholinergic Drugs

14/1 Receptors for acetylcholine are discussed in Mishnia et al., *Nature* 313, 364–369 (1985); Ridel, *Nature* 324, 68–70 (1986); Dixon, *Nature* 326, 73–77 (1987) and Davis, *Nature* 326, 760–765 (1987).

15 Anti-Asthma Drugs

15/1 *Asthma*, Office of Health Economics, London, 1976, provides an overview.

15/2 A useful review is "Recent Advances in Anti-Allergic and Anti-Asthmatic Drugs," E. Lunt in *Progress in Pharmaceutical Research*, K. R. H. Woolridge, Ed., Blackwell, Oxford, 1982.

15/3 The leukotrienes have attracted much attention. Three references are E. J. Corey et al., *J. Am. Chem. Soc.* 102, 4278 (1980); *Biochem. Biophys. Res. Commun.* 92, 946 (1980), and S. Okuyama et al., *Chem. Pharm. Bull.* 30, 2453 (1982).

16 Parkinson's Disease

16/1 There are Office of Health Economics booklets about *Parkinson's Disease*, London, 1974, and *Huntingdon's Chorea*, London, 1980.

16/2 For early work on the clinical use of L-dopa, see G. C. Cotzias et al., *N. Engl. J. Med.* 276, 374 (1967); 280, 337 (1969). The asymmetric synthesis is reported by W. S. Knowles et al., U. S. Patent 4005127 to Monsanto Chemical Co., 25 Jan. 1977.

16/3 For the fermentation route to L-dopa, see U.S. Patent 3,791,924 (2/12/74) to Ajinimoto, and A. Aikinia et al., *Chem. Brit.*, July 1987, p. 645. The effect of L-dopa on sleeping sickness victims was described vividly in *Awakenings*, O. W. Sacks, Duckworth, New York, 1973.

17 Miscellaneous Drugs

17/1 The L-thyroxine synthesis is derived from U.S. Patents 2889363-4 and 2803654.

17/2 A review of the antiacne retinoids is provided by C. E. Orfanos et al., *Drugs* 34, 459–503, (1987), and two research articles are to be found in the same journal, 26, 9 (1983), and 28, 6 (1984).

17/3 The syntheses of most of the vitamins are given in Wittcoff and Reuben (33).

18 Tropical Diseases

18/1 See *Medicines, Health and the Poor World*, (3/2). The Chinese antimalarial Qinghaosu is discussed in *Brit. Med. J.* 284, 767 (1982).

18/2 The development of anthelmintic drugs was reviewed by A. A. Baklien, *Chemistry in Australia*, May 1979.

18/3 H. C. Richards presents "Oxamniquine—A Drug for the Third World," in *Chem. Brit.*, Nov. 1985, p. 1001, and has a similar account in Roberts and Price (35).

19 Prostaglandins

19/1 Several prostaglandin references up to 1977 were given in Wittcoff and Reuben (33), but these have really been incorporated into the comprehensive review in Burger (20).

19/2 Many synthetic routes not discussed in Burger are to be found in *The Synthesis of Prostaglandins*, A. Mitra, Wiley, New York, 1977, and *Prostaglandin Research*, P. Crabbe, Ed., Academic Press, New York, 1977. See also *Total Synthesis of Prostaglandins*, *Synthesis*, Vol. 1, U. Axen, J. E. Pike, and W. Schneider, J. W. ApSimon, Eds., Wiley, New York, 1973. The Nobel prize award to Bergstrom, Samelsson, and Vane was reported in the *New York Times*, 12 Oct. 1982.

19/3 We have drawn heavily on "Prostaglandins and the Arachidonic Acid Cascade," N. A. Nelson, R. C. Kelly, and R. A. Johnson, *Chem. Eng. News*, 16 Aug. 1982, p. 30, an excellent article whose predictions have largely been borne out.

19/4 Manufacture of thromboxane antagonists is discussed by S. Lee, *Chem. Ind.*, 6 Apr. 1987, p. 223.

20 Antiviral Drugs

20/1 There is an excellent article on virus therapy by C. E. Hoffmann in *CHEMTECH* 8, 726 (1978). The prospects for acyclovir are discussed by D. McInnes, *New York Times*, 19 June 1983, p. F8. Bromovinyl deoxyuridine was described in *Chem. Brit.* 1981, p. 396.

20/2 The problem of drugs to combat herpes 2 was featured in the *New York Times*, 26 May 1983, and the *London Observer*, 2 May 1982, but has, of course, been very widely exposed. See also "Treatment of Herpes virus Infection," M. S. Hirsch and M. S. Schooley, *N. Eng. J. Med.* 309, 963, (1983).

20/3 All the recent literature features AIDS. We have found useful R. Dagani, *Chem. Eng. News*, 29 June 1987 and 8 December 1986, p. 7; "Viruses," *Time*, 3 November 1986, p. 42; and "Synthetic Antiviral Agents," R. K. Robins, Chem. Eng. News 27 Jan. 1986, p. 28.

21 Anticancer Drugs

21/1 The estimate that 80% of cancers are environmentally produced was put forward by R. Doll and R. Peto, *J. Nat. Cancer Inst.* 66, 1191 (1981). A subsequent report by the National Research Council, with recommendations, was summarized in *Chem. Eng. News*, 21 June 1982, p. 4.

21/2 A progress report on cancer treatment appeared in the *Mayo Clinic Health Letter*, March 1986, and was the source of Figure 21.3.

21/3 A more chemically oriented account of cancer is given in "Chemicals and Cancer," K. Vaughan, *Chemistry International*, 1980, p. 11. An equally old book is *The Anticancer Drugs*, W. E. Pratt and R. W. Ruddon, OUP, New York, 1979, but progress in this area has been slow and the material is still valid.

21/4 Discussion of platinum based anticancer agents will be found in P. J. Sadler, *Chem. Brit.*, March 1982 p. 182; S. J. Berners-Price and P. J. Sadler, *Chem. Brit.*, June 1987, p. 541; C. F. J. Barnard, M. J. Cleave, and P. C. Hydes, *Chem. Brit.*, November, 1986, p. 1001; and J. L. van der Veer and J. Reedijk, *Chem. Brit.*, August, 1988, p. 775.

21/5 The use of interferon in cancer therapy is discussed by P. Hersey, *Aust. NZ J. Med.* 16, 425 (1986), and D. Goldstein and J. Laszlo, *Cancer Res.* 46, 4315–4329, (1986).

22 Orphan Drugs

22/1 The story of triethylene tetramine dihydrochloride for Wilson's disease was told by J. M. Walshe in *Orphan Drugs*, F. E. Karch, Ed., Dekker, New York, 1982. The FDA go-ahead for various orphan drugs was reported in *Chem. Eng. News*, 4 October 1982, p. 13, and 25 October 1982, p. 24.

22/2 Some of the effects of the Orphan Drugs Act are reviewed in *Chem. Mkt. Rep.*, 4 February 1985, p. 7 and *Medical World*, 8 September 1986.

23 Innovations and Issues

23/1 Table 23.1 is based on the lists given in the end-of-year issues of SCRIP for 1986, 87 and 88.

23/2 A useful if somewhat dated article is "Brain Chemistry", M. B. Krassner, *Chem. Eng. News*, 25 Aug. 1983.

23/3 Easily the best book critical of the pharmaceutical industry (because it offers suggestions as well as criticism) is *Corporate Crime in the Pharmaceutical Industry*, J. Braithwaite, Routledge, London, 1984.

23/4 Over-prescribing is discussed from the physicians viewpoint by D. Ryde, *General Practitioner*, 16 Jan. 1981, p. 48; *The Practitioner* 224, 235 (1980) and 225, 283 (1981).

23/5 A valuable update on the commercial output of the burgeoning biotechnology industry is given in "Biotechnology Industry moving Pharmaceutical Products to Market", R. M. Baum, *Chem. Eng. News*, 20 July 1987, p. 11. Developments in peptide research are reviewed in "Small Peptides—New Targets for Drug Research", A. S. Dutta, *Chem. Brit.*, Feb. 1989, p. 159.

Index